AQA
GCSE Mathematics

Glyn Payne   Ron Holt   Mavis Rayment   Ian Robinson

Higher

www.heinemann.co.uk

Inspiring generations

Heinemann Educational Publishers
Halley Court, Jordan Hill, Oxford OX2 8EJ
Part of Harcourt Education

Heinneman is the registered trademark of Harcourt Education Limited
© Harcourt Education Ltd 2006

First published 2006

10 09 08 07 06
10 9 8 7 6 5 4 3 2 1

British Library Cataloguing in Publication Data is available from the British Library on request.

10-digit ISBN: 0 435210 459
13-digit ISBN: 978 0 435210 45 8

Edited by Katherine Pate and Jon Billam
Designed by Phil Leafe
Typeset by Tech-Set Ltd, Gateshead, Tyne and Wear

Original illustrations © Harcourt Education Limited, 2006

Illustrated by Adrian Barclay and Mark Ruffle

Cover design by mccdesign

Printed in the United Kingdom by Pindar Graphics

Cover photo: Alamy Images ©

Consultants
Andrew Darbourne, Jackie Fairchild, Margaret Hayhurst

Acknowledgements
Harcourt Education Ltd would like to thank those schools who helped in the development and trialling of this course.

The author and publisher would like to thank the following individuals and organisations for permission to reproduce photographs:

Rex Features p35; Photolibrary.com p81; Corbis pp232, 236; Getty Images/PhotoDisc pp232, 285; Kobal Collection/MGM/EON/Maidment, Jay p284; Chrissie Martin p288; iStockPhoto pp290, 335; Harcourt Education Ltd/Tudor Photography p291; iStockPhoto/Gloria-Leigh Logan p293.

AQA material is reproduced by permission of the Assessment and Qualifications Alliance. AQA take no responsibility for the accuracy of questions or answers in this book.

The map on page 298 is reproduced by permission of The Ordnance Survey. © Crown Copyright. All rights reserved.

Every effort has been made to contact copyright holders of material reproduced in this book. Any omissions will be rectified in subsequent printings if notice is given to the publishers.

There are links to relevant websites in this book. In order to ensure that the links are up-to-date, that the links work, and that sites are not inadvertently linked to sites that could be considered offensive, we have made the links available on the Heinemann website at www.heinemann.co.uk/hotlinks. When you access the site, the express code is 0459P.

Tel: 01865 888058   www.heinemann.co.uk   www.tigermaths.co.uk

About this book

AQA GCSE Mathematics has been written to meet the requirements of the National Curriculum and provides full coverage of the new two-tier AQA Syllabus which is to be first examined in June 2008.

The book is geared to examination success and is suitable for students in Years 10 and 11. It can be used as a classroom textbook with teacher input, or as a self-help guide for students, using the comment boxes and hints that are a key feature of the book.

The chapters on Number are headed in purple, Algebra chapters in yellow, Shape, Space and Measures chapters in green and Handling Data chapters in brown. This will help you to find related chapters quickly.

Each chapter consists of an introduction outlining the scope of the chapter followed by examples with explanatory notes. Helpful comments and hints feature throughout and key words are highlighted.

Exercises give students plenty of practice and are structured to provide a clear progression path. Icons indicate whether use of a calculator is advised.

At the end of each chapter, exam style questions are included, followed by a summary of key points. The summary sections include a guide to the examination grade. In addition, there are two examination papers at the end of the book, designed to reflect the current AQA papers.

Using and Applying Mathematics (UAM) questions test the ability to solve problems and explain and justify methods of solution. They place extra demands on students. These questions are flagged by the UAM icon.

Proof questions, either in the form of proof by counter example or by more formal methods, now occur on all examination papers. Chapter 31 shows students exactly how to tackle this type of question.

The authors hope that this textbook will make Mathematics more accessible for a wide range of students and will lead to a greater understanding and enjoyment of the subject!

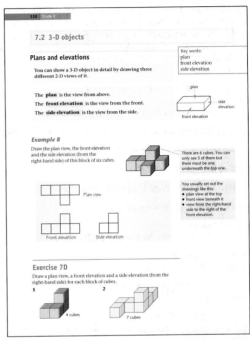

You can find a guide to help you link the specification objectives to this student book at www.tigermaths.co.uk.

Contents

This chapter will show you how to:

✔ find squares, cubes, square roots and cube roots
✔ write a number as a product of its prime factors
✔ find the highest common factor (HCF)
✔ find the lowest common multiple (LCM)

1.1 Square numbers, cube numbers, square roots and cube roots

Key words:
square
power
index
cube
square root
cube root

Squares and cubes

A **square** number is what you get when you multiply a whole number by itself.

$4 \times 4 = 16$
$15 \times 15 = 225$

The area is 225 square units.

16 and 225 are examples of square numbers.

Square numbers always have an odd number of factors.

The factors of 16 are

  1   2   4   8   16

Notice that there are pairs of factors except for the square root on its own in the middle.

4×4 can be written as 4^2.
The ^{2} is called a **power** (or **index**) and you read this as

   '4 squared' or 'the square of 4' or '4 to the power of 2'.

A **cube** number is what you get when you multiply a whole number by itself, then by itself again.

$2 \times 2 \times 2 = 8$
$10 \times 10 \times 10 = 1000$

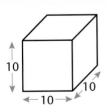

The volume is 1000 cubic units.

8 and 1000 are examples of cube numbers.

$2 \times 2 \times 2$ can be written as 2^3.
The ^{3} is a power (or index) and you read this as

   '2 cubed' or 'the cube of 2' or '2 to the power of 3'.

Your calculator has a key for squaring numbers

... and a key for cubing numbers or

Square roots and cube roots

The inverse of squaring is finding the square root .

You say the square root of 16 is 4, because $4^2 = 16$.
This is written as $\sqrt{16} = 4$.
Since $(-4) \times (-4) = 16$ then $(-4)^2 = 16$.
So -4 is also a square root of 16.

> This is called the negative square root.

> Any *positive* number has *two* square roots, a positive square root and a negative square root.

The inverse of cubing is finding the cube root .

You say the cube root of 8 is 2, because $2^3 = 8$.
This is written as $\sqrt[3]{8} = 2$.
When you cube a positive number, such as 2, the answer is positive.

$2 \times 2 \times 2 = 8$.

When you cube a negative number, such as -2, the answer is negative.

$(-2) \times (-2) \times (-2) = -8$.

So $\sqrt[3]{8} = 2$ and $\sqrt[3]{-8} = -2$

> $(-2) \times (-2) = +4$
> then $(+4) \times (-2) = -8$

> The cube root of a positive number is positive;
> the cube root of a negative number is negative.

Your calculator has a key for square roots

> The square root key on a calculator will give you the positive square root.

... and a key for cube roots

 or

Exercise 1A

1 Work out the following, using only the positive square root in parts **(d)**, **(e)** and **(f)**.

   (a) $3^2 + 4^3$ **(b)** $5^3 - 6^2$ **(c)** $8^2 \times 2^3$

   (d) $\sqrt{25} + \sqrt{100}$ **(e)** $\sqrt{64} \div \sqrt{4}$ **(f)** $\dfrac{\sqrt{81} \times 10^2}{\sqrt{9}}$

2 Work out each of the following, where possible, and then write a comment about the signs.

(a) 4^2 $(-4)^2$ 5^2 $(-5)^2$ 10^2 $(-10)^2$

(b) 2^3 $(-2)^3$ 3^3 $(-3)^3$ 5^3 $(-5)^3$

(c) $\sqrt{16}$ $\sqrt{-16}$ $\sqrt{25}$ $\sqrt{-25}$

(d) $\sqrt[3]{8}$ $\sqrt[3]{-8}$ $\sqrt[3]{64}$ $\sqrt[3]{-64}$

> Is there a difference between -4^2 and $(-4)^2$?

3 True or False?

(a) 144 is a square number.

(b) The cube of 10 is 100.

(c) The square root of -25 is -5.

(d) The difference between two consecutive square numbers is odd.

(e) An even number cubed is always even.

4 Evaluate the following.

(a) $6.2^2 + 3.8^3 - (-4.9)^2$ (b) $(9.2 \times 5.8)^2$

(c) $\sqrt{1.19 + 4.1}$ (d) $(-8)^2 + 5.4^3$

(e) $\sqrt[3]{216} + 2^3$ (f) $\sqrt{49} + 3^2$

> Write down all possible answers.

5 729 is a square number (27^2) and also a cube number (9^3).

(a) Find two more numbers like this.

(b) Write down a rule for finding these numbers.

1.2 Writing a number as the product of its prime factors

> **Key words:**
> prime factor
> index form
> factor tree

Any factor of a number which is also a prime number is called a prime factor .

> A prime number is a number with exactly two factors, itself and 1:
> 2, 3, 5, 7, 11, 13, 17, 19, 23, ...

For example, the factors of 10 are 1, 2, 5 and 10. 2 and 5 are also prime numbers, so you say that 2 and 5 are prime factors of 10, or that (written as a product of its prime factors) $10 = 2 \times 5$.

You can write any number as the product of its prime factors.

> Remember that to find a **product** you must multiply.

1 Start with the first prime number, 2. Divide your number by 2 (if it is possible to do so) as many times as you can.

2 When you can no longer divide by 2, try dividing by the next prime number, 3. Do this as many times as you can.

3 Then try dividing by 5 as many times as you can.

4 Then by 7, then by 11, ... and so on until you have an answer of 1.

> You do not have to divide by the prime numbers in this order, but it will help you to remember where you are.
>
> The method will work whatever order you choose.

Your original number is the product of all the primes you have divided by.

Example 1

Write 84 as the product of its prime factors.

$$
\begin{array}{r|r}
2 & 84 \\
2 & 42 \\
3 & 21 \\
7 & 7 \\
& 1
\end{array}
$$

$84 = 2 \times 2 \times 3 \times 7$

or

$84 = 2^2 \times 3 \times 7$

> You are often asked to write your answer in **index form**.
> So $84 = 2^2 \times 3 \times 7$

Example 2

Write 1980 as the product of its prime factors, giving your answer in index form.

$$
\begin{array}{r|r}
2 & 1\,980 \\
2 & 990 \\
3 & 495 \\
3 & 165 \\
5 & 33 \\
11 & 3 \\
& 1
\end{array}
$$

$1980 = 2 \times 2 \times 3 \times 3 \times 5 \times 11$

or

$1980 = 2^2 \times 3^2 \times 5 \times 11$

You can also write a number as the product of its prime factors, using **factor trees**.

1 Divide your number by any factor, prime or otherwise, and write the factor pair on 'branches' from the original number,

2 Keep dividing the factors in this way until you only have prime numbers at the ends of the branches.

3 The answer is the product of the primes on the branches.

Example 3

Write (a) 84 and (b) 1980 as the products of their prime factors.
Give your answers in index form.

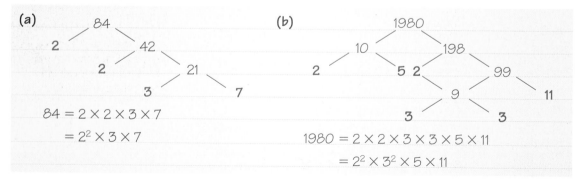

(a)
$84 = 2 \times 2 \times 3 \times 7$
$ = 2^2 \times 3 \times 7$

(b)
$1980 = 2 \times 2 \times 3 \times 3 \times 5 \times 11$
$ = 2^2 \times 3^2 \times 5 \times 11$

Exercise 1B

1 Write each number as a product of prime factors.
Give your answers in index form.

    (a) 20   (b) 72   (c) 45   (d) 144

    (e) 540   (f) 1000   (g) 3150   (h) 1323

2 Which of these numbers are prime factors of 264?

        12   3   5   11   7   4

3 Which of the following numbers are prime factors of both 72 and 90?

        2   5   12   18   9   3   6

4 (a) List the prime factors of 96.

    (b) List the prime factors of 60.

    (c) Write out the common prime factors of 96 and 60.

1.3 Highest common factor (HCF)

Key words:
highest common factor

The **highest common factor** (HCF) of two or more numbers is the largest number that divides into all of them exactly.

To find the HCF, write each number as the product of its prime factors.

Once you have done this, pick out the prime factors which are common to all the numbers. Multiply the common prime factors to get the highest common factor.

Example 4

Find the highest common factor of 54, 72 and 90.

2)54	2)72	2)90
3)27	2)36	3)45
3) 9	2)18	3)15
3) 3	3) 9	5) 5
1	3) 3	1
	1	

18 divides 3 times into 54
             4 times into 72
             5 times into 90

You can tell this by looking at the factors that are left once you have taken out the common factors.

The factors in common to all three are **2**, **3** and **3**.

So the HCF of 54, 72 and 90 is **2** × **3** × **3** = 18.

Exercise 1C

1 Find the highest common factor of:

    (a) 48 and 80   **(b)** 360 and 540

    (c) 120 and 336   **(d)** 525 and 90

    (e) 168, 315 and 525   **(f)** 400, 216 and 280.

2 Match two or three numbers from the first column with their HCF from the second column. Numbers can be used more than once.

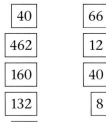

3 Which of the following pairs have no common prime factors?

    (a) 154 and 225   **(b)** 195 and 392   **(c)** 170 and 357

4 A jeweller has two lengths of gold chain. One is 672 cm long and one 720 cm long. He wants to cut them into pieces the same length to make necklaces, which are as long as possible. He wants to use all the chain, so what length should he cut them into?

1.4 Lowest common multiple (LCM)

Key words:

lowest common multiple

The **lowest common multiple** (LCM) of two or more numbers is the smallest number that is a multiple of all of them.

To find the LCM, write each number as the product of its prime factors, in the same way as for the HCF.

Once you have done this, pick out the highest power of each of the prime factors in the lists and multiply them to get the LCM.

These are the same numbers as in Example 4 so the lists of the prime factors have been taken from that example.

You must write each number as the product of its prime factors before you can work out the LCM.

Example 5

Find the lowest common multiple of 54, 72 and 90.

$$54 = 2 \times 3 \times 3 \times 3 \quad\quad = 2 \times 3^3$$
$$72 = 2 \times 2 \times 2 \times 3 \times 3 = 2^3 \times 3^2$$
$$90 = 2 \times 3 \times 3 \times 5 \quad\quad = 2 \times 3^2 \times 5$$

The LCM of 54, 72 and 90 is $2^3 \times 3^3 \times 5 = 1080$.

The highest power of each of the prime factors is shown in red.

$1080 = 20 \times 54$
$1080 = 15 \times 72$
$1080 = 12 \times 90$
1080 is the smallest number in the times tables of 54, 72 and 90.

Exercise 1D

1 Find the least common multiple of

(a) 16 and 12   (b) 32 and 40   (c) 27 and 54

(d) 11, 14 and 15   (e) 40, 120 and 35   (f) 36, 60 and 54.

'Least common multiple' means the same as 'lowest common multiple'.

UAM **2** A parrot squawks every 14 seconds. A frog croaks every 26 seconds. They both make a noise at the same time. After how long do they both make a noise at the same time again? Give your answer in minutes and seconds.

UAM **3** Pat is waiting for the Paddington bus which runs every 15 minutes. Ron is waiting for the Euston bus which runs every 18 minutes. They decide to wait until both buses come together before they get on. What is the longest they would have to wait?

4 Work out:

(a) $\frac{5}{12} + \frac{4}{15}$ (b) $\frac{29}{72} - \frac{7}{48}$ (c) $2\frac{5}{18} + 3\frac{7}{24}$

Find the LCM of the denominators.

UAM **5** Three bells on a tower ring at the same time. One rings every 12 seconds, another every 18 seconds and a third every 20 seconds. How long will it be before all three ring together?

Examination questions

1 (a) Write 28 as a product of its prime factors. Give your answer in index form.

(b) Find the least common multiple (LCM) of 28 and 42.   *(4 marks)*

AQA, Spec A, 1H, November 2004

2 (a) a and b are prime numbers.

$ab^3 = 54$

Find the value of a and b.

(b) Find the highest common factor (HCF) of 54 and 135.   *(5 marks)*

AQA, Spec A, 1H, June 2003

3 (a) You are given that $2x^3 = 250$.

Find the value of x.

(b) Write 75 as the product of its prime factors.   *(3 marks)*

AQA, Spec A, 1H, November 2003

4 Ahmed, Bill and Kath are counting beats. Ahmed strikes a gong every 6 beats, Bill hits a drum every 5 beats and Kath rings a bell every 3 beats. All three start playing at the same time. How many beats is it before they next play at the same time?   *(2 marks)*

Summary of key points

Powers and roots

A *square number* is what you get when you multiply a whole number by itself.

The inverse of squaring is finding the *square root*.

Any *positive* number has *two* square roots, a positive square root and a negative square root.

A *cube number* is what you get when you multiply a whole number by itself, then by itself again.

The inverse of cubing is finding the *cube root*.

The cube root of a positive number is positive; the cube root of a negative number is negative. For example $\sqrt[3]{8} = 2$ and $\sqrt[3]{-8} = -2$.

Writing a number as the product of its prime factors (grade C)

Any factor of a number which is also a prime number is called a *prime factor*. For example, the factors of 10 are 1, 2, 5 and 10 and of these both 2 and 5 are prime numbers, so you say that 2 and 5 are prime factors of 10.

You can write any number as the product of its prime factors. Divide the number by the prime numbers 2, 3, 5, 7, ... etc. in order until you have an answer of 1. Your answer is the product of the numbers you have divided by.

For example, $84 = 2 \times 2 \times 3 \times 7$ or $84 = 2^2 \times 3 \times 7$ (in index form).

You can also write a number as the product of its prime factors using 'factor trees'.

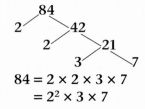

$$84 = 2 \times 2 \times 3 \times 7$$
$$= 2^2 \times 3 \times 7$$

Highest Common Factor (HCF) (grade C)

The highest common factor of two or more numbers is the highest number that divides into all of them exactly.

To find the HCF of two or more numbers, first write the numbers as products of their prime factors. Then pick out the factors which are common to all the numbers and multiply to get the HCF.

Lowest Common Multiple (LCM) (grade C)

The lowest common multiple of two or more numbers is the smallest number that is a multiple of all of them.

To find the LCM of two or more numbers, first write them as products of their prime factors, then pick out the highest power of each prime factor and multiply them to get the LCM.

2 Fractions

This chapter will show you how to:

✔ **compare fractions**
✔ **add, subtract, multiply and divide fractions**
✔ **find a fraction of a quantity**
✔ **solve fraction problems**
✔ **find reciprocals**

2.1 Writing fractions with a common denominator

Key words:
equivalent fractions
simplest form

$$\frac{3}{4} = \frac{6}{8} = \frac{12}{16}$$

These are called **equivalent fractions** .

There is a connection between these three fractions:

$$\frac{3}{4} \overset{\times 2}{\underset{\times 2}{=}} \frac{6}{8} \overset{\times 2}{\underset{\times 2}{=}} \frac{12}{16}$$

$$\frac{3}{4} \overset{\times 4}{\underset{\times 4}{=}} \frac{12}{16}$$

You could also write:

$$\frac{12}{16} \overset{\div 2}{\underset{\div 2}{=}} \frac{6}{8} \overset{\div 2}{\underset{\div 2}{=}} \frac{3}{4}$$

$$\frac{12}{16} \overset{\div 4}{\underset{\div 4}{=}} \frac{3}{4}$$

When there is no number that divides exactly into the numerator and denominator, a fraction is in its **simplest form** (in this case $\frac{3}{4}$).

You can find equivalent fractions by multiplying or dividing numerator and denominator by the same value.

Example 1

Find three more fractions which are equivalent to $\frac{2}{9}$.

$$\frac{2 \times 2}{9 \times 2} = \frac{4}{18}$$

$$\frac{2 \times 3}{9 \times 3} = \frac{6}{27}$$

$$\frac{2 \times 10}{9 \times 10} = \frac{20}{90}$$

Any choice of multiplier will give an equivalent fraction as long as it is applied to both the numerator *and* the denominator.

Example 2

Write these fractions in their simplest form:

(a) $\frac{8}{10}$ **(b)** $\frac{36}{48}$ **(c)** $\frac{40}{72}$

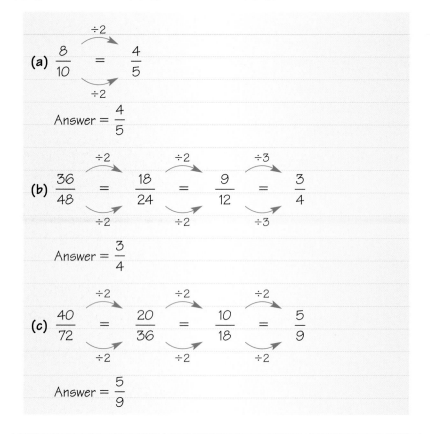

(a) $\frac{8}{10} = \frac{4}{5}$ ÷2 ÷2

Answer $= \frac{4}{5}$

(b) $\frac{36}{48} = \frac{18}{24} = \frac{9}{12} = \frac{3}{4}$ ÷2 ÷2 ÷3

Answer $= \frac{3}{4}$

(c) $\frac{40}{72} = \frac{20}{36} = \frac{10}{18} = \frac{5}{9}$ ÷2 ÷2 ÷2

Answer $= \frac{5}{9}$

You could work out **(b)** more quickly if you spotted that both 36 and 48 divide exactly by 4 or 6 or 12.

In **(c)**, you might spot that both 40 and 72 divide exactly by 4 or 8.

You can see that even if you do not spot the quickest way, you will get the correct answer if you follow the rules.

Exercise 2A

1 Write each fraction in its simplest form:

(a) $\frac{6}{8}$ **(b)** $\frac{12}{20}$ **(c)** $\frac{10}{15}$ **(d)** $\frac{9}{30}$

(e) $\frac{40}{48}$ **(f)** $\frac{16}{24}$ **(g)** $\frac{60}{72}$ **(h)** $\frac{27}{45}$

2 Luke ate $\frac{15}{30}$ of a pie and Katy ate $\frac{8}{32}$. Write these fractions in their simplest form.

3 In a garden $\frac{2}{12}$ is lawn, $\frac{3}{9}$ is flower beds and $\frac{9}{18}$ is a vegetable plot.

   (a) Write each fraction in its simplest form.

   (b) Draw a rectangle split into 6 equal parts to represent the garden. Shade in and label the fractions for the lawn, flower beds and vegetable plot.

2.2 Comparing fractions

You will need to know:

● how to write equivalent fractions

> **Key words:**
> lowest common multiple (LCM)
> lowest common denominator

To compare fractions, first convert them into equivalent fractions with the same denominator.

Example 3

Put these fractions in order, smallest first, $\frac{1}{3}, \frac{7}{20}, \frac{3}{10}$.

LCM of 3, 20 and 10 is 60.

$$\frac{1}{3} \xlongequal{\times 20} \frac{20}{60}$$

$$\frac{7}{20} \xlongequal{\times 3} \frac{21}{60}$$

$$\frac{3}{10} \xlongequal{\times 6} \frac{18}{60}$$

The order is $\dfrac{3}{10}, \dfrac{1}{3}, \dfrac{7}{20}$.

> The quickest way to find a common denominator is to multiply the two original denominators together.

> The **lowest common denominator** is the LCM of the denominators.

> 60 is the smallest number you can use.
> Any number in the 3, 10 and 20 'times-tables' would have worked.

> Alternatively, change the fractions to decimals first.
> $\frac{1}{3} = 0.333...$
> $\frac{7}{20} = 0.35$
> $\frac{3}{10} = 0.3$

> You can write this as
> $\frac{3}{10} < \frac{1}{3} < \frac{7}{20}$.

Exercise 2B

1 Put these fractions, in order, smallest first:

(a) $\frac{1}{4}, \frac{1}{2}, \frac{3}{8}$ (b) $\frac{3}{5}, \frac{1}{2}, \frac{4}{10}$ (c) $\frac{5}{6}, \frac{1}{3}, \frac{3}{4}$

2 Change these to equivalent fractions, then write them using $>$ or $<$:

(a) $\frac{3}{5} \quad \frac{5}{8}$ (b) $\frac{1}{3} \quad \frac{3}{10}$ (c) $\frac{5}{6} \quad \frac{4}{9}$

(d) $\frac{3}{4} \quad \frac{5}{6}$ (e) $\frac{3}{5} \quad \frac{5}{7}$ (f) $\frac{1}{4} \quad \frac{2}{5}$

> $<$ means 'less than'
> $>$ means 'greater than'

3 Put in order from smallest to largest:

(a) $\frac{7}{10}, \frac{1}{4}, \frac{3}{5}$ (b) $\frac{5}{6}, \frac{2}{3}, \frac{7}{12}$ (c) $\frac{5}{8}, \frac{7}{10}, \frac{17}{40}$ (d) $\frac{57}{100}, \frac{13}{25}, \frac{14}{20}$

2.3 Adding and subtracting fractions

You will need to know:
- **how to write equivalent fractions**
- **how to handle improper fractions and mixed numbers**

You can only *add* and *subtract* fractions when they have the *same denominator*.

When fractions do not have the same denominator you will need to use equivalent fractions to change one or both fractions so that the denominators are the same.

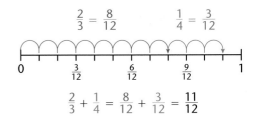

$$\frac{2}{3} = \frac{8}{12} \qquad \frac{1}{4} = \frac{3}{12}$$

$$\frac{2}{3} + \frac{1}{4} = \frac{8}{12} + \frac{3}{12} = \frac{11}{12}$$

Example 4

Work out: (a) $\frac{2}{3} + \frac{4}{5}$ (b) $4\frac{3}{4} - 1\frac{4}{5}$

(a) $\frac{2}{3} + \frac{4}{5}$

$= \frac{10}{15} + \frac{12}{15}$

$= \frac{22}{15}$

$= 1\frac{7}{15}$

(b) $4\frac{3}{4} - 1\frac{4}{5}$

$= \frac{19}{4} - \frac{9}{5}$

$= \frac{95}{20} - \frac{36}{20}$

$= \frac{59}{20}$

$= 2\frac{19}{20}$

Look at the denominators, 3 and 5, and change both into a denominator of 15 by using equivalent fractions.

Write the mixed numbers as improper fractions then use a denominator of 20 to find equivalent fractions.

Always give your answer as a mixed number.

Exercise 2C

1 Work out:

(a) $\frac{5}{6} + \frac{2}{3}$

(b) $\frac{4}{5} - \frac{3}{4}$

(c) $1\frac{5}{8} + 3\frac{1}{6}$

(d) $\frac{2}{3} + 1\frac{2}{5} + \frac{7}{10}$

(e) $4\frac{2}{7} - 3\frac{4}{5}$

(f) $1\frac{3}{4} + 3\frac{1}{3} - 2\frac{5}{6}$

(g) $3\frac{1}{4} + 2\frac{1}{6}$

(h) $2\frac{5}{9} - 1\frac{4}{5}$

(i) $1\frac{7}{8} + 4\frac{3}{10} - 3\frac{3}{5}$

2.4 Finding a fraction of a quantity

You need to be able to do calculations such as finding $\frac{3}{5}$ of £80.

> In mathematics 'of' means multiply, so the calculation can be written as
>
> $$\frac{3}{5} \text{ of } £80 = \frac{3}{5} \times 80 = \frac{3 \times 80}{5} = \frac{240}{5} = 48$$

Another way to do this is to work out $\frac{1}{5}$ of £80 first, then multiply by 3, because $\frac{3}{5} = 3 \times \frac{1}{5}$.

$$\frac{1}{5} \text{ of } 80 = \frac{1}{5} \times 80 = \frac{80}{5} = 16$$

So $\quad \frac{3}{5}$ of £80 = $3 \times £16 = £48$

You can see that both methods give £48 for the answer. You can choose whichever one you prefer.

Example 5

A factory employs 216 people.
$\frac{5}{12}$ of them are men.

(a) How many men work in the factory?

(b) How many women work in the factory?

$$\frac{5}{12} \text{ of } 216 = \frac{5}{12} \times 216 = \frac{5 \times 216}{12} = \frac{1080}{12} = 90$$

(a) 90 men work in the factory.

(b) So the number of women = $216 - 90$

$$= 126.$$

Exercise 2D

1 I spend $\frac{1}{6}$ of my pocket money on a magazine and $\frac{3}{4}$ of what is left on a CD. What fraction of my pocket money have I spent altogether?

2 (a) What is the area of $\frac{2}{5}$ of a circle with a radius of $3\frac{1}{2}$ cm? Use $\frac{22}{7}$ for π.

(b) I join this to the end of a square with side length $3\frac{1}{3}$ cm. What is the total area of the new shape?

Area of a circle = πr^2.

3 I spend $\frac{3}{8}$ of my pocket money on sweets, $\frac{2}{5}$ of the remainder on a magazine and $\frac{1}{2}$ of what is left on a ticket to the cinema. If I have £1.50 left, how much was my pocket money?

2.5 Multiplying fractions

To *multiply* two fractions you *multiply the numerators*, and *multiply the denominators*.

You then need to write the answer in its simplest form.

Example 6

Work out: **(a)** $\frac{4}{7} \times \frac{3}{5}$ **(b)** $2\frac{2}{7} \times \frac{3}{8}$

(a) $\frac{4}{7} \times \frac{3}{5}$

$= \frac{4 \times 3}{7 \times 5}$

$= \frac{12}{35}$

(b) $2\frac{2}{7} \times \frac{3}{8}$

$= \frac{16}{7} \times \frac{3}{8}$

$= \frac{\overset{2}{\cancel{16}}}{7} \times \frac{3}{\underset{1}{\cancel{8}}}$

$= \frac{2 \times 3}{7 \times 1}$

$= \frac{6}{7}$

Exercise 2E

1 Find the value of:

(a) $\frac{4}{7} \times \frac{3}{4}$

(b) $3\frac{1}{3} \times 1\frac{4}{5}$

(c) $\frac{6}{11} \times 7\frac{1}{3}$

(d) $2\frac{1}{3} \times 4\frac{5}{7}$

(e) $1\frac{2}{5} \times 1\frac{7}{18} \times 1\frac{13}{14}$

(f) $\frac{2}{3} \times \frac{4}{8} \times 3\frac{4}{11}$

2.6 Reciprocals

Key words:
reciprocal
invert

Look at these calculations:

$$\frac{10}{2} = 5 \qquad\qquad 10 \times \frac{1}{2} = 5$$
$$10 \times 2 = 20 \qquad 10 \div \frac{1}{2} = 20$$

2 and $\frac{1}{2}$ are examples of **reciprocals** .

Dividing by 2 is the same as multiplying by $\frac{1}{2}$.

Multiplying by 2 is the same as dividing by $\frac{1}{2}$.

> **The reciprocal of a number is the value obtained when it is divided into 1.**

$$2 \times \frac{1}{2} = 1 \qquad 3 \times \frac{1}{3} = 1 \qquad 10 \times \frac{1}{10} = 1$$

> **When you multiply a number by its reciprocal you always get 1.**

Zero has no reciprocal because division by zero is not defined.

The reciprocal of a number can be found using the $\boxed{x^{-1}}$ or $\boxed{1/x}$ key on your calculator.

For example, $\boxed{7}$ $\boxed{x^{-1}}$ $= 0.14285714...$

which is the decimal equivalent of $\frac{1}{7}$

Negative powers are explained in Chapter 13.

The reciprocal of a fraction such as $\frac{2}{3}$ is $\frac{3}{2}$ since $\frac{2}{3} \times \frac{3}{2} = \frac{6}{6} = 1$

> **To find the reciprocal of a fraction you **invert** the fraction (turn it upside down).**

The reciprocal of any fraction m/n is n/m where m and n are integers.

Example 7

Find the reciprocal of

(a) 8   (b) $\frac{1}{9}$ (c) $5\frac{1}{2}$ (d) 0.625

(a) $\frac{1}{8}$ Check: $8 \times \frac{1}{8} = \frac{8}{8} = 1$

(b) 9 Check: $\frac{1}{9} \times 9 = \frac{9}{9} = 1$

(c) $5\frac{1}{2} = \frac{11}{2}$

    so reciprocal $= \frac{2}{11}$ Check: $\frac{11}{2} \times \frac{2}{11} = \frac{22}{22} = 1$

(d) 1.6 Check: $0.625 \times 1.6 = 1$

Change mixed numbers into improper fractions before you find the reciprocal.

Use the $\boxed{x^{-1}}$ key on your calculator.

Exercise 2F

1 Find the reciprocals of the following:

(a) 4   (b) $\frac{1}{6}$ (c) 15   (d) $\frac{1}{20}$

(e) $\frac{3}{4}$ (f) $\frac{2}{9}$ (g) $1\frac{2}{5}$ (h) $4\frac{1}{3}$

(i) $2\frac{3}{7}$ (j) 2.5   (k) 0.4   (l) 1.125

(m) $6\frac{5}{13}$ (n) $12\frac{4}{11}$

2 Put the *reciprocals* of these numbers in order of size, smallest first.

$$1.25,\ 1\tfrac{5}{8},\ 1\tfrac{2}{7},\ 2,\ 1.2,\ 1\tfrac{1}{3},\ 1\tfrac{5}{6},\ 2.1,\ 1.5,\ 1\tfrac{4}{11}$$

2.7 Dividing fractions

Key words:
invert

To *divide* two fractions you **invert** (turn upside down) *the fraction you are dividing by* and *change the division sign to a multiplication sign.*

$$\text{So}\quad \frac{1}{3} \div \frac{2}{5} = \frac{\frac{1}{3}}{\frac{2}{5}} = \frac{\frac{1}{3} \times \frac{5}{2}}{\frac{2}{5} \times \frac{5}{2}} = \frac{\frac{1}{3} \times \frac{5}{2}}{1} = \frac{1}{3} \times \frac{5}{2} = \frac{5}{6}$$

This method also works when you are dividing by a whole number or when the division involves mixed numbers. Change mixed numbers to improper fractions first.

> You turn the *2nd* fraction upside down. Then you follow the method for multiplying fractions.

Example 8

Work out: (a) $1\frac{3}{4} \div \frac{5}{6}$ (b) $\frac{1}{7} \times \frac{3}{4} \div \frac{4}{7}$

(a) $1\frac{3}{4} \div \frac{5}{6}$

$$= \frac{7}{4} \div \frac{5}{6}$$

$$= \frac{7}{4} \times \frac{6}{5}$$

$$= \frac{42}{20}$$

$$= \frac{21}{10}$$

$$= 2\frac{1}{10}$$

> Write $1\frac{3}{4}$ as an improper fraction.
>
> Invert the *2nd* fraction and multiply.

> Write the answer as a mixed number in its simplest form.

continued ▶

(b) $\dfrac{1}{7} \times \dfrac{3}{4} \div \dfrac{4}{7}$

$$= \dfrac{3}{28} \div \dfrac{4}{7}$$

$$= \dfrac{3}{28} \times \dfrac{7}{4}$$

$$= \dfrac{3 \times \overset{1}{\cancel{7}}}{\underset{4}{\cancel{28}} \times 4}$$

$$= \dfrac{3}{16}$$

Exercise 2G

1 Calculate:

(a) $\dfrac{5}{6} \div \dfrac{2}{3}$

(b) $3\dfrac{1}{7} \div 1\dfrac{3}{8}$

(c) $1\dfrac{2}{7} \div 1\dfrac{1}{2}$

(d) $\dfrac{8}{9} \div \dfrac{7}{3}$

(e) $\dfrac{3}{10} \times 2\dfrac{1}{4} \div 1\dfrac{1}{5}$

(f) $3\dfrac{1}{7} \times 2\dfrac{4}{5} \div 1\dfrac{1}{10}$

(g) $\left(\dfrac{3}{4} \times \dfrac{8}{9}\right) + \left(1\dfrac{3}{5} \div 2\dfrac{2}{15}\right)$

2.8 Solving fractions problems

You will need to be able to apply your skills in handling fractions when you are asked to solve problems. You will need to identify which operator $(+, -, \times, \div)$ is appropriate.

Example 9

John eats $\dfrac{2}{5}$ of a bar of chocolate. Linda eats $\dfrac{4}{9}$ of what remains.
What fraction of the bar of chocolate have they eaten between them?

John eats $\dfrac{2}{5}$ of the bar, which leaves $1 - \dfrac{2}{5} = \dfrac{3}{5}$

Linda eats $\dfrac{4}{9}$ of $\dfrac{3}{5}$ $\quad = \dfrac{4}{9} \times \dfrac{3}{5} = \dfrac{4 \times 1}{3 \times 5} = \dfrac{4}{15}$

Total amount eaten $= \dfrac{2}{5} + \dfrac{4}{15}$

$$= \dfrac{6 + 4}{15}$$

$$= \dfrac{10}{15}$$

$$= \dfrac{2}{3}$$

Find out what fraction is left after John has eaten some chocolate *before* you work out the fraction that Linda eats.

Subtraction, multiplication and addition are all needed.

Exercise 2H

1 A rectangular path is $3\frac{1}{8}$ metres long and $\frac{4}{5}$ metres wide.
What is its area?

2 I eat $\frac{1}{4}$ of a pizza and give two friends $\frac{1}{6}$ of the remainder each.
How much do I have left?

3 A cylinder with a mass of $2\frac{2}{3}$ kg and pistons with a mass of $2\frac{3}{4}$ kg
are put into a wooden box with a mass of $\frac{5}{6}$ kg. The mail company
charges £2 per kg for packages up to 4 kg and then charges £1 for
every $\frac{1}{4}$ kg over that. How much will the package cost to post?

4 **(a)** How many pieces of wood $\frac{3}{5}$ m long can I cut from a piece
$2\frac{1}{4}$ m long?

(b) How much is left after I have cut the $\frac{3}{5}$ m lengths possible?

5 $\frac{2}{5}$ of my garden is lawn and $\frac{1}{3}$ of the
rest is a flower bed.
How much of my garden is left for
a vegetable plot?

UAM **6** The distance between Newborough and Castletown by the
motorway is $29\frac{1}{3}$ miles. Using a B road the distance is $14\frac{4}{5}$ miles.
My car uses $\frac{1}{5}$ litre of petrol per mile, whichever route is used.
How much more petrol will I use if I take the motorway?

7 Find the height (h) of a cuboid with length (l) $1\frac{4}{5}$ m, width (w)
$1\frac{1}{3}$ m and a total volume of 6 m³.

> Remember:
> Volume $= l \times w \times h$
>
> So $h = \dfrac{\text{volume}}{l \times w}$

Examination questions

1 **(a)** There are 40 sweets in a bag.
Ben eats $\frac{1}{8}$ of the 40 sweets.
Jerry eats $\frac{1}{5}$ of the 40 sweets.
What fraction of the sweets do Ben and Jerry eat together?

(b) Work out $\frac{2}{5} - \frac{3}{8}$

(5 marks)
AQA, Spec A, 1H, November 2004

2 Linda uses $\frac{3}{5}$ of a tin of paint to paint a fence panel.
What is the **least** number of tins she needs to paint 8 fence panels?   *(3 marks)*
AQA, Spec A, 2I, November 2004

Summary of key points

Equivalent fractions

Equivalent fractions have the same value. To find equivalent fractions, multiply or divide numerator and denominator by the same number.

For example, $\dfrac{2}{3} = \dfrac{2 \times 5}{3 \times 5} = \dfrac{10}{15}$

Comparing fractions

First convert them into equivalent fractions with the same denominator.

Adding and subtracting fractions

You can only *add* and *subtract* fractions when they have the *same denominator*.

When fractions do not have the same denominator, use equivalent fractions to change one or both fractions so that the denominators are the same.

Fraction of a quantity

A calculation such as $\frac{3}{5}$ of 80 can be written as $\dfrac{3 \times 80}{5}$ or $\dfrac{80}{5} \times 3$.

Reciprocals (grade C)

The reciprocal of a number is the value obtained when it is divided into 1.

For example: The reciprocal of 3 is $\frac{1}{3}$

When you multiply a number by its reciprocal you always get 1.

For example: $3 \times \frac{1}{3} = 1$

To find the reciprocal of a fraction, you invert the fraction.

For example: The reciprocal of $\frac{2}{3}$ is $\frac{3}{2}$ since $\frac{2}{3} \times \frac{3}{2} = \frac{6}{6} = 1$

Multiplying and dividing fractions (grade D)

To *multiply* two fractions you *multiply the numerators*, and *multiply the denominators*.

You must then write the answer in its simplest form.

To *divide* two fractions you *invert* (turn upside down) *the fraction you are dividing by* and *change the division sign to a multiplication sign*.

So $\dfrac{1}{3} \div \dfrac{2}{5} = \dfrac{1}{3} \times \dfrac{5}{2} = \dfrac{5}{6}$

This method works when you are dividing by a whole number or when the division involves mixed numbers. Change mixed numbers to improper fractions first.

3 Percentages

This chapter will show you how to:
- ✔ convert between fractions, decimals and percentages
- ✔ solve percentage problems
- ✔ solve compound percentage problems
- ✔ interpret index numbers

3.1 Converting between fractions, decimals and percentages

You will need to know:
- about place value
- how to multiply and divide by 10, 100, 1000
- how to write equivalent fractions

To change a *percentage* to a *fraction or a decimal*, divide by 100.

For example, $43\% = \dfrac{43}{100} = 0.43$

> Remember: a percentage of a quantity shows how many parts of a hundred that quantity is. $50\% = \frac{50}{100}$

Example 1

Write these percentages as fractions in their simplest form.

(a) 58%   (b) $12\frac{1}{2}\%$

(a) 58%

$= \dfrac{58}{100}$

$= \dfrac{29}{50}$

(b) $12\frac{1}{2}\%$

$= \dfrac{12.5}{100}$

$= \dfrac{25}{200}$

$= \dfrac{1}{8}$

> Multiply numerator and denominator by 2 to convert to the numerator to a whole number.

To change a *fraction or a decimal* to a *percentage*, multiply by 100.

For example, $\dfrac{2}{5} = \left(\dfrac{2}{5}\right) \times 100\% = 40\%$

and $0.725 = (0.725 \times 100)\% = 72.5\%.$

Example 2

Write these fractions as percentages:

(a) $\frac{5}{8}$ (b) $\frac{31}{40}$

(a) $\frac{5}{8}$

$= \frac{5}{8} \times 100\%$

$= \frac{5}{8_2} \times 100^{25}$

$= \frac{125}{2}\%$

$= 62.5\%$

(b) $\frac{31}{40}$

$= \frac{31}{40} \times 100\%$

$= \frac{31}{40_2} \times 100^5$

$= \frac{155}{2}\%$

$= 77.5\%$

> This type of question could appear on the non-calculator paper.

To change a *decimal* to a *fraction*, use the place value of the *last* significant figure, as the denominator of the fraction.

For example, $0.24 = \frac{24}{100} = \frac{6}{25}$ (in its simplest form).

Example 3

Write these decimals as fractions:

(a) 0.48   (a) 0.037   (c) 2.175

(a) 0.48

$= \frac{48}{100}$

$= \frac{12}{25}$

(b) 0.037

$= \frac{37}{1000}$

(c) 2.175

$= 2\frac{175}{1000}$

$= 2\frac{7}{40}$

> Write (a) and (c) in their simplest form.
> (b) cannot be simplified.

Exercise 3A

1 Write the following as fractions in their simplest form.

(a) 15%   (b) 36%   (c) 0.4   (d) 0.78   (e) 0.125

2 Write the following as percentages and place them in order from smallest to largest.

$2\frac{1}{10}$ $2\frac{7}{100}$ 2.4 $2\frac{8}{25}$ 2.34

3 Copy and complete the following table of equivalent fractions, decimals and percentages.

Percentage	Fraction	Decimal
60%		
		0.72
	$\frac{5}{8}$	
		1.025
$12\frac{1}{2}\%$		

> Write the fractions in their simplest form.

4 A group of students were trying to find the proportion of a certain chemical in different soils. They each tested a sample and got these results.

$$0.438 \qquad \frac{9}{20} \qquad 42\% \qquad 0.4 \qquad 43\tfrac{3}{4}\%$$

Place these in order, from smallest to largest.

3.2 Using percentages to compare amounts

You can use percentages to compare quantities, such as working out which of two or more amounts is the greater, or which of two or more offers is the 'best buy'.

Example 4

Village A has a population of 1540 of which $\frac{3}{20}$ are under 16 years of age. Village B has a population of 1250 of which 18% are under 16 years of age. Which village has the greater number of people under 16 years of age?

Village A

Number of people under 16 $= \frac{3}{20}$ of 1540

$$= \frac{3}{20} \times 1540$$

$$= \frac{3 \times 1540}{20}$$

$$= 231$$

Village B

Number of people under 16 $= 18\%$ of 1250

$$= \frac{18}{100} \times 1250$$

$$= \frac{18 \times 1250}{100}$$

$$= 225$$

So Village A has the greater number of people under 16.

To compare, use fractions or percentages to work out the amounts.

Exercise 3B

1 Which one is the best offer?
Explain your answer.

2 Which weekend break is cheaper?

3 A special offer on breakfast cereal was shown on two different
boxes originally priced at £1.72 for 144 g.
One stated '25% extra free' the other said '$\frac{1}{4}$ off the price'.
Is this the same offer or is one a better buy?

3.3 Writing one quantity as a percentage of another

To write one quantity as a percentage of another:

**Write the first quantity as a fraction of the second quantity.
Then multiply by 100 to convert to a percentage.**

For example, 18 as a percentage of 45 $= \dfrac{18}{45} \times 100\% = 40\%$.

When you sit a test, your mark is often written as a fraction.

For example, if you score 17 out of 25 it would be written as $\frac{17}{25}$.

You convert this to a percentage by multiplying by 100.

So $\dfrac{17}{25} = \dfrac{17}{25} \times 100\% = 68\%$

Check:
$\frac{17}{25} = 17 \div 25$
$\quad = 0.68$
$\quad = 68\%$

Example 5

A secretary can type 60 words per minute, but makes an average of
3 errors in every 60 words. What percentage of the words contain errors?

$\dfrac{3}{60}$ words contain errors

so $\dfrac{3}{60} = \dfrac{3}{60} \times 100\% = 5\%$.

Exercise 3C

1 Ramat gets the following scores in his tests.
Write them as percentages.

Subject	English	Maths	French	History	Geography
Score	22	40	17	34	43
Total marks	40	75	30	65	80

For questions 1, 2 and 3, give your answers to the nearest whole percentage.

2 In my class there are 4 left-handed people. If there are 28 people in the class, what percentage are left-handed?

3 In a Linkgo set there are 14 blue blocks, 20 green and 12 red. Write the number of red blocks as a percentage of the set.

4 A model is 8 cm long. The width of a wheel is 25 mm. Write the width of the wheel as a percentage of the length of the model.

Take care to make all the units the same.

5 The following table shows the number of people in a drama club.

	Male	Female
Adult	17	15
Child	12	18

(a) What percentage of the club are male?

(b) What percentage of the club are girls?

(c) Write the number of adults as a percentage of the number of people in the club.

Give your answers to the nearest whole percentage.

6 A trainee secretary makes an average of 10 errors in every 80 words. What percentage of the words contain errors?

7 A factory produces 500 medical instruments in a day but on average 15 of these are faulty. What percentage is faulty?

3.4 Adding and subtracting a percentage of a quantity

You will often be asked to increase (or decrease) a quantity by a given percentage. There are several ways to do this. Two methods are shown below.

Method A
1 Work out the actual increase (or decrease).
2 Add to (or subtract from) the original amount.

Method B
1 Add the % increase to 100%
  (Subtract the % decrease from 100%)
2 Convert this % to a decimal.
3 Multiply the original amount by this decimal.

Example 6

Liz earns £150 per week.
Next week she is due to receive a pay rise of 6%.
What will be her new weekly pay?

Method A

Pay rise $= 6\%$ of £150

$\qquad = \dfrac{6}{100} \times 150$

$\qquad = £9$

Liz's new weekly pay

$\qquad = £150 + £9$

$\qquad = £159$

Method B

Increase $= 6\%$

New % $= 100\% + 6\%$

$\qquad = 106\%$

$\qquad = 1.06$

Liz's new weekly pay

$\qquad = 1.06 \times £150$

$\qquad = £159$

> Divide by 100 to convert a % to a decimal.

Example 7

Toby reduces his weight by 8% from 95 kg.
What is his new weight?

Method A

Weight loss $= 8\%$ of 95 kg

$\qquad = \dfrac{8}{100} \times 95 \text{ kg}$

$\qquad = 7.6 \text{ kg}$

Toby's new weight

$\qquad = 95 \text{ kg} - 7.6 \text{ kg}$

$\qquad = 87.4 \text{ kg}$

Method B

Decrease $= 8\%$

New % $= 100\% - 8\%$

$\qquad = 92\%$

$\qquad = 0.92$

Toby's new weight

$\qquad = 0.92 \times 95 \text{ kg}$

$\qquad = 87.4 \text{ kg}$

> Divide by 100 to convert a % to a decimal.

Exercise 3D

1 Work out the following:

(a) Increase 30 by 5% (b) Increase 58 by 12%

(c) Decrease 3400 ml by 18% (d) Decrease 480 by 32%

(e) Increase £135 by $7\frac{1}{2}$% (f) Decrease 890 ml by 8.4%

For each answer, give the new adjusted amount.

2 Firefighters are given a 4% pay rise. If John earned £420 per week, how much will he earn after the pay rise?

3 There has been a 6% decrease in the number of reported thefts in Walton this past year. There were 350 reported thefts last year. How many were there this year?

4 I pay £28 for my ticket and Peter pays £17.60.
How much will we have to pay after the increase?

All fares to increase by $7\frac{1}{2}$%

5 A new car costs £8450. After two years the value of the car will have decreased by 43%. How much will the car be worth?

6 Sales of a magazine, costing £1.95, decreased by 11% this week. They sold 4300 copies last week.
How much money have they lost on their sales this week?

7 Jane started her new job earning £12 000 per year, increasing by 3% after 3 months. Milly started on £11 500 per year, increasing by 5% after 3 months.
Who has the greater salary after 3 months?

8 Ali weighed 84 kg. He lost $4\frac{1}{2}$% of his body weight when he started running but then put 2% of his new weight back on.

(a) How much was Ali's lowest weight?

(b) How much does Ali weigh now?

9 In 2000 the number of pairs of breeding sparrows was estimated as 200 000. This decreased by 38% in 2001. In 2002 there was a slight recovery with an increase of 4%.
How many breeding pairs were estimated at the end of 2002?

10 Tom, Tina and Tracy each invested £250. Tom's money increased in value by $6\frac{3}{4}$%. Tina's money decreased by $2\frac{1}{2}$% and Tracy's money increased by $4\frac{1}{4}$%. How much does each have now?

Value Added Tax (VAT)

VAT is added to many items that you buy and to bills for services. This is an example of percentage increase.

The rate of VAT is $17\frac{1}{2}$%, which means that you will pay an extra $17\frac{1}{2}$% on top of the cost of the item you buy.

Example 8

A digital camera is advertised for sale at £240 (excluding VAT). How much will you have to pay?

'Excluding' means that VAT must be added on to the advertised price.

Method A

VAT $= 17\frac{1}{2}$% of £240

$= \dfrac{17.5}{100} \times 240$

$= £42$

Cost of digital camera

$= £240 + £42$

$= £282$

Method B

Increase $= 17\frac{1}{2}$%

New % $= 100\% + 17\frac{1}{2}\%$

$= 117\frac{1}{2}\%$

$= 1.175$

Cost of digital camera

$= 1.175 \times £240$

$= £282$

$117\frac{1}{2} \div 100 = 117.5 \div 100$
$= 1.175$
This is your multiplier to work out the % increase.

The VAT makes quite a difference to the price!

Exercise 3E

1 Work out **(a)** the VAT and **(b)** the total cost of each item.
   (i) a DVD player at £130 (ex. VAT)
   (ii) a phone at £39.99 (ex. VAT)

Give your answers to the nearest penny.

2 What is the total cost of a TV sold for £126 + VAT?

3 Which car is the more expensive?

4 A meal for four costs £73.58 plus VAT. If the four people decide to share the bill equally, how much will each pay?

5 The cost of a scooter is £375 + VAT. Jas decides to pay by credit. He pays an initial deposit of 15% and then 12 monthly payments of £35. How much more does Jas pay by buying on credit?

6 Which method is cheaper? Find how much can be saved using the cheaper method.

£199 + VAT

BUY NOW!

or

22% Deposit
Plus
12 monthly payments
of £19.75

3.5 Percentage change

Key words:
percentage change

You will need to write one quantity as a percentage of another when you are asked to work out **percentage change** .

$$\text{Percentage change} = \frac{\text{actual change}}{\text{original amount}} \times 100\%.$$

Always work out the actual change first. Don't forget that it is the *original* amount that goes on the bottom of the fraction.

Example 9

The original cost of a DVD player is £225. In a sale it is reduced to £180. What is the percentage decrease in the price?

Actual decrease in price = £225 − £180

$= £45$

$$\text{Percentage decrease} = \frac{\text{actual decrease}}{\text{original amount}} \times 100\%$$

$$= \frac{45}{225} \times 100\%$$

$$= 20\%$$

These examples are typical of the questions you can expect on your GCSE exam papers.

Example 10

The average attendance at a team's home games last season was 55 000. This year the average attendance is 58 575. What percentage increase is this?

Actual increase = 58 575 − 55 000

$= 3575$

First find the difference.

$$\text{Percentage increase} = \frac{\text{actual increase}}{\text{original amount}} \times 100\%$$

Then divide by the original amount.

$$= \frac{3575}{55\,000} \times 100\%$$

$$= 6.5\%$$

Write the answer as a percentage.

Exercise 3F

1 What is the percentage decrease in the price of the coat (originally £95)?

2 Jamie was given a pay rise. He used to earn £12 000 and now earns £12 400. What was the percentage increase in his pay?

3 The number of car thefts in Brigton rose from 483 in 2002 to 504 in 2003. What percentage increase is this?

4 The number of students taking a particular subject at A level has dropped from 4015 to 3984.

   (a) What is the percentage reduction?

   (b) If there is the same percentage reduction the following year, how many students will take the A level then?

5 The cost of a holiday has risen from £556 to £604.50.

   (a) What is the percentage increase?

   (b) Owing to a fuel crisis, the holiday company puts a 15% surcharge on the new cost of the holiday.
   What is the total percentage increase of the holiday now?

6 The Matis tribe living in the Amazon basin numbered 350. The numbers fell to 198 one month after contact with diseases from the outside world.

   (a) What was the percentage reduction in numbers?

   (b) The following year their number decreased by 4%. What is the total percentage reduction since the Matis came into contact with the outside world?

Percentage profit and loss

When you are dealing with money problems you will often meet the terms **profit** and **loss**, depending on whether you make money or lose money on the sale of an item.

You also need to know the terms **cost price**, the amount you pay for an item, and **selling price**, the amount you sell it for.

The formula for percentage change now becomes:

$$\text{Percentage profit (loss)} = \frac{\text{actual profit (loss)}}{\text{cost price}} = 100\%$$

> **Key words:**
> profit
> loss
> cost price
> selling price

> The cost price is the *original* price of the item and goes on the bottom of the fraction.

Example 11

A second-hand car salesman buys a car for £2500 then sells it for £3200. What is his percentage profit?

> Actual profit = £3200 − £2500
>
> = £700
>
> Percentage profit = $\dfrac{\text{actual profit}}{\text{cost price}} \times 100\%$
>
> = $\dfrac{700}{2500} \times 100\% = 28\%$

Example 12

Peter bought a new motorbike for £5450 but traded it in for another a year later. He got a trade-in value of £4578. What percentage loss is this?

> Actual loss = £5450 − £4578
>
> = £872
>
> Percentage loss = $\dfrac{\text{actual loss}}{\text{cost price}} \times 100\%$
>
> = $\dfrac{872}{5450} \times 100\% = 16\%$

Index numbers

An index number compares one number (often an average price) with another. The UK retail price index in December 2005 was 194.1. This compares with a base price of 100 in 1987. This means that average prices increased by 94.1% between 1987 and 2005.

> The index number is a percentage of the base, but the percentage sign is left out.

Example 13

The index for gold was 190 in January 2006, compared with a base of 100 in January 2001.

(a) What was the percentage increase in gold prices in that period?

(b) Gold cost £170 per ounce in January 2001. What was the price in January 2006?

> (a) 190 − 100 = 90
>
> Percentage increase = 90%
>
> (b) £170 × $\frac{190}{100}$ = £323

Exercise 3G

1 Ahmed sold a CD player for £30. He had bought it for £35. What was his percentage loss?

2 Sarah restores skateboards. She bought one for £5 and sold it for £18.25. What was her percentage profit?

3 For each of the following, find the percentage profit or loss.

Cost price	£50	£34	£6	£73.50	£125	£3500
Selling price	£65	£24	£7.50	£69.50	£84	£3635

4 A sack of pet food costs £15.50. If you buy two sacks the shop only charges £13.25 per sack. What is the percentage discount per sack when you buy two sacks?

> Discount is an amount taken off the price. It is worked out in the same way as a profit or loss.

5 Amy has a ticket for a rock concert. It cost her £32.50 and she wants to sell it for £45. What will be her percentage profit?

6 A DVD was sold for £14.99. It cost the shop £9 to buy. What was the percentage profit?

7 The house price index was quoted as 106.2 in February 2006, compared with a base index of 100 in February 2005.
  (a) What was the percentage increase?
  (b) If a house cost £190,000 in February 2006, what was its likely value 12 months earlier?

> Give your answer to the nearest £1000.

3.6 Reverse percentages

In some problems you are told the amount *after* a percentage increase or decrease and you have to work out the *original* amount.

Remember that the *original* amount in any calculation is always taken to be *100%* and the final amount will be a given percentage above or below 100%, depending on whether there has been an increase or a decrease.

If it is a percentage increase,
... **add** the % change to 100% then convert this % to a decimal.

If it is a percentage decrease,
... **subtract** the % change from 100% then convert this % to a decimal.

Set up your answer with this information and you will see that a simple division is all you need to do.

Example 14

Calculate the original amount.

Original amount	% change	Final amount
£?	10% increase	£66

Original (100%) + 10% = 110%

110% as a decimal = 1.10

Original amount × 1.10 = Final amount

Original amount × 1.10 = £66

$$\text{Original amount} = \frac{£66}{1.10}$$

$$= £60$$

The original amount has been increased by 10%.

Always multiply the original amount by the decimal.

The answer to questions of this type is always found by doing a division.

Do **not** make the mistake of working out the given percentage of the *final* amount and then subtracting.

Example 15

Sarah bought a new washing machine from a store offering all items at 20% off. If she paid £344, what was the original price of the washing machine?

100% − 20% = 80%

80% as a decimal = 0.80

Original price × 0.80 = Sale price

Original price × 0.80 = £344

$$\text{Original price} = \frac{£344}{0.80}$$

$$= £430$$

The original price has been reduced by 20%.

Check: 20% of
£430 = 0.20 × £430 = £86
£430 − £86 = £344

Exercise 3H

1 A computer is reduced by 12% in a sale. It now costs £334.40. What was the original price of the computer?

2 After a pay rise of 4%, Kim earned £14 560. What did she earn before the pay rise?

3 A mobile phone was priced at £135 including VAT. What would be the price before VAT was added?

Assume VAT = 17.5%

4 A spring is stretched by 32% to a length of 22 cm. What length was it before it was stretched?

Give your answer to the nearest millimetre.

5 A seal pup increases its body weight by $4\frac{1}{2}$% in one day. It now weighs 6.27 kg. What did it weigh yesterday?

6 A shop advertises 'We will pay your VAT'. How much would I pay for a bike priced at £87.50 inclusive of VAT?

7 The total number of pupils in school this year is 1034. This is a $5\frac{1}{2}$% increase on last year.

   (a) How many pupils were there last year?

   (b) Next year they expect an increase in pupils of 3%. How many pupils will there be then?

8 The value of a car depreciates by 17% in its first year.

   (a) If it is worth £12 035 after one year, what was its original price?

   (b) The original price included VAT. How much was the car *excluding* VAT?

> Depreciates means that its value goes down.

9 After paying a 12% deposit on a house, I needed a mortgage for £76 120. The estate agent charged me $2\frac{1}{4}$% of the *full* cost of the house for a survey of the property. How much did I have to pay for the survey?

10 During heating, a metal rod expands by $1\frac{1}{4}$%.

   (a) If the rod measures 40.5 cm when hot, what was its length originally?

   (b) After 10 minutes, it remains 0.5% longer than its original length. What is the length of the rod at this time?

3.7 Repeated percentage change

> Key words:
> compound interest
> invest

Examples of repeated percentage change include **compound interest** and population changes.

When you **invest** money the interest you earn is calculated using compound interest.

> The amount you pay back on a loan is also calculated using compound interest.

In compound interest
● the rate of interest is fixed
● the amount of interest you receive each year is *not* the same
● the interest is calculated on the sum of money you invested in the first place *plus* any interest you have already received.

Example 16

Venetia puts £800 into an account paying 5% p.a. compound interest. How much does she have in her account after two years?

> p.a. stands for 'per annum' meaning 'each year'.

At the start of year 1 she has £800

Interest at the end of year 1 = 5% of £800 = 0.05 × £800

= £40

> Notice the decimal version of 5% (0.05). It can be easier to use than $\frac{5}{100}$.

At the end of year 1 she has £800 + £40 = £840

Interest at the end of year 2 = 5% of £840 = 0.05 × £840

= £42

At the end of year 2 she has £840 + £42 = £882

Venetia has £882 in her account after 2 years.

> Notice how you use the sum of money at the end of one year to calculate the interest earned at the end of the next year.

Another way to calculate the amount of money you will have if you invest it at compound interest is as follows:

> This method is quick and easy to use.

1 Add the rate of interest to 100%.
2 Convert this % to a decimal.
3 Multiply the original amount of money by this decimal as many times as the number of years given in the question.

> You *must* remember to add the rate of interest to 100% before you convert to the decimal you will use as a multiplier.

In Example 16, Venetia invested £800 at a rate of interest of 5% p.a. for two years.

Step 1   100% + 5% = 105%

Step 2   105% = 1.05

Step 3   £800 × 1.05 × 1.05 = £800 × $(1.05)^2$ = £882

> Notice that this method gives you the final amount of money, not the interest.

These ideas can also be used to solve problems which do not involve money, but any quantity that is increasing or decreasing.

> In Example 17 the number of whales reduces by 8% of the number *at the start of each year*.
>
> Each % calculation is done on a *different* number so you must use the same methods as for compound interest problems.

Example 17

There are 5000 whales of a certain species, but scientists think that their numbers are reducing by 8% each year.

Estimate how many of these whales there will be in three years.

> The whales are *reducing* in number so you must *subtract* the % from 100%.

Numbers are reducing by 8% each year.

So 100% − 8% = 92% and 92% = 0.92 as a decimal

Number of whales in 3 years' time

= 5000 × 0.92 × 0.92 × 0.92

= 5000 × $(0.92)^3$

= 3893.44

Estimated number of whales   = 3900

> Using the power button on your calculator speeds up the calculation.

> You cannot give 3893.44 as your answer!!!
> 3893, 3890 and 3900 are all acceptable as estimates.

Example 18

Tony invests £500 in a building society. The rate of interest is fixed at 3.6% compound interest. How much money will he have if he leaves it in the building society for 10 years?

Rate of interest = 3.6%

$100\% + 3.6\% = 103.6\% = 1.036$

After 10 years Tony will have £500 $\times$ $(1.036)^{10}$

$= £712.1435717\ldots$

$= £712.14$

> The money will increase so add 3.6% to 100%.

> Using the power button is much quicker than doing £500 $\times$ 1.036 $\times$ 1.036 $\times$ 1.036 $\times$ 1.036 $\times$ ... (10 times altogether).

Exercise 3I

1 Amy invests £135 for two years at a rate of 4% p.a. compound interest.
   How much will she have at the end of the two years?

2 A new car cost £12 000. Each year the value of the car depreciates by $8\frac{1}{2}\%$. What will the car be worth at the end of three years?

3 The population of robins is increasing in parts of Britain. One estimate is that a population of 2000 pairs is increasing at a rate of $6\frac{1}{2}\%$ each year. How many pairs of robins will there be at the end of five years?

4 Sami has a choice of ways to invest the £250 legacy left by her grandfather. She can invest it at $4\frac{1}{4}\%$ p.a. for 5 years or she can invest it for 2 years at 5% p.a. followed by 3 years at $3\frac{1}{2}\%$ p.a. Which one is best?

5 The seal population in Scotland is estimated to decline at the rate of 16% each year owing to pollution. In 2001, 3000 seals were counted. How many years will it take for the population to fall below 1000?

6 A company leases a car at £260 per month. In the leasing agreement, the price is reduced by $2\frac{1}{2}\%$ each month after the first 6 months.
   (a) Work out the difference between the amount paid in the 6th and 7th months.
   (b) How much will the car cost in total in the first 9 months?
   (c) Work out the total amount paid up to the end of the second year.

7 Jane earns £21 000 a year. This increases by 6% each year.
How long will it take her to earn over £25 000 per year?

8 Mamet earns £14 250 per year. His contract says this will
increase each year in line with the annual rate of inflation.
An estimate for the annual rate of inflation is $3\frac{1}{2}$ %.
  (a) Find his salary at the end of three years.
  (b) Find his monthly pay at the end of five years.
  (c) If the rate of inflation is $5\frac{1}{4}$ % instead of $3\frac{1}{2}$ %, how much
      more is his monthly pay at the end of five years?

> Compare the monthly pay for the two rates of inflation.

9 How long will it take each of the following investments to reach
£1 000 000?
  (a) £200 000 invested at a rate of 15%
  (b) £150 000 invested at a rate of 19%

10 If £7400 grows to £10 873 in five years, what is the rate of
compound interest?

> Use original amounts $\times n^5$
> = final amount,
> where n = compound
> interest + 100%.

Examination questions

1 A special savings account earns 10% per year compound interest.
  (a) Jill invests £2500 in the special account.
      How much will she have in her account after 2 years?
  (b) James also invests in the special account.
      After earning interest for one year, he has £1320 in his account.
      How much money did James invest? *(6 marks)*
      AQA, Spec A, 1H, November 2003

2 In 2003 the State Pension was increased by 2% to £78.03.
What was the State Pension before this increase? *(3 marks)*
AQA, Spec A, 2H, June 2003

3 (a) During 2003 the average wage earned by some factory workers in
      Barnsley rose from £350 to £372.
      What was the percentage increase?
  (b) During 2003 the number of people out of work in Barnsley fell by 8%.
      At the end of the year there were 2576 people out of work in Barnsley.
      How many people were out of work at the beginning of the year? *(6 marks)*
      AQA, Spec A, 2H, November 2004

Summary of key points

Fractions, decimals and percentages (up to grade C)

To change a *percentage* to a *fraction or a decimal*, divide by 100.

For example, $43\% = \dfrac{43}{100} = 0.43.$

To change a *fraction or a decimal* to a *percentage*, multiply by 100.

For example, $\dfrac{2}{5} = \left(\dfrac{2}{5} \times 100\right)\% = 40\%$

and $0.725 = (0.725 \times 100)\% = 72.5\%.$

To change a *decimal* to a *fraction*, use the place value of the *last* significant figure, as the denominator of the fraction.

For example, $0.24 = \dfrac{24}{100} = \dfrac{6}{25}$ (in its simplest form).

To *compare amounts*:
Work out the fraction or percentage of the amounts to be compared.
For example, which is greater, $\frac{3}{20}$ of 1540 or 18% of 1250?

$\frac{3}{20}$ of $1540 = \dfrac{3 \times 1540}{20} = 231$

18% of $1250 = \dfrac{18 \times 1250}{100} = 225$

So $\frac{3}{20}$ of 1540 is the greater amount.

To write *one quantity as a percentage of another*:
Write the first quantity as a fraction of the second quantity.
Then multiply by 100 to convert to a percentage.

For example,   18 as a percentage of $45 = \dfrac{18}{45} \times 100 = 40\%.$

To *increase* or *decrease* a quantity *by a given percentage*:

Method A: Work out the actual increase or decrease then add to or subtract from the original.

Method B: Add the % to 100% or subtract the % from 100%. Convert this % to a decimal and multiply it by the original.

For example, to increase £20 by 15%:

115% of £$20 = 1.15 \times$ £$20 =$ £$23.$

Value Added Tax (VAT) (grade D)

VAT at the rate of $17\frac{1}{2}\%$ is added to the cost of many items that you buy.

Percentage change (grade C)

To calculate *percentage change*:

$$\text{Percentage change} = \frac{\text{actual change}}{\text{original amount}} \times 100\%.$$

To calculate *percentage profit or loss*:
The formula for percentage change can be adapted to:

$$\text{Percentage profit (loss)} = \frac{\text{actual profit (loss)}}{\text{cost price}} \times 100\%.$$

where the actual profit (or loss) is the difference between the cost price (what you pay for an item) and the selling price (what you sell the item for).

Reverse percentages (grade B)

Remember that the *original* amount in any calculation is always taken to be *100%* and the final amount will be a given percentage above or below 100%, depending on whether there has been an increase or a decrease.

If it is a percentage *increase … add* the % to 100% then convert this % to a decimal.

If it is a percentage *decrease … subtract* the % from 100% then convert this % to a decimal.

Repeated percentage change (grade C)

Examples of repeated percentage change include compound interest and population changes.

1 Add the rate of change (or interest) to 100% (or subtract if it is a decrease).

2 Convert this % to a decimal.

3 Multiply the original amount by this decimal as many times as the number of years.

This chapter will show you how to:

✔ recognise angles of different types
✔ discover angle properties
✔ use bearings to describe directions
✔ investigate angle properties of triangles, quadrilaterals and other polygons

4.1 Angles

An angle is a measure of turn.
Angles are usually measured in degrees.
A complete circle (or full turn) is 360°.

The minute hand of a clock turns through 360° between 1400 (2 pm) and 1500 (3 pm).

Key words:
acute
right angle
obtuse
reflex
line segment

Angle	Picture	Properties
Acute		An angle between 0° and 90° is an acute angle.
Right angle		An angle of 90°, a $\frac{1}{4}$ turn, is a right angle.
Obtuse		An angle between 90° and 180° is an obtuse angle.
Straight line		An angle of 180°, a $\frac{1}{2}$ turn, is a straight line.
Reflex		An angle between 180° and 360° is a reflex angle.

A right angle is usually marked with this symbol: ⌐

You can describe angles in three different ways:

1 'Trace' the angle using capital letters. Write a 'hat' symbol over the middle letter: $A\hat{B}C$

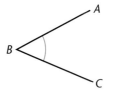

> *AB, BC, PQ* and *QR* are called **line segments** .

2 Use an angle sign or write the word 'angle': $\angle PQR$ or angle *PQR*

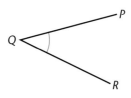

> The letter on the point of the angle always goes in the middle.

3 Use a single letter:

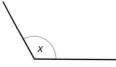

Measuring angles

You use a **protractor** to measure angles accurately.

Follow these instructions carefully.

1 Estimate the angle first, so you don't mistake an angle of 30°, say, for an angle of 150°.

2 Put the centre point of the protractor exactly on top of the angle's point.

3 Place one of the 0° lines of the protractor directly on top of one of the angle 'arms'. If the line isn't long enough, extend it so that it reaches beyond the edge of the protractor.

4 Measure from the 0°, following the scale round the edge of the protractor. If you are measuring from the *left hand* 0°, use the *outside* scale. If you are measuring from the *right hand* 0°, use the *inside* scale.

5 On the correct scale, read the size of the angle in degrees, where the other 'arm' cuts the edge of the protractor.

> Use your estimate to help you choose the correct scale.

Measuring from the left-hand 0°.

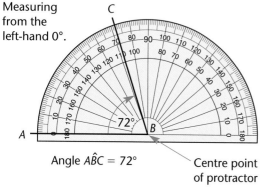

Angle $A\hat{B}C = 72°$

Centre point of protractor

Measuring from the right-hand 0°.

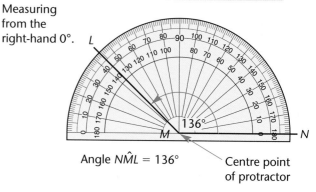

Angle $N\hat{M}L = 136°$

Centre point of protractor

6 To measure a reflex angle (an angle that is bigger than 180°), measure the acute or obtuse angle, and subtract this value from 360°.

Angle properties

Key words:
vertically opposite
perpendicular

You need to know these angle facts:

Angles on a straight line add up to 180°.

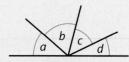

These angles lie on a straight line,
so $a + b + c + d = 180°$.

Angles around a point add up to 360°.

These angles make a full turn,
so $p + q + r + s + t = 360°$.

Vertically opposite angles are equal.

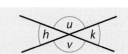

In this diagram $h = k$ and $u = v$.

Perpendicular lines intersect at 90° and are marked with the right angle symbol.

Equal angles are shown by matching arcs:

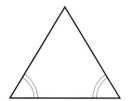

Example 1

Calculate the size of the angles r, s and t, giving reasons.

(a)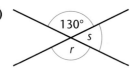

(b)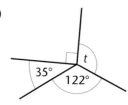

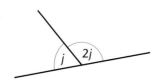 means 90°.

(a) $r = 130°$ (vertically opposite)

$s + 130° = 180°$ (angles on a straight line)

$s = 180° - 130° = 50°$

(b) $t + 122° + 35° + 90° = 360°$ (angles around a point)

$t = 360° - 122° - 35° - 90°$

$t = 113°$

Exercise 4A

Calculate the size of the angles marked with letters.

1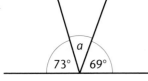

2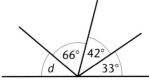

3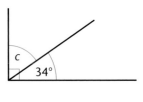

4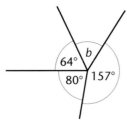

5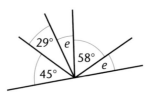

6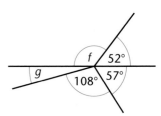

7

8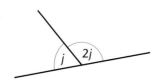

9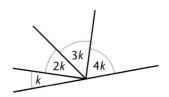

Angles in parallel lines

Parallel lines are the same distance apart all along their length. You can use arrows to show that lines are parallel.

A straight line that crosses a pair of parallel lines is called a **transversal** .

A transversal creates pairs of equal angles.

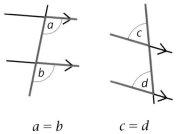

$$a = b \qquad c = d$$

a and b $\Big\}$ are **corresponding** angles.
c and d

The lines make an F shape.

$$e = f \qquad g = h$$

e and f $\Big\}$ are **alternate** angles.
g and h

The lines make a Z shape.

In this diagram the two angles are not equal, j is obtuse and k is acute.

The two angles lie on the *inside* of a pair of parallel lines.

They are called **co-interior** angles or allied angles.

Co-interior angles add up to 180°.

$$j + k = 180°$$

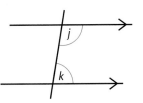

Corresponding angles are equal.
Alternate angles are equal.
Co-interior angles add up to 180°.

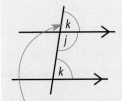

This angle = k
(corresponding angles)
$j + k = 180°$ (angles
on a straight line).

Example 2

Calculate the size of the angles marked with letters.

(a)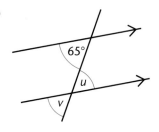

(b)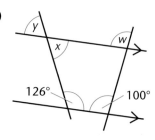

(a) Method A	**(b)** w = 100° (corresponding)
u = 65° (alternate)	x + 126° = 180°
v = 65° (vertically opposite)	(co-interior angles)
Method B	x + 126° − 126° = 180° − 126°
v = 65° (corresponding)	x = 54°
u = 65° (vertically opposite)	y = 54° (vertically opposite)

You could use Method A or Method B.

Exercise 4B

Calculate the size of the angles marked with letters.

1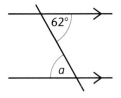

2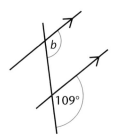

3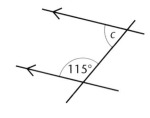

4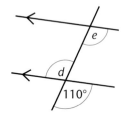

5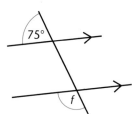

6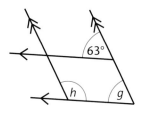

7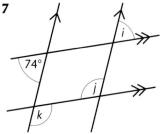

8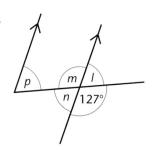

9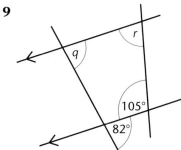

4.2 Three-figure bearings

Key words:
bearing

You can use compass points to describe a direction.

Another method is to use three-figure **bearings** .

> **A three-figure bearing gives a direction in degrees.**
> **It is an angle between 0° and 360°.**
> **It is always measured from the North in a clockwise direction.**

The diagram shows the cities of Manchester and Sheffield.

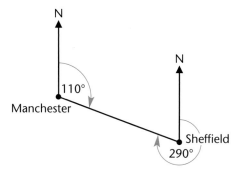

The bearing of Sheffield from Manchester is 110°.

The bearing of Manchester from Sheffield is 290°.

A bearing must always be written with *three figures*.

So for angles less than 100°, include a zero at the front, e.g. 090°.

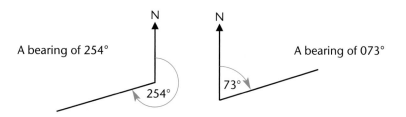

You write 073° because the bearing must have three figures.

You need to be able to measure bearings accurately using a protractor.

Your answer needs to be within 2° of the correct value.

Example 3

Write down

(a) the bearing of Q from P

(b) the bearing of P from Q.

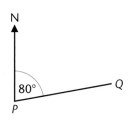

(a) 080°

(b)

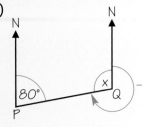

If you stand at P and face North, you need to turn through 80° to face Q.

Draw in the North line at Q.

This is the bearing of P from Q.

The two North lines are parallel, so

$80° + x = 180°$ (co-interior angles)

$80° - 80° + x = 180° - 80°$

$x = 100°$

Bearing of P from Q + x = 360° (angles around a point)

Bearing of P from Q + 100° = 360°

Bearing of P from Q = 260°.

Exercise 4C

1 For each diagram:
  (i) write down the bearing of Q from P
  (ii) work out the bearing of P from Q.

(a)
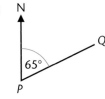

(b)
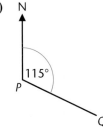

(c)
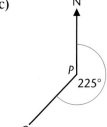

(d)
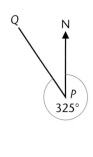

2 Draw accurate diagrams to show these three-figure bearings.
  (a) 036°   **(b)** 145°   **(c)** 230°   **(d)** 308°
  (e) 074°   **(f)** 256°   **(g)** 348°   **(h)** 115°

3 The diagram shows a triangle *LMP*.

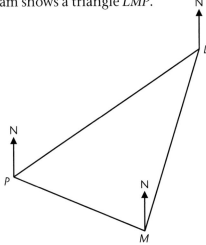

Use a protractor to measure these bearings:

(a) *L* from *P*   (b) *M* from *P*   (c) *L* from *M*

(d) *P* from *L*   (e) *P* from *M*   (f) *M* from *L*

4.3 Triangles

Types of triangle

Key words:
scalene
isosceles
equilateral
right-angled

Triangle	Picture	Properties
scalene		The three sides are different lengths. The three angles are different sizes.
isosceles		Two equal sides. *AB = AC* Two equal angles ('base angles') angle *ABC* = angle *ACB*.
equilateral		All three sides are equal in length. All angles 60°.
right-angled		One of the angles is a right angle (90°). Angle *XZY* = 90°.

Triangle properties 1

The sum of the three angles of a triangle is 180°. To show this:

1 Draw a triangle on a piece of paper.

2 Mark each angle with a different letter or shade them different colours.

3 Tear off each corner and place them next to each other on a straight line.

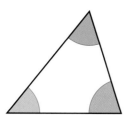

You are not *proving* this result, you are simply showing that it is true for the triangle you drew.

You will see that the three angles fit exactly onto the straight line. So the three angles add up to 180°.

You can prove this for all triangles using facts about alternate and corresponding angles.

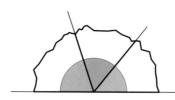

For the triangle *ABC*:

● Extend side *AC* to point *E*.

● From *C* draw a line *CD* parallel to *AB*.

Let $\angle BCD = x$ and $\angle DCE = y$.

$$x = b \text{ (alternate angles)}$$
$$y = a \text{ (corresponding angles)}$$
$$c + x + y = 180° \text{ (angles on a straight line at point } C)$$

which means that $c + b + a = 180°$.

This *is* a proof since it is true for general angles *a*, *b*, *c* and not just for one particular triangle.

The sum of the angles of a triangle is 180°.

Example 4

Calculate the size of the angles marked with letters.

(a)

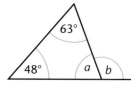

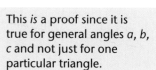

$a = 180° - 63° - 48°$
    (angle sum of △)
$a = 69°$
$b = 180° - 69°$
    (angles on a straight line)
$b = 111°$

△ means 'triangle'.

(b)

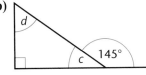

$c = 180° - 145°$

(angles on a straight line)

$c = 35°$

$d = 180° = 90° - 35°$

(angle sum of △)

$d = 55°$

(c)

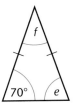

$e = 70°$ (base angles isosceles △)

$f = 180° - 70° - 70°$

(angle sum of △)

$f = 40°$

The triangle has two equal sides. It is isosceles.

Exercise 4D

Calculate the size of the angles marked with letters.

1

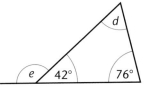

2

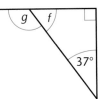

3

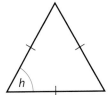

4

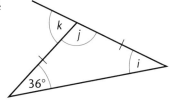

5

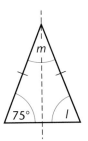

6

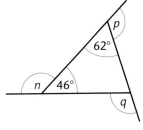

Key words:
interior
exterior

Triangle properties 2

The angles inside a triangle are called **interior** angles.

An **exterior** angle is formed by extending one of the sides of the triangle.

Angle *BCE* in this diagram is an exterior angle.

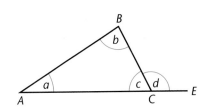

Let angle $BCE = d$

then $\qquad c + d = 180°$ (angles on a straight line)

but $\qquad c + b + a = 180°$ (angle sum of △ABC)

which means that $\qquad d = a + b$

In a triangle, the exterior angle is equal to the sum of the two opposite interior angles.

This is a *proof*.

Sometimes you need to calculate the size of 'missing' angles before you can calculate the ones you want.
You can label the extra angles with letters.

This helps make your explanations clear.

Example 5

Calculate angle *i*.

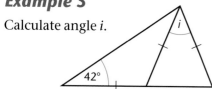

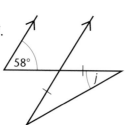

$x = 42°$

   (base angles isosceles △ 1)

$y = 42° + 42°$

   (exterior angle of △ 1)

$y = 84°$

$z = 84°$

   (base angles isosceles △ 2)

$i = 180° - 84° - 84°$

   (angle sum △ 2)

$i = 12°$

Label the 'missing' angles *x*, *y* and *z*.

Label the two triangles 1 and 2.

Example 6

Calculate angle *j*.

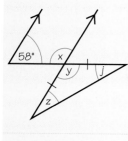

$x = 180° - 58°$ (co-interior angles, parallel lines)

$x = 122°$

$y = 122°$ (vertically opposite)

$z + j = 180° - 122°$ (angle sum of △)

$z + j = 58°$

$z = j$ (base angles isosceles △)

so both *z* and $j = \frac{1}{2}$ of 58° (or 58° ÷ 2)

$j = 29°$

Label the 'missing' angles *x*, *y* and *z*.

58 ÷ 2 = 29.

Exercise 4E

Calculate the size of the angles marked with letters.

Copy the diagram and label any 'missing' angles.

1

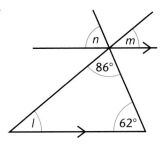

2

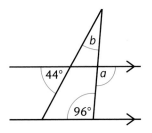

3

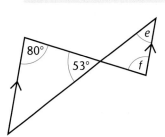

4

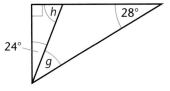

5

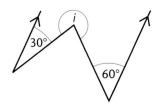

6

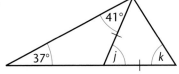

7

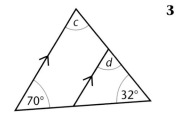

8

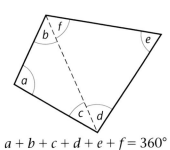

4.4 Quadrilaterals and other polygons

Quadrilaterals

A **quadrilateral** is a 2-dimensional 4-sided shape.

Key words:
quadrilateral
bisect
diagonal
adjacent

This quadrilateral can be divided into two triangles:

In each triangle, the angles add up to 180°, so $a + b + c = 180°$

and $d + e + f = 180°$

The six angles from the two triangles add up to 360°.

$$a + b + c + d + e + f = 360°$$

You can divide *any* quadrilateral into 2 triangles.

$a, b + f, e, c + d$ are the angles of the quadrilateral.

In any quadrilateral the sum of the interior angles is 360°.

Once again, we have *proved* the result.

Some quadrilaterals have special names and properties.

Quadrilateral	Picture	Properties
square		Four equal sides. All angles 90°. Diagonals **bisect** each other at 90°.
rectangle		Two pairs of equal sides. All angles 90°. **Diagonals** bisect each other.
parallelogram		Two pairs of equal and parallel sides. Opposite angles equal. Diagonals bisect each other.
rhombus		Four equal sides. Two pairs of parallel sides. Opposite angles equal. Diagonals bisect each other at 90°.
trapezium		One pair of parallel sides.
kite		Two pairs of **adjacent** sides equal. One pair of opposite angles equal. The longer diagonal bisects the shorter diagonal at 90°.

Bisect means 'cut in half'.

Lines with equal arrows are parallel to each other.

Adjacent means 'next to'.

Exercise 4F

1 Write down the names of all the quadrilaterals with these properties:

(a) diagonals which cross at 90°

(b) all sides are equal in length

(c) only one pair of parallel sides

(d) two pairs of equal angles but not all angles equal

(e) all angles are equal

(f) two pairs of opposite sides are parallel

(g) only one diagonal bisected by the other diagonal

(h) two pairs of equal sides but not all sides equal

(i) the diagonals bisect each other

(j) at least one pair of opposite sides are parallel

(k) diagonals equal in length

(l) at least two pairs of adjacent sides equal.

Polygons

A **polygon** is a 2-dimensional shape with many sides and angles. Here are some of the most common ones:

Pentagon (5 sides)   Hexagon (6 sides)   Octagon (8 sides)

 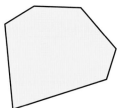 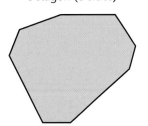

A polygon with all of its sides the same length and all of its angles equal is called **regular**.

Regular pentagon   Regular hexagon   Regular octagon

 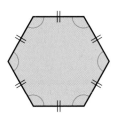 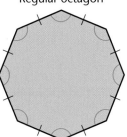

The sum of the exterior angles of any polygon is 360°.

> **Key words:**
> polygon
> regular
> vertex
> vertices

> Polygon means 'many angled'.

You can explain this result like this.

If you 'walk round' the sides of this hexagon until you get back to where you started, you complete a full turn or 360°.

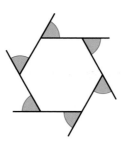

Exterior angle

The exterior angles represent your 'turn' at each corner, so they must add up to 360°.

On a regular hexagon, all six exterior angles are the same size.

For a regular hexagon:

$$\text{exterior angle} = \frac{360°}{6}$$
$$= 60°$$

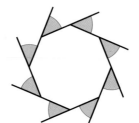

$$\frac{360°}{6} = \frac{360°}{\text{number of sides}}$$

For a regular octagon:

$$\text{exterior angle} = \frac{360°}{8} = 45°$$

A regular octagon has 8 equal exterior angles.

For a regular polygon:

$$\textbf{Exterior angle} = \frac{\textbf{360°}}{\textbf{number of sides}}$$

$$\textbf{Number of sides} = \frac{\textbf{360°}}{\textbf{exterior angle}}$$

Once you know the exterior angle, you can calculate the interior angle.

At each **vertex** the exterior and interior angles lie next to each other (adjacent) on a straight line.

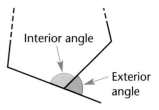

Interior angle

Exterior angle

In a polygon, each pair of interior and exterior angles adds up to 180°.

In a regular polygon all the exterior angles are equal. So all the interior angles are equal.

So, for any polygon

$$\text{interior angle} = 180° - \text{exterior angle}$$

For example:

- regular hexagon   interior angle = 180° − 60° = 120°
- regular octagon   interior angle = 180° − 45° = 135°

When a polygon is *not* regular the interior angles could all be different sizes. You can find the *sum* of the interior angles.

These diagrams show how you can divide any polygon into triangles.

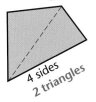
4 sides
2 triangles

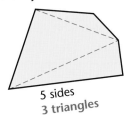
5 sides
3 triangles

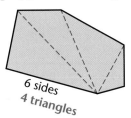
6 sides
4 triangles

Draw dotted lines from the **vertices** .

The table shows you how to calculate the sum of the interior angles for polygons with up to 10 sides.

Name of polygon	Number of sides	Number of triangles	Sum of interior angles
Triangle	3	1	$1 \times 180° = 180°$
Quadrilateral	4	2	$2 \times 180° = 360°$
Pentagon	5	3	$3 \times 180° = 540°$
Hexagon	6	4	$4 \times 180° = 720°$
Heptagon	7	5	$5 \times 180° = 900°$
Octagon	8	6	$6 \times 180° = 1080°$
Nonagon	9	7	$7 \times 180° = 1260°$
Decagon	10	8	$8 \times 180° = 1440°$

You can see that the number of triangles is always 2 less than the number of sides of the polygon.

The sum of the interior angles for each polygon is given by the formula

$(n - 2) \times 180°$ **where n is the number of sides of the polygon.**

You should learn this formula.

Example 7

Calculate the size of the exterior and interior angles of a regular polygon with 20 sides.

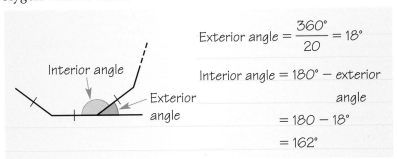

Interior angle

Exterior angle

$$\text{Exterior angle} = \frac{360°}{20} = 18°$$

Interior angle $= 180° -$ exterior angle

$= 180 - 18°$

$= 162°$

Exterior angle of regular

$$\text{polygon} = \frac{360°}{\text{number of sides}}$$

Example 8

Calculate the size of the angles marked with letters.

(a)
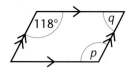

$p = 118°$ (opposite angles of parallelogram are equal)

$q = 62°$ (p and q are co-interior angles, so $p + q = 180°$)

> You could use the fact that the sum of the angles of a quadrilateral is 360° to calculate q.

(b)
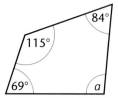

Sum of the interior angles of a pentagon (5 sides) $= 3 \times 180° = 540°$.

$r = 540° - (95° + 110° + 124° + 73°)$

$r = 138°$

> $(n - 2) \times 180°$ with $n = 5$

Exercise 4G

Calculate the size of the angles marked with letters.

1
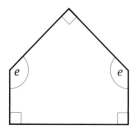

2
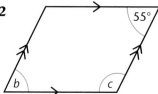

3

4

5
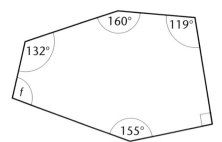

6
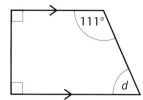

7

8
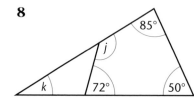

9
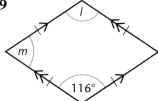

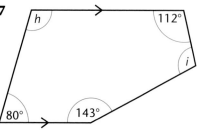

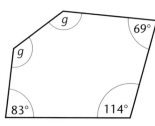

10

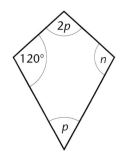

11

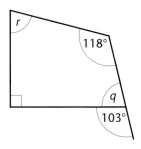

12

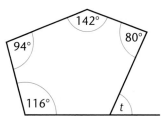

13

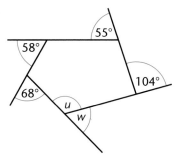

14

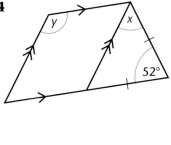

15 Calculate the interior angle of a regular polygon with 18 sides.

16 Can a regular polygon have an interior angle of 130°?
Explain your answer.

Examination questions

1 *ABCDEF* is a hexagon.

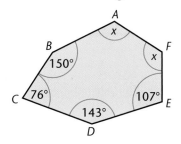

Work out the value of *x*. *(3 marks)*

2 The diagram shows part of a regular polygon.
Each interior angle is 150°.

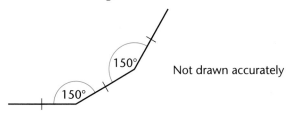

Not drawn accurately

Calculate the number of sides of the polygon. *(3 marks)*

3 The diagram shows a regular octagon.

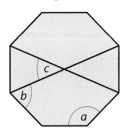

Calculate the size of angles *a*, *b* and *c*.

(3 marks)

Summary of key points

Angle properties (up to grade D)

Angles on a straight line add up to 180°.

$a + b + c = 180°$

Angles around a point add up to 360°.

$d + e + f + g = 360°$

Vertically opposite angles are equal.

$d = g$

$f = e$

Corresponding angles are equal.

$j = k$

Alternate angles are equal.

$l = k$

Co-interior angles add up to 180°.

$l + m = 180°$

Perpendicular lines intersect at 90° and are marked with a right angle symbol:

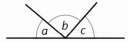

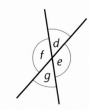

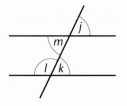

Three-figure bearings (up to grade D)

A three-figure bearing gives a direction in degrees.
It is an angle between 0° and 360°and is always measured
from the North **in a** *clockwise* **direction.**

This diagram shows that the bearing of the ship
from the submarine is 248°.

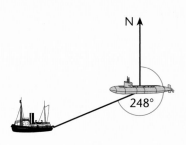

Triangle properties

The sum of the angles of a triangle is 180º

$$a + b + c = 180°$$

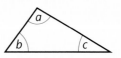

The exterior angle is equal to the sum of the two opposite interior angles.

$$z = x + y$$

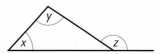

Quadrilateral and polygon properties (grades D/C)

The sum of the interior angles of a quadrilateral is 360°.

$$a + b + c + d = 360°$$

A polygon with all its sides the same length and all of its angles equal is called *regular*.

Regular hexagon

The sum of the exterior angles of any polygon is 360°.

$$p + q + r + s + t = 360°$$

In a *regular* polygon

$$\text{Exterior angle} = \frac{360°}{\text{number of sides}}$$

For example, for a regular polygon with 10 sides,

$$\text{Exterior angle} = \frac{360°}{10} = 36°$$

$$\text{Number of sides} = \frac{360°}{\text{exterior angle}}$$

For example, the exterior angle of a regular polygon is 40°.

$$\text{Number of sides} = \frac{360°}{40°} = 9$$

In any polygon the interior angle and the exterior angle add up to 180°

$$x + y = 180°$$

The sum of the interior angles of a polygon with *n* sides is given by the formula $(n - 2) \times 180°$.

For example ...

This is a pentagon, so $n = 5$

Sum of interior angles $= 3 \times 180° = 540°$

Hence $a + b + c + d + e = 540°$

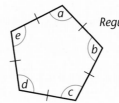

Regular pentagon

This chapter will show you how to:

✔ solve simple equations
✔ solve equations that have negative, decimal or fractional answers
✔ solve equations combining two or more operations
✔ use equations to solve problems
✔ solve simultaneous equations
✔ write and solve inequalities, and represent solutions on a number line

In algebra, letters are used to represent numbers.

In a statement such as $x + 3 = 12$ you must find a value of x that makes this a true statement.

This is called 'solving an equation' and the value of x is the solution of the equation.

5.1 Solving simple equations

A **simple equation** is one involving an unknown letter. In this section you will consider only **linear equations** (equations that have no letters with powers) involving only one operation. The operation might be addition, subtraction, multiplication or division by a whole number.

Key words:
simple equation
linear equation
subject

> **To solve linear equations, treat both sides of the equation in the same way – what you do to one side of the equation you must do to the other.**

If you add 6 to one side of an equation you must add 6 to the other side. If you divide one side by 2 you must divide the other side by 2 as well.

Example 1

Solve the equation $p - 4 = 9$.

$$p - 4 = 9$$
$$p - 4 + 4 = 9 + 4$$
$$p = 13$$

The inverse of -4 is $+4$, so add 4 to both sides of the equation.

The aim, always, is to leave the letter (the **subject**) on its own on one side of the equation.

Example 2

Solve the equation $7 = c + 2$.

$$7 = c + 2$$
$$7 - 2 = c + 2 - 2$$
$$5 = c$$
$$c = 5$$

Subtract 2 from both sides.

You usually write the answer with the subject on the left-hand side of the equals sign.

Remember $5 = c$ is the same as $c = 5$.

Example 3

Solve the equation $4a = 12$.

$$4a = 12$$
$$\frac{4a}{4} = \frac{12}{4}$$
$$a = 3$$

$4a$ means $4 \times a$. The inverse process is divide by 4, so divide both sides of the equation by 4.

Example 4

Solve the equation $\dfrac{d}{5} = 8$.

$$\frac{d}{5} = 8$$
$$\frac{d}{5} \times 5 = 8 \times 5$$
$$d = 40$$

$\dfrac{d}{5}$ means the same as $d \div 5$.

The inverse of $\div 5$ is $\times 5$, so multiply both sides by 5 to make d the subject of the equation.

Exercise 5A

1 Solve

(a) $h + 4 = 6$ (b) $b + 1 = 12$ (c) $x + 9 = 10$

(d) $k - 5 = 2$ (e) $y - 9 = 11$ (f) $a - 20 = 30$

(g) $5 + f = 32$ (h) $40 + w = 120$ (i) $14 = 8 + n$

2 Solve

(a) $3b = 9$ (b) $5t = 25$ (c) $7p = 21$

(d) $4g = 28$ (e) $9x = 36$ (f) $8n = 24$

(g) $10m = 130$ (h) $6y = 360$ (i) $7k = 49$

(j) $20 = 5h$ (k) $64 = 8r$ (l) $32 = 4a$

3 Solve

(a) $\dfrac{t}{2} = 12$ (b) $\dfrac{y}{6} = 3$ (c) $\dfrac{f}{4} = 9$ (d) $\dfrac{r}{3} = 11$ (e) $\dfrac{q}{6} = 6$ (f) $\dfrac{g}{9} = 8$

(g) $\dfrac{s}{2} = 56$ (h) $\dfrac{b}{5} = 30$ (i) $\dfrac{n}{8} = 1$ (j) $4 = \dfrac{m}{5}$ (k) $16 = \dfrac{v}{3}$ (l) $100 = \dfrac{u}{9}$

4 Solve

(a) $8t = 48$ (b) $4 + p = 7$ (c) $\dfrac{r}{7} = 56$ (d) $15 = \dfrac{m}{4}$ (e) $45 = x - 35$

(f) $g - 100 = 550$ (g) $\dfrac{s}{12} = 5$ (h) $9d = 99$ (i) $42 = 7j$

5.2 Equations involving two operations

Remember that you treat both sides of the equation in the same way
– what you do to one side of the equation you must do to the other.

Example 5

Solve the equation $5g - 4 = 36$.

$$5g - 4 = 36$$
$$5g - 4 + 4 = 36 + 4$$
$$5g = 40$$
$$\frac{5g}{5} = \frac{40}{5}$$
$$g = 8$$

Using inverse operations

$$\begin{array}{cc} \times 5 & -4 \\ g \to 5g \to 5g - 4 \end{array}$$

$$\begin{array}{cc} \div 5 & +4 \\ 8 \leftarrow 40 \leftarrow 36 \end{array}$$

The order of the inverse operations is very important. You must add 4 before you divide by 5.

Example 6

Solve the equation $\dfrac{k}{4} + 3 = 7$.

$$\frac{k}{4} + 3 = 7$$

$$\frac{k}{4} + 3 - 3 = 7 - 3$$

$$\frac{k}{4} = 4$$

$$\frac{k}{4} \times 4 = 4 \times 4$$

$$k = 16$$

Look for the inverse operations. The inverse of $\div 4$ is $\times 4$, and the inverse of $+3$ is -3.

You must subtract 3 from both sides before you multiply both sides by 4.

Example 7

Solve the equation $\dfrac{12 - 2n}{3} = 2$.

$$\frac{12 - 2n}{3} = 2$$

$$\frac{12 - 2n}{3} \times 3 = 2 \times 3$$ — Begin by multiplying both sides of the equation by 3.

$$12 - 2n = 6$$

$$12 - 2n + 2n = 6 + 2n$$ — Adding $2n$ to both sides means that the value of $2n$ remains positive.

$$12 = 6 + 2n$$

$$12 - 6 = 6 - 6 + 2n$$ — Subtract 6 from both sides then divide both sides by 2.

$$6 = 2n$$

$$3 = n$$

$$n = 3$$

Exercise 5B

1 Solve the following equations.

(a) $3w + 2 = 14$ (b) $2u + 9 = 19$ (c) $5t - 4 = 36$
(d) $9m - 2 = 25$ (e) $8d + 3 = 19$ (f) $7g + 6 = 69$
(g) $4s + 5 = 17$ (h) $8b + 6 = 6$ (i) $20a - 30 = 70$

2 Solve

(a) $\dfrac{r}{3} + 2 = 5$ (b) $\dfrac{h}{4} - 5 = 2$ (c) $\dfrac{f}{3} - 8 = 2$

(d) $\dfrac{x}{2} + 4 = 5$ (e) $\dfrac{d}{10} - 10 = 1$ (f) $\dfrac{a}{3} + 7 = 10$

3 Solve the following equations.

(a) $\dfrac{3x - 2}{11} = 2$ (b) $\dfrac{6t + 5}{7} = 5$ (c) $\dfrac{4w + 9}{7} = 7$

(d) $\dfrac{14 - 3a}{2} = 4$ (e) $\dfrac{30 + 10n}{5} = 14$ (f) $\dfrac{22 + 8q}{9} = 6$

(g) $5 = \dfrac{4y - 20}{4}$ (h) $\dfrac{e - 4}{2} = -5$ (i) $-11 = \dfrac{6p - 4}{2}$

5.3 Equations involving fractions, decimals and negative values

You will need to know:
- how to convert between fractions and decimals

Equations often have solutions that are negative.

Sometimes the solutions are fractions or decimals.

Unless the question asks for a particular type of answer, you can give your solution either as a fraction or decimal.

Example 8

Solve $9t + 7 = 19$ giving your answer as a mixed number.

$$9t + 7 = 19$$
$$9t + 7 - 7 = 19 - 7$$
$$9t = 12$$
$$\frac{9t}{9} = \frac{12}{9}$$
$$t = \frac{12}{9}$$
$$t = 1\frac{3}{9}$$
$$t = 1\frac{1}{3}$$

Note the order of the inverse operations.

You must subtract 7 from both sides before dividing both sides by 9.

A mixed number has a whole number part and a fractional part.

$\frac{3}{9} = \frac{1}{3}$ (in its lowest terms).

Example 9

Solve $5x - 6 = 3$.

$$5x - 6 = 3$$
$$5x - 6 + 6 = 3 + 6$$
$$5x = 9$$
$$\frac{5x}{5} = \frac{9}{5}$$
$$x = \frac{9}{5}$$
$$x = 1.8$$

Add 6 to both sides before dividing both sides by 5.

You can give your answer as 1.8 or $1\frac{4}{5}$.

Example 10

Solve the equation $2u + 7 = 1$.

$$2u + 7 = 1$$
$$2u + 7 - 7 = 1 - 7$$
$$2u = -6$$
$$\frac{2u}{2} = \frac{-6}{2}$$
$$u = -3$$

You must subtract 7 from both sides before dividing both sides by 2.

Dividing a negative number by a positive number gives a negative answer.

Example 11

Solve the equation $\dfrac{d}{2} - 4 = -5$.

$$\frac{d}{2} - 4 = -5$$

$$\frac{d}{2} - 4 + 4 = -5 + 4$$

$$\frac{d}{2} = -1$$

$$\frac{d}{2} \times 2 = -1 \times 2$$

$$d = -2$$

You must add 4 to both sides before multiplying both sides by 2.

Multiplying a negative number by a positive number gives a negative answer.

Exercise 5C

1 Solve the following equations, giving your answer as a mixed number.

(a) $2b - 4 = 5$ (b) $7c + 4 = 13$ (c) $3y + 6 = 16$

(d) $7j - 13 = 6$ (e) $3g - 2 = 3$ (f) $8p - 2 = 10$

2 Solve the following equations, giving your answer as a decimal.

(a) $8k - 3 = 7$ (b) $4e - 5 = 1$ (c) $10i - 10 = 12$

(d) $5s + 9 = 10$ (e) $4m - 2 = 5$ (f) $2a + 5 = 20$

3 Solve these equations.

(a) $3k + 8 = 2$ (b) $5f + 12 = 7$ (c) $6m + 20 = 2$

(d) $10g - 5 = -55$ (e) $2w - 1 = -5$ (f) $4s - 4 = -12$

(g) $3u + 6 = -12$ (h) $5f + 13 = 3$ (i) $4v + 23 = 3$

(j) $19 + 2x = 7$ (k) $12 - 6a = 18$ (l) $3n + 7 = -8$

4 Solve the following equations.

(a) $\dfrac{m}{2} + 3 = 1$ (b) $\dfrac{q}{3} + 7 = 6$ (c) $\dfrac{x}{5} - 3 = -5$

(d) $\dfrac{b}{8} - 7 = -10$ (e) $20 + \dfrac{e}{4} = 17$ (f) $\dfrac{t}{6} + 10 = 4$

5.4 Equations with unknowns on both sides

Key words:
unknown

Equations sometimes have the **unknown** (letter) on both sides of the equation. To solve these equations you often need to carry out more than two operations.

Example 12

Solve the equation $5a - 2 = 2a + 4$.

$$5a - 2 = 2a + 4$$
$$5a - 2a - 2 = 2a - 2a + 4$$
$$3a - 2 = 4$$
$$3a - 2 + 2 = 4 + 2$$
$$3a = 6$$
$$a = 2$$

$5a$ on the left-hand side is greater than $2a$ on the right-hand side. In this case it is easier to subtract $2a$ from both sides, keeping the unknown as the subject on the left-hand side.

Adding 2 then dividing by 3 completes the solution.

Exercise 5D

1 Solve the following equations.

  (a) $3x + 4 = 2x + 6$ (b) $5u + 9 = 3u + 17$

  (c) $r + 2 = 5r - 10$ (d) $7p + 11 = 6p + 16$

  (e) $4w - 6 = 3w + 1$ (f) $6b - 2 = 2b + 14$

2 Solve the following equations.

  (a) $4a + 6 = 2a + 2$ (b) $7m + 10 = 4m + 1$

  (c) $8d - 2 = 3d - 12$ (d) $y - 5 = 6y + 15$

  (e) $7 - 6t = 10 - 4t$ (f) $13 - 5r = 3r + 29$

3 Solve

  (a) $5x - 13 = 3x - 8$ (b) $4z + 24 = 9z - 12$

  (c) $15m - 2 = 7 - 12m$ (d) $17h + 35 = 5h - 7$

  (e) $11k + 3 = 30 + 5k$ (f) $20c - 12 = 28c + 16$

5.5 Using equations to solve problems

Key words:
formulae

You can use equations to solve problems.

You will need to read the problem carefully then set up an equation and solve it using the methods introduced earlier.

Example 13

The length of a rectangular garden is 10 m more than its width.
The perimeter of the garden is 48 m. Find the width of the garden.

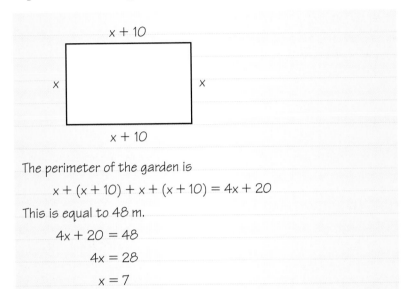

The perimeter of the garden is

$$x + (x + 10) + x + (x + 10) = 4x + 20$$

This is equal to 48 m.

$$4x + 20 = 48$$
$$4x = 28$$
$$x = 7$$

The width of the garden is 7 m.

> It is often useful to sketch a diagram and label it.

> Use the letter x to represent the width of the garden. The length is 10 m longer than the width so it must be $x + 10$.

> Check by putting $x = 7$ into your original equation.
> $7 + 17 + 7 + 17 = 48$ ✓

Exercise 5E

1 The length of a rectangular field is three times its width. If the perimeter of the field is 360 m find the width of the field.

> Let the width of the field be x.

2 The sum of three consecutive whole numbers is 276. Find the three numbers.

> Let the smallest of these numbers be x.
> Then write the other two in terms of x.

3 Sarah is two years older than Susan, who is seven years older than Stephen. If their combined age is 61 years find the age of each person.

> Let the age of youngest be x.

4 The perimeter of the triangle shown below is 22 cm.

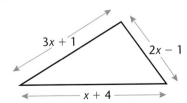

Formulate an equation in terms of x and solve it to find the lengths of the sides of the triangle.

5 Trebling a number gives the same answer as adding 12 to it. What is the number?

Let the number be x.

6 AB is a straight line.

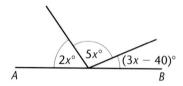

Find the value of x and the angles shown.

7 The opposite angles of a parallelogram are $(2a + 40)°$ and $(3a - 10)°$. Find the two angles.

8 A cricket bat costs £15 more than a tennis racket. If the total price for a cricket bat and tennis racket is £81 what is the cost of the cricket bat?

Let the price of the tennis racket be £x.

5.6 Simultaneous equations

Key words:
simultaneous equations
method of elimination
coefficient

A linear equation such as $3x + 2 = 14$ has a unique solution.

In this case it is $x = 4$. This is the only value of x that satisfies the equation.

These equations each contain two unknowns (x and y):

$$2x + y = 11 \qquad (1)$$
$$3x - y = 14 \qquad (2)$$

Values of x and y that satisfy equation (1) may not satisfy equation (2).

For example, $x = 4$ and $y = 3$ satisfy equation (1) but they do not satisfy equation (2).

There are values of x and y that satisfy both equations at the same time ... you might have spotted that they are $x = 5$ and $y = 1$.

An equation of a straight line is known as a linear equation. Each equation has an infinite number of coordinate pairs that satisfy it. Chapter 16 looks at linear equations and straight line graphs in a lot more detail. It includes finding points of intersection of straight lines.

Solving two linear equations at the same time is called solving **simultaneous equations**.

You can solve simultaneous equations by the **method of elimination**. This eliminates one of the unknowns by adding or subtracting the equations, provided they have the same amount of either x's or y's.

You need to see that the coefficients of either x or y are the same (ignoring the sign).

The **coefficient** is the number in front of the x or y.

Example 14

Solve the simultaneous equations

$$2x + y = 11$$
$$3x - y = 14.$$

	$2x + y = 11$	(1)
	$3x - y = 14$	(2)
(1) + (2)	$5x = 25$	
so	$x = 5$	

Notice that the coefficients of y are 1 and -1 so **adding** the two equations will eliminate the y terms.

Substitute $x = 5$ in equation (1)

$$2 \times 5 + y = 11$$
$$10 + y = 11$$

so $\quad y = 1$

When you have found the first solution, substitute it into one of the equations to work out the other solution.

It doesn't matter which equation you choose.

Check in (2): $3 \times 5 - 1 = 15 - 1 = 14$ ✓

Check that the solutions you have found work for the other equation.

The solutions are $x = 5$ and $y = 1$

Example 15

Solve the simultaneous equations

$$2x + 5y = 27$$
$$2x - 3y = 3.$$

When the coefficients (of x in this example) are exactly the same, subtracting the equations will eliminate one unknown.

	$2x + 5y = 27$	(1)
	$2x - 3y = 3$	(2)
(1) − (2)	$8y = 24$	
so	$y = 3$	

Take great care when subtracting.
$$5y - (-3y)$$
$$= 5y + 3y$$
$$= 8y$$

Substitute $y = 3$ in equation (1)

$$2x + 5 \times 3 = 27$$
$$2x + 15 = 27$$
$$2x = 27 - 15$$
$$2x = 12$$

so $\quad x = 6$

Substitute $y = 3$ into one of the equations.

Check in (2) $2 \times 6 - 3 \times 3 = 12 - 9 = 3$ ✓

The solutions are $x = 6$ and $y = 3$

Check that the solutions you have found work for the equation you did not choose before.

The rule is:

If the *x*'s or *y*'s have the same numerical value but differ in sign ... ADD

If the *x*'s or *y*'s have the same numerical value and the same sign ... SUBTRACT

Sometimes you need to multiply one of the equations so that the coefficients of one of the terms become the same, as in the next example.

Example 16

Solve the simultaneous equations

$$4g + 6h = 2$$
$$3g + 12h = 9$$

	$4g + 6h = 2$	(1)
	$3g + 12h = 9$	(2)
(1) × 2 gives	$8g + 12h = 4$	(3)
	$3g + 12h = 9$	(2)
(3) − (2)	$5g = -5$	
so	$g = -1$	

The coefficients of g and h in (1) and (2) are not the same. To make the coefficient of h the same you multiply equation (1) by 2 giving a new equation (3).

Replace equation (1) with equation (3).

Do not alter equation (2).

Substitute $g = -1$ in equation (1)

$$4 \times (-1) + 6h = 2$$
$$-4 + 6h = 2$$
$$6h = 2 + 4$$
$$6h = 6$$
$$h = 1$$

Solve equations (2) and (3) by the method you used in the previous examples.

You can still substitute in equation (1) even though you replaced it with equation (3) to do the elimination.

Check in equation (2)

$$3 \times -1 + 12 \times 1 = -3 + 12 = 9 \ \checkmark$$

Always check your solutions using the other equation.

The solutions are $g = -1$ and $h = 1$

You may have to multiply both equations before you can use the method of elimination. This is shown in Example 17.

Example 17

Solve the simultaneous equations

$2x - 5y = 13$
$3x + 2y = 10.$

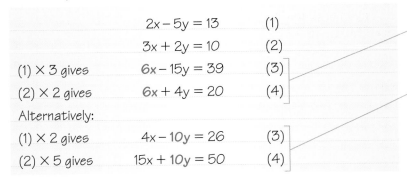

	$2x - 5y = 13$	(1)
	$3x + 2y = 10$	(2)
(1) $\times$ 3 gives	$6x - 15y = 39$	(3)
(2) $\times$ 2 gives	$6x + 4y = 20$	(4)
Alternatively:		
(1) $\times$ 2 gives	$4x - 10y = 26$	(3)
(2) $\times$ 5 gives	$15x + 10y = 50$	(4)

These equations can now be subtracted to eliminate the x terms.

These equations can now be added to eliminate the y terms.

Whichever of these you choose you will obtain the solutions $x = 4$, $y = -1$.

Example 18

The sum of two numbers is 60 and the difference between them is 26. What are the two numbers?

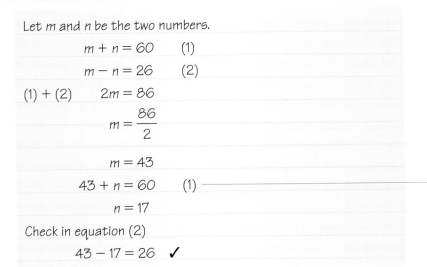

Let m and n be the two numbers.

$m + n = 60$ (1)
$m - n = 26$ (2)

(1) + (2) $2m = 86$

$$m = \frac{86}{2}$$

$m = 43$

$43 + n = 60$ (1)

$n = 17$

Check in equation (2)

$43 - 17 = 26$ ✓

First express the problem, using letters for the unknown numbers.

Substitute in equation (1).

Example 19

At a theme park, the cost for 4 adults and 5 children is £156, and the cost for 2 adults and 3 children is £86. How much is each ticket?

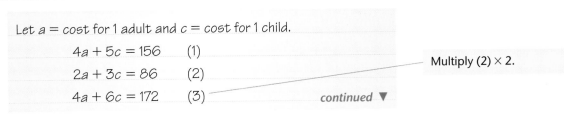

Let a = cost for 1 adult and c = cost for 1 child.

$4a + 5c = 156$ (1)
$2a + 3c = 86$ (2)
$4a + 6c = 172$ (3)

Multiply (2) $\times$ 2.

continued ▼

$$c = 172 - 156 \quad \text{Subtract (1) from (3).}$$
$$c = 16$$
$$2a + 3 \times 16 = 86 \quad (2)$$
$$2a = 86 - 48 = 38$$
$$a = 19$$

An adult ticket costs £19 and a child's ticket £16.

Exercise 5F

1 Solve the simultaneous equations.

(a) $3a + 2b = 14$
$a - 2b = 2$

(b) $5x - 3y = 21$
$3x + 3y = 3$

(c) $-3b + 4c = 2$
$3b - 7c = 1$

2 Solve the simultaneous equations.

(a) $4s - 2t = 14$
$3s - 2t = 9$

(b) $6u - 3v = 18$
$u - 3v = 8$

(c) $2p - 5q = 13$
$2p - 2q = 4$

3 Find the values of the two unknowns in the following.

(a) $x + 3y = 10$
$3x + 3y = 12$

(b) $4a + 2b = 7$
$2a + 2b = 4$

(c) $3p + 6q = 9$
$3p + q = -1$

4 Solve the following simultaneous equations.

(a) $2x + y = 5$
$x + 3y = 5$

(b) $3a + b = 5$
$a - 3b = 15$

(c) $x + 2y = 12$
$3x - 4y = 4$

(d) $5p - 3q = 1$
$-15p + 7q = 1$

(e) $3e - 2f = 5$
$5e - 6f = 3$

(f) $4x - 5y = 7$
$x - 3y = 7$

5 Solve these simultaneous equations.

(a) $2x - 3y = 3$
$5x + 2y = 17$

(b) $4x + 3y = 20$
$3x + 5y = 26$

(c) $5a - 2b = 9$
$3a + 7b = -11$

(d) $4g + 3h = 15$
$5g + 2h = 24$

(e) $2w + 5z = 43$
$9w - 4z = 8$

(f) $2x + 2y = 2$
$5x + 3y = -1$

6 Find two numbers whose sum is 116 and difference is 34.

7 The cost of 3 apples and 4 bananas is £1.17 and the cost of 5 apples and 2 bananas is £1.11. Find the cost of each type of fruit.

8 Solve these simultaneous equations.

(a) $3x - 2y = 0$
$5x - 7y = -22$

(b) $2a + 3b = 5$
$5a - 2b = -16$

(c) $5g + 3h = 23$
$2g + 4h = 12$

(d) $3w + 2z = 7$
$2w - 3z = -4$

(e) $5x - 7y = 27$
$3x - 4y = 16$

(f) $4x - 0.5y = 12.5$
$5x + 1.5y = 13.5$

5.7 Writing inequalities

Expressions in which the right- and left-hand sides are not equal are called **inequalities** . There are four inequality symbols that you need to know.

$<$ means **less than**

$>$ means **greater than**

$\leqslant$ means **less than or equal to**

$\geqslant$ means **greater than or equal to**

Example 20

Write the correct inequality sign between these pairs of numbers.

(a) $21, 56$ (b) $-7, 6$ (c) $-7, -10$

(a) 21 is less than 56 $21 < 56$

(b) -7 is less than 6 $-7 < 6$

(c) -7 is greater than -10 $-7 > -10$

The thinner end $<$ points towards the smaller number.

The thicker end $>$ points towards the larger number.

Example 21

Write down the integer values of the following letters that makes the inequality true.

(a) $n > 5$ and $n < 12$ (b) $b > 2$ and $b \leqslant 7$ (c) $x \geqslant -4$ and $x < 3$

An integer is a whole number (negative or positive). Zero is also an integer.

(a) $n > 5$ and $n < 12$

$n = 6, 7, 8, 9, 10$ and 11

(b) $b > 2$ and $b \leqslant 7$

so $b = 3, 4, 5, 6$ and 7

(c) $x \geqslant -4$ and $x \leqslant 3$

so $x = -4, -3, -2, -1, 0, 1$ and 2

The values must be greater than 5 but less than 12.

The values must be greater than 2 but less than or equal to 7.

The values must be greater than or equal to -4 but less than 3.

Inequalities can also be shown on a number line.

The answers to Example 21 can be shown on the following number lines.

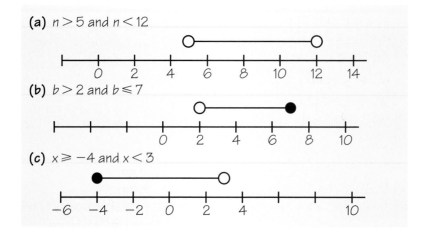

(a) $n > 5$ and $n < 12$

(b) $b > 2$ and $b \leqslant 7$

(c) $x \geqslant -4$ and $x < 3$

Use the symbol $\bigcirc$ when the end value is *not* included.

Use the symbol $\bullet$ when the end value *is* included.

Inequalities can also be combined. The three inequalities in Example 21 can be written as follows.

(a) $5 < n < 12$

(b) $2 < b \leqslant 7$

(c) $-4 \leqslant x < 3$

In **(a)** you say 'n is greater than 5 but less than 12'.

Or that 'n lies strictly between 5 and 12'.

Exercise 5G

1 Place the correct inequality sign between these pairs of numbers.

(a) $5.5, 6.4$ (b) $12.7, 12.6$ (c) $0, 0.01$ (d) $1112, 1121$

2 Write down whether these statements are True (T) or False (F).

(a) $5 \leqslant 4$ (b) $23 > 22.99$ (c) $0.01 < 0.001$

(d) $89 = 88$ (e) $45 \geqslant 44$ (f) $99.9 > 99.89$

3 Write down the whole number values of the letter that satisfy the following inequalities.

(a) $b > 4$ and $b < 9$ (b) $r \leqslant 12$ and $r > 7$

(c) $p > 0$ and $p \leqslant 5$ (d) $k \leqslant 100$ and $k > 94$

(e) $x > -2$ and $x \leqslant 3$ (f) $q \geqslant -9$ and $q < -4$

4 Write down the values (to 1 decimal place) that satisfy the following inequalities.

(a) $h > 3.2$ and $h < 4.1$ (b) $t \leqslant 21.2$ and $t > 20.7$

(c) $y \leqslant 0.9$ and $y \geqslant 0.3$ (d) $m \geqslant 78.9$ and $m < 79.5$

5 Write each of the inequalities in questions **3** and **4** as a combined inequality.

6 Draw the following inequalities on separate number lines.
All number lines should be between -4 and 12.

   (a) $x > 4$ **(b)** $t > 2$ and $t < 11$

   (c) $u \leqslant 12$ and $u > 6$ **(d)** $w \leqslant 9$ and $w > 0$

   (e) $2 < j \leqslant 8$ **(f)** $8 \leqslant s < 12$

   (g) $-3 \leqslant m < 6$ **(h)** $-4 \leqslant d \leqslant 0$

5.8 Solving inequalities

Key words:
solution set
double inequality

Finding the solution of an inequality is very similar to finding the solution of an equation, and the same rules of algebra apply.

There is usually more than one solution when you solve an inequality and you need to state all possible solutions.

You call this the **solution set** .

Example 22

Solve $4n \geqslant 20$ where n is an integer.

$$4n \geqslant 20$$

$$\frac{4n}{4} \geqslant \frac{20}{4}$$

$$n \geqslant 5$$

Divide both sides by 4.

So n can have the values 5, 6, 7 , 8, ... This is called the solution set.

Example 23

Solve $3m - 6 < 9$ where m is an integer.

$$3m - 6 < 9$$

$$3m - 6 + 6 < 9 + 6$$

$$3m < 15$$

$$\frac{3m}{3} < \frac{15}{3}$$

$$m < 5$$

Add 6 to both sides of the inequality and then divide both sides by 3.

m can take any value in the set 4, 3, 2, 1, 0, -1, -2, ...

Example 24

Solve the inequality $5p \leqslant 2p + 12$ where p is an integer.

$$5p \leqslant 2p + 12$$
$$5p - 2p \leqslant 2p - 2p + 12$$
$$3p \leqslant 12$$
$$p \leqslant 4$$

Subtract $2p$ from both sides then divide both sides of the inequality by 3.

Example 25

Solve $4 < 2m \leqslant 16$ where m is an integer.

This is a **double inequality** because it has inequality signs at both ends.

Method A $4 < 2m$ and $2m \leqslant 16$

$$\frac{4}{2} < \frac{2m}{2} \qquad \frac{2m}{2} \leqslant \frac{16}{2}$$

$$2 < m \qquad m \leqslant 8$$

The solution is $2 < m \leqslant 8$

Split the inequality into two parts.

m can take the values 3, 4, 5, 6, 7 and 8.

Method B $4 < 2m \leqslant 16$

$$\frac{4}{2} < \frac{2m}{2} \leqslant \frac{16}{2}$$

$$2 < m \leqslant 8$$

Divide throughout by 2.

Exercise 5H

1 Solve these inequalities.

(a) $6m \leqslant 18$ (b) $2v > 3$ (c) $10x \leqslant 24$

(d) $4w \leqslant -16$ (e) $9a > -81$ (f) $7t < -21$

2 Solve the following inequalities, where each letter has integer values.

(a) $2x + 5 > 9$ (b) $8s + 3 < 19$ (c) $5u - 7 \leqslant 18$

(d) $4a - 22 \geqslant 18$ (e) $6q - 7 > 13$ (f) $12 + 4r \leqslant 18$

(g) $3n + 7 \geqslant 1$ (h) $7p + 25 \leqslant 4$ (i) $10y + 50 > 10$

3 Solve the following inequalities. All the letters represent integer values. State the solution set clearly.

(a) $8q > 5q + 12$ (b) $3c \leqslant c + 20$

(c) $12v \leqslant 7v + 100$ (d) $5n \geqslant 8n - 6$

(e) $3x + 14 \leqslant 5x$ (f) $7g \leqslant 2 + 5g$

(g) $4m + 10 \leqslant 2m$ (h) $7w + 12 > 4w$

(i) $5k - 6 \leqslant 2k + 3$ (j) $3t + 25 \geqslant 8t - 15$

(k) $6f + 9 > 12 + 3f$ (l) $14y - 23 \leqslant 5y + 4$

Always rearrange so that the letter term is **positive**.

4 Write down the solution set for each of the following inequalities. State the solution set clearly.

(a) $6 < 2z < 20$

(b) $21 \leqslant 7c < 28$

(c) $6 < 3x \leqslant 18$

(d) $200 < 10b \leqslant 150$

(e) $0 < 9h < 81$

(f) $3 < 4t \leqslant 10$

(g) $-6 < 3p \leqslant 3$

(h) $-25 \leqslant 10q \leqslant -5$

(i) $3 \leqslant 2t + 7 \leqslant 15$

(j) $-7 < 3y - 10 < 26$

(k) $-13 < 4r + 5 < 15$

(l) $6 \leqslant 5d + 18 \leqslant 30$

Inequalities with a negative unknown

So far you have solved inequalities such as $-6 < 3p \leqslant 3$ where the unknown is *positive*. You need to be aware of an extra rule to follow, if dealing with *negative* unknowns, as in the next example.

Example 26

Solve the inequality $9 < 15 - 2y$.

$$9 < 15 - 2y$$
$$9 + 2y < 15 - 2y + 2y$$
$$9 + 2y < 15$$
$$9 + 2y - 9 < 15 - 9$$
$$2y < 6$$
$$y < \frac{6}{2}$$
$$y < 3$$

Add $2y$ to both sides then subtract 9 from both sides.

Always rearrange so that the letter term is on its own on one side of the inequality and is **positive**.

Check a number less than 3 e.g. $y = 2$ $RHS = 15 - 2 \times 2$
$= 15 - 4$
$= 11$
which is greater than 9. ✓

Look what might happen if you leave the y term negative.

$$9 < 15 - 2y$$
$$9 - 15 < 15 - 2y - 15$$
$$-6 < -2y$$
$$\frac{-6}{-2} < y$$
$$3 < y$$

This is a contradiction of the first result (which is correct, as you can see from the check).

What has gone wrong?

The problem is in the last two lines of the solution (highlighted in blue).

When you divide (or multiply) by a negative number you must always remember to **reverse the inequality sign**. So the last two lines of the solution should be:

$$\frac{-6}{-2} > y$$

$3 > y$ which is the same solution as before.

Look at these examples,

$5 < 8$

multiply both sides by -2

$-10 > -16$

... notice that the sign is now reversed.

$100 > 60$

divide both sides by -20

$-5 < -3$

... as before, the sign is now reversed.

Exercise 5I

Solve these inequalities.

1 $14 - 3x > 5$ **2** $123 < 15 - 4y$ **3** $13 - 5z \leqslant -7$

4 $7 - 3h \geqslant 22$ **5** $16 - 2k \leqslant 7$ **6** $10 - 4t > 1$

7 $12 > 27 - 2d$ **8** $9 - 6e \geqslant 24$ **9** $26 \leqslant 14 - 5f$

Examination questions

1 Solve the equations

     (a) $\frac{x}{3} = 12$ **(b)** $y + 6 = 13$ **(c)** $8c - 5 = 11$ *(5 marks)*

2 Solve the equations

     (a) $6r + 2 = 8$ **(b)** $7s + 2 = 5s + 3$ **(c)** $\frac{x}{4} = 8$ **(d)** $\frac{12 - y}{3} = 5$ *(8 marks)*

                                                      AQA, Spec A, 2I, November 2003

3 (a) Solve the equation $\frac{23 - 2x}{5} = 3$.

     (b) Solve the inequality $3x + 8 < 29$.   *(5 marks)*

                                                      AQA, Spec A, 2H, June 2003

4 Solve the simultaneous equations

         $2x + 3y = 9$

         $3x + 2y = 1.$

Do not use trial and improvement.
You **must** show your working.

                                                   (4 marks)

                                                        AQA, Spec A, 1H, June 2003

5 (a) Solve the inequality
$2x + 3 \geqslant 1$.

(b) Write down the inequality
shown by the following diagram.

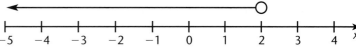

(c) Write down all the integers that satisfy both inequalities shown
in parts **(a)** and **(b)**.

(4 marks)
AQA, Spec A, 2H, June 2004

Summary of key points

Solving equations (grade D/C to B)

Linear equations can be solved by treating both sides of the equation in the same way
– what you do to one side of the equation you must do to the other.

The solutions to equations may be positive or negative whole numbers (integers) or
fractions or decimals.

Equations sometimes have the unknown (letter) on both sides of the equation. To
solve these equations you often need to carry out more than two operations.

You can use equations to solve problems.
You will need to read the problem carefully then set up an equation and solve it.

Simultaneous equations (grade B)

Solving two linear equations at the same time is called solving simultaneous equations.

The rule is:

If the x's or y's have the same numerical value **but differ in sign** ..... **ADD**

If the x's or y's have the same numerical value **and the same sign** ... **SUBTRACT**

Inequalities (grade C/B)

An expression in which the right-hand side and left-hand side are not equal is called
an *inequality*. Finding the solution of an inequality is very similar to finding the
solution of an equation, and the same rules of algebra apply. There is usually more
than one solution when you solve an inequality and you need to state all possible
solutions. You call this the *solution set*.

You need to know the following inequality symbols

$<$ means *less than*

$>$ means *greater than*

$\leqslant$ means *less than or equal to*

$\geqslant$ means *greater than or equal to*

A *number line* can be used to show the solutions to an inequality.

Use the symbol $\bigcirc$ when the end value is *not* included.

Use the symbol $\bullet$ when the end value *is* included.

6 Measurement and approximations

Number 4

This chapter will show you how to:

✔ convert between metric units of measurement and handle metric/imperial conversions
✔ handle the compound measures of speed and density
✔ round numbers
✔ estimate answers to calculations by using approximations
✔ understand limits of accuracy
✔ calculate absolute error and percentage error

6.1 Conversions

You will need to know:
● the metric and imperial units of length, capacity and mass.

Conversions between metric units

Metric units of length:
10 mm = 1 cm   100 cm = 1 m
1000 mm = 1 m   1000 m = 1 km

Metric units of mass:
1000 mg = 1 g   1000 g = 1 kg
1000 kg = 1 tonne (t)

Metric units of capacity:
100 c*l* = 1 *l*   1000 m*l* = 1 *l*
1 cm³ = 1 cc = 1 m*l*
1000 cm³ = 1 *l*   1000 cc = 1 *l*

1 *l* = 1 litre
1 c*l* = 1 centilitre

Example 1

Change
(a) 8 km into cm   (b) 3 t into kg   (c) 72 000 c*l* into *l*.

(a) 8 km = 8 × 1000 m = 8000 m
8000 m = 8000 × 100 cm = 800 000 cm
(b) 3 t = 3 × 1000 kg = 3000 kg
(c) 72 000 c*l* = 72 000 ÷ 100 *l* = 720 *l*

Exercise 6A

1 Change these lengths into the units given.

   (a) 3.9 metres → millimetres   **(b)** 5.4 metres → centimetres

   (c) 12.3 cm → mm   **(d)** 4050 m → km

   (e) 6125 mm → m   **(f)** 303 mm → cm

2 Change these masses into the units given.

   (a) 5.05 tonnes → kilograms   **(b)** 8200 grams → kg

   (c) 0.56 kg → grams   **(d)** 16 120 kg → tonnes

3 Change these capacities into the units given

   (a) 7450 ml → l **(b)** 80 cl → l

   (c) 30.4 cl → ml **(d)** 0.53 l → cl

4 Find the missing values.

   (a) 7008 mm = __m   **(b)** 95 ml = __cl

   (c) 0.04 kg = __g   **(d)** 0.005 km = __m

   (e) 5050 kg = __tonnes   **(f)** 200 ml = __l

5 The masses of 4 boxes are given as 40.3 kg, 8780 g, 0.64 kg and 0.018 t. What is the total mass in tonnes?

6 The lengths achieved by 3 jumping frogs were 2.83 m, 0.002 km and 2798 mm. Change the lengths into metres and place in order from longest to shortest.

Conversions between metric and imperial units

You need to be able to convert between metric and imperial units.

Metric	Imperial
* 8 km	5 miles
* 30 cm	1 foot
* 1 litre	$1\frac{3}{4}$ pints
* 4.5 litres	1 gallon
* 1 kg	2.2 pounds
2.5 cm	1 inch
1 m	39 inches
25 g	1 ounce

You need to remember those marked* for your exam.
The others are useful.

These conversions are approximate.

Example 2

The marathon is run over a distance of 26 miles.
How many kilometres is this?

5 miles $\approx$ 8 km

26 miles $\approx \dfrac{8 \times 26}{5}$ km

≈ 41.6 km

$\approx$ means 'approximately equal to'.

25 miles $\approx$ 40 km
1 mile $\approx \frac{8}{5}$ km = 1.6 km
So   26 miles $\approx$ 40 + 1.6
     = 41.6 km

Exercise 6B

1 Change these into the units given.

(a) 2 inches $\rightarrow$ cm
(b) 7.5 cm $\rightarrow$ inches
(c) 2 feet $\rightarrow$ cm
(d) 35 cm $\rightarrow$ inches
(e) 15 miles $\rightarrow$ km
(f) 32 km $\rightarrow$ miles
(g) 2.5 miles $\rightarrow$ km
(h) 2 km $\rightarrow$ miles

2 Change these into the units given.

(a) $3\frac{1}{2}$ pints $\rightarrow$ litres
(b) 5 litres $\rightarrow$ pints
(c) 7 gallons $\rightarrow$ litres
(d) 13.5 litres $\rightarrow$ gallons
(e) 2.25 litres $\rightarrow$ gallons
(f) 4.2 gallons $\rightarrow$ litres

3. Change these into the units given.

(a) 3 kilograms $\rightarrow$ pounds
(b) $6\frac{1}{2}$ kg $\rightarrow$ pounds
(c) 11 pounds $\rightarrow$ kg
(d) 53 pounds $\rightarrow$ kg
(e) $8\frac{1}{2}$ ounces $\rightarrow$ grams
(f) 340 grams $\rightarrow$ ounces

4 Change these into the units given.

(a) $5\frac{1}{4}$ inches $\rightarrow$ cm
(b) 3.8 cm $\rightarrow$ inches
(c) 14 lbs $\rightarrow$ kg
(d) 8.4 litres $\rightarrow$ gallons
(e) 0.6 kg $\rightarrow$ pounds
(f) 14.8 miles $\rightarrow$ km
(g) $2\frac{1}{2}$ pints $\rightarrow$ litres
(h) 23.4 km $\rightarrow$ miles

5 A shop sells smoothies in 2 pint bottles or $1\frac{1}{2}$ litre cartons.
Which one holds the most?

6 The maximum mass for a suitcase on HJ Airlines is 22 kg.
My case weighs 49.8 pounds. Will I be able to take it?

Conversions of areas and volumes

Key words:
square
cubic

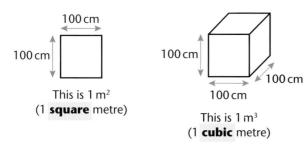

This is 1 m² (1 **square** metre)

This is 1 m³ (1 **cubic** metre)

1 m = 100 cm
so   **1 m² = 100 cm × 100 cm = 10 000 cm²**
and   **1 m³ = 100 cm × 100 cm × 100 cm = 1 000 000 cm³**

Exercise 6C

1 Copy and complete the following.

(a) $3\,m^2 = \underline{\quad}\,cm^2$ (b) $5.2\,m^3 = \underline{\quad}\,cm^3$

(c) $70\,500\,cm^2 = \underline{\quad}\,m^2$ (d) $\frac{1}{2}\,m^3 = \underline{\quad}\,cm^3$

(e) $0.4\,m^2 = \underline{\quad}\,cm^2$ (f) $300\,000\,cm^3 = \underline{\quad}\,m^3$

2 How many

(a) mm^2 in $1\,cm^2$ (b) cm^2 in $1\,km^2$

(c) square feet in a square yard?

1 yard = 3 feet

3 How many

(a) m^3 in $1\,km^3$ (b) mm^3 in $1\,m^3$

(c) cubic inches in 1 cubic foot?

12 inches = 1 foot

4 A lake covers an area of 3.25 million square metres. What is its area in square kilometres?

5 A rectangular carpet measures 420 cm by 380 cm.

(a) What is the area of the carpet in cm^2?

(b) What does the carpet measure in square metres?

(c) How many whole square metres should I buy?

(d) If the carpet costs £10.50 per m^2, what is the total cost?

6 A rectangular container for a poisonous spider measures 85 cm by 65 cm by 45 cm.

(a) What is the volume of the container?

(b) The rules on keeping this kind of spider say it must have at least $\frac{1}{4}\,m^3$ of space. Is this container suitable?

6.2 Rounding

Key words:
degree of accuracy

Why is it necessary to use a rounded or approximate answer rather than an exact answer?

'The land area of the UK is 94 251 square miles.'
How can the land area of the UK be measured so accurately?
Do you really need to know the exact area?
You could say:
'The land area of the UK is about 90 000 square miles.'

You can give the land area to a greater **degree of accuracy** by rounding to the nearest 1000 or 100 or 10. The answers are 94 000, 94 300 and 94 250 respectively, but it is unlikely you would want it to be any more accurate than 94 000.

Remember that when a number falls *exactly* halfway between two limits you always round *upwards*.
So 1465 (exactly halfway between 1460 and 1470) rounds up to 1470, to the nearest 10.

Rounding to a given number of decimal places

You are often asked to give an answer to a number of decimal places.

The final answer must have only as many decimal places as the question asks for, no more and no less.

Follow these simple rules:
1 Count the number of places you want, to the right of the decimal point.
2 Look at the next digit.
   If it is 5 or more you need to round up the digit in the previous decimal place.
   If it is 4 or less you leave the previous decimal digit as it is.

Example 3

Write 8.14973 to **(a)** 1 d.p.   **(b)** 2 d.p. **(c)** 3 d.p.

(a) $8.14973 = 8.1$ (1 d.p.)

The *next* digit is 4 so the 1 is *unchanged*.

(b) $8.14973 = 8.15$ (2 d.p.)

The *next* digit is 9 so the 4 must be *rounded up* to 5.

(c) $8.14973 = 8.150$ (3 d.p.)

The *next* digit is 7 so the 9 must be *rounded up*.
Since this would make it 10, the 4 needs to change to 5.

You need to keep the 0 in **(c)** because you must have 3 d.p.
The answers of 8.15 in **(b)** and 8.150 in **(c)** are not the same since 8.15 is to an accuracy of 2 d.p. and 8.150 is to an accuracy of 3 d.p.

Exercise 6D

1 Round the following to 1 decimal place.

    (a) 5.83   **(b)** 7.39   **(c)** 2.15   **(d)** 5.681

    (e) 4.332   **(f)** 15.829   **(g)** 11.264   **(h)** 17.155

    (i) 145.077   **(j)** 521.999

2 Round to 2 decimal places.

    (a) 3.259   **(b)** 6.542   **(c)** 0.877   **(d)** 0.031

    (e) 11.055   **(f)** 4.007   **(g)** 3.899   **(h)** 2.3093

    (i) 0.0009   **(j)** 5.1299

3 Round to 3 d.p.

    (a) 1.2546   **(b)** 5.2934   **(c)** 4.1265

    (d) 0.0007   **(e)** 0.000 08

4 Round 15.1529 to

    (a) 1 decimal place   **(b)** 2 d.p.

    (c) 3 d.p.   **(d)** the nearest whole number.

5 In a race the time for the winner was given as 15.629 seconds.
Write this to **(a)** the nearest second **(b)** 1 d.p. **(c)** 2 d.p.
The time for the second place was 15.634 seconds.

    (d) Round this as you did for **(a)**, **(b)**, and **(c)** above.
      What do you notice?

Rounding to a given number of significant figures

> **Key words:**
> significant figure

Significant figures are 'important' figures and the most significant figure in a number is the one with the greatest place value.

> The most significant figure is the *first non-zero* figure reading from the left.

Follow these simple rules:

1 Start from the most significant figure and count the number of figures that you want.

2 Look at the next digit.
If it is 5 or more you need to round up the previous digit.
If it is 4 or less you leave the previous digit as it is.

3 Put in any zeros needed to locate the decimal point and indicate place value.

> This is *very* important.

Example 4

Round 26 818 to

(a) 1 significant figure (1 s.f.)   **(b)** 2 s.f.

> **(a)** 26 818 = 30 000   (1 s.f.)

> **(b)** 26 818 = 27 000   (2 s.f.)

> The *next* digit is 6 so the 2 must be *rounded up* to 3. You *must* include zeros to locate the decimal point and indicate the place value of the 3.

> The *next* digit is 8 so the 6 must be *rounded up* to 7. Either 30 000 or 27 000 would be a sensible approximation of 26 818.

Example 5

Write 0.007 038 6 to

(a) 1 s.f.   **(b)** 2 s.f.

> **(a)** 0.007 038 6 = 0.007

> **(b)** 0.007 038 6 = 0.0070

> The most significant figure is the *first non-zero* figure reading from the left, 7. The *next* digit is 0 so the 7 is *unchanged*.

> The *next* digit is 3 so the 0 is *unchanged*.
> You *must* keep the 0 after the 7 because it is the second significant figure.

In both of these examples, the zeros after the decimal point but before the 7 locate the decimal point and indicate the place value of the 7.

Exercise 6E

1 Round the following to 1 significant figure:
   (a) 2862   **(b)** 19 473   **(c)** 257 394
   (d) 84 693   **(e)** 4.86

2 Round to 2 s.f.
   (a) 6842   **(b)** 32 841   **(c)** 153 945
   (d) 267 149   **(e)** 5.32

3 Round the following to 1 s.f.
   (a) 0.037   **(b)** 0.042   **(c)** 0.0059
   (d) 0.0035   **(e)** 0.509

4 Round to 2 s.f.
   (a) 0.574   **(b)** 0.000 382   **(c)** 0.001 92
   (d) 0.203   **(e)** 0.001 99

5 The population of a town is given as 293 465.
   What is this rounded to
   (a) 1 s.f.   **(b)** 2 s.f.?

6 The length of a boat is given as 43.562 m.
   What is this rounded to
   (a) 2.s.f   **(b)** 3.s.f.?

6.3 Compound measures

You will need to know:
● how to convert times between 'hours' and 'hours and minutes'

Speed

Key words:
speed
distance
time

Speed is a measurement of how fast something is travelling.
It involves two other measures, **distance** and **time** .

If you travel from Newcastle to Leeds, a distance of 100 miles, in 2 hours, you have travelled at an average speed of 50 miles per hour. To calculate a speed you divide a distance by a time.
There are three formulae connecting speed, distance and time:

Your speed would not be constant for the whole journey so the speed calculated is an **average** speed.

$$\text{Speed} = \frac{\text{distance}}{\text{time}} \qquad \text{Time} = \frac{\text{distance}}{\text{speed}}$$

$$\text{Distance} = \text{speed} \times \text{time}$$

This triangle will help you to remember the formulae.

Speed can be measured in many different units.
The most common are:
metres per second (m/s),
kilometres per hour (km/h or kph)
centimetres per second (cm/s),   miles per hour (mph)

If you cover up the one you want to find, you are left with the formula you need. For example, if you cover up T, you are left with $\frac{D}{S}$.
$T = \frac{D}{S}$ or Time $= \frac{\text{distance}}{\text{speed}}$

The word 'per' means divide. Kilometres per hour (km/h) means you divide a distance in km by a time in hours.

When you do calculations involving speed, distance and time, you must make sure that you are consistent with the units.
For example, if you want to calculate a speed in *metres per second*, you must have the distance in *metres* and the time in *seconds*.
If you divide a distance in *miles*, by a speed in *miles per hour*, the time taken will automatically be in *hours*.

Example 6

Jo travelled a distance of 180 miles. The journey took $4\frac{1}{2}$ hours.
What was her average speed for the journey?

$$\text{Average speed} = \frac{\text{total distance travelled (miles)}}{\text{total time taken (hours)}}$$
$$= \frac{180}{4.5} \text{ mph}$$
$$= 40 \text{ mph}$$

Distance in miles divided by time in hours gives speed in miles per hour.

Example 7

Simon averaged a speed of 80 km/h for a journey of 208 km.
How long did the journey take?

Distance in km divided by speed in km/h gives time in hours.

$$\text{Time} = \frac{\text{Distance}}{\text{Speed}} = \frac{208}{80} = 2.6$$

0.6 hours = 0.6 × 60 minutes = 36 minutes

Time taken for journey = 2 hours 36 minutes

You must *not* read this as 2 hours 6 minutes.
It is 2.6 hours.
To convert the decimal part to minutes you multiply by 60.
Always do this to change part of one hour to minutes.

Example 8

Ian drove for 2 hours 45 minutes at an average speed of 48 mph.
How far did he travel?

2 hours 45 minutes = $2\frac{3}{4}$ hours = 2.75 hours

Distance = Speed × Time

= 48 × 2.75

= 132 miles

This is an example where the units are not consistent. The time of 2 hours 45 minutes must be written in hours only.
45 minutes is $\frac{3}{4}$ of an hour, the time will be 2.75 hours. (Divide by 60 to change minutes to hours.)

Exercise 6F

1 Change the following decimal times into hours and minutes.

(a) 2.1 hours   (b) 3.8 hours

(c) 5.65 hours   (d) 4.7 hours

Change the following into hours, giving your answers to 2 d.p.

(e) 3 h 15 min   (f) 6 h 25 min

(g) 4 h 8 min   (h) 12 min

Many calculators have a D°M'S' button which can be used for hours, minutes and seconds, e.g.
2.64 hours = 2 h 38 min 24 sec.

2 Find the average speed of a car which travelled 245 miles in 4 hours 10 minutes.

3 What distance will a car travel in 3 hours 40 minutes, assuming an average speed of 75 km/h?

4 How long will it take to do a journey of 730 miles at 68 mph?

5 John cycled to his friend's house 4 miles away which took 25 minutes. He and his friend walked back round the lake which took 3 hours 10 minutes at an average speed of $3\frac{1}{2}$ mph.

(a) What was John's average speed on his cycle ride?

(b) How much further was his walk than his cycle ride?

(c) What was the average speed for the whole journey including the cycle ride and the walk?

6 What is the speed 7 m/s in kilometres per hour?

7 The journey to my holiday hotel was in three parts. I drove at an average speed of 55 mph to the airport which took 1 hour 10 minutes. The flight was 1200 miles in 2 hours 36 minutes and the bus travelled at 35 mph to the hotel 25 miles away.

(a) What was the total travelling time?

(b) What was the total distance travelled?

(c) What was the average speed for the whole journey?

8 My dog can run at a speed of 8 km/h. Pete's cat runs 30 metres in 20 seconds. Which one is faster?

9 Ali is on a 5000 m run. He runs the first 150 m in 22 seconds. For the next 3500 m he maintains an average speed of 7 m/s. He slows his speed to 4 m/s over the next 500 m. For the next 750 m he picks up speed and runs at a speed of 6 m/s, putting in a final burst to run the last 100 m in 12 seconds.

(a) What was Ali's time for the race?

(b) What was his average speed for the 5000 m in m/sec?

(c) What was Ali's average speed in km/h?

10 Sally is in training for a triathlon which involves cycling, swimming and running.
The cycling takes $2\frac{1}{2}$ hours at an average speed of 35 km/h. The swimming involves crossing a lake 1200 m wide which takes 20 minutes, followed by a 1500 m cross country run, which she completes at an average speed of 6 km/h.

(a) What is her speed on the swimming section?

(b) What is the total distance she travels?

(c) What is her average speed for the whole triathlon?

Density

Key words:
density
mass
volume

Density is defined as **mass** per unit **volume**.
It is the mass (usually given in grams) of one unit of volume of material (usually given in cm³).
This means that density is calculated by dividing the mass of an object by its volume.
There are three formulae connecting density, mass and volume:

$$\text{Density} = \frac{\text{mass}}{\text{volume}} \qquad \text{Volume} = \frac{\text{mass}}{\text{density}}$$

$$\text{Mass} = \text{density} \times \text{volume}$$

This triangle will help you to remember the formulae.
You use it like the one for speed, distance and time.

You need to be consistent with the units that you use, just as you did in the speed, distance and time calculations. Density will usually be in g/cm³. Volume will then be in cm³. Mass will then be calculated in g.

Example 9

A piece of wood weighs 124 g and has a volume of 140 cm³.
What is the density of the wood?

$$\text{Density} = \frac{mass}{volume} = \frac{124}{140}$$

$$= 0.886$$

$$\text{Density} = 0.886 \; g/cm^3 \; (3 \; s.f.)$$

> Measurements in g, kg, lbs or oz are strictly measurements of mass, although in everyday speech we often refer to them as weights.

Example 10

A measuring cylinder contains a liquid whose density is 1.18 g/cm³.
The volume of the liquid is 0.35 *l*.
What is the mass of the liquid?

$$0.35 \; \ell = 0.35 \times 1000 \; cm^3 = 350 \; cm^3$$

$$\text{Mass} = density \times volume = 1.18 \; g/cm^3 \times 350 \; cm^3 = 413 \; g$$

> You need to change the units of volume from litres to cm³.
> 1 *l* = 1000 cm³

Exercise 6G

1 Calculate the missing measurements in the following table.

	Mass	Density	Volume
(a)	24 g	1.6 g/cm³	
(b)		2.05 g/cm³	40 cm³
(c)	28 g		13.3 cm³
(d)		3.2 g/cm³	1.5 litres
(e)	5.7 kg	1.24 g/cm³	

> Change the units first.

> Round your answers to 3 s.f.

2 A solid gold bar measures 5.5 cm by 3.4 cm by 2.6 cm. The density of gold is 19.3 g/cm³. What is the mass of the bar?

3 A cube of lead with side 8.4 cm has a mass of 6.757 kg. What is the density of lead?

4 The cuboid has a density of 6.2 g/cm³.
  (a) Find its volume.
  (b) Two sides of the cuboid are 5.4 cm and 4.1 cm. How long is the third side of the cuboid?

5 The density of a type of card is 1274 kg/m³.
  (a) What is the volume of 850 kg of card?
  (b) What is the mass of card with a volume of 1.75 m³?
  Give your answers to 3 s.f.

6 What is the volume of a concrete block that has a mass of 50 kg and a density of 1600 kg/m³? Give your answer in cm³.

7 A sculpture in a museum has a volume of about 3 m³. The material from which it is made has a density of 11.4 g/cm³. What is the approximate mass of the sculpture? Give your answer in kg.

8 A ball bearing has a radius of 2.3 cm. It is made from material with a density of 6.3 g/cm³. What is its mass?

> The formula for finding the volume of a sphere is $\frac{4}{3}\pi r^3$

9 A gold chain is made from 1.4 cm³ of gold. The density of gold is 19.3 g/cm³. Gold chains are priced by weight at £23.18 per gram. How much does the gold chain cost?

 10 A cube has a mass of 2.6 kg and is made from wood with a density of 5.2 g/cm³. What is the length of the side of the cube?

6.4 Approximations and error

Estimating using approximations

When you do a calculation using your calculator, do you always believe the answer? What if you press a wrong button? Would you realise what you had done?
You can check your answer using **approximations** .

> **Key words:**
> approximation
> estimate

> To **estimate** an approximate answer:
>
> **1** Round all of the numbers in the calculation to 1 significant figure.
>
> **2** Do the calculation using these approximations.

> This is a skill you will need. Some examination questions ask you to show how to get estimates to calculations by using approximations.

Example 11

Use approximations to estimate the answers to these calculations.
(a) 6.89×3.375
(b) $38.45 \div 7.56$
(c) $\dfrac{181.03 \times 0.48}{8.641}$
(d) $\dfrac{6.85^2}{2.19 \times 11.7}$

(a) $6.89 \times 3.375 \approx 7 \times 3 = 21$

(b) $38.45 \div 7.56 \approx 40 \div 8 = 5$

(c) $\dfrac{181.03 \times 0.48}{8.641} \approx \dfrac{200 \times 0.5}{9} = \dfrac{100}{9} \approx 11$

(d) $\dfrac{6.85^2}{2.19 \times 11.7} \approx \dfrac{7^2}{2 \times 10} = \dfrac{49}{20} \approx 2.5$

The approximate answers in Example 11 are close to the answers of the original calculations.

(a) $6.89 \times 3.375 = 23.25375$ Approximate answer $= 21$

(b) $38.45 \div 7.56 = 5.0859...$ Approximate answer $= 5$

(c) $\dfrac{181.03 \times 0.48}{8.641} = 10.0560...$ Approximate answer $= 11$

(d) $\dfrac{6.85^2}{2.19 \times 11.7} = 1.8312...$ Approximate answer $= 2.5$

> All of these are acceptable because the approximations are sensible and make the calculations easy.
> This is the most important thing to remember when using approximations to estimate answers.

Exercise 6H

1 Use approximations to estimate the answers to the following:

(a) 7.63×8.36

(b) $15.28 \div 4.85$

(c) $(3.62 + 7.17) \times 8.57$

(d) $\dfrac{4\,287 \times 4\,862}{37 \times 53}$

(e) $\dfrac{8.63 \times 4.95}{(2.7)^2}$

(f) $\dfrac{7.53 - 2.86}{1.35 + 1.44}$

2 Calculate the actual answers in question 1 and check to see if your approximations were reasonable.

3 A company bought 2815 kg of cloves to crush and package. The cloves cost £17.85 per kg. Estimate the total cost.

4 A supermarket sells a box of 460 pears for £38 in total. Estimate how much each pear cost.

5 The cost of a school trip per pupil is £4.85 for tickets, £2.15 for transport, plus £1.12 for meals. If there are 224 pupils going on the trip, estimate the total cost.

6 My car, on average, does 8.5 miles to the litre over a distance of 348 miles. Work out approximately

(a) how many litres of petrol were used

(b) total cost if petrol cost 89p per litre.

7 A wall will need 127 bricks for each row and will be 17 bricks high. A brick has a mass of 3.4kg.

(a) Estimate the number of bricks needed.

(b) What is the approximate total mass of the bricks?

8 Given that $\dfrac{0.416 \times 87.5}{3.206} = 11.354$ work out $\dfrac{41.6 \times 0.0875}{32.06}$

> Find a connnection between the given expression and the one you have to work out. Where will the decimal point go in the answer?

9 Given that $\dfrac{5.473}{0.26 \times 9.78} = 2.1524$ work out $\dfrac{0.5473}{26 \times 0.978}$

10 Given that $\dfrac{(2.736)^2}{0.000451} = 16598$ work out $\dfrac{(27.36)^2}{0.0451}$

Accuracy of measurement

You need to be familiar with two kinds of measurement.
Discrete measure is for quantities that can be counted.
For example, the number of spectators at a football match was 60 499.

Continuous measure can take any values. The accuracy of the measure depends on the measuring instrument.
For example, if you measure your height as 174 cm, it may not be exactly 174 cm. It is likely to be 174 cm *to the nearest centimetre*. This means that the true height lies between 173.5 cm and 174.5 cm.
You can write this as: $173.5 \text{ cm} \leqslant \text{height} < 174.5 \text{ cm}$

In the same way, if your mass is 68 kg to the nearest kilogram, it will be nearer to 68 kg than to either 67 kg or 69 kg.
The true mass is $67.5 \text{ kg} \leqslant \text{mass} < 68.5 \text{ kg}$

> **Any measurement given *to the nearest whole unit* may be inaccurate by up to *one half unit in either direction*.**
> The minimum value is known as the **lower bound**.
> The maximum value is known as the **upper bound**.

> **Key words:**
> lower bound
> upper bound

> If the attendance was 60 000 correct to the nearest thousand, then we can write $59\,500 \leqslant$ attendance $< 60\,500$. With a discrete measure, there is an actual maximum value – in this case 60 499.

> Notice that the $\leqslant$ sign is used for the 'smaller' value but the $<$ sign is used for the 'larger' value.
> You will see the reason for this at the end of this section.

Example 12

A tube of toothpaste is 10 cm long, to the nearest half centimetre. What are the upper and lower bounds for the length?

Upper bound $= 10.25$ cm

Lower bound $= 9.75$ cm

Exercise 6I

1 Give the range of possible values for the following measurements.

(a) 4 cm (to the nearest cm)   (b) 45 kg (to the nearest kg)

(c) 173 g (to the nearest g)   (d) 340 mm (to the nearest mm)

(e) 3.6 m (to 1 d.p.)   (f) 12.3 l (to 1 d.p.)

(g) 122.5 g (to 1 d.p.)   (h) 0.6 cm (to 1 d.p.)

(i) 7.24 km (to 2 d.p.)   (j) 9.67 min (to 2 d.p.)

(k) 4.92 g (to 2 d.p.)   (l) 7.02 ml (to 2 d.p.)

2 Give the possible range of values of the following:

(a) 7 m   (b) 7.0 m   (c) 7.00 m   (d) 7.000 m

(e) Explain why the answers are not all the same.

Example 13

A rectangle has sides of 8 cm and 5 cm, each measured to the nearest centimetre.
Find the upper and lower bound for the area.

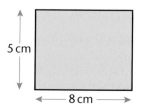

> 8 cm means a length in the range 7.5 cm ≤ length < 8.5 cm
>
> 5 cm means a width in the range 4.5 cm ≤ width < 5.5 cm
>
> The rectangle could be as small as 7.5 cm by 4.5 cm
>
> So the smallest possible area = 7.5 × 4.5 cm²
>
> = 33.75 cm²
>
> Lower bound of the area of the rectangle = 33.75 cm²
>
> The area must be less than 8.5 × 5.5 = 46.75 cm²
>
> Upper bound of the area of the rectangle = 46.75 cm²

Note that the rectangle cannot be 46.75 cm² in area. Any value up to, but not including, this is possible.

The lower bound of the area and the upper bound of the area are *not* an equal distance from the nominal area 8 × 5 = 40 cm². The lower bound is 6.25 cm² less but the upper bound is 6.75 cm² more.

This idea can be extended to values which are given to *any* degree of accuracy, not just to the nearest whole number.

Example 14

$v = u + at$ is a formula used to calculate the final velocity (v) of an object when you know its initial velocity (u), its acceleration (a) and the time (t) over which the acceleration takes place.

If $u = 4.6$ m/s, $a = 1.8$ m/s² and $t = 7.5$ s, all correct to 1 d.p., calculate the greatest possible value of the final velocity.

Acceleration is measured in m/s² (metres per second per second).

> 4.6 m/s must lie between 4.55 m/s and 4.65 m/s
>
> 1.8 m/s² must lie between 1.75 m/s² and 1.85 m/s²
>
> 7.5 s must lie between 7.45 s and 7.55 s
>
> So   max v = 4.65 + (1.85 × 7.55)
>
> = 4.65 + 13.9675
>
> = 18.6175
>
> = 18.6 (to 1 d.p.)

Note that if a number is given correct to 1 d.p. (to the nearest 0.1) it must lie 0.05 either side of its nominal value. For example, 4.6 will be 4.6 ± 0.05 = 4.55 to 4.65.

The maximum value of the product $a \times t$ occurs when both a and t take their maximum value.

Example 14 shows you how to handle a product of terms.

The maximum value is obtained when each of the terms in the product takes its maximum value.

The minimum value would be found when each term takes its minimum value.

Example 15

You want to cut 53 cm from a piece of wood 120 cm long, both measurements given to the nearest cm.
What is the greatest possible length of the remaining piece?

120 cm lies between 119.5 cm and 120.5 cm.

53 cm lies between 52.5 cm and 53.5 cm.

(120.5 cm − 52.5 cm) will give the largest possible answer.

So the greatest possible length of the remaining piece

$$= 120.5 − 52.5 \text{ cm}$$

$$= 68 \text{ cm}$$

Maximum − Minimum

You can see that
119.5 − 53.5 would give
the *least* possible length of
the remaining piece
ie. Minimum − Maximum.

For division, you use one upper bound and one lower bound to obtain the greatest and least answers, in the same way.

These results are summarised as:

Consider two quantities *A* and *B*.

Maximum $(A + B) = \max A + \max B$
Minimum $(A + B) = \min A + \min B$

Maximum $(A \times B) = \max A \times \max B$
Minimum $(A \times B) = \min A \times \min B$

Maximum $(A − B) = \max A − \min B$
Minimum $(A − B) = \min A − \max B$

Maximum $(A ÷ B) = \max A ÷ \min B$
Minimum $(A ÷ B) = \min A ÷ \max B$

Note that for *addition* and *multiplication* ...
MAX = max and max
MIN = min and min

... but for *subtraction* and *division* ...
MAX = max and min
MIN = min and max

When questions involve a combination of these operations you must take care to perform the appropriate calculation at each stage.

Example 16

This formula works out the value of the acceleration, a, for an object which increases its velocity from u to v in travelling a distance s.

$$a = \frac{v^2 - u^2}{2s}$$

It is a formula similar to the one used in Example 14.

If $v = 12.4$ m/s and $u = 3.8$ m/s (both correct to 1 d.p.) and $s = 14$ m (correct to the nearest metre), calculate the least possible value of a. (The units of a will be m/s^2.)

The lower and upper bounds of each of the quantities are:

$$12.35 \leqslant v < 12.45$$

$$3.75 \leqslant u < 3.85$$

$$13.5 \leqslant s < 14.5$$

$a = (v^2 - u^2) \div (2s)$ which is a division

So $\min a = \min(v^2 - u^2) \div \max(2s)$

But $(v^2 - u^2)$ is a subtraction

So $\min(v^2 - u^2) = \min v^2 - \max u^2$

$$= (\min v)^2 - (\max u)^2$$

Final calculation is:

$$\min a = \frac{(\min v)^2 - (\max u)^2}{\max(2s)}$$

$$= \frac{12.35^2 - 3.85^2}{2 \times 14.5}$$

$$= \frac{137.7}{29}$$

$$= 4.74827\ldots \text{ m/s}^2$$

Do not attempt to do the calculation without first breaking it down into smaller steps.

You can then see what are the appropriate values to use for each bit of the numerator and for the denominator.

The final answer has been left in its exact form as seen on your calculator display.

You may be asked to give the final answer to a certain degree of accuracy.

Exercise 6J

1 If $a = 3.2$ cm, $b = 6.7$ cm and $c = 4.3$ cm, all correct to the nearest millimetre, find the least possible value for each of these calculations.

    (a) $a + b + c$ **(b)** $b \times a$ **(c)** $b - c$ **(d)** $c \div a$

2 A rectangle is measured as 3.6 cm by 7.2 cm correct to 1 d.p.

    (a) Find the upper and lower bound for each measurement.

    (b) Find the smallest possible area of the rectangle.

3 Sami runs a race in 24 minutes and Peter runs it in 21 minutes, both measured to the nearest minute. What is the greatest possible difference in their times?

4 The lawn shown is measured to the nearest metre. What is the difference between the greatest and the smallest possible areas?

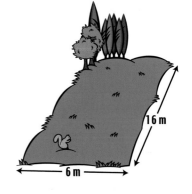

5 Sally wins a 100 m race in 15.2 seconds to the nearest tenth of a second. The length of the track is measured to the nearest metre.
  (a) What is the possible range of times that Sally ran?
  (b) What is the possible range of lengths for the race?
  (c) What is the fastest possible speed of her run?

6 A cube has a volume of 72 cm³, to the nearest cm³.
  Find the range of possible values for the length of the side.

7 A rectangular block with a mass of 7.84 kg measures 6.4 cm by 3.7 cm by 2.8 cm to 1 d.p.
  (a) Find the greatest and least possible volume of the block.
  (b) Find the range of possible mass for the block.
  (c) Find the greatest possible density of the material.

8 A car travelled for 32.4 miles at an average speed of 34.8 mph.
  Find the least amount of time in minutes for the journey.

The upper bound

Why is the ⩽ sign used for the lower bound but the < sign used for the upper bound?

Look at the value for u given in Example 16, $u = 3.8$.
You wrote $3.75 \leqslant u < 3.85$

The convention used in mathematics is that when a number ends in '5' it is rounded upwards.
If you were asked to write 3.85 correct to 1 d.p. you would write 3.9.
If you were asked to write 3.84999... correct to 1 d.p. the correct response is to write 3.8.

This is a recurring decimal where the 9s go on forever. You will meet them again in Chapter 26.

But let
$$x = 3.84999...$$

then $10x = 38.4999...$
so $9x = 10x - x = 38.4999... - 3.84999...$
which gives $9x = 34.65$
$$x = 34.65 \div 9$$
$$x = 3.85$$

This shows that 3.84999... and 3.85 are *the same*!

So, although 3.85 rounds up to 3.9 it is also the upper bound of 3.8 (it seems to be connected to both!)
This is why the < sign is always used for the upper bound.

Error

In Chapter 3 you learnt about percentage error. This section will show you a more structured way to find error.

Absolute error is the difference between the measured value and the **nominal value** of a quantity.

The **percentage error** is obtained by writing the absolute error as a percentage of the nominal value.

$$\text{Percentage error} = \frac{\text{absolute error}}{\text{nominal value}} \times 100\%$$

The nominal value is the value that the quantity is *supposed* to have. You can think of it as the *accurate* value (the value it would have if there were no error).

Example 17

A book shelf is meant to be 80 cm long but measures 79.7 cm. Calculate the absolute error and the percentage error.

Absolute error = nominal value − measured value

$$= 80 - 79.7 = 0.3 \text{ cm}$$

$$\text{Percentage error} = \frac{\text{absolute error}}{\text{nominal value}} \times 100\%$$

$$= \frac{0.3}{80} \times 100\%$$

$$= 0.375\%$$

The absolute error would be the same if the measurement had been 80.3 cm ... it does not matter whether the difference is +0.3 cm or −0.3 cm.

Example 18

$s = \frac{1}{2}gt^2$ is a formula that calculates the distance (s) an object has fallen when it is dropped vertically downwards for a given time (t). The acceleration due to gravity is g, which is constant at 9.8 m/s². If $t = 5.3$ s, correct to 2 s.f., calculate the maximum percentage error in the value of the distance fallen, s (in metres).

The lower and upper bounds of t are

$$5.25 \text{ s} \leqslant t < 5.35 \text{ s}$$

Least possible distance $= \frac{1}{2} \times 9.8 \times (5.25)^2$

$$= 135.05625 \text{ m}$$

Greatest possible distance $= \frac{1}{2} \times 9.8 \times (5.35)^2$

$$= 140.25025 \text{ m}$$

continued ▶

You need to break the question down into stages.

First find the lower and upper bounds of the time.

Then use *both* of these values in the formula for the distance fallen ... you cannot tell which one will give the maximum percentage error.

Nominal distance $= \frac{1}{2} \times 9.8 \times (5.3)^2$

$\qquad\qquad\qquad = 137.641\,m$

Absolute error for 'least' distance

$\qquad = 137.641 - 135.05625 = 2.58475\,m$

Absolute error for 'greatest' distance

$\qquad = 140.25025 - 137.641 = 2.60925\,m$

... and this one is the maximum absolute error

So maximum percentage error

$$= \frac{\text{Maximum absolute error}}{\text{Nominal distance}} \times 100$$

$$= \frac{2.60925}{137.641} \times 100$$

$$= 1.895692\ldots\%$$

$$= 1.90\% \;(3\,s.f.)$$

> You need to compare each answer to the nominal distance, working out the percentage error for the appropriate one.

Exercise 6K

1 A path is meant to be 120 m long but when checked measures 121.8 m. Calculate the absolute error and the percentage error.

2 The mass of a crate is measured as 23.6 kg. It should have been 24 kg. Find the percentage error.

3 A rectangle is measured correct to the nearest cm.

    (a) Find the largest possible area.

    (b) Find the percentage error from the area of the given measurements.

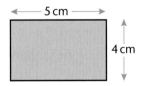

4 $\dfrac{3.82 \times 4.47}{4.41}$

John finds an approximate answer to this calculation by rounding the values to 1 significant figure. He then uses a calculator to find the answer to 2 d.p. What is the percentage error of his approximation?

5 I buy two pieces of curtain material 2.4 m long, correct to 1 d.p.

   (a) What are the upper and lower bounds for each piece?

   (b) If I put two pieces together, what are the maximum and minimum lengths that I could get?

What is

   (c) the maximum absolute error?

   (d) the maximum percentage error for the total length?

6 A ball has been measured with a diameter of 14.6 cm, correct to 1 d.p. Using the minimum possible value, calculate the percentage error in the volume of the ball.

Volume of a sphere = $\frac{4}{3}\pi r^3$

where r = radius = $\dfrac{\text{diameter}}{2}$.

7 A cube has a side of 6 cm, to the nearest cm.

What is the largest percentage error of **(a)** the area of a face?

   (b) the volume of the cube?

6 cm

UAM 8 3 litres of *Tropical* orange juice has a mass of 3.8 kg, to 1 d.p.

   (a) Find the range of the density of this orange juice.

   (b) To meet the description 'orange juice' the density should be not less than 1.3 g/cm³. What is the greatest possible percentage error from 1.3 g/cm³, that *Tropical* orange juice may have?

Give your answer to 3 s.f.

Examination questions

1 Hannah, Gemma and Jo use their calculators to work out the value of $\dfrac{28.78}{4.31 \times 0.47}$

Hannah gets 142.07, Gemma gets 14.207 and Jo gets 3.138
Use approximations to show which one of them is correct.
You **must** show your working.

(3 marks)
AQA, Spec A, 1H, November 2004

2 A boy runs 50 metres at a speed of 5 m/s.
Both values are measured to an accuracy of one significant figure.
What is the least possible time taken?

(3 marks)
AQA, Spec A, 1H, November 2003

3 Use approximations to estimate the value of $\dfrac{316 \times 4.03}{0.198}$

You **must** show your working.

(3 marks)
AQA, Spec A, 1H, June 2004

Summary of key points

Metric units of length:

10 mm = 1 cm
1000 mm =1 m
100 cm = 1 m
1000 m = 1 km

Metric units of mass:

1000 mg = 1 g
1000 g = 1 kg
1000 kg = 1 tonne (1 t)

Metric units of capacity:

100 cl = 1000 ml = 1 l
1 cm^3 = 1 cc = 1 ml
1000 cm^3 = 1000 cc = 1 l

Conversions between metric and imperial units (approximate)

Metric	8 km	30 cm	1 litre	4.5 litres	1 kg
Imperial	5 miles	1 foot	$1\frac{3}{4}$ pints	1 gallon	2.2 pounds

Conversions of areas and volumes (grade D/C)

1 m = 100 cm   so   1 m^2 = 100 cm × 100 cm = 10 000 cm^2
1 m^3 = 100 cm × 100 cm × 100 cm = 1 000 000 cm^3

Estimating using approximations (grade C)

To estimate an approximate answer:
1 Round all of the numbers to 1 s.f.
2 Work out the answer using these rounded numbers.

Compound measures (grade D/C)

$$\text{Speed} = \frac{\text{distance}}{\text{time}}$$ Distance = speed × time $$\text{Time} = \frac{\text{distance}}{\text{speed}}$$

$$\text{Density} = \frac{\text{mass}}{\text{volume}}$$ Mass = density × volume $$\text{Volume} = \frac{\text{mass}}{\text{density}}$$

Accuracy of measurement (grade C)

Any measurement given *to the nearest whole unit* may be inaccurate by up to *one half unit in either direction*.
The minimum value is the lower bound. The maximum value is the upper bound.

Accuracy of measurement (grade A)

Maximum$(A + B)$ = max A + max B Minimum$(A + B)$ = min A + min B
Maximum$(A \times B)$ = max A × max B Minimum$(A \times B)$ = min A × min B
Maximum$(A - B)$ = max A − min B Minimum$(A - B)$ = min A − max B
Maximum$(A \div B)$ = max A ÷ min B Minimum$(A \div B)$ = min A ÷ max B

Absolute error and percentage error (grade A)

Absolute error is the difference between the measured value and the nominal value.
$$\text{Percentage error} = \frac{\text{absolute error}}{\text{nominal value}} \times 100\%$$

This chapter will show you how to:

✔ find areas of compound shapes
✔ calculate the circumference and area of a circle
✔ find the arc length and area of a sector
✔ calculate the surface area and volume of prisms, pyramids, cones and spheres
✔ calculate the dimension of a formula

7.1 Perimeter and area

> **Key words:**
> perimeter
> area

The **perimeter** of a shape is the sum of the lengths of all its sides.

The **area** is the amount of space inside a shape.

Simple shapes

	Shape	Perimeter	Area
Rectangle	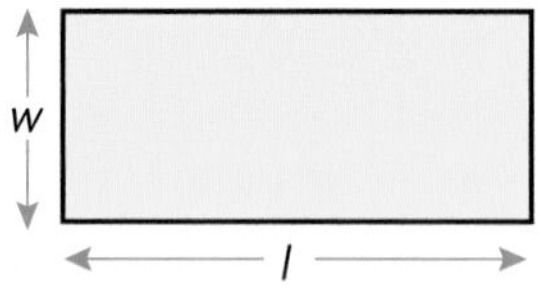	$2(l + w)$	$l \times w$
Triangle	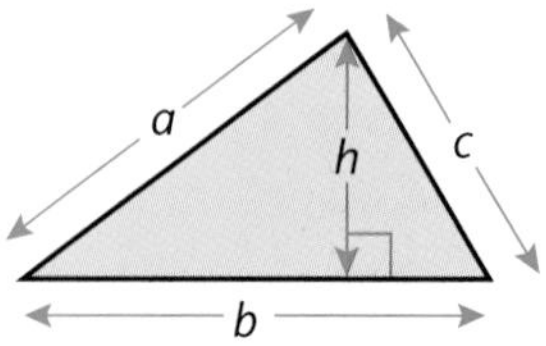	$a + b + c$	$\frac{1}{2} \times b \times h$
Parallelogram	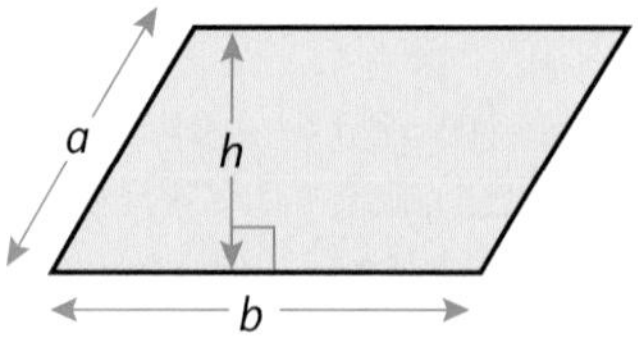	$2(a + b)$	$b \times h$ Note: You must remember to use the **perpendicular** height for the area, not the slant height.
Trapezium	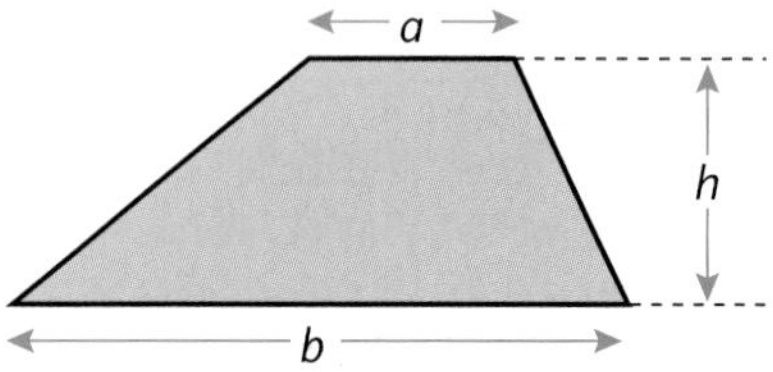	sum of all four sides	$\frac{1}{2} \times (a + b) \times h$ This is often remembered as $\frac{1}{2} \times$ sum of parallel sides $\times$ distance between them

Compound shapes

A compound shape is made from simple shapes.
To find the area of a compound shape, first split it into simple
shapes. Then use the formulae for the areas of these basic shapes
and add together.

Example 1

Find the perimeter and area of
this compound shape.

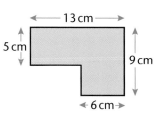

Split the shape into 2 rectangles A and B:

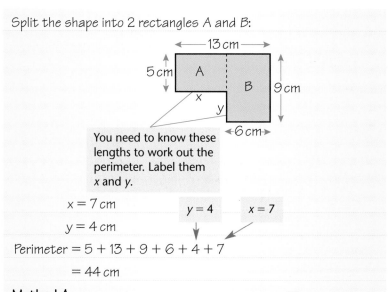

You need to know these
lengths to work out the
perimeter. Label them
x and y.

$x = 7\,cm$

$y = 4\,cm$

$y = 4$ $x = 7$

Perimeter $= 5 + 13 + 9 + 6 + 4 + 7$

$\qquad\quad = 44\,cm$

Method A

Area $=$ area of A $+$ area of B

$\quad = (5 \times 7) + (9 \times 6)$

$\quad = 35 + 54$

$\quad = 89\,cm^2$

Method B

Area of large rectangle $= 13 \times 9$

$\qquad\qquad\qquad\quad = 117\,cm^2$

Area of E $= 4 \times 7 = 28\,cm^2$

Area of shape $=$ area of large

$\qquad$ rectangle $-$ area of E

$\qquad = 117 - 28 = 89\,cm^2$

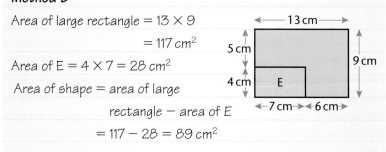

You could split it like this
instead:

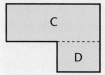

You would get the same
answers.

Work out the missing lengths
Total width of shape $= 13$
So $x + 6 = 13$
Total height of shape $= 9$
So $y + 5 = 9$

Start at the bottom left
corner and work clockwise.

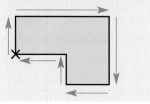

Brackets first.

Add a small rectangle E to
'fill in' the missing part.

Example 2

Find the area of this shape.

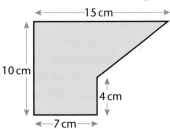

Total width of shape = 15

So $x + 7 = 15$

                        $x = 8$

Total height of shape = 10

So $y + 4 = 10$

                        $y = 6$

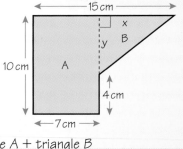

Split the shape into rectangle A and triangle B.

Find the lengths marked x and y before you can find the area of the triangle.

Area = rectangle A + triangle B

     $= (10 \times 7) + (\frac{1}{2} \times 6 + 8)$

     $= 70 + 24 = 94\,cm^2$

Area of triangle $= \frac{1}{2} \times h \times b$

Example 3

For each of these shapes you are given the area.
Calculate the value of the unknown length x in each case.

(a) Area = 35 cm²

(b) Area = 27 cm²

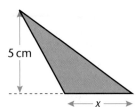

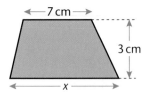

(a) Area of triangle $= \frac{1}{2} \times b \times h$

     $35 = \frac{1}{2} \times x \times 5$

     $70 = 5x$

     $x = 14\,cm$

(b) Area of trapezium $= \frac{1}{2} \times (a + b) \times h$

     $27 = \frac{1}{2} \times (x + 7) \times 3$

     $54 = 3(x + 7)$

     $18 = x + 7$

     $x = 11\,cm$

Exercise 7A

Find the area of the following shapes.

1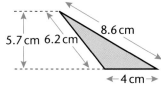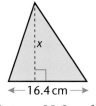

2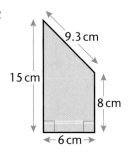

3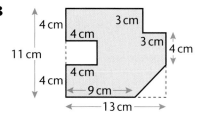

4 (a) Find the area of this shape, giving your answer in terms of x.

    (b) Find the perimeter.

    (c) If the perimeter is 16 cm, to 2 s.f., what is the value of x?

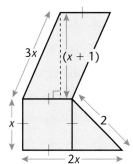

5 Find the height (x), in each of the following

(a)

Area = 98.2 cm²

(b)

Area = 33.8 cm²

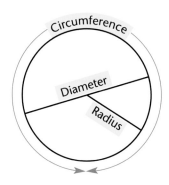

Circumference of a circle

The diameter (d) is twice the length of the radius (r).

> **Circumference = πd**   or   **Circumference = $2\pi r$**

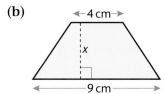

Key words:
circumference
diameter
radius

Area of a circle

- Draw a circle of radius 8 cm.

- Mark off points on the circumference at 30° intervals.

- Draw the 12 sectors.

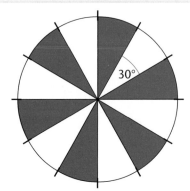

Circumference
$= 2\pi r$
$= 2 \times \pi \times 8$
$= 50.265...$
$= 50$ cm (to the nearest cm)

- Cut out the sectors and arrange them like this:

    This is approximately the shape of a rectangle.

    The width of the 'rectangle' is 8 cm.

    The length of the 'rectangle' is approximately 25 cm.

    So the area of the 'rectangle' $\approx 25 \times 8 = 200$ cm^2

    The formula for the circumference of the circle is $2\pi r$, so the length of the 'rectangle' $= \pi r$.

    The width of the 'rectangle' is r.

    So the area of the 'rectangle' = length × width = $\pi r \times r = \pi r^2$

The radius of the original circle.

$\frac{1}{2}$ the circumference of the original circle.

$\approx$ means 'is approximately equal to'.

This is the same as the area of the original circle.

$$\text{Area of a circle } A = \pi r^2 \quad \text{or} \quad \frac{\pi d^2}{4}$$

For a circle of radius 8 cm: $A = \pi \times 8^2 = 201.06...$ cm^2

This is very close to the approximate answer of 200 cm^2.

Example 4

The distance around the edge of a circular pond is 10.5 m.

Calculate the radius of the pond.

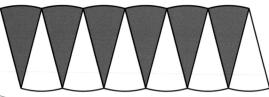

$$C = \pi d$$
$$10.5 = \pi \times d$$
$$\frac{10.5}{\pi} = d$$
$$d = 3.342...$$
$$r = \frac{3.342...}{2} = 1.67 \text{ m (3 s.f.)}$$

$C = 10.5$

Dividing by π rearranges the formula to make d the subject.

Example 5

A garden is in the shape of a rectangle with two semicircles, one on the length of the rectangle and one on the width, as shown in the diagram.
Calculate the area of the garden, giving your answer to the nearest square metre.

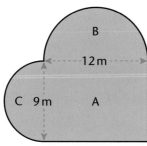

Area of rectangle $A = 12 \times 9$

$\qquad = 108 \text{ m}^2$

Area of semicircle $B = \frac{1}{2} \times \pi \times 6^2$

$\qquad = 56.548\ldots$

Area of semicircle $C = \frac{1}{2} \times \pi \times 4.5^2$

$\qquad = 31.808\ldots$

Total area $= 108 + 56.548\ldots + 31.808\ldots$

$\qquad = 196.35\ldots$

$\qquad = 196 \text{ m}^2 \text{ (to nearest m}^2)$

Area of semicircle
$= \frac{1}{2}$ area of circle.
Diameter of B = 12 m, so radius = 6 m.
Diameter of C = 9 m, so radius = 4.5 m.

If you rounded each of the three answers to the nearest whole number and then added them your answer would be
$108 + 57 + 32 = 197 \text{ m}^2$
(which is incorrect).
You should round at the *end* of the calculation.

Exercise 7B

1 The following shape is made up of a semicircle with a diameter of 10 cm, and a quarter circle with a radius of 3 cm.

Find   **(a)** the area, to 2 d.p.
       (b) the perimeter, to 3 s.f.

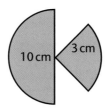

2 A tyre has a circumference of 70 cm. What is the radius of the wheel it fits? Give your answer to 1 d.p.

3 (a) A circular lawn covers an area of 28.27 m². What is the radius of the lawn to 2 s.f.?

(b) The lawn is cut in half and changed to give a semicircle on top of a square. What is the perimeter of the lawn now?

Sectors and arcs

Key words:
sector
arc
minor sector
major sector

A **sector** is part of a circle formed by drawing two radii. The distance between the points where the radii meet the circumference is the **arc** length.

The diagram shows two sectors. The smaller one is called the **minor sector** and the larger one the **major sector**.

The length of the arc of the smaller sector is called the **minor arc** and the length of the larger one is called the **major arc**.

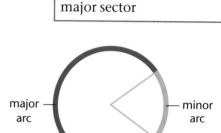

A sector is a fraction of the whole circle. The arc length and the sector area both depend on the size of the angle at the centre of the circle.

If the angle at the centre of the circle is θ, the arc length and the sector area are calculated as follows:

$$\textbf{Arc length} = \frac{\theta}{360} \times \textbf{circumference} = \frac{\theta}{360} \times 2\pi r$$

$$\textbf{Sector area} = \frac{\theta}{360} \times \textbf{area of whole circle} = \frac{\theta}{360} \times \pi r^2$$

Example 6

Find the arc length and area of a sector of a circle of radius 6 cm if the angle of the sector is 75°. Give your answers to 3 s.f.

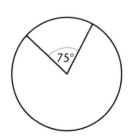

$$\text{Arc length} = \frac{75}{360} \times 2 \times \pi \times 6$$

$$= 7.85 \text{ cm (3 s.f.)} \quad or \quad 2.5\pi \text{ cm}$$

$$\text{Sector area} = \frac{75}{360} \times \pi \times 6^2$$

$$= 23.5619\ldots$$

$$= 23.6 \text{ cm}^2 \text{ (3 s.f.)} \quad or \quad 7.5\pi \text{ cm}^2$$

This question could be on the non-calculator paper or the calculator paper.

On a non-calculator paper, give the answer in terms of π.
On a calculator paper give it to the degree of accuracy required.

Example 7

The length of the minor arc of a circle is 3π cm and the length of the major arc is 15π cm.

Find   (a) the radius of the circle

(b) the angle of the minor sector

(c) the area of the minor sector, giving your answer in terms of π.

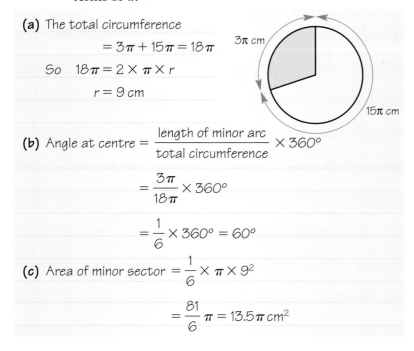

(a) The total circumference

$$= 3\pi + 15\pi = 18\pi$$

So $18\pi = 2 \times \pi \times r$

$$r = 9 \text{ cm}$$

(b) Angle at centre $= \dfrac{\text{length of minor arc}}{\text{total circumference}} \times 360°$

$$= \frac{3\pi}{18\pi} \times 360°$$

$$= \frac{1}{6} \times 360° = 60°$$

(c) Area of minor sector $= \dfrac{1}{6} \times \pi \times 9^2$

$$= \frac{81}{6} \pi = 13.5\pi \text{ cm}^2$$

The question could have simply asked for the area of the minor sector, with none of the structure seen here.

You would then have to find an appropriate method.

This would make it UAM1 (problem solving).

Exercise 7C

1 Find the arc length and area of each of the following sectors. Give answers to 2 d.p.

(a)

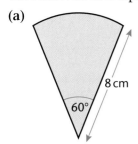

(b)

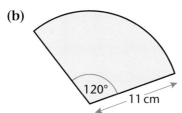

(c)

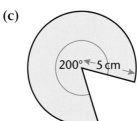

2 Find the angle at the centre of a sector with an arc length of 7.85 cm and a radius of 15 cm. Give your answer to the nearest degree.

3 The length of a minor arc of a circle is 5π cm and the length of the major arc is 7π cm. Find the area of the minor sector.

Example 7 will help.

7.2 3-D objects

Plans and elevations

Key words:
plan
front elevation
side elevation

You can show a 3-D object in detail by drawing three different 2-D views of it.

The **plan** is the view from above.

The **front elevation** is the view from the front.

The **side elevation** is the view from the side.

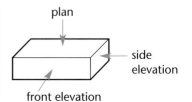

Example 8

Draw the plan view, the front elevation and the side elevation (from the right-hand side) of this block of six cubes.

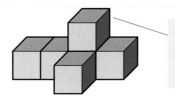

There are 6 cubes. You can only see 5 of them but there must be one underneath the top one.

You usually set out the drawings like this:
• plan view at the top
• front view beneath it
• view from the right-hand side to the right of the front elevation.

Plan view

Front elevation   Side elevation

Exercise 7D

Draw a plan view, a front elevation and a side elevation (from the right-hand side) for each block of cubes.

1

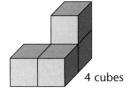

4 cubes

2

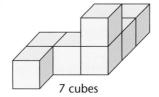

7 cubes

3

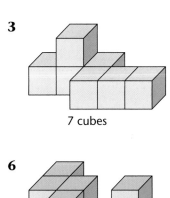

7 cubes

4

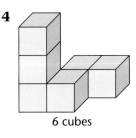

6 cubes

5

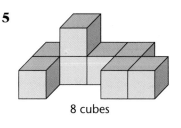

8 cubes

6

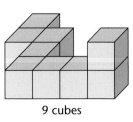

9 cubes

7

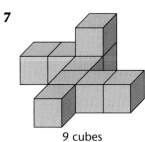

9 cubes

8

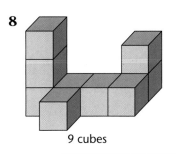

9 cubes

Planes of symmetry

You can also draw 3-D objects on squared paper.

Here are three diagrams of a cuboid. In each one you can see a **plane of symmetry** (shaded red).

> **A plane of symmetry divides a 3-D object into two equal halves, where one half is the mirror image of the other.**

It is the 3-D equivalent of a line of symmetry.

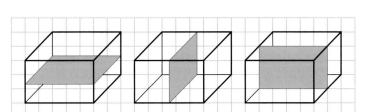

Example 9

Copy this 3-D object on squared paper.
Show all its planes of symmetry.
Draw a separate diagram for each plane
of symmetry.

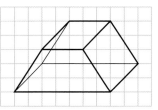

This object has two planes of symmetry. If you made a horizontal cut, the top 'half' would be smaller than the bottom 'half', so the cut would *not* be a plane of symmetry.

Exercise 7E

Copy these 3-D objects on squared paper.

Show all their planes of symmetry.

Draw a separate diagram for each plane of symmetry.

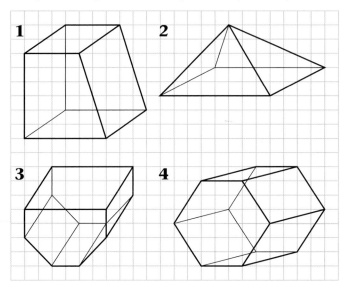

7.3 Volume and surface area

Prisms

Key words:
cross-section

A prism is a 3-D shape with a uniform **cross-section** .

The cross-sectional area is the area of the end of the shape which runs throughout its length.

In these examples of prisms, the area of the cross-section is shaded.

To find the volume of a prism multiply the cross-sectional area by the length of the prism.

Volume = area of cross-section × length

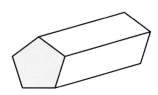

The surface area of a prism is found by adding the areas of all of its faces.

Cuboids

A cuboid is an example of a prism.
Volume = area of cross-section × length

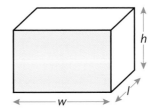

> **Volume of cuboid = width × height × length**
> $$= w \times h \times l$$

The surface area of a cuboid is found by adding together the areas of all six faces.

So for the cuboid shown in the diagram,

Surface area $= 2 \times (l \times w + l \times h + w \times h)$
$= 2(lw + lh + wh)$

Cubes

A cube is a special case of a cuboid.
Since all its sides are the same length, say x
Volume of cube $= x^3$.
Surface area of cube $= 6x^2$.

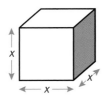

Example 10

Find the volume of this prism. All lengths are given in cm.

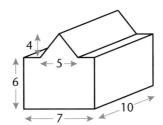

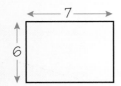

The area of the end of the prism must be found first. Split it up into a rectangle and a triangle, as shown.

Area of rectangle $= 7 \times 6 = 42$
Area of triangle $= \frac{1}{2} \times 5 \times 4 = 10$
Area of cross-section $= 42 + 10 = 52$
Volume of prism $= 52 \text{ cm}^2 \times 10 \text{ cm} = 520 \text{ cm}^3$

Exercise 7F

1 Find the volume of the following prisms. All measurements are in centimetres.

(a)

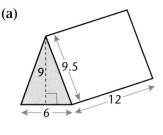

(b)

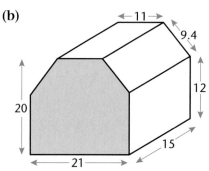

(c)

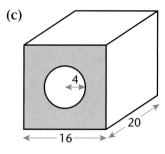

2 Find the surface area of the prisms in question 1.

3 The skip shown is to be used to remove a rectangular mound of soil 6 m by 4 m by 2 m. How many times will it have to be emptied to remove all the soil from the site?

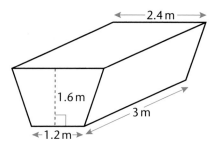

Cylinders

A cylinder is a prism whose cross-section is a circle.

Volume of a cylinder = area of cross-section × height

$$V = \pi r^2 h$$

The curved surface area = circumference of base × height

$$\text{CSA} = 2\pi r h$$

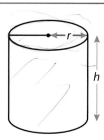

The **total surface area** of a cylinder is the area of the curved surface plus the areas of the two ends.

The ends are circles so each has an area of πr^2.

So the total surface area $= 2\pi r h + 2\pi r^2$
$$\text{TSA} = 2\pi r (h + r)$$

This factorised version is quick and easy to use.

Example 11

A thin cylindrical tube is 2 metres long and has a diameter of 4.6 cm. Calculate its total surface area and its volume.

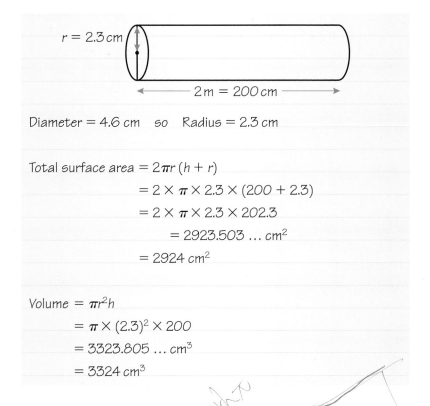

Always sketch a diagram to help you answer the question.

Diameter = 4.6 cm   so   Radius = 2.3 cm

Beware! You have mixed units and they need to be the same. It is best to work in centimetres.

Total surface area = $2\pi r (h + r)$

$\qquad\qquad = 2 \times \pi \times 2.3 \times (200 + 2.3)$

$\qquad\qquad = 2 \times \pi \times 2.3 \times 202.3$

$\qquad\qquad\quad = 2923.503 \ldots \text{ cm}^2$

$\qquad\qquad = 2924 \text{ cm}^2$

Volume = $\pi r^2 h$

$\qquad = \pi \times (2.3)^2 \times 200$

$\qquad = 3323.805 \ldots \text{ cm}^3$

$\qquad = 3324 \text{ cm}^3$

Use the π button on the calculator.

Note that the answers are given to a sensible degree of accuracy.

Example 12

A cylindrical container holds 8.5 litres of liquid when full.
The height of the container is 24 cm.
Calculate the radius of the base.

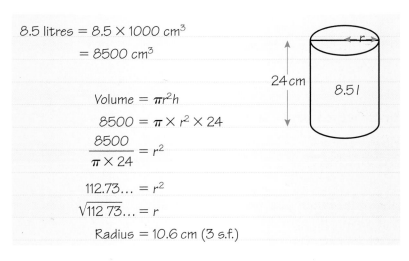

8.5 litres = $8.5 \times 1000 \text{ cm}^3$

$\qquad\qquad = 8500 \text{ cm}^3$

You must convert litres into cm^3 since the height is given in cm.

Volume = $\pi r^2 h$

$\qquad 8500 = \pi \times r^2 \times 24$

$\qquad \dfrac{8500}{\pi \times 24} = r^2$

$\qquad 112.73\ldots = r^2$

$\qquad \sqrt{112.73\ldots} = r$

$\qquad\qquad \text{Radius} = 10.6 \text{ cm (3 s.f.)}$

Note that you need to rearrange the formula to make r^2 the subject. Do this by dividing by π and by 24.

Exercise 7G

1 Find the volume of the following cylinders.

(a)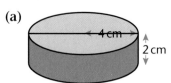

(b)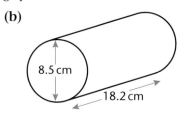

Take care with units.

(c)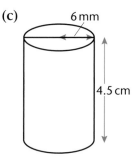

2 Find the surface area for each of the cylinders in question 1.

3 The annulus shown has an outside diameter of 27.6 cm, an inside diameter of 19.6 cm and is 45 cm long.

(a) Find its volume.

(b) It is made out of plastic with a density of 8.4 g/cm³. What is its mass?

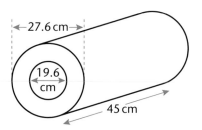

Pyramids and cones

Volume

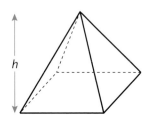 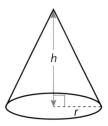

The volume of a cone or a pyramid is given by the formula

$$\text{Volume} = \tfrac{1}{3} \times \text{area of base} \times \boxed{\textbf{perpendicular height}}$$

So for a cone of base radius r and perpendicular height h,

$$\text{Volume} = \tfrac{1}{3}\pi r^2 h$$

Surface area

The surface area of a pyramid with a base which is square, rectangular, triangular or the shape of any polygon, is found by adding together the areas of the faces of the pyramid.

The surface area of a cone is slightly different since the cone is made up of a flat, circular base and a curved surface.

The area of the curved surface of the cone is given by the formula
Curved surface area = $\pi \times$ radius of base $\times$ **slant height**

$$= \pi r l$$

where r is the radius of the base of the cone and l is the slant height of the cone.

> So, the **total** surface area of a cone
>
> = area of base + curved surface area
>
> = $\pi r^2 + \pi r l$
>
> = $\pi r (r + l)$

There is more about cones in Chapter 22.

It is quicker to use this factorised version.

Example 13

Calculate the volume of this pyramid which has a square base of side 10 cm and a perpendicular height of 15 cm.

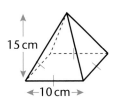

15 cm

←10 cm→

Volume = $\frac{1}{3} \times$ area of base $\times$ perpendicular height

$= \frac{1}{3} \times 10 \times 10 \times 15$

$= 10 \times 10 \times 5$

$= 500 \ cm^3$

Example 14

A cone has a volume of 200 cm³.
The radius of the base is 4 cm.
Calculate the perpendicular height of the cone.

Volume = $\frac{1}{3} \pi r^2 h$

$200 = \pi \times 4^2 \times h$

$\dfrac{200 \times 3}{\pi \times 4^2} = h$

Height of cone = 11.9 cm (3 s.f.)

Rearranging the formula to make h the subject.

Exercise 7H

1 Find the volume of the following objects.

(a)

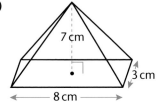

(b)

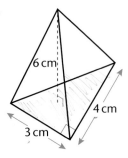

(c)

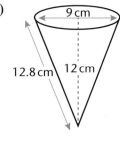

2 Find the surface area of **1(c)**.

3 A square based pyramid has a perpendicular height of 7 cm and a volume of 84 cm³. What is the length of a side of the base?

Spheres

The volume and surface area of a sphere of radius r are given by the formulae

Volume $= \frac{4}{3}\pi r^3$ **Surface area $= 4\pi r^2$**

Example 15

A sphere has a radius of 9 cm. Calculate its volume and surface area.

Volume $= \frac{4}{3}\pi r^3$

$\qquad = \frac{4}{3} \times \pi \times 9^3 \quad$ cm³

$\qquad = \frac{4}{3} \times \pi \times 729 \quad$ cm³

$\qquad = 972\pi$ cm³

$\qquad = 3050$ cm³ (3 s.f.)

Surface area $= 4\pi r^2$

$\qquad = 4 \times \pi \times 9^2 \quad$ cm²

$\qquad = 4 \times \pi \times 81 \quad$ cm²

$\qquad = 324\pi$ cm²

$\qquad = 1020$ cm² (3 s.f.)

On the non-calculator paper you could leave your answers as a multiple of π.

Example 16

A solid metal cone of base radius 10 cm and perpendicular height
15 cm is melted down and recast into a solid sphere.
Calculate the surface area of the sphere.

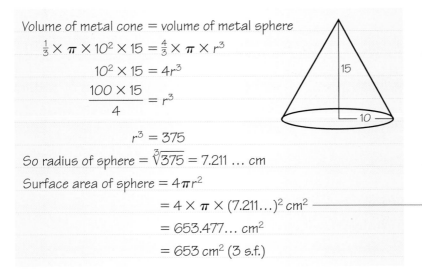

Volume of metal cone = volume of metal sphere

$$\tfrac{1}{3} \times \pi \times 10^2 \times 15 = \tfrac{4}{3} \times \pi \times r^3$$

$$10^2 \times 15 = 4r^3$$

$$\frac{100 \times 15}{4} = r^3$$

$$r^3 = 375$$

So radius of sphere = $\sqrt[3]{375}$ = 7.211 ... cm

Surface area of sphere = $4\pi r^2$

$$= 4 \times \pi \times (7.211...)^2 \, cm^2$$

$$= 653.477... \, cm^2$$

$$= 653 \, cm^2 \ (3 \ s.f.)$$

First equate the two
volumes to find the radius
of the sphere.

Cancel the factors of 3 and
the π terms.

Keep the full calculator
display for the radius to
ensure accuracy.

Exercise 7I

1 Find the volume of spheres with

   (a) radius 12 cm   **(b)** diameter 45.2 cm

2 Find the surface area of the spheres in question 1.

UAM **3** A hemisphere of clay with a radius of 16 cm is remodelled into a
cylindrical pipe.

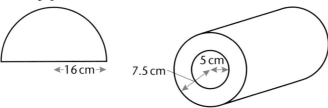

   If the outer radius of the pipe is 7.5 cm and the inner radius 5 cm,
how long is the pipe if all the clay is used?

7.4 Dimension theory

In this chapter you have been working out **lengths** (perimeter
and circumference, for example), **areas** (of rectangles and circles
for example) and **volumes** (of cuboids and spheres, for example).

Lengths are measured in units such as cm or m. You say they have
dimension = 1.

Key words:
lengths
areas
volumes
dimension

Areas are measured in square units (e.g. cm² or m²). You say they have dimension = 2.

Volumes are measured in cubic units (e.g. cm³, m³ or litres). You say they have dimension = 3.

> Remember that
> 1 litre = 1000 cm³.

Here are some formulae for calculating lengths:

Perimeter of a rectangle $\qquad P = 2(l + b)$

l and b are both lengths so when added together still give a length.
So $(l + b)$ has dimension = 1.
2 is simply a number and has no dimension.
So $2(l + b)$ has dimension = 1.

> When you add or subtract quantities of the same dimension, the answer has the same dimension too.

Circumference of a circle $\qquad C = \pi d$

π has no dimension and d has dimension = 1
So πd ($\pi \times d$) has dimension = 1.

> Notice that the dimension of the answer is 1 because you are multiplying a length by a number which still gives a length.

> **Lengths have dimension = 1.**

Here are some formulae for calculating areas:

Area of a triangle $\qquad A = \frac{1}{2}bh$

b and h both have dimension = 1 and $\frac{1}{2}$ has no dimension.
So $\frac{1}{2} \times b \times h$ has dimension = 2.

Curved surface area of cylinder $\qquad CSA = 2\pi rh$

r has dimension = 1 so r^2 ($r \times r$) has dimension = 2.
π has dimension = 0
So $2\pi r^2$ has dimension = 2.

> When you multiply two or more quantities you *add* the dimensions.
> For example
> length (cm) × length (cm) = area (cm²)
> dim 1 × dim 1 = dim 2

Area of a circle $\qquad A = \pi r^2$

r and h both have dimension = 1, π and 2 both have dimension = 0.
So $2\pi rh$ has dimension = 2.

> **Areas have dimension = 2.**

Here are some formulae for calculating volumes:

Volume of cube $\qquad V = x^3$

x has dimension = 1
So x^3 ($x \times x \times x$) has dimension = 3.

> Dimension = 3 when you multiply *three* quantities of dimension = 1 or multiply a quantity with dimension 2 by a quantity with dimension 1.

Volume of cylinder $\qquad V = \pi r^2 h$

π has dimension = 0, r^2 has dimension = 2 and h has dimension = 1.
So $\pi r^2 h$ has dimension = 3.

> **Volumes have dimension = 3.**

If you know these rules you can always work out what a formula represents. The rules will prevent you making mistakes such as confusing the formulae for the circumference and the area of a circle.

$(C =) 2\pi r$ has dimension = 1 so represents a length.
$(A =) \pi r^2$ has dimension = 2 so represents an area.

You can use dimension theory to work out whether a formula represents a length, an area, a volume or none of these.

Example 17

In each of these formulae w, x, y and z represent lengths. State whether each formula represents a length, an area, a volume or none of these.

(a) wx

(b) $2y + 5z - 7x$

(c) $3y + w^3 - xz$

(d) $\pi x(y^2 + wz)$

(d) $\dfrac{5xw^2}{y}$

(a) w has dimension $= 1$ and x has dimension $= 1$.

Multiplying them gives dimension $= 2$.

So wx represents an area.

(b) $2y + 5z - 7x$:   2, 5 and 7 have dimension $= 0$.

y, z and x all have dimension $= 1$.

They are added /subtracted so still have dimension $= 1$.

So $2y + 5z - 7x$ represents a length.

(c) $3y + w^3 - xz$:   3 has dimension $= 0$.

y has dimension $= 1$, w^3 has dimension $= 3$,

xz has dimension $= 2$.

The quantities all have different dimensions. So the formula does not represent a length, area or volume.

> You cannot add/subtract quantities with different dimensions.

(d) $\pi x(y^2 + wz)$: π has dimension $= 0$.

Inside the bracket, y^2 and wz both have dimension $= 2$.

They are added so still have dimension $= 2$.

Outside the bracket, x has dimension $= 1$.

Multiplying gives dimension $= 3$.

So $\pi x(y^2 + wz)$ represents a volume.

(e) $\dfrac{5xw^2}{y}$:   5 has dimension $= 0$.

x has dimension $= 1$, w^2 has dimension $= 2$,

So multiplying gives $5xw^2$ dimension $= 3$.

$5xw^2$ is the numerator, and has dimension $= 3$.

The denominator, y, has dimension $= 1$.

So $\dfrac{5xw^2}{y}$ has dimension $3 - 1 = 2$.

The formula represents an area.

> When you divide two quantities you *subtract* the dimensions.

Exercise 7J

1 In each of these formulae x, y and z represent lengths. State whether each formulae represents a length, an area, a volume or none of these. Give reasons for your answer.

(a) πy^2 (b) yz^2 (c) $xy + yz + xz$

(d) $\pi x^2 y^2$ (e) $3xy + 7xyz$ (f) πz

2 p is the length 7 cm, q is the length 3.8 cm and t is the length 4.9 cm. Select the volume formula from those listed below and use it to find the volume required.

(a) $p(q + t) + q^2$ (b) $p(pq) + t(pq)$ (c) $\pi q(q + t + p)$

3 If $7t(p^2)$ is an area formula, what can you say

(a) about p (b) about t ?

Examination questions

1 The first diagram shows a cylindrical block of wood of diameter 24 cm and height 10 cm. It is cut into six equal prisms as shown.
One of the prisms is shown in the second diagram.

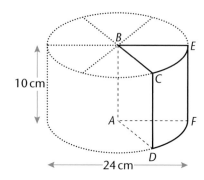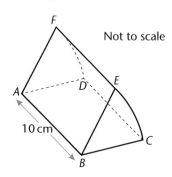

(a) Calculate the area of sector *BEC*, the cross-section of the prism. Give your answer in terms of π.

(b) Calculate the area of *CDFE*, the curved surface of the prism. Give your answer in terms of π.

(c) Calculate the volume of the prism. Give your answer in terms of π.

(8 marks)
AQA, Spec A, 1H, November 2004

Summary of key points

Rectangle

Perimeter $= 2(l + w)$ Area $= l \times w$

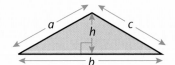

Triangle (grade D)

Perimeter $= a + b + c$ Area $= \frac{1}{2} \times b \times h$

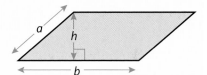

Parallelogram (grade D)

Perimeter $= 2(a + b)$ Area $= b \times h$

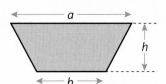

Trapezium (grade D)

Perimeter $=$ sum of all four sides Area $= \frac{1}{2} \times (a + b) \times h$

Circle (grade D/C)

Circumference $= \pi d$ or Circumference $= 2\pi r$

Area $= \pi r^2$ or Area $= \dfrac{\pi d^2}{4}$

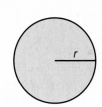

Sectors (grade A)

Arc length $= \dfrac{\theta}{360} \times 2\pi r$ Sector area $= \dfrac{\theta}{360} \times \pi r^2$

Prism (grade D/C)

A prism is a 3-D shape with a uniform cross-section.
Volume $=$ area of cross-section $\times$ length
Volume of a cuboid $=$ length $\times$ width $\times$ height $= l \times w \times h$
The surface area of a prism is found by adding the areas of all of its faces.

Plans and elevations (grade D)

Any 3-D shape can be shown in detail by drawing three different 2-D views of it.
The plan is the view from above. The front elevation is the view from the front.
The side elevation is the view from the side.

Plane of symmetry (grade D)

A plane of symmetry divides a three-dimensional shape into
two equal halves, where one half is the mirror image of the other.

Cylinder (grade C)

A cylinder is a prism whose cross-section is a circle.
Volume of a cylinder = area of cross-section × height
Curved surface area = circumference of base × height
Total surface area = CSA + areas of ends = $2\pi rh + 2\pi r^2 = 2\pi r(h + r)$

$$V = \pi r^2 h$$
$$\text{CSA} = 2\pi rh$$

Pyramids and cones (grade A/A*)

Volume of pyramid = $\frac{1}{3}$ × area of base × perpendicular height
Volume of cone = $\frac{1}{3}\pi r^2 h$
Curved surface area of cone = πrl
Total surface area of cone = $\pi r^2 + \pi rl = \pi r(r + l)$

$$r^2 + h^2 = l^2$$

Spheres (grade A/A*)

Volume of sphere = $\frac{4}{3}\pi r^3$ Surface area = $4\pi r^2$

Dimension theory (grade B)

You can use dimension theory to work out whether a formula represents a
length, an area, a volume, or none of these. Lengths have dimension = 1,
areas have dimension = 2, and volumes have dimension = 3.
Numbers have dimension = 0.

This chapter will show you how to:

✔ use a straight edge and compasses to construct
  - a triangle given all three sides
  - the mid-point and perpendicular bisector of a line segment
  - the perpendicular from a point to a line
  - the perpendicular from a point on a line
  - the bisector of an angle
✔ construct angles of 60° and 90°
✔ understand, interpret and solve problems using simple loci
  (including scale drawing and bearings)

8.1 Constructions

You will need to know:
- how to use a pair of compasses
- how to measure accurately

> **Key words:**
> construction
> arc

Standard **constructions** use only a straight edge (ruler) and a pair of compasses to draw accurate diagrams.

Your drawings must be accurate.

> NO ARCS, NO MARKS!

When you use compasses you must leave the construction **arcs** as evidence that you have used the correct method.

> Arcs are parts of a curve drawn with compasses.

8.2 Construct a triangle given all three sides

> **Key words:**
> intersect

Example 1 shows the method for constructing a triangle given the lengths of all three sides.

Example 1

Construct a triangle *ABC* where *AB* = 6 cm, *AC* = 3 cm and *BC* = 5 cm.

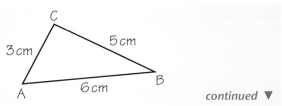

continued ▼

> It helps to draw and label a sketch first.

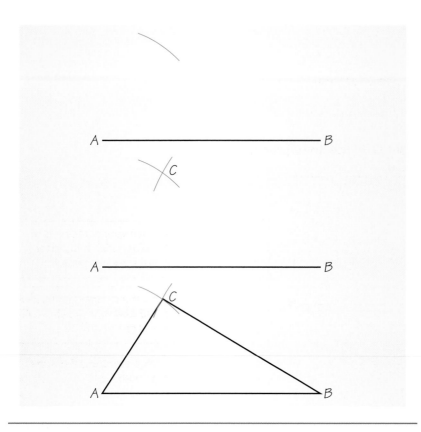

Use your ruler to draw the longest side 6 cm long. Label it *AB*.

Open your compasses to a radius of 3 cm. Put the point on *A* and draw an arc in the space above line *AB*.

Open your compasses to a radius of 5 cm. Put the point on *B* and draw an arc to intersect the first arc. Where the arcs intersect is C.

Point *C* is 3 cm from *A* and 5 cm from *B*.

Join *C* to *A* and *B* to complete the triangle.

Remember: don't rub out the arcs!

Exercise 8A

Construct triangles *ABC* with these measurements:

1 *AB* = 8 cm, *AC* = 6 cm, *BC* = 5 cm.

2 *AB* = 9 cm, *AC* = 8.5 cm, *BC* = 4 cm.

3 *AB* = 7.5 cm, *AC* = 10 cm, *BC* = 4.5 cm.

4 *AB* = 6 cm, *AC* = 8 cm, *BC* = 10 cm.

5 Equilateral triangle *ABC* with side 6 cm.

6 Equilateral triangle *ABC* with side 9.5 cm.

In an equilateral triangle all sides are the same length.

8.3 Constructing perpendiculars

Key words:
bisect
line segment
perpendicular
mid-point

Construct the perpendicular bisector of a line segment

To **bisect** means to cut in half.

A straight line has infinite length so you will be looking at a finite part of it – a **line segment** .

line segment

A **perpendicular** bisector:

- cuts a line segment in half
- is perpendicular (at 90°) to the line segment.

Example 2

Construct the perpendicular bisector of the line segment *AB*.

A ——————————— B

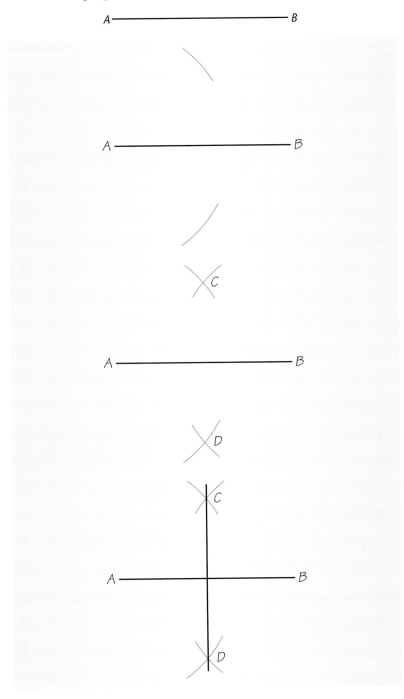

Open your compasses to a radius which is just over half the length of *AB*.

Put the compass point on *A* and draw one arc above *AB* and one below.

Keep the radius the same for both arcs.

Keeping the radius the same, put your compass point on *B* and draw two more arcs to intersect the first pair.
Label these points *C* and *D*.

Join *CD*. This line is the perpendicular bisector of *AB*.

CD crosses *AB* at the **mid-point** of *AB*.

CD makes an angle of 90° with *AB*.

Why does this work?
We have made a rhombus.

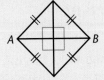

Look at the properties of a rhombus, page 52.

Exercise 8B

Draw six line segments of different lengths. Label the ends *A* and *B*. Construct the perpendicular bisector for each line segment.

Check by measuring that your perpendicular bisector passes through the mid-point of the line segment.

Check, using a protractor, that the angle between the two lines is 90°.

Construct the perpendicular from a point to a line

Example 3

Construct the line which is perpendicular to *AB* and passes through point *P*.

The line from *P* to *AB* will meet *AB* at 90°.

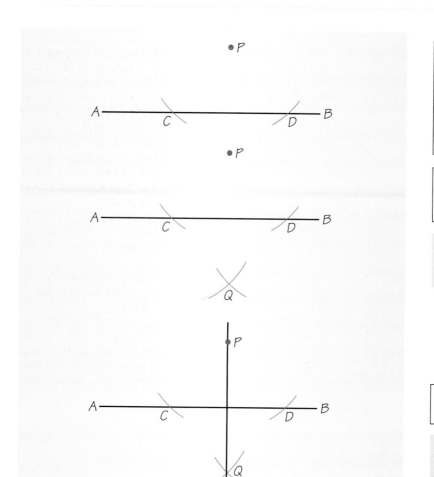

Open your compasses to a radius slightly longer than the distance of *P* from the line. Put the compass point on *P* and draw an arc to intersect *AB* twice. Label these points *C* and *D*.

From *C* and *D*, draw arcs of the same radius to intersect below *AB*. Label this point *Q*.

This need not be the same radius as the first arc you drew from *P*.

Join *PQ*. This line is the perpendicular from *P* to *AB*.

PQ is at 90° to *AB* and passes through *P*. So shape *PDQC* is a kite.

Exercise 8C

Draw six line segments AB with point P, similar to this:

For each, construct the perpendicular from the point P to the line segment AB.

Construct the perpendicular from a point on a line

This time the point is *on* the line and you construct the perpendicular to the line through that point.

> You can use this construction to construct an angle of 90°. Another construction for an angle of 90° is shown in Example 7.

Example 4

Construct the line perpendicular to AB from point P, which lies on AB.

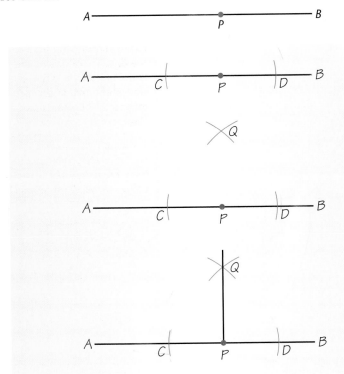

> Put your compass point on P and draw two arcs, of a fairly small radius (about 1.5 cm), to intersect AB on either side of P. Label these points C and D.

> From C and D draw arcs of equal radius to intersect in the space above AB. Label this point Q.

> Make this radius larger than the previous one.

> Join PQ. This line is the perpendicular from the point P.

> PQ passes through P and is at 90° to AB. Look at Example 14 to see why this works.

Exercise 8D

Draw six line segments AB.
Mark a point P on each line segment.
Construct the perpendicular from point P for each.

8.4 Constructing angles

Key words:
bisector

Construct the bisector of an angle

The **bisector** of an angle divides an angle into two equal parts.

Sometimes this is called an angle bisector.

Example 5

Construct the bisector of angle *A*.

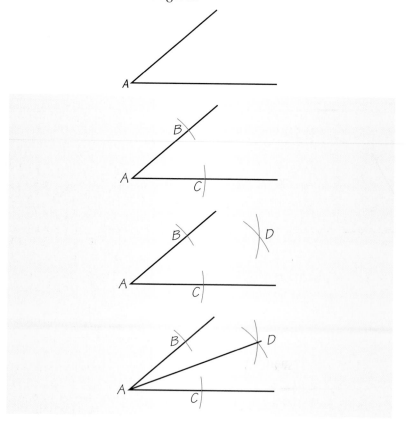

Open your compasses to about 2 cm. Put your compass point on *A* and mark off two small arcs, one on each arm of the angle. Label these points *B* and *C*.

From *B* and *C* draw two arcs of equal radius to intersect in the space between the arms of the angle. Label this point of intersection *D*.

Join *AD*. This line is the bisector of angle *A*.

AD divides angle A into two equal parts. See Example 15.

Exercise 8E

Draw six angles of different sizes. Construct the angle bisector for each of them.

Check the accuracy of your constructions by measuring the angles with a protractor.

Construct angles of 60° and 90°

To construct an angle of 60° you can use the method for constructing an *equilateral* triangle.

Remember: an equilateral triangle has three angles of 60°.

Example 6

Construct an angle of 60° at point *P* on line segment *AB*.

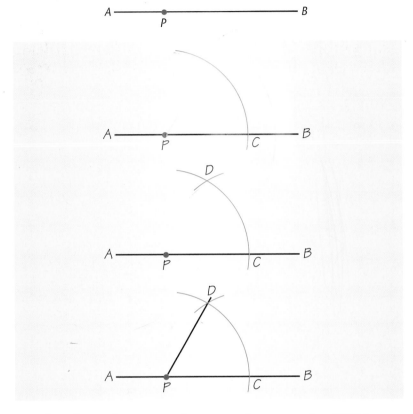

Open your compasses to a radius of about 3 cm.
Put the compass point on *P* and draw an arc which starts just below the line *AB* on one side of *P* and ends almost above *P*.
Label point *C*, where this arc intersects the line *AB*.

Keeping the radius the same, from point *C* draw an arc to intersect the first arc at point *D*.

Join *DP*. Angle *DPC* is 60°.

This construction is very similar to the first steps of the one for 90° in Example 7.

Can you see that △*DPC* is equilateral?

Example 4 showed one way of constructing an angle of 90°, by drawing the perpendicular bisector. Example 7 shows another.

Example 7

Construct an angle of 90° at point *P* on line segment *AB*.

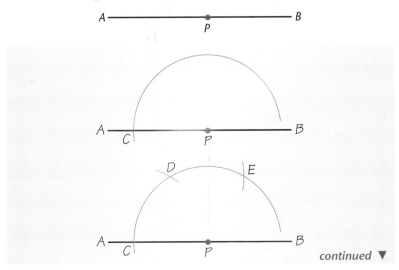

Open your compasses to a radius of about 3 cm.
Put the compass point on *P* and draw an arc which starts just below the line *AB* on one side of *P* and ends fairly close to the line on the other side of *P*.
Label point *C*, where this arc intersects *AB*.

Keeping the radius the same, from point *C* draw an arc to cut the first arc at point *D*.
Keeping the radius the same, draw an arc from point *D* to cut the first arc at point *E*.

Keep the radius the same for the first three arcs.

continued ▼

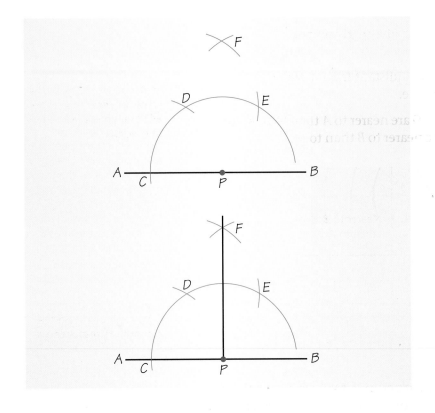

From points *D* and *E*, draw arcs of equal radius to intersect in the space above them. Label this point *F*.

This radius need not be the same as the first one.

Join *FP*. This line makes a 90° angle with line *AB*.

∠FPB = ∠FPA = 90°

If you are asked to construct a 90° angle you can use the method you prefer.

Exercise 8F

Draw six line segments *AB*.
Mark point *P* on each line segment.
Construct a 60° angle at *P* for each line.

Repeat for angles of 90°.

8.5 Understanding and using loci

Key words:
locus (loci)
equidistant

The line *CD* is the perpendicular bisector of the line *AB*.

All the points on the line *CD* are exactly the same distance from *A* as they are from *B*. Some of them are shown on the diagram.

CD is an example of a **locus**

(plural **loci**).

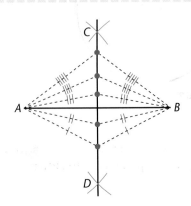

For more on the perpendicular bisector, see Example 2.

A locus is a set of points that obey a given rule.

For *CD* the rule is 'all points equidistant from *A* and *B*'. All the points on line *CD* obey this rule.

All the points to the left of *CD* are nearer to *A* than to *B*. All the points to the right of *CD* are nearer to *B* than to *A*.

Equidistant means 'the same distance from'. You will often see it in locus questions.

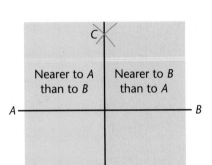

CD is the locus of points equidistant from *A* and *B*.

The locus of points which are equidistant from two fixed points is the perpendicular bisector of the two fixed points.

Example 8

What is the locus of points which are always 2 cm from a fixed point *P*?

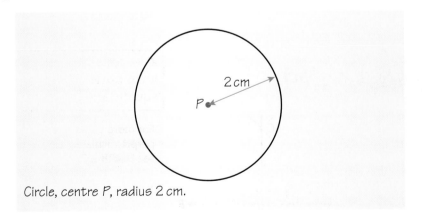

Circle, centre *P*, radius 2 cm.

All points on this circle obey this rule.

All points *inside* the circle are less than 2 cm from *P*. All points *outside* the circle are more than 2 cm from *P*.

The locus of points which are a fixed distance from a fixed point is a circle, with the fixed point as the centre.

Example 9

What is the locus of points which are equidistant from the two lines *AB* and *AC*?

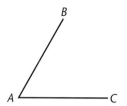

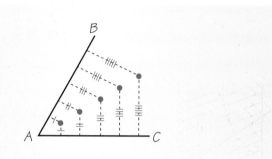

The locus of points which are equidistant from two fixed lines is the angle bisector of the two fixed lines.

> Plot some points equidistant from *AB* and *AC*.
> If you join the points you get a straight line – the angle bisector.

> To construct the angle bisector follow the steps in Example 5.

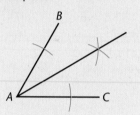

Example 10

Construct the locus of points which are exactly 2 cm from the fixed line segment *AB*.

A ——————— B

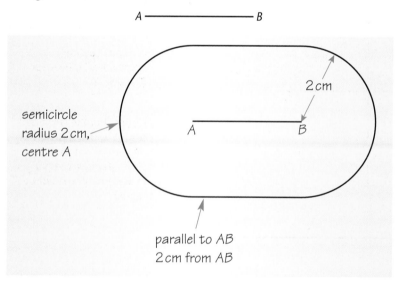

semicircle radius 2 cm, centre A

parallel to AB
2 cm from AB

> Using a ruler, mark points 2 cm from different points on *AB*.

> This shows the locus you need to construct. The locus will be a 'racetrack' shape around the outside of the line segment *AB*.

> It can help to sketch a locus before you construct it.

The locus of points which are a fixed distance from a line segment *AB* is a 'racetrack' shape. The shape has two lines parallel to *AB* and two semicircular ends.

For locus questions:
- **think about the points**
- **make a sketch**
- **construct the locus accurately using standard constructions.**

Example 11

In $\triangle ABC$, $AB = 4$ cm, $AC = 6$ cm and $BC = 5$ cm.
Shade the region inside the triangle where
the points are

(a) less than 3 cm from B and

(b) nearer to C than A.

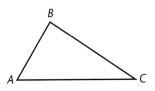

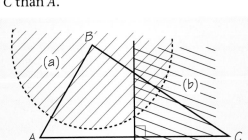

not to scale

First make a sketch.

(a) Points less than 3 cm
from B are in a circle
centre B, radius 3 cm.

(b) Points nearer to C than
A are to the right of the
perpendicular bisector
of AC.

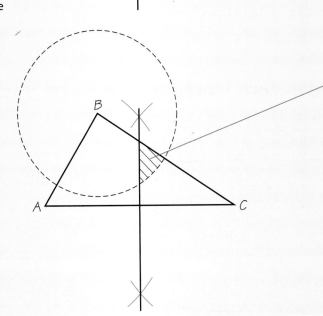

The points that satisfy **(a)**
and **(b)** are in this region.

Construct the diagram
accurately.

For constructing a
perpendicular bisector see
Example 2.

Notice that the circle is
drawn with a dashed line
because points on the
circumference are not
included in the required
region.

Exercise 8G

Use accurate constructions to
answer these questions.

1 In $\triangle ABC$, $AB = 5$ cm,
$AC = 7$ cm and $BC = 6$ cm.
Copy the diagram and shade
the region where points are:

(a) less than 4 cm from C and

(b) nearer to A than B.

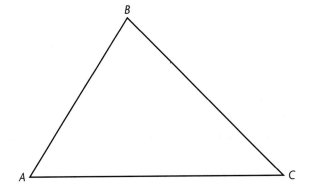

UAM **2** In △*PQR*, *PQ* = 6 cm, *PR* = 8 cm and *QR* = 4 cm.
Copy the diagram and shade the region where points are:

(a) nearer to *R* than *P*   and   **(b)** closer to *PR* than *QR*.

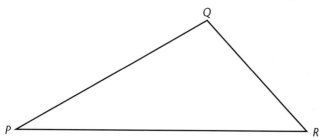

UAM **3** Main roads *AB*, *AC* and *BC* connect towns *A*, *B* and *C*.
AB = 16 miles, *AC* = 14 miles and *BC* = 12 miles.
A new leisure centre is to be built so that it is

(a) closer to road *AC* than it is to road *AB*

(b) between 8 miles and 10 miles from *B*.

Copy the diagram and construct the region where the leisure
centre will be built.

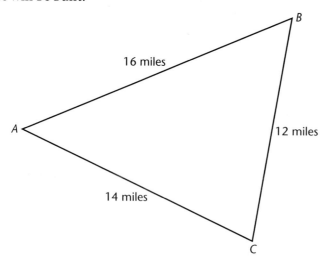

Use a scale of 1 cm for
2 miles.

8.6 Using bearings and scale

Example 12

The positions of two ships *A*
and *B* are shown in the diagram.

Ship *A* sails on a bearing of 070°.
Ship *B* is at anchor.

Mark clearly all points where
ship *A* is within 10 km of ship *B*.

North

Look back at bearings in
Section 4.2 if you need
some help.

Use a scale of 1 cm for 5 km.

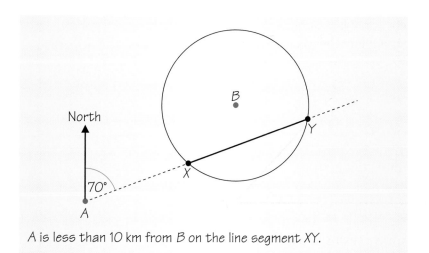

Draw in the line for the direction that *A* sails on, using a protractor for the angle.

The locus of points 10 km from *B* is a circle centre *B*. On the diagram using the scale 1 cm for 5 km, this is a circle with radius 2 cm.

A is less than 10 km from B on the line segment XY.

Example 13

The diagram shows a triangular field *ABC* where *AB* = 60 m, *AC* = 80 m and *BC* = 70 m. Some treasure is buried in the field.

It is between bearings of 040° and 060° from *A*, closer to *CA* than to *CB* and less than 40 m from *B*.

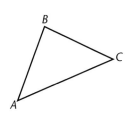

Using a scale of 1 cm for 10 m, make accurate constructions to find the area in which the treasure lies. Shade the region.

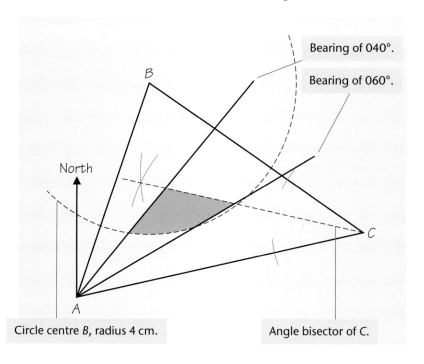

Circle centre *B*, radius 4 cm.

Angle bisector of *C*.

Construct the triangle. Draw in the bearings using a protractor.

Bearing of 040°.

Bearing of 060°.

Closer to *CA* than to *CB*. Draw the angle bisector of angle *C*, then choose the correct side of it (closer to *CA*). Draw it dashed because points on the bisector are not included in the region.

Less than 40 m from *B* means *inside* a circle of radius 4 cm, centre *B*.

Exercise 8H

1 The positions of two ships *A* and *B* are shown in the diagram. Ship *A* starts 20 km from B and sails on a bearing of 110° while ship B remains stationary. Copy the diagram and mark clearly, with a line *XY*, all points where ship *A* is within 12 km of ship *B*.

North

A

B

Use a scale of 1 cm for 4 km for your diagram.

2 *A*, *B* and *C* are places on Treasure Island. *AB* = 200 m, *AC* = 160 m and *BC* = 120 m. The hidden treasure is

(a) nearer to *C* than *B*

(b) closer to *AB* than to *AC*

(c) less than 70 m from *C*.

Using a scale of 1 cm for 20 m, construct the triangle and the region where the treasure is hidden.

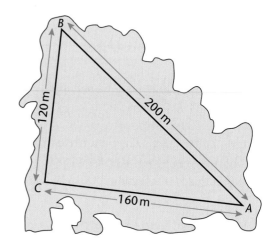

8.7 Verification of construction theorems

Key words:
congruent

Congruent shapes are shapes which are identical. This means they have exactly the same shape and are exactly the same size. There are four different cases to consider.

In Chapter 19 you will find a detailed explanation of congruent triangles.

Triangles are congruent if they have:
1 Three sides equal (side, side, side – SSS)

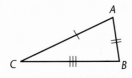

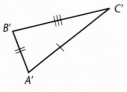

2 Two sides equal and the included angle the same (side, angle, side – SAS)

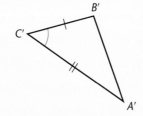

3 Two angles the same and a corresponding side equal (angle, side, angle – ASA, or angle, angle, side – AAS)

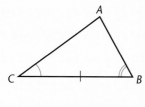
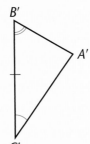

or

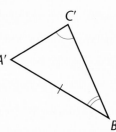

BC and *B′C′* are the corresponding sides.

AB and *A′B′* are the corresponding sides.

Corresponding side means 'the sides in the same position in both triangles'.

4 A right angle, a hypotenuse and a corresponding side equal (right angle, hypotenuse, side – RHS)

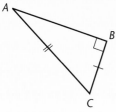

The hypotenuse is the longest side of a right-angled triangle. It is always opposite the right angle.

BC and *B′C′* are the corresponding sides.

These results can be used to verify some of the standard straight edge and compass constructions that you have seen in this chapter. The first case (three sides equal, or SSS) is particularly useful.

Example 14

Construct the perpendicular from a point *P* on a line *AB*. (This is the construction you saw in Example 4.)

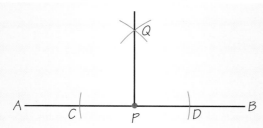

Putting the compass point on *P* and marking off points *C* and *D* means that *PC* = *PD*. Putting the compass point on *C* and *D* and marking off arcs which intersect at *Q* means that *CQ* = *DQ*. *PQ* is a side in both of the triangles *CPQ* and *DPQ*. This means that triangles *CPQ* and *DPQ* are congruent (SSS).

Verification

Since these two triangles are congruent, angle CPQ = angle DPQ.

But these are angles adjacent to each other on a straight line (CD) so they must add up to 180°.

You can see that angles CPQ and DPQ must both be 90°, which verifies the fact that PQ must be perpendicular to AB.

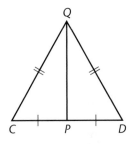

Example 15

Construct the bisector of an angle.
(This is the construction you saw in Example 5.)

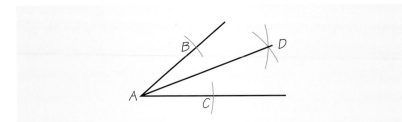

Putting the compass point on A and marking off points B and C means that $AB = AC$. Putting the compass point on B and C and marking off arcs which intersect at D means that $BD = CD$. AD is a side in both of the triangles ABD and ACD. This means that triangles ABD and ACD are congruent (SSS).

Verification

Since these two triangles are congruent, angle BAD = angle CAD which verifies the fact that the line AD bisects angle BAC.

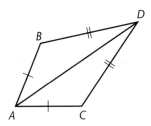

Examination questions

1 A is due North of B.
The bearing of C from A is 115°.
The bearing of C from B is 075°.

North

Mark the position of C on the diagram.

(3 marks)
AQA, Spec A, 1I, November 2003

2 The map below shows three boats, *A*, *B* and *C*, on a lake.
Along one edge of the lake there is a straight path.

Treasure lies at the bottom of the lake.

The treasure is:
    between 150 m and 250 m from *B*,
    nearer to *A* than *C*,
    more than 100 m from the path.

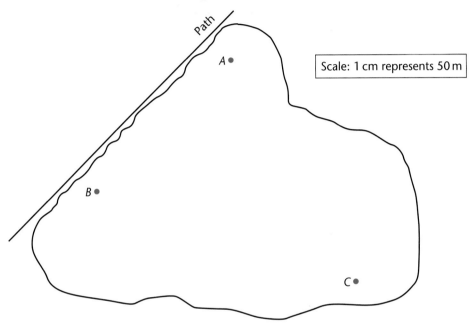

Scale: 1 cm represents 50 m

Using a ruler and compasses only, shade the region in which the treasure lies.
You **must** show clearly all your construction arcs.

(5 marks)

AQA, Spec A, 1I, June 2003

Summary of key points

Constructions

Constructions must be drawn using *only* a straight edge (ruler)
and pair of compasses.

Leave in the construction arcs as evidence that you have used
the correct method.

Construction of a triangle given all three sides (grade E)

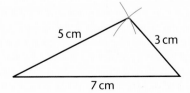

5 cm 3 cm

7 cm

Construction of the perpendiculars (grade C)

Construction of angle of 60° (grade C)

Construction of the bisector of an angle (grade C)

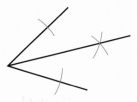

Locus (grade C)

A locus is a set of points that obey a given rule.

The locus of points which are equidistant from two fixed points is the perpendicular bisector of the two fixed points.

The locus of points which are a fixed distance from a fixed point is a circle.

The locus of points which are equidistant from two fixed lines is the angle bisector of the two fixed lines.

The locus of points which are a fixed distance from a line segment is a 'racetrack' shape. The shape has two lines parallel to the line segment and two semi-circular ends.

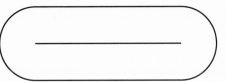

For locus questions:
- think about the points
- make a sketch
- construct the locus accurately using standard constructions.

Loci involving bearings and scale drawing (grade E/D)

You can measure bearings with a protractor.

Congruent triangles (grade C)

Triangles are congruent if they have:

1 Three sides equal (SSS)
2 Two sides equal and the included angle the same (SAS)
3 Two angles the same and a corresponding side equal (ASA or AAS)
4 A right angle, a hypotenuse and a corresponding side equal (RHS)

This chapter will show you how to:

✔ identify different types of data
✔ construct frequency tables for discrete and grouped data
✔ create a questionnaire based on a hypothesis
✔ use random and representative sampling techniques
✔ design and use two-way tables for discrete and grouped data

9.1 Types of data

Key words:
primary data
secondary data
qualitative data
quantitative data
discrete data
continuous data

Primary data is information you collect directly yourself, for example from questionnaires.

Secondary data is information that you get from existing records, for example newspapers, magazines, the internet.

Qualitative data contains descriptive words, for example a colour (red, green), or an activity (climbing, sailing), or a location (London, Paris). It is sometimes called categorical data.

Quantitative data contains numbers, such as temperatures, masses, areas, lengths, time, number of TVs or cars.

There are two types of quantitative data:

1 Discrete data can only have exact values.

Discrete data is 'countable'.
Discrete data examples:
- Shoe size $5, 5\frac{1}{2}, 8$
- Scores on a dice $4, 2, 6$
- Goals scored in a match $0, 2, 3$

> Shoe sizes can only be whole or half numbers.
> '8' is not possible.
> You can't score $2\frac{1}{2}$ goals!

2 Continuous data can take any value in a particular range.

Continuous data examples:
- Mass   72 kg, 15.3 g, 5 lbs
- Temperature   −4°C, 25°C, 100°C
- Length   800 m, 300 000 km, 2.6 mm

> 5 lbs is measured to the nearest pound.
> 2.6 mm is measured to the nearest tenth of a millimetre.

Continuous data cannot be measured exactly. The accuracy depends on the measuring instrument, for example, ruler or thermometer, and the values are often approximate.

Exercise 9A

1 State whether each of the following sets of data are quantitative or qualitative.

(a) Height *Quantitative*   (b) Age *Quantitative*

(c) Eye colour *Qualitative*   (d) Place of birth *Qualitative*

(e) Distance between home and school *Quantitative*

(f) Time spent travelling to work *Quantitative*

2 State which of the following sets of data are discrete and which are continuous.

(a) Cost *Discrete*   (b) Population *Discrete*   (c) Time *Discrete*

(d) Weight *Continuous*   (e) Area *Continuous*   (f) IQ score *Discrete*

3 State whether each source will give primary or secondary data.

(a) Collecting data from a traffic survey *Primary*

(b) Downloading data from the internet *Secondary*

(c) Using data from the 2001 Census *Secondary*

(d) Using data found in a newspaper *Secondary*

(e) Giving people a questionnaire *Primary*

9.2 The data handling cycle

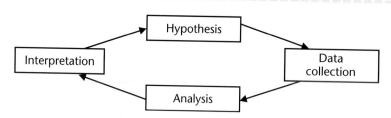

> **Key words:**
> data handling cycle
> hypothesis
> analysis

When you carry out an investigation, you will follow the data handling cycle . Start with a hypothesis . This is simply a statement that you want to test.

For example 'The vowels *a, e, i, o* and *u* all occur with the same frequency in written English' is a hypothesis.

You then collect the data and carry out an **analysis** of your data. This means arranging your data in a way that enables you to make sense of it. Putting data into a bar chart, pie chart, frequency diagram or box plot helps you to analyse it. You will also need to do calculations, such as finding the mean or median.

Finally, interpret or draw conclusions from your data – say what you have found out. This interpretation might lead to an alternative hypothesis.

Key words:
tally chart
frequency
frequency table
frequency distribution

9.3 Data collection

You will need to know:
- how to record and count data using tally marks
- what the frequency represents

When you have a large amount of data, start by organising it in a table. Use a **tally chart** to record each piece of data. Find the **frequency** (how often something occurs) by adding together the tally marks.

The complete table showing the tally marks and the frequency is called a **frequency table** or **frequency distribution** .

Example 1

Here is the list of vowels in the first four sentences of a book. Construct a frequency table for this data.

i, a, o, e, a, i, i, e, i, i, o, a, i, a, a, e, a, a, i, e

e, a, i, o, e, i, i, o, a, i, a, a, i, e, o, i, a, i, a, a

i, i, i, i, e, a, e, u, o, o, e, i, a, u, i, e, e, o, o, e

e, u, e, e, a, a, i, a, u, i, e, a, u, i, e

Vowel	Tally	Frequency
a	ℍℍ ℍℍ ℍℍ ℍℍ	20
e	ℍℍ ℍℍ ℍℍ |||	18
i	ℍℍ ℍℍ ℍℍ ℍℍ |||	23
o	ℍℍ ||||	9
u	ℍℍ	5

To see how often each vowel occurs, put the results into a tally chart like this.

Group the tally marks into sets of five by drawing the fifth line through the previous four.

Seven looks like ℍℍ || rather than |||||||.

Key words:
class interval

Class intervals

When you are dealing with continuous data that is widely spread, you can group the results together in regular intervals called classes or **class intervals** .

Example 2

The heights of 31 pupils in a Year 10 class were measured to the nearest centimetre.

> 161, 153, 168, 171, 143, 161, 165, 156, 147, 158
> 161, 160, 180, 173, 149, 155, 164, 167, 173, 159
> 151, 163, 162, 174, 166, 157, 170, 161, 156, 181
> 168

Put these heights into a tally chart using class intervals of 140–149 cm, 150–159 cm, etc.

Height (cm)	Tally	Frequency
140–149	|||	3
150–159	||||| |||	8
160–169	||||| ||||| |||	13
170–179	|||||	5
180–189	||	2
	Total	31

When data is grouped like this in a class interval, you need to remember that information about the individual heights is lost.

Check the total in the frequency column is the same as the original number of values in the actual data list.

Class intervals are groupings of quantitative data. When you group data, you should use between five and ten class intervals.

Class intervals are usually of equal widths.

You can also write class intervals using the symbols < for 'less than' and ≤ for 'less than or equal to' . For example, the data from Example 2 could be shown as in the table below.

Height h (cm)	Tally	Frequency
$140 \leqslant h < 150$	|||	3
$150 \leqslant h < 160$	||||| |||	8
$160 \leqslant h < 170$	||||| ||||| |||	13
$170 \leqslant h < 180$	|||||	5
$180 \leqslant h < 190$	||	2
	Total	31

Let h be the symbol for the height in cm.

The first class interval $140 \leqslant h < 150$ means: include any height greater than or equal to 140 cm but less than 150 cm. It does not include 150 cm.

Key words:
data-capture sheet

Data-capture sheets

A pre-prepared table like the one shown in Example 2 is called a
data-capture sheet . These are useful for collecting and ordering
data before choosing how to present your results.

Exercise 9B

1 The list below shows the marks gained by 29 pupils in a test.

22 27 18 23 25 19 17 20

23 20 18 22 20 19 24 21

16 19 18 22 14 19 18 21

20 19 25 25 21

Copy and complete the frequency table below.

Mark	Tally	Frequency
1–10		
11–20	ＷＴＩＬＴＩＬＴＩ	16
21–30	ＩＬＴＩＬＴＩＩＩ	13
	Total	29

2 The masses of twenty students in the sixth form were recorded.

46 kg   55 kg   53 kg   75 kg   62 kg

59 kg   84 kg   63 kg   68 kg   74 kg

78 kg   60 kg   57 kg   61 kg   48 kg

81 kg   73 kg   67 kg   64 kg   77 kg

Copy and complete the frequency table below using five equal
class intervals.

Mass m (kg)	Tally	Frequency
$40 \leqslant m < 50$	｜｜	2
$50 \leqslant m < 60$	｜｜｜｜	4
$60 \leqslant m < 70$	ＨＴＩＩ	7
$70 \leqslant m < 80$	ＨＴ	5
$80 \leqslant m < 90$	｜｜	2
	Total	20

3 During a school cross-country run the following times were recorded (rounded to the nearest minute).

15	18	23	27	17	21	24	19	20	28
33	26	24	19	23	28	25	19	17	31
38	24	28	21	24	26	26	19	18	27
17	26	21	23	26	19	18	26	24	27
26	16	24	28	29	19	18	33	23	29

Construct a grouped frequency table showing the tally marks and frequencies and using class intervals 10–14, 15–19, 20–24, 25–29, 30–34, 35–40.

4 The following amounts of money were collected for a charity:

£86   £89   £51   £61   £11   £56   £82   £87   £4   £89

£43   £93   £31   £20   £61   £3   £65   £3   £56   £84

£38   £80   £42   £64   £66   £22   £3   £34

Construct a frequency table using five equal class intervals beginning with $0 \leqslant m < 20$, where m means the amount of money collected in £s.

5 Construct a data-capture sheet for throwing a six-sided fair dice. The sheet should include headings for tally marks and frequency.

6 A traffic survey is being conducted on the colours of cars passing a road junction. Construct a data-capture sheet to record this data.

9.4 Questionnaires and sampling techniques

Key words:
survey
questionnaire
hypothesis
biased
pilot study

A **survey** collects primary data. One way of collecting primary data is to use a **questionnaire** .

Questionnaire guidelines:

- Be clear about what you want to find out. A survey should test your theory or **hypothesis** . Think about what data you need to collect.

- Ask specific questions, for example, *what music do you like? how much do you earn? what is your favourite sport?*

- Keep your questions short and simple. Tick boxes give a clear choice of replies.

A traffic survey collects data on traffic.

A questionnaire makes sure you ask everyone in the survey for the same information.

Are you:

Male? ☐   Female? ☐

This question has a clear choice of **two** options. They include all possible answers.

What is your favourite sport?

Hockey ☐   Football ☐   Gymnastics ☐

Rugby ☐   Athletics ☐   Other ☐

There are six choices here. 'Other' covers all the choices not in the first five.

- Avoid questions that are vague or could be misunderstood.

Do you use the internet:

Sometimes? ☐   Occasionally? ☐   Often? ☐

These words mean different things to different people, so avoid using them.

Instead you could ask:

Do you use the internet:

Never? ☐   Once a day? ☐   Twice a day? ☐   3 times or more a day? ☐

- Avoid using personal or embarassing questions.

Do you have a boyfriend? ☐   girlfriend? ☐

- Avoid asking questions that are **biased** :

Do you agree that television is bad for you?

Yes ☐   No ☐

This question suggests that the answer is *Yes*. A biased or 'leading question' encourages people to give a particular answer.

- Do not ask too many questions.

If your questionnaire is too long, people won't want to answer it.

It is sensible to test a questionnaire on a few people first to find out if it works. This is called a **pilot study** . It may show that your questionnaire needs improving.

Questionnaire tips:
- Do not ask for names (personal information).
- Make sure that the questionnaire is easy to complete quickly (tick boxes are best).
- Make sure that the questionnaire can be photocopied easily. You may need lots.
- Design a data-capture sheet. You will need to copy the questionnaire answers on to this.

Populations and samples

The **population** is the whole group that a survey is interested in.

The definition of the population will depend on the investigation you have chosen. A population might be all the students in a school, or all the dandelion plants on the school field, for example.

When the population is very large, collecting the data would take too long or cost too much. Instead you could collect data from part of the population – a **sample** . You must select your sample so that it is **representative** of the population.

For example, in a school there are 1000 pupils. You could choose to sample 10% of the population for a survey, using one of a range of sampling methods.

Random sampling

A **random sample** allows every member of the population an equal chance of being selected.

You could do this by:
- putting everyone's name into a hat and drawing out 100 names.
- giving each name a number and selecting 100 numbers using a random number generator.

Systematic sampling

Systematic sampling involves selecting a member of the population at regular intervals.

So for a 10% sample, select every 10th pupil from an *unordered* list of all the pupils in the school. The first pupil would be chosen by generating a random number between 1 and 10. For example, if your random number was 7, you would select the 7th pupil, the 17th, then the 27th, and so on until you reached number 997 on the list.

Key words:
population
sample
representative
sampling fraction
random sample
systematic sampling
stratified sampling
bias

A sample size of 10% is large enough to represent the views of the whole school.

The fraction to be sampled is known as the **sampling fraction**. For example, sampling 150 out of a population of 1200 gives a sampling fraction of $\frac{150}{1200} = \frac{1}{8}$

You might also meet...
cluster sampling... when the population is divided into groups (clusters) and a number of these clusters are chosen at random. Observations from the chosen clusters are selected at random. **Quota sampling**... often used in market research. Interviewers are given a number of people (a quota) to interview. They might ask 20 men and 20 women about their TV viewing habits. It is easy to do but the sample is not random.

Stratified sampling

Stratified sampling ensures that you get a fair proportion of samples from each group within the population.

For example, the numbers in each year group in a school are:

Year group	7	8	9	10	11
Number of pupils	250	200	250	200	100

Your sample should reflect the numbers in each year group.
This means that in your sample there should be

- equal numbers of Year 7 and Year 9 pupils
- equal numbers of Year 8 and Year 10 pupils
- twice as many Year 8 or Year 10 pupils as the number from Year 11

For a 10% sample, take 10% from each year group.

Year group	7	8	9	10	11
Number of pupils in sample	25	20	25	20	10

Choose the sample from each year group by random sampling.

Years 7 and 9 both have 250 pupils.

Years 8 and 10 both have 200 pupils.

200 pupils in Years 8 and 10 but only 100 pupils in Year 11.

In a stratified sample:

- **The sample should reflect the number in each group.**
- **Take the same fraction or percentage of each group.**
- **Select the sample from each group using random sampling.**

You might also decide that the proportion of boys and girls should be reflected in the sample.

If there are twice as many boys as girls in a particular year group then the sample from that year group should have approximately twice as many boys as girls in it.

Example 3

A survey of 2371 science students is carried out at a university.

The percentages of students studying each science is shown in the table.

Subject	Biology	Chemistry	Geology	Physics
%	33	24	19	24

A 10% stratified sample is chosen. Write down the number of students from each category that they should sample.

The total number of students in the stratified sample is.

10% of 2371 = $\frac{10}{100}$ × 2371 = 237.1 = 237

Number of biology students = 237 × $\frac{33}{100}$ = 78.21 = 78

Number of chemistry students = 237 × $\frac{24}{100}$ = 56.88 = 57

Number of geology students = 237 × $\frac{19}{100}$ = 45.03 = 45

Number of physics students = 237 × $\frac{24}{100}$ = 56.88 = 57

Always round the numbers to the nearest whole number.

Check your total is correct.
78 + 57 + 45 + 57 = 237

Bias

Bias means that your results do not give an accurate picture of the whole population. Bias might arise because:

- you have not correctly identified the population
- you have not chosen a representative sample of the population
- you have not designed a survey or questionnaire carefully.

Exercise 9C

1 A whole school survey is being conducted in a school containing roughly equal numbers of boys and girls. For each of the following methods, say whether the results would be a random sample or if they would be biased.

(a) Taking all surnames on a school list beginning with H.

(b) Taking only 10 boys in each year group.

(c) Putting all the names into a box and picking out 50 without looking.

(d) Asking an equal number of boys and girls from each year group.

(e) Taking every 20th name from the alphabetical school list.

(f) Only asking 10 girls and 10 boys in Years 7, 9 and 11.

2 A secondary school has the following number of students in each year group.

Year group	7	8	9	10	11
Number of students	241	256	234	263	251

Taking a 5% stratified sample, write down how many students from each year group should be in the sample.

3 Describe how you would collect data to find out if the following statements are true. You should include a questionnaire and details about your sample.

- Most students in school carry a mobile phone.
- More boys than girls eat crisps during the school day.
- People born in summer months are more likely to choose adventurous sports.

4 (a) Explain what is meant by **(i)** a random sample
(ii) a systematic sample
(iii) a stratified sample.

(b) This table shows the age distribution of the members of a gym.

Age (years)	Under 16	16–30	31–45	46–65	Over 65
Number of people	86	107	130	94	63

A stratified sample of 50 is planned.
(i) What percentage is this of the total membership?
(ii) Calculate the number of people that should be sampled for each age group.

9.5 Two-way tables

Key words:
two-way tables

Two-way tables are similar to frequency tables but show two or more types of information at the same time.

Bus timetables, league tables, school performance tables and holiday brochure prices are often two-way tables.

Example 4

In a survey of 32 pupils, 6 girls walked to school, 10 boys went by bus, and 4 boys cycled. Of the remaining 11 girls, only 1 cycled to school. No-one travelled by any other method.

(a) Draw a two-way table to show this information.

(b) Complete the table and find
(i) how many girls went by bus?
(ii) how many pupils walked to school?

(a)

	Walked	Cycled	Bus	Total
Boys		4	10	
Girls	6	1		17
Total				32

6 girls walked and there were 11 remaining
6 + 11 = 17

The total number of boys is 32 − 17 = 15.

The number of boys that walked is 15 − (4 + 10) = 1.

(b)

	Walked	Cycled	Bus	Total
Boys	1	4	10	15
Girls	6	1	10	17
Total	7	5	20	32

The number of girls who came to school by bus is 17 − (6 + 1) = 10.

The total number who walked, cycled and came by bus can now be completed.

(i) 10 girls went by bus.

(ii) 7 pupils walked to school.

Check the totals are correct:
7 + 5 + 20 = 32.

Example 5

The following two-way table shows the heights of pupils in five Year 11 classes.

Height h (cm)	Frequency				
	11A	**11B**	**11C**	**11D**	**11E**
$140 \leqslant h < 150$	3	6	4	5	4
$150 \leqslant h < 160$	6	5	7	5	6
$160 \leqslant h < 170$	10	9	11	8	12
$170 \leqslant h < 180$	7	6	4	9	5
$180 \leqslant h < 190$	4	5	4	3	5

This is an example of a two-way table using continuous data for the height column.

(a) How many pupils are taller than or equal to 180 cm?

(b) How many pupils are in class 11C?

(c) How many pupils took part in the survey?

(d) What percentage of pupils are taller than or equal to 160 cm but less than 180 cm? (Give your answer to the nearest whole number.)

Height h (cm)	Frequency					
	11A	**11B**	**11C**	**11D**	**11E**	**Total**
$140 \leqslant h < 150$	3	6	4	5	4	22
$150 \leqslant h < 160$	6	5	7	5	6	29
$160 \leqslant h < 170$	10	9	11	8	12	50
$170 \leqslant h < 180$	7	6	4	9	5	31
$180 \leqslant h < 190$	4	5	4	3	5	21
Totals	30	31	30	30	32	153

Find the totals across and down by adding more columns and rows to the table. The final total in the bottom right-hand corner should be the same by adding the final row and final column.

(a) 21 pupils taller than or equal to 180 cm.

(b) 30 pupils in 11C.

(c) 153 pupils took part.

(d) $\dfrac{(50 + 31)}{153} \times 100 = 53\%$.

Example 6

The table below shows part of a holiday brochure giving the cost of a fly–drive holiday. The prices shown are per person in £s.

Group	Number of days						Extra night
	2	3	4	5	6	7	
5/6 adults sharing	170	178	185	190	193	196	25
4 adults sharing	173	184	190	197	199	205	25
3 adults sharing	179	192	202	213	220	227	25
2 adults sharing	179	192	202	213	220	227	25
Child	148	148	148	148	148	148	25

(a) Find the cost of a 3-day holiday for 4 adults and 3 children.

(b) What is the cost of a fly–drive holiday for 2 adults and 2 children for 10 days?

(a) A 3-day holiday for 4 adults costs 4 × £184 = £736

For 3 children costs 3 × £148 = £444

Total cost of the holiday = £736 + £444 = £1180

> Use the figures in the '3 days' column. Read the rows for 4 adults and 1 child, don't forget to multiply by 3.

(b) 2 adults for 10 days = (2 × £227) + (2 × 3 × £25) = £604

2 children for 10 days = (2 × £148) + (2 × 3 × £25) = £446

Total cost of holiday = £604 + £446 = £1050

> Work out the cost for 7 days then add on 3 extra nights. Read the rows for 2 adults and 1 child. Multiply the cost for 1 child by 2.

Exercise 9D

1 In a school survey of 50 boys and 50 girls, 41 boys were right-handed and only 6 girls were left-handed. Copy the two-way table below and complete it to show this information. Use the table to work out an estimate of the percentage of pupils in the school who are left-handed.

	Left-handed	Right-handed	Total
Girls	6		
Boys			
Total			

2 In the 2001 Census the male population of Poynton (Central) was 3522 and the number of females in Poynton (West and East) was 3898. The population of Poynton (Central) was 6792 and the total population of Poynton was 13 433.
Construct a two-way table to show this information.
Complete the table.
What is the total percentage of females in Poynton?
(Give your answer to the nearest whole number.)

3 The following table gives the KS3 English SATs results for a local school.

		Level					
		3	4	5	6	7	8
English	Boys	11	28	34	31	15	1
	Girls	4	20	36	43	22	5

(a) Copy the table and extend it to find the totals in each row and each column

(b) How many pupils took the KS3 English SATs test?

(c) What percentage of boys achieved a level 5 or higher?
Give your answer to 1 d.p.

(d) What percentage of girls achieved a level 7?
Give your answer to 1 d.p.

4 The table shows the distances between some major French cities in km.

Bordeaux

870	Calais					
658	855	Grenoble				
649	1067	282	Marseille			
804	1222	334	188	Nice		
579	292	565	776	931	Paris	
244	996	536	405	560	706	Toulouse

The table shows that the distance between Calais and Paris is 292km.
Find:

(a) the distance between Bordeaux and Marseille

(b) the distance between Toulouse and Grenoble

(c) the total distance travelled from Paris to Calais to Bordeaux and back to Paris.

5 Using the two-way table in Example 6 on page 155, find the cost of:

(a) a 5-day holiday for 3 adults and 4 children

(b) an 8-day holiday for 6 adults and no children.

6 In a local fun-run event the following results were recorded by all the competitors. Times are given to the nearest minute.

Age a (years)	Time t (minutes)			
	$0 \leqslant t < 10$	$10 \leqslant t < 20$	$20 \leqslant t < 30$	$30 \leqslant t < 40$
$5 \leqslant a < 10$	0	1	12	36
$10 \leqslant a < 15$	2	46	59	27
$15 \leqslant a < 20$	7	65	37	13

This is an example of a two-way table using continuous data for both the columns and the rows.

(a) How many runners finished in under 20 minutes?

(b) How many runners were aged 15 years or over?

(c) How many runners took part in the fun-run altogether?

(d) What percentage of runners completed the course in under 10 minutes? Give your answer to the nearest whole number.

Examination questions

1 The ages to the nearest year of 20 people taking part in a fun run are recorded below

     5   22   34   14   19   16   21   28   30   14

    44   12   18   24   7   32   16   8   13   9

Complete a grouped frequency table with intervals beginning 1–10.

How many people were aged between 11 and 20?   *(3 marks)*

2 The two-way table shows the number of desks and the number of chairs in each office in a building.

		Number of chairs			
		1	2	3	4
Number of desks	1	3	3	3	2
	2	1	5	2	3
	3	0	0	1	2

(a) How many offices have 4 chairs?

(b) How many offices have the same number of chairs as desks?   *(3 marks)*

3 Jane reads in a magazine that there is a link between the number of children in a family and the number of holidays taken.

(a) Design a two-way table to record the number of holidays and the number of children in a sample of families.

(b) Complete your two-way table by inventing data for six families.   *(4 marks)*

4 The manager of a cinema wants to find out how often teenagers attend the cinema. He uses a questionnaire.

(a) Here is part of the questionnaire.

> **Question** How often do you attend the cinema?
>
> **Response** Sometimes ☐   Occasionally ☐
>
> Regularly ☐

Write down two criticisms of his response section.

(b) Design a replacement question with a response section for the question in part **(a)**.

(c) Explain how the manager could distribute 50 questionnaires randomly to pupils from a school of 1000 pupils.

(5 marks)

5 Chandri wants to survey pupils in her school about their reading habits.

(a) Write a question that would help Chandri to investigate how often pupils in her school read for pleasure. Include a response section.

(b) There are 1000 pupils in Chandri's school. Chandri samples 50 pupils at random and asks them to complete her survey. She finds that 16 of the pupils in the sample read comics. Estimate the number of pupils in the school who read comics.

(4 marks)
AQA, Spec A, 1I, June 2003

6 (a) State two conditions that must be satisfied when collecting data for a stratified sample.

(b) A small village has a population of 400. The population is classified by age as shown in the table below.

Age (years)	0–12	13–24	25–40	41–60	61+
Number of people	35	58	125	103	79

A stratified sample of 50 is planned. Calculate the number of people that should be sampled from each age group.

(4 marks)
AQA, Spec A, 1H, June 2004

Summary of key points

Types of data (grade D)

Primary data is information you collect yourself. Secondary data is information that you get from existing records.

Qualitative data contains descriptive words.
Quantitative data contains numbers and can be discrete or continuous.

Discrete data can only have exact values and is countable.
Continuous data can have any value in a certain range.

The data handling cycle

When you carry out an investigation, you will follow the data handling cycle. Start with a hypothesis.

Collecting data (grade D/C/B/A/A*)

Class intervals are used to group quantitative data. When you group your data, use between 5 and 10 class intervals.

A data-capture sheet is used to record primary or secondary data.

Questionnaire tips:
- Do not ask for names (personal information).
- Make sure that the questionnaire is easy to complete quickly.
- Make sure that the questionnaire can be photocopied easily.
- Design a data-capture sheet. You will need to copy the questionnaire answers on to this.

Sampling

The population is the whole group that a survey is interested in.

A random sample is one in which each member of the population is equally likely to be selected.

Systematic sampling involves selecting members of the population at regular intervals. For example, selecting every 10th item from an unordered list.

Stratified sampling divides the population into groups.

- The sample should reflect the number in each group.
- Take the same fraction or percentage of each group.
- Select the sample from each group using random sampling.

Two-way tables

Two-way tables are used to show two pieces of information at the same time. For example, timetables show times and places.

This chapter will show you how to:

✔ draw bar charts, pie charts, line graphs and stem-and-leaf diagrams
✔ construct scatter graphs, identify correlation and find the line of best fit
✔ construct frequency polygons and frequency diagrams

10.1 Frequency diagrams

Key words:
bar chart

Frequency diagrams, such as **bar charts**, can show patterns or trends in data.

In a bar chart, the bars can be either vertical or horizontal. They must be of equal width.

Bar charts can be used for quantitative or qualitative data.

Example 1

The table shows the frequency of vowels occurring in the first four lines of a book.

Vowel	a	e	i	o	u
Frequency	20	18	23	9	5

This is qualitative data.

Draw a bar chart for this data.

Choose a sensible scale.

Plot frequency on the vertical axis.

The height of each bar represents the frequency.

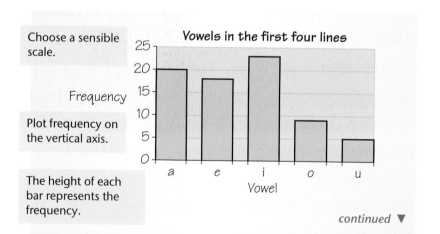

Give your bar chart a title.

Leave gaps between the bars.

Label the axes.

continued ▼

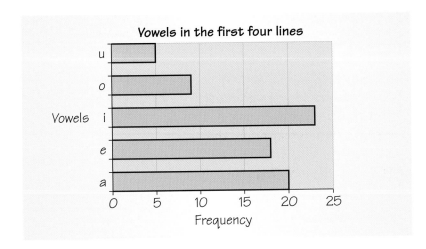

Vowels in the first four lines

You could plot this bar chart with horizontal bars.

The length of each bar represents the frequency.

There are still gaps between the bars.

Compound bar charts

Key words:
compound bar chart

You can use **compound bar charts** to make comparisons.

Example 2

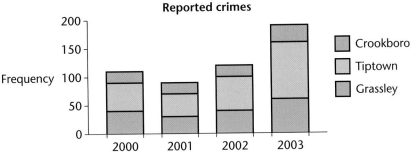

Reported crimes

- Crookboro
- Tiptown
- Grassley

Use the bar chart to say whether these statements are true or false:

(a) Tiptown has the highest number of reported crimes.

(b) Crimes have increased for all three towns in 2003.

(c) Grassley's reported crimes fell by one half from 2000 to 2001.

(d) Crookboro's crime levels stayed roughly the same between 2000 and 2002.

(a) True

(b) True

(c) False

(d) True

When you draw a bar chart make sure that you:
- label the horizontal and vertical axes clearly
- give the chart a title
- use a sensible scale to show all the information clearly
- shade or colour the bars
- leave spaces between the bars.

Vertical line graphs

Vertical line graphs can also be used to show discrete data. You can construct a vertical line graph in the same way as a bar chart, using a thick line instead of a bar.

Example 3

The number of goals scored in hockey matches during one season is recorded in the table. Draw a vertical line graph for the data.

Number of goals	0	1	2	3	4
Number of matches	5	7	3	2	2

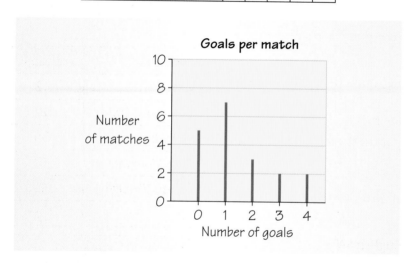

Frequency diagrams for continuous data

For grouped, continuous data you will need to draw a frequency diagram similar to a bar chart, but with no gaps between the bars.

Example 4

Draw a frequency table to show the heights of 31 sunflowers.

Height (cm)	Frequency
$140 \leqslant h < 150$	3
$150 \leqslant h < 160$	8
$160 \leqslant h < 170$	13
$170 \leqslant h < 180$	5
$180 \leqslant h < 190$	2

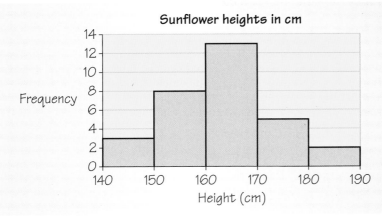

For continuous data there are no gaps between the bars.

The width of each bar is the same as the class interval.

A frequency diagram often shows the 'spread' of the data.

The horizontal axis must have a continuous scale.

Exercise 10A

1 The following frequency table shows the results of a crisp manufacturer's survey to find the most popular flavour.

(a) Draw a bar chart to display this.

(b) How many people took part in the survey?

Flavour	Frequency
Plain	12
Cheese and Onion	16
Smoky Bacon	33
Prawn Cocktail	13
Salt and Vinegar	21
Beef	5

2 (a) The table gives the KS3 English SATs results for a local school. Plot a compound bar chart for this data.

(b) Comment on the levels achieved by boys and girls.

(c) Are there any advantages in displaying this data as a compound bar chart instead of two separate bar charts?

		Level					
		3	4	5	6	7	8
English	**Boys**	11	28	34	31	15	1
	Girls	4	30	36	43	22	5

You used this data in Exercise 9D question 3.

3 A survey of the most common birds in the UK gave the following results:

Bird	Number (millions of pairs)
Blackbird	4.7
Blue tit	3.5
Chaffinch	5.8
Robin	4.5
Sparrow	3.8
Wood pigeon	2.4
Wren	7.6

Draw a bar chart to represent this information.

4 At a holiday show, families were asked how many holidays they had taken last year. Draw a vertical line graph to show this data.

Number of holidays	Number of families
0	2
1	14
2	17
3	8
4	1

5 Draw a frequency diagram to show the following information.

Length, x (cm)	Frequency
$0 < x \leqslant 5$	6
$5 < x \leqslant 10$	11
$10 < x \leqslant 15$	8
$15 < x \leqslant 20$	5

6 The table below gives the age range of the members of a local leisure centre.

Draw a frequency diagram to show the spread of ages.

Age	Frequency
$0 \leqslant \text{age} < 10$	23
$10 \leqslant \text{age} < 20$	45
$20 \leqslant \text{age} < 30$	56
$30 \leqslant \text{age} < 40$	36
$40 \leqslant \text{age} < 50$	49
$50 \leqslant \text{age} < 60$	32
$60 \leqslant \text{age} < 70$	16

Key words:
pie chart
sector
key

10.2 Pie charts

Pie charts show how data is shared or divided.

The sections of the pie chart are called **sectors**.

To draw a pie chart:

- First calculate the angle for each sector.
- Draw the angles using a protractor.
- Give the pie chart a **key** to explain what each sector represents.

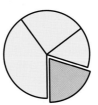

A sector is like a slice of the pie.

Example 5

Thirty people were asked how they travelled to work.
The results were:

Mode of travel	Frequency
Walk	7
Bus	9
Car	13
Cycle	1

Draw a pie chart to show this information.

360° represents 30 people

So $\dfrac{360°}{30} = 12°$ represents 1 person

$7 + 9 + 13 + 1 = 30$

Mode of travel	Frequency	Sector angle calculation	Angle
Walk	7	7 × 12°	84°
Bus	9	9 × 12°	108°
Car	13	13 × 12°	156°
Cycle	1	1 × 12°	12°
Total	30	Total angle	360°

Check that the total frequency equals the number of people and that the angles add up to 360°.

continued ▶

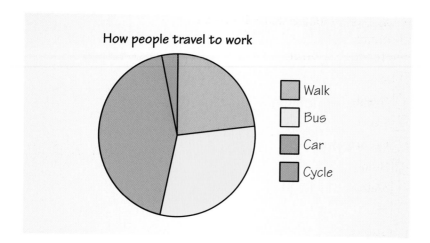

Example 6

This pie chart shows how a family spends its money in a week.
The amount spent on food is £120.
How much do they spend on each of the other items?

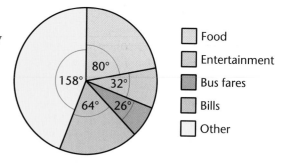

Food has a sector of 80° and the amount for food is £120.

80° represents £120

1° represents $\frac{£120}{80}$ = £1.50

Entertainment = 32 × £1.50 = £48

Bus fares = 26 × £1.50 = £39

Bills = 64 × £1.50 = £96

Other = 158 × £1.50 = £237

Check: £120 + £48 + £39 + £96 + £237 = £540

and 360 × £1.50 = £540

> Work out how much an angle of 1° represents.

> If the sector angles were not labelled, you could measure them using a protractor.

> The whole pie chart is 360° so the total of all the amounts of money must be 360 × £1.50.

Exercise 10B

1 Just before a general election, a small survey asked how people intended to vote.
Here are the results:

Draw a pie chart to show the results of this survey.

Conservative	60
Green	24
Labour	72
Liberal Democrat	56
Others	28

2 540 students were asked which was their favourite school subject.
The results are shown in this pie chart.
Work out the number who voted for each subject.

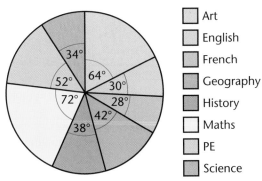

	Art
	English
	French
	Geography
	History
	Maths
	PE
	Science

Use Example 6 to help you.

3 This pie chart shows what Year 11 students did in the year after
their GCSE exams.
135 students went to college.
Work out the number of students in each of the other categories.

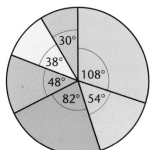

	Stayed in the 6th form
	Went to college
	Got a full-time job
	On a training scheme
	Out of work
	Other

4 A packet of Brand A breakfast cereal showed the following
nutritional information:

Ingredient	Protein	Carbohydrate	Fat	Fibre
Amount per 100 g of cereal	15 g	62.5 g	10 g	12.5 g

(a) If 360° represents 100 g of cereal, what angle represents 1 g?

(b) Calculate the angle of the sector for each ingredient.

(c) Draw a pie chart to show this information.

5 A different cereal, Brand B, has ingredients in these proportions:

   400 g carbohydrate
   150 g protein
   120 g fibre
    50 g fat

(a) Copy and complete the calculation and table below.

$$1 \text{ g of ingredient} = \frac{360°}{\text{Total weight of cereal}} = \frac{360}{\boxed{}} \text{g} = 0.\boxed{}°$$

Ingredient	Amount in g	Sector angle calculation	Angle
Carbohydrate	400	$400 \times 0.\square°$	
Protein	150	$150 \times 0.\square°$	
Fibre	120	$120 \times 0.\square°$	
Fat	50	$50 \times 0.\square°$	
Total		Total angle	360°

(b) Draw a pie chart to show this information.

(c) Compare your pie charts for questions 4 and 5.
Which brand could claim it has
   (i) less carbohydrate   **(ii)** less fat?

6 A government's expenditure one year was analysed as follows:

Area	£ (billions)
General Public Services	7
Defence	19
Public Order and Safety	7
Education	20
Health	19
Social Security	50
Housing	8
Transport	4

Draw a pie chart to show this expenditure.

10.3 Time series line graphs

Key words:
line graph
time series

Line graphs show trends in the data. You can only draw line graphs for continuous data.

In a line graph, the points plotted are joined with straight lines.

Line graphs often show how something changes over time. These are called **time series**.

Example 7

The table shows the maximum temperature (in °C) in Paris in each month.

Month	J	F	M	A	M	J	J	A	S	O	N	D
Max temp (°C)	6	7	11	14	18	21	24	24	21	15	9	7

Draw a time series line graph to show this data.

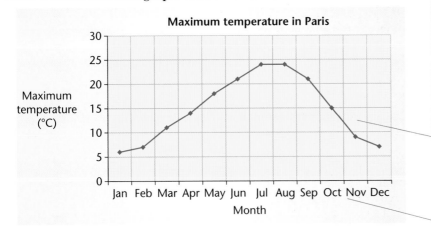

Temperature (°C) and time (months) are continuous data, so we can draw a line graph.

Join the points with straight lines.

This line shows the trend from October to November was a fall in temperature. It does not show actual temperature values.

For a graph showing changes over time, always put the time on the horizontal axis.

Exercise 10C

1 The table shows the amount of rainfall during the first week in February.

Draw a line graph to display this data.

Day	Rainfall (mm)
1	1.5
2	2.1
3	3.2
4	1.4
5	2.3
6	0.6
7	0.4

2 The maximum temperature (°C) and amount of rainfall (mm) recorded over a 12-month period in Santa Cruz (Bolivia) are shown in the table.

Draw two line graphs, one for temperature and one for rainfall, to show how the temperature and rainfall vary throughout the year.

Month	Temperature	Rainfall
Jan	29	200
Feb	30	125
Mar	29	122
Apr	27	120
May	25	95
June	23	80
July	24	65
Aug	26	55
Sept	27	68
Oct	28	105
Nov	30	121
Dec	30	162

10.4 Stem-and-leaf diagrams

Key words:
stem-and-leaf

A **stem-and-leaf** diagram keeps the original data and gives a 'picture' of the spread of the data.

Here are the scores out of 50 for a general knowledge quiz:

32	16	46	31	27	29	36	41	22	44
37	28	34	42	33	25	37	43	20	37
32	41	26	38	46	18	20	30	38	

You can write the data using the 'tens' digit as the stem and the 'units' digit as the leaf.

1|6 represents 16

```
1 | 6 8
2 | 7 9 2 8 5 0 6 0
3 | 2 1 6 7 4 3 7 7 2 8 0 8
4 | 6 1 4 2 3 1 6
```

Stem

4|4 represents 44   Leaves

Key: 1|6 means 16.

The key shows how to read the values.

You need to write the diagram with the leaves in ascending order. Write the leaves in neat columns, keeping all the numbers in line.

```
1 | 6 8
2 | 0 0 2 5 6 7 8 9
3 | 0 1 2 2 3 4 6 7 7 7 8 8
4 | 1 1 2 3 4 6 6
```

The total number of values here is 29 and this is the same as in the original list.

Check that you have not missed out any numbers. Make sure the total number of values is the same as in the original list.

Example 8

The lengths of 50 nails made in a factory were measured to the nearest tenth of a cm. The results were:

15.1	9.7	12.0	10.2	14.9	15.6	10.8	11.0	15.7	10.9
11.4	17.1	10.1	15.2	15.7	10.4	14.7	16.3	15.0	11.2
9.9	14.6	10.5	15.3	16.2	10.0	16.2	13.9	15.4	15.2
10.6	9.7	15.0	16.1	11.7	10.9	15.2	14.8	8.8	14.6
16.0	15.8	11.0	14.8	15.6	9.5	10.7	14.9	15.2	11.1

(a) Draw a stem-and-leaf diagram for these results.

(b) How many nails were less than 10 cm long?

(a)

8	8
9	5 7 7 9
10	0 1 2 4 5 6 7 8 9 9
11	0 0 1 2 4 7
12	0
13	9
14	6 6 7 8 8 9 9
15	0 0 1 2 2 2 2 3 4 6 6 7 7 8
16	0 1 2 2 3
17	1

Key: 8|8 represents 8.8

(b) 5 nails were less than 10 cm long.

Use the 'whole numbers' as the stem and the 'decimal parts' as the leaves.

Write the 'leaves' in ascending order.

Check: There are 50 pieces of data in the diagram, which is the same number as in the original list.

Count the 'leaves' on the 8 and 9 stems.

Exercise 10D

7,6,18,15, 19,9, 11,17, 18,14, 12,17,19

11,12,14,15,16,17,17

11,12,14,15,16,17,18,18,19,19,19

1 The temperature in °C is recorded in 20 towns on one day.

23	16	18	21	15	24	21	19	19	11
17	18	14	22	23	9	12	17	20	19

Copy and complete the stem-and-leaf diagram for this data.

0	9
1	1 2 4 5 6 7 7 8 9 9 9 9
2	0,1 1 2 3 3 4

Key: 1|3 represents 13°C.

2 Here are students' test scores:

9	43	33	26	37	12	18	19	25	32
14	29	43	33	37	31	29	40	17	

(a) Draw a stem-and-leaf diagram for these scores.

(b) How many students scored less than 13? 2

Find where 13 would be on the diagram.

3 A number of people were asked how many driving lessons they had taken before passing their test.
The results are shown below.

8	21	21	14	18	41	35	12	17
14	32	29	38	25	20	34	13	19

(a) How many people were asked?

(b) Draw a stem-and-leaf diagram for this data.

4 The following list gives the number of litres of fuel (given to the nearest litre) sold at a garage one morning.

```
26  31  18  44  37  30  29  32  35  40  20
51  15  36  30  25  40  30  34  27  20  35
45  38  40  42  28  12  39  44  35  30  24
36  33  43  23  46  38  42  50
```

(a) Draw a stem-and-leaf diagram for these results.

(b) How many customers bought more than 40 litres of fuel?

10.5 Scatter graphs

Key words:
scatter graph
correlation
line of best fit
positive correlation
negative correlation
no correlation

Scatter graphs help you to compare two sets of data. They show if there is a connection or relationship called **correlation** between the two quantities plotted.

Sometimes scatter graphs are called *scatter diagrams* or *scattergrams*.

The table shows the masses and heights of 10 men registered at a gym.

For a scatter graph, plot the height along the *x*-axis (horizontal) and the mass along the *y*-axis (vertical).

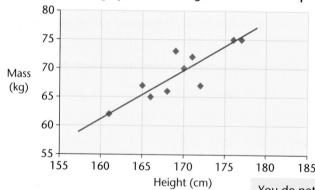

Scatter graph to show height–mass relationship

Height (cm)	Mass (kg)
166	65
169	73
172	67
161	62
177	75
171	72
168	66
165	67
170	70
176	75

The scatter graph suggests that the taller they are the heavier they are.

You do not need to start the axes at zero. Find the smallest and largest values in each set of data to help you decide on the scale. The straight line shows the line of best fit.

A good way to show this is by drawing a straight line through, or as close to, as many points as possible. This line is called the **line of best fit** .

Try to have equal numbers of points above and below the line of best fit.

You draw it 'by eye', using a ruler.

Here the line of best fit slopes from bottom left to upper right. This is called a **positive correlation** .

Positive correlation means as one variable gets bigger, so does the other.

Example 9

The following table shows how the fuel consumption (in litres per 100 km) changes as the speed of a car increases.

(a) Plot a scatter graph for this data.

(b) Draw in a line of best fit.

(c) Comment on the correlation.

Speed (kph)	Fuel consumption (litres/100 km)
20	9.6
30	8.7
40	8.0
50	6.8
60	6.0
70	5.5
80	4.5

(a)(b)

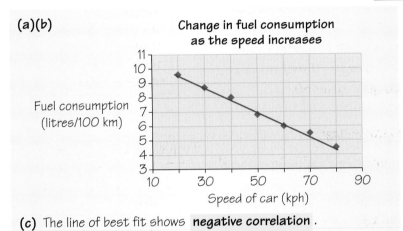

As the speed *increases* the fuel consumption *decreases*.

(c) The line of best fit shows **negative correlation**.

Example 10

This table shows the results of a recent geography test (as a percentage) and students' hand span measurements (in cm).

(a) Plot a scatter graph for this data.

(b) Comment on the correlation shown.

Test result (%)	Hand span (cm)
32	15
85	19
54	21
47	16
41	23
36	14
29	18
57	17
67	20
60	21

(a)

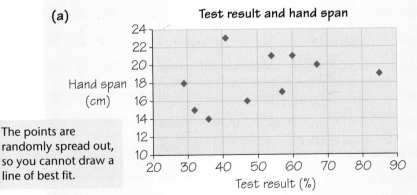

The points are randomly spread out, so you cannot draw a line of best fit.

(b) There is no linear connection between the test results and the hand span measurement. There is **no linear correlation**.

There is no *linear* relationship because the points do not lie on or near a *line*.

Exercise 10E

1 Here are four sketches of scatter diagrams.

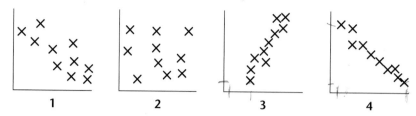

| 1 | 2 | 3 | 4 |

Which ones show:
(a) positive correlation
(b) negative correlation
(c) zero correlation?

> No correlation is sometimes called zero correlation.

2 What type of correlation would you expect if you drew a scatter graph of the following?
(a) The football league position against the number of goals conceded.
(b) A person's earnings against their height.
(c) The number of ice creams sold against the temperature during the day.
(d) The marks gained in a mock GCSE exam against those gained in the actual GCSE exam.
(e) The size of a car engine against the amount of fuel used by that engine.

3 In a science experiment, one end of a metal bar is heated.
The results show the temperature at different points along the metal bar.

(a) Draw a scatter graph to show this data. Let the x-axis represent the position between 0 and 10 cm, and the y-axis represent the temperature between 0 and 100°C.

Position (cm)	Temperature (°C)
1	15.6
2	17.5
3	36.6
4	43.8
5	58.2
6	61.6
7	64.2
8	70.4
9	98.8

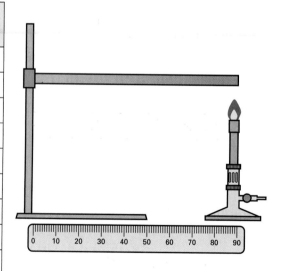

(b) Draw the line of best fit.
(c) Comment on the correlation shown.

4 The table shows the number of female
competitors taking part in each Olympic
Games from 1948 to 1984.

(a) Plot a scatter graph for this data.

Put 'Year' on the
horizontal axis.

(b) Draw the line of best fit.

(c) State what type of correlation you see.

(d) Describe the trend or pattern.

Year	Female competitors
1948	385
1952	518
1956	384
1960	610
1964	683
1968	781
1972	1070
1976	1251
1980	1088
1984	1620

5 The table below shows the mean annual temperature (in °C)
for 10 major cities and their latitude.

City	Mean annual temperature (°C)	Latitude (degrees)
Mumbai	32	19
Kolkata	26	22
Dublin	12	53
Hong Kong	26	22
Istanbul	18	41
London	12	51
Oslo	10	60
New Orleans	21	30
Paris	15	49
St Petersburg	7	60

Latitude describes position
on the globe, as degrees
North from the Equator.

(a) Plot the latitude along the *x*-axis (horizontal) from 0 to
70 degrees.
Plot the mean temperature along the *y*-axis (vertical)
between 0 and 40°C.

(b) What type of correlation does the graph show?

(c) What happens to the temperature as you move further North
from the Equator?

6 The table below shows the exam results for twelve students in maths and science.

(a) Draw a scatter graph to show this information.

Put the maths mark along the x-axis from 10 to 100.

(b) Draw the line of best fit.

(c) What type of correlation is shown?

(d) Estimate the science mark for a student who gained 38 marks in the maths exam.

Maths mark	Science mark
42	38
83	58
29	23
34	17
45	30
47	35
55	47
74	55
61	36
59	50
53	37
77	63

10.6 Frequency polygons

Key words:
frequency polygon
mid-point values

You can display grouped data in a **frequency polygon**. You draw a frequency diagram and then use straight lines to join the middle values (**mid-point values**) of the tops of the rectangles for each class interval.

A frequency polygon shows patterns or trends in the data more clearly than a frequency diagram.

Example 10

The frequency table shows the heights of 50 boys and 50 girls.

Draw a frequency polygon for the boys' heights.

Height (cm)	Frequency (boys)	Frequency (girls)
$155 \leqslant h < 160$	2	5
$160 \leqslant h < 165$	6	9
$165 \leqslant h < 170$	14	22
$170 \leqslant h < 175$	19	11
$175 \leqslant h < 180$	8	3
$180 \leqslant h < 185$	1	0

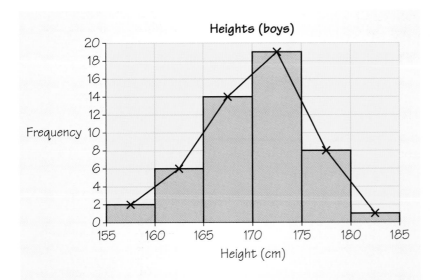

The frequency diagram for the boys' data is shown left. Join the mid-points of each bar with straight lines.

Frequency always goes on the vertical axis.

The straight lines form the frequency polygon.

Height (cm)	Frequency (boys)	Frequency (girls)	Mid-point values
$155 \leqslant h < 160$	2	5	157.5
$160 \leqslant h < 165$	6	9	162.5
$165 \leqslant h < 170$	14	22	167.5
$170 \leqslant h < 175$	19	11	172.5
$175 \leqslant h < 180$	8	3	177.5
$180 \leqslant h < 185$	1	0	182.5

You could draw the frequency polygon without drawing the frequency diagram first.
First extend the table and work out the mid-point values.

For each class interval, add together the boundary values and divide by 2.
The mid-point value of $155 \leqslant h < 160$ is
$\dfrac{155 + 160}{2} = 157.5$.

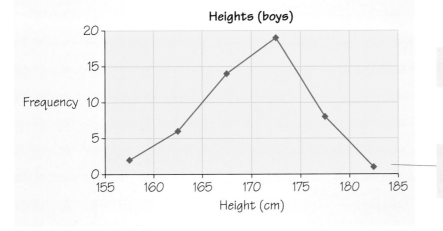

Plot each frequency against its mid-point value.

This is the point (1, 182.5). It represents 1 boy in the range $180 \leqslant h < 185$.

You could draw a frequency diagram and frequency polygon for the girls' data in the same way.

Exercise 10F

1 Work out the mid-point value for each of these class intervals:

    (a) 160 cm–170 cm   **(b)** $16\,\text{kg} \leqslant m < 18\,\text{kg}$

    (c) $3s \leqslant t < 7s$ **(d)** $15\,\text{m}^2$–$20\,\text{m}^2$

    (e) $100\,\text{m}\ell < v \leqslant 200\,\text{m}\ell$

2 (a) Draw a frequency polygon for the girls' data in Example 11.

    (b) Compare your polygon with the boys' polygon. What do you notice?

Copy the axes for the boys' data.

3 The table shows the total catches for 32 anglers in a fishing competition.

Catch c (kg)	Frequency
$0.05 < c \leqslant 0.55$	6
$0.55 < c \leqslant 1.05$	6
$1.05 < c \leqslant 1.55$	9
$1.55 < c \leqslant 2.05$	4
$2.05 < c \leqslant 2.55$	3
$2.55 < c \leqslant 3.05$	4

    (a) Copy and complete this extended table to show the mid-point values.

    (b) Draw a frequency polygon for this data.

Catch c (kg)	Frequency	Mid-point values
$0.05 < c \leqslant 0.55$	6	0.3
$0.55 < c \leqslant 1.05$	6	0.8
$1.05 < c \leqslant 1.55$	9	1·3
$1.55 < c \leqslant 2.05$	4	1·8
$2.05 < c \leqslant 2.55$	3	2·3
$2.55 < c \leqslant 3.05$	4	2·8

4 The temperature was recorded over a period of 60 days.

    (a) Copy and extend the table with a column showing the mid-point values.

    (b) Draw a frequency polygon for this data.

Temperature t (°C)	Frequency
$0 < t \leqslant 4$	6
$4 < t \leqslant 8$	14
$8 < t \leqslant 12$	11
$12 < t \leqslant 16$	15
$16 < t \leqslant 20$	8
$20 < t \leqslant 24$	3
$24 < t \leqslant 28$	3

5 A box holds 100 tomatoes. The mass of each tomato was measured and recorded in this frequency table.

Draw a frequency polygon to show this data.

Mass of tomato m (g)	Frequency
$35 < m \leqslant 40$	7
$40 < m \leqslant 45$	23
$45 < m \leqslant 50$	30
$50 < m \leqslant 55$	26
$55 < m \leqslant 60$	14

10.7 Histograms

Key words:
histogram
class intervals
frequency density

Data that is grouped and continuous can be displayed as a **histogram** .

- As the data is continuous there are no gaps between the bars.
- The bars can be different widths (to represent different **class intervals**).
- The *areas* of the bars are proportional to the frequencies they represent.
- The vertical axis shows the **frequency density** .
- The horizontal axis must be a continuous scale.

The class interval is the difference between the two ends points of the data. The class interval is also called the class width.

The frequency density is given by

$$\text{frequency density} = \frac{\text{frequency}}{\text{class interval}}.$$

The shape of a histogram gives a measure of how the data is distributed or spread.

Another word for spread is *dispersion*.

Example 11

Construct a histogram for this data on the heights of students in a survey.

Height (cm)	Frequency
$155 \leqslant h < 160$	2
$160 \leqslant h < 165$	6
$165 \leqslant h < 170$	14
$170 \leqslant h < 175$	19
$175 \leqslant h < 180$	8
$180 \leqslant h < 185$	1

The class interval for $155 \leqslant h < 160$ is $160 - 155 = 5$.

Height (cm)	Frequency	Frequency density
$155 \leqslant h < 160$	2	$2 \div 5 = 0.4$
$160 \leqslant h < 165$	6	$6 \div 5 = 1.2$
$165 \leqslant h < 170$	14	$14 \div 5 = 2.8$
$170 \leqslant h < 175$	19	$19 \div 5 = 3.8$
$175 \leqslant h < 180$	8	$8 \div 5 = 1.6$
$180 \leqslant h < 185$	1	$1 \div 5 = 0.2$

Extend the table by including a column for calculating frequency density.

The class intervals are all 5.

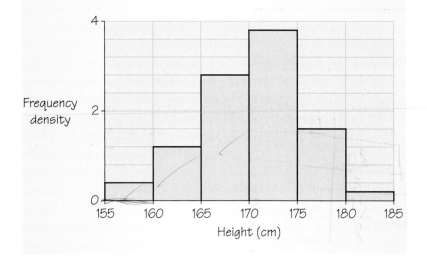

The data is continuous, so there are no gaps between the bars.

Since the class intervals are all equal, the heights of the bars are proportional to the frequency as well as the areas being proportional to the frequency.

Example 12

The masses of tomatoes in a mixed bag are given in the following frequency table.

Mass m (grams)	Frequency
$10 \leqslant m < 20$	10
$20 \leqslant m < 25$	15
$25 \leqslant m < 30$	12
$30 \leqslant m < 40$	42
$40 \leqslant m < 60$	50
$60 \leqslant m < 90$	27

Draw a histogram for this data.

Mass m (grams)	Frequency	Class interval (grams)	Frequency density
$10 \leqslant m < 20$	10	10	10/10 = 1.0
$20 \leqslant m < 25$	15	5	15/5 = 3.0
$25 \leqslant m < 30$	12	5	12/5 = 2.4
$30 \leqslant m < 40$	42	10	42/10 = 4.2
$40 \leqslant m < 60$	50	20	50/20 = 2.5
$60 \leqslant m < 90$	27	30	27/30 = 0.9

In this example, the class intervals are *not* equal. Extend the table to include the class interval.
This helps reduce errors in calculating the frequency density.

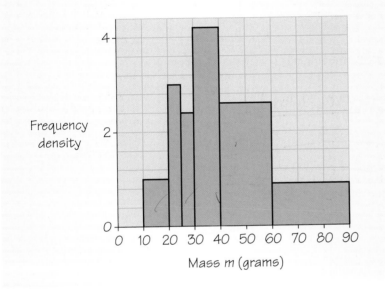

In this example, the **area** of each rectangle represents the frequency.

Exercise 10G

1 The frequency table below shows the salary structure for a small company. Use this information to draw a histogram.

Salary, s (£)	Frequency
$5000 \leqslant s < 10\,000$	2
$10\,000 \leqslant s < 18\,000$	4
$18\,000 \leqslant s < 24\,000$	2
$24\,000 \leqslant s < 35\,000$	6
$35\,000 \leqslant s < 50\,000$	1

2 In a fishing competition the total catches for each angler were recorded as shown in the frequency table below.

Catch (kg)	Frequency
0 to less than 0.5	2
0.5 to less than 0.9	6
0.9 to less than 1.5	4
1.5 to less than 2.2	1
2.2 to less than 2.9	4
2.9 to less than 3.8	2
3.8 to less than 5.4	1

Draw a histogram to display this data.

3 The age structure for members of a local leisure centre is shown below.

Age, a (years)	Frequency
$0 \leqslant a < 5$	4
$5 \leqslant a < 11$	18
$11 \leqslant a < 16$	36
$16 \leqslant a < 19$	24
$19 \leqslant a < 22$	41
$22 \leqslant a < 35$	49
$35 \leqslant a < 50$	32
$50 \leqslant a < 65$	14
$65 \leqslant a < 90$	3

Draw a histogram to represent the age structure of the centre.

4 The following histogram shows the amount of sponsor money collected from a class of 32 students.

Construct a frequency table for this data to show how many students are involved in each class interval.

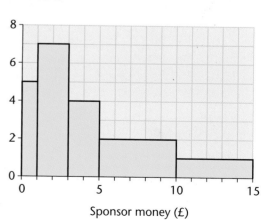

Examination questions

1 The number of cars passing through a set of traffic lights each time they
go on green is recorded.

12   15   23   20   18   16   27   9   10
19   22   26   14   11   8   4   12   23

Copy and complete the stem-and leaf-diagram, including a key, to represent the data.

```
0 | 8, 9, 4                        (3)          1 | 2  means  12
1 | 0, 1, 2, 2, 4, 5, 6, 8, 9      (9)
2 | 0, 2, 3, 3, 7,                 (5)
```

(3 marks)

AQA, Spec B, 1I (part question), March 2003

2 The time, in minutes, that seven
teenagers spent using their computer
and spent watching TV on one day
is recorded in the table.

Time spent using computer (minutes)	10	20	30	40	45	55	60
Time spent watching TV (minutes)	50	40	45	40	30	30	20

  (a) Plot these data as a scatter graph.

  (b) Draw a line of best fit on your
scatter graph.

  (c) Describe the relationship shown in the scatter graph.

Quite strong negative

(4 marks)

AQA, Spec B, 1I, June 2003

3 Ali and Bob are telephone sales assistants. The length and frequency of telephone calls made
by them during one day are shown in the table.

Length, t (minutes)	Frequency (Ali)	Frequency (Bob)
$0 < t \leqslant 2$	22	12
$2 < t \leqslant 4$	40	14
$4 < t \leqslant 6$	16	40
$6 < t \leqslant 8$	12	32
$8 < t \leqslant 10$	4	2

  (a) Draw frequency polygons for this data.

  (b) Write down two comparisons between the lengths of Ali's and Bob's calls.   *(4 marks)*

4 Batteries are tested by
putting them into toys
and seeing how long
they last.

Here are the results of
60 tests.

Draw a histogram to
show this information.

Time, t (minutes)	Frequency
$500 \leqslant t < 600$	8
$600 \leqslant t < 700$	15
$700 \leqslant t < 750$	10
$750 \leqslant t < 950$	18
$950 \leqslant t < 1150$	9

0,08
0,15
0,5
0,09
0,045

(3 marks)

AQA, Spec A, 1H, November 2003

Summary of key points

Bar charts (grade E/D)

Bar charts can be used to show patterns or trends in the data. You draw a bar chart with gaps between bars and the bars can be vertical or horizontal. The bars must be equal widths.

For grouped, continuous data, draw a frequency diagram similar to a bar chart, but with no gaps between the bars.

Pie charts (grade E/D)

These show how data is shared or divided. All pie chart sector angles add up to 360°. A key shows what each sector represents.

Time series line graphs (grade D)

These show trends in continuous data. They are drawn using straight lines between the data points.

Stem-and-leaf diagrams (grade D)

These keep the original data but allow you to form a picture of the spread of the data.

Scatter graphs (grade D/C)

These help you to compare two sets of data to find a connection or correlation. A line drawn to pass as closely as possible to all the points plotted is called the line of best fit.

Frequency polygons (grade C/B)

A frequency polygon is formed by straight lines between the mid-point values found in frequency diagrams. They show trends and patterns in the data and are used with continuous data.

Histograms (grade A/A*)

For a histogram

- There are no gaps between the bars since the data is continuous.
- The area of each bar is proportional to the frequency.
- The vertical axis represents the frequency density.
- The horizontal axis must be a continuous scale.

$$\text{Frequency density} = \frac{\text{frequency}}{\text{class interval}}$$

This chapter will show you how to:

✔ expand and simplify expressions with brackets
✔ solve equations and inequalities involving brackets
✔ factorise by removing a common factor
✔ expand two brackets

11.1 Expanding brackets

You will need to know how to:
- multiply positive and negative numbers
- add and subtract negative numbers
- collect like terms

When multiplying algebraic terms remember that:

$$x \times 3 = 3 \times x = 3x$$
$$y \times y = y^2$$
$$gh = g \times h = h \times g = hg$$

More complicated multiplications can also be simplified.

Example 1

Simplify $3f \times 4g$.

$$3f \times 4g = 3 \times f \times 4 \times g$$
$$= 3 \times 4 \times f \times g$$
$$= 12 \times fg$$
$$= 12fg$$

To multiply algebraic terms: multiply the numbers, then multiply the letters.

Multiplying a bracket

You can work out 6×34 by thinking of 34 as $30 + 4$.

$$6 \times 34 = 6 \times (30 + 4)$$
$$= 6 \times 30 + 6 \times 4$$
$$= 180 + 24$$
$$= 204$$

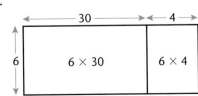

Brackets are often used in algebra:

$6(x + 4)$ means $6 \times (x + 4)$

As in the 6×34 example, you have to multiply each term inside the brackets by 6:

$$6(x + 4) = 6 \times x + 6 \times 4$$
$$= 6x + 24$$

It is like working out the area of a rectangle that has length $x + 4$ and width 6:

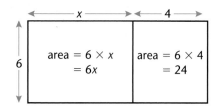

Total area $= 6(x + 4)$
$$= 6 \times x + 6 \times 4$$
$$= 6x + 24$$

You find the total area by adding the area of the two smaller rectangles.

When you do this it is called **expanding the brackets**.

It is also known as **removing the brackets** or **multiplying out the brackets**.

When you remove the brackets you must multiply each term inside the brackets by the term outside the bracket.

Example 2

Simplify the following by multiplying out the brackets:

(a) $5(a + 6)$ (b) $2(x - 8)$ (c) $3(2c - d)$

(a) $5(a + 6) = 5 \times a + 5 \times 6$
$$= 5a + 30$$

You must multiply each term inside the bracket by the term outside the bracket.

(b) $2(x - 8) = 2 \times x - 2 \times 8$
$$= 2x - 16$$

(c) $3(2c - d) = 3 \times 2c - 3 \times d$
$$= 6c - 3d$$

$3 \times 2c = 3 \times 2 \times c = 6 \times c = 6c$

Exercise 11A

1 Simplify these expressions.

(a) $2 \times 5k$ (b) $3 \times 6b$ (c) $4a \times 5$

(d) $3a \times 2b$ (e) $4c \times 3d$ (f) $x \times 5y$

2 Expand the brackets to find the value of these expressions. Check your answers by working out the brackets first:

(a) $2(50 + 7)$ (b) $5(40 + 6)$ (c) $6(70 + 3)$

(d) $3(40 - 2)$ (e) $7(50 - 4)$ (f) $8(40 - 3)$

> Remember: you must multiply each term inside the bracket by the term outside the bracket.
>
> A common mistake is to forget to multiply the second term.

3 Remove the brackets from the following:

(a) $5(p + 6)$ (b) $3(x + y)$ (c) $4(u + v + w)$

(d) $2(y - 8)$ (e) $7(9 - z)$ (f) $8(a - b + 6)$

4 Expand the brackets in these expressions:

(a) $3(2c + 6)$ (b) $5(4t + 3)$ (c) $2(5p + q)$

(d) $3(2a - b)$ (e) $6(3c - 2d)$ (f) $7(2x + y - 3)$

(g) $6(3a - 4b + c)$ (h) $2(x^2 + 3x + 2)$ (i) $4(y^2 - 3y - 10)$

> Remember:
> $3 \times 2c = 3 \times 2 \times c = 6c$

5 Write down the 6 pairs of cards which show equivalent expressions:

$4(x + 2y)$	$4x + 2y$	$2(4x + y)$	$4(2x - y)$
A	B	C	D
$8x - 8y$	$4x + 8y$	$8(x - y)$	$2x - 8y$
E	F	G	H
$8x + 2y$	$2(x - 4y)$	$2(2x + y)$	$8x - 4y$
I	J	K	L

You can use the same method for expressions that have an *algebraic* term instead of a *number* term outside the bracket.

Example 3

Expand the brackets in these expressions:

(a) $a(a + 4)$ (b) $x(2x - y)$ (c) $3t(t^2 + 1)$

(a) $a(a + 4) = a \times a + a \times 4$
             $= a^2 + 4a$

> Remember: $a \times a = a^2$

(b) $x(2x - y) = x \times 2x - x \times y$
              $= 2x^2 - xy$

> $x \times 2x = x \times 2 \times x = 2 \times x \times x = 2x^2$
>
> Remember: $x \times y = xy$

(c) $3t(t^2 + 1) = 3t \times t^2 + 3t \times 1$
               $= 3t^3 + 3t$

> $3t \times t^2 = 3 \times t \times t \times t = 3t^3$

Exercise 11B

Expand the brackets in these expressions:

1 $b(b + 4)$ **2** $a(5 + a)$ **3** $k(k - 6)$

4 $m(9 - m)$ **5** $a(2a + 3)$ **6** $g(4g + 1)$

7 $p(2p + q)$ **8** $t(t + 5w)$ **9** $m(m + 3n)$

10 $x(2x - y)$ **11** $r(4r - t)$ **12** $a(a - 4b)$

13 $2t(t + 5)$ **14** $3x(x - 8)$ **15** $5k(k + l)$

16 $3a(2a + 4)$ **17** $2g(4g + h)$ **18** $5p(3p - 2q)$

19 $3x(2y + 5z)$ **20** $r(r^2 + 1)$ **21** $a(a^2 + 3)$

22 $t(t^2 - 7)$ **23** $2p(p^2 + 3q)$ **24** $4x(x^2 + x)$

> Remember:
> $3x \times 4x = 4 \times 3 \times x \times x$
> $= 12x^2$

Adding and subtracting expressions with brackets

> Key words:
> like terms

Adding

To add expressions with brackets, expand the brackets first, then collect **like terms** to simplify your answer.

> Collecting like terms means adding all the terms in x, all the terms in y and so on.

Example 4

Expand then simplify these expressions:

(a) $3(a + 4) + 2a + 10$ **(b)** $3(2x + 5) + 2(x - 4)$

(a) $3(a + 4) + 2a + 10 = 3a + 12 + 2a + 10$
$$= 3a + 2a + 12 + 10$$
$$= 5a + 22$$

> Expand the brackets first. Then collect like terms.

(b) $3(2x + 5) + 2(x - 4) = 6x + 15 + 2x - 8$
$$= 6x + 2x + 15 - 8$$
$$= 8x + 7$$

> Expand both sets of brackets first.

Subtracting

If you have an expression like $-3(2x - 5)$, multiply both terms in the brackets by -3.

$$-3 \times 2x = -6x \quad \text{and} \quad -3 \times -5 = 15$$

So $-3(2x - 5) = -3 \times 2x + -3 \times -5$
$$= -6x + 15$$

> Remember:
> $^-\text{ve} \times {}^+\text{ve} = {}^-\text{ve}$
> $^-\text{ve} \times {}^-\text{ve} = {}^+\text{ve}$
> When you multiply:
> unlike signs give $-$
> like signs give $+$

Example 5

Expand these expressions:

(a) $-2(3t + 4)$ (b) $-3(4x - 1)$

(a) $-2(3t + 4) = -2 \times 3t + -2 \times 4$

$\qquad\qquad\quad = -6t + -8$

$\qquad\qquad\quad = -6t - 8$

$-2 \times 3 = -6 \quad -2 \times 4 = -8$

(b) $-3(4x - 1) = -3 \times 4x + -3 \times -1$

$\qquad\qquad\quad = -12x + 3$

$-3 \times 4 = -12 \quad -3 \times -1 = +3$

Example 6

Expand then simplify these expressions:

(a) $3(2t + 1) - 2(2t + 4)$ (b) $8(x + 1) - 3(2x - 5)$

(a) $3(2t + 1) - 2(2t + 4) = 6t + 3 - 4t - 8$

$\qquad\qquad\qquad\qquad\quad = 6t - 4t + 3 - 8$

$\qquad\qquad\qquad\qquad\quad = 2t - 5$

Remember to multiply both terms in the second bracket by -2.

(b) $8(x + 1) - 3(2x - 5) = 8x + 8 - 6x + 15$

$\qquad\qquad\qquad\qquad\quad = 8x - 6x + 8 + 15$

$\qquad\qquad\qquad\qquad\quad = 2x + 23$

Expand the brackets first. Remember that $-3 \times -5 = +15$. Then collect like terms.

Exercise 11C

Expand these expressions:

1 $-2(2k + 4)$ **2** $-3(2x + 6)$ **3** $-5(3n - 1)$

4 $-4(3t + 5)$ **5** $-3(4p - 1)$ **6** $-2(3x - 7)$

Expand then simplify these expressions:

7 $3(y + 4) + 2y + 10$ **8** $2(k + 6) + 3k + 9$

9 $4(a + 3) - 2a + 6$ **10** $3(t - 2) + 4t - 10$

11 $3(2y + 3) + 2(y + 5)$ **12** $4(x + 7) + 3(x + 4)$

13 $3(2x + 5) + 2(x - 4)$ **14** $2(4n + 5) + 5(n - 3)$

15 $3(x - 5) + 2(x - 3)$ **16** $4(2x - 1) + 2(3x - 2)$

17 $3(2b + 1) - 2(2b + 4)$ **18** $4(2m + 3) - 2(2m + 5)$

19 $2(5t + 3) - 2(3t + 1)$ **20** $5(2k + 2) - 4(2k + 6)$

21 $8(a + 1) - 3(2a - 5)$ **22** $2(4p + 1) - 4(p - 3)$

23 $5(2g - 4) - 2(4g - 6)$ **24** $2(w - 4) - 3(2w - 1)$

25 $x(x + 3) + 4(x + 2)$ **26** $x(2x + 1) - 3(x - 4)$

11.2 Solving equations involving brackets

Equations sometimes involve brackets.

When dealing with equations involving brackets, you usually expand the brackets first.

Then use the methods you learnt in Chapter 5 for solving equations.

Example 7

Solve $4(c + 3) = 20$.

Method A

$$4(c + 3) = 20$$
$$4c + 12 = 20$$
$$4c + 12 - 12 = 20 - 12$$
$$4c = 8$$
$$\frac{4c}{4} = \frac{8}{4}$$
$$c = 2$$

Expand the bracket by multiplying both terms inside the bracket by the term outside the bracket.

This is now like the equations in Chapter 5. You must subtract 12 from both sides before dividing both sides by 4.

Method B

$$4(c + 3) = 20$$
$$c + 3 = \frac{20}{5}$$
$$c + 3 = 5$$
$$c = 5 - 3$$
$$c = 2$$

Since 4 divides exactly into 20 you can divide both sides by 4 first.

Example 8

Solve $2(3p - 4) = 7$.

$$2(3p - 4) = 7$$
$$6p - 8 = 7$$
$$6p - 8 + 8 = 7 + 8$$
$$6p = 15$$
$$\frac{6p}{6} = \frac{15}{6}$$
$$p = 2.5$$

Expand the bracket.

You must add 8 to both sides before dividing both sides by 6.

Exercise 11D

1 Solve these equations:

  (a) $4(g + 6) = 32$ **(b)** $7(k + 1) = 21$ **(c)** $5(s + 10) = 65$

  (d) $2(n - 4) = 6$ **(e)** $3(f - 2) = 24$ **(f)** $6(v - 3) = 42$

  (g) $4(m - 3) = 14$ **(h)** $2(w + 7) = 19$

2 Solve these equations:

  (a) $4(5t + 2) = 48$ **(b)** $3(2r + 4) = 30$ **(c)** $2(2b + 2) = 22$

  (d) $2(3w - 6) = 27$ **(e)** $3(4x - 2) = 24$ **(f)** $5(2y + 11) = 40$

  (g) $6(2k - 1) = 36$ **(h)** $3(2a - 13) = 18$

When two brackets are involved, expand both brackets then collect like terms before solving.

Like terms are terms of the same kind. In Example 9 there are only m terms and number terms.

Example 9

Solve $2(2m + 10) = 12(m - 1)$.

$$2(2m + 10) = 12(m - 1)$$
$$4m + 20 = 12m - 12$$
$$20 + 12 = 12m - 4m$$
$$32 = 8m$$
$$m = 4$$

Expand the brackets on both sides of the equation and collect like terms.

Collect ms on the RHS because $12m$ on the RHS is greater than $4m$ on the LHS. This keeps the m term positive.

When an equation involves fractions, it can be transformed into an equation without fractions by multiplying all terms by the LCM of the numbers in the denominators.

LCM means Lowest Common Multiple.

Example 10

Solve the equation $\dfrac{x + 17}{4} = x + 2$.

$$\frac{x + 17}{4} = x + 2$$
$$\frac{4(x + 17)}{4} = 4(x + 2)$$
$$x + 17 = 4x + 8$$
$$17 - 8 = 4x - x$$
$$9 = 3x$$
$$x = 3$$

Multiply both sides by 4, collect like terms and then finally divide by 3.

Note the use of brackets.

Collect the xs on the RHS and the numbers on the LHS. $4x$ on the RHS is greater than x on the LHS.

Example 11

Solve the equation $\dfrac{x-6}{3} = \dfrac{x+4}{5}$.

$$\frac{x-6}{3} = \frac{x+4}{5}$$

$$\frac{15(x-6)}{3} = \frac{15(x+4)}{5}$$

$$5(x-6) = 3(x+4)$$

$$5x - 30 = 3x + 12$$

$$5x - 3x = 12 + 30$$

$$2x = 42$$

$$x = 21$$

Look at the denominators. 3 and 5 have a LCM of 15 so multiply both sides of the equation by 15.

Note the use of brackets. Always put them in when you multiply in this way.

Then solve using the method shown in Example 9.

Example 12

Solve the equation $\dfrac{2x+3}{6} + \dfrac{x-2}{3} = \dfrac{5}{2}$.

$$\frac{6(2x+3)}{6} + \frac{6(x-2)}{3} = \frac{6(5)}{2}$$

$$2x + 3 + 2(x-2) = 15$$

$$2x + 3 + 2x - 4 = 15$$

$$4x - 1 = 15$$

$$4x = 16$$

$$x = 4$$

The LCM here is 6.

Note the use of brackets.

The most common mistake is to forget to multiply **all** terms by the LCM.

This means the number on the RHS as well as the terms on the LHS.

Exercise 11E

1 Solve these equations.

(a) $2a + 4 = 5(a - 1)$ (b) $3(d - 2) = 2d - 1$

(c) $5(x + 3) = 11x + 3$ (d) $12p + 3 = 3(p + 7)$

(e) $4t + 3 = 3(2t - 3)$ (f) $3b - 4 = 2(2b - 7)$

(g) $8(3g - 1) = 15g + 10$ (h) $3(2k + 6) = 17k + 7$

(i) $2(y + 5) = 3y + 12$ (j) $5r + 3 = 4(2r + 3)$

2 Solve the following equations by expanding both brackets.

(a) $2(b + 1) = 8(2b - 5)$ (b) $5(4a + 7) = 3(8a + 9)$

(c) $6(x - 2) = 3(3x - 8)$ (d) $5(2p + 2) = 6(p + 5)$

(e) $9(3s - 4) = 5(4s - 3)$ (f) $4(10t - 7) = 3(6t - 2)$

(g) $4(2w + 2) = 2(5w + 7)$ (h) $3(3y - 2) = 7(y - 2)$

Use Example 9 to help.

3 Solve these equations.

(a) $\dfrac{d+3}{5} = 3 - d$

(b) $\dfrac{6y-5}{5} = y + 3$

(c) $\dfrac{3x-1}{8} = x - 2$

(d) $\dfrac{6+a}{2} = a + 4$

(e) $\dfrac{c-8}{4} = c + 1$

(f) $\dfrac{10-b}{3} = 12 + b$

Use Example 10 to help.

4 Solve these equations.

(a) $\dfrac{x+1}{3} = \dfrac{x-1}{4}$

(b) $\dfrac{2x-1}{3} = \dfrac{x}{2}$

(c) $\dfrac{3x+1}{5} = \dfrac{2x}{3}$

(d) $\dfrac{x+3}{2} = \dfrac{x-3}{5}$

(e) $\dfrac{x+2}{7} = \dfrac{3x+6}{5}$

(f) $\dfrac{8-x}{2} = \dfrac{2x+2}{5}$

Use Example 11 to help.

5 Solve

(a) $\dfrac{x+1}{2} + \dfrac{x+2}{5} = 3$

(b) $\dfrac{x+2}{4} + \dfrac{x+1}{7} = 3$

(c) $\dfrac{3x+2}{5} + \dfrac{x+2}{3} = 2$

(d) $\dfrac{3x-1}{5} - \dfrac{x+2}{3} = \dfrac{1}{5}$

(e) $\dfrac{x-3}{4} - \dfrac{x+3}{2} = 1$

(f) $\dfrac{2x+5}{4} - \dfrac{x+4}{3} = 2$

Use Example 12 to help.

11.3 Solving inequalities involving brackets

Inequalities can involve brackets.

Expand the brackets first, then use the methods you learned for solving inequalities in Chapter 5.

Remember that there is usually more than one answer when you solve an inequality and you need to state all possible values of the solution set.

Example 13

Solve these inequalities:

(a) $9 \leqslant 3(y-1)$

(b) $3(2x-5) > 2(x+4)$

(a) $9 \leqslant 3(y-1)$

        $9 \leqslant 3y - 3$

    $9 + 3 \leqslant 3y$

       $12 \leqslant 3y$

         $4 \leqslant y$

(b) $3(2x-5) > 2(x+4)$

      $6x - 15 > 2x + 8$

   $6x - 2x > 8 + 15$

         $4x > 23$

          $x > \frac{23}{4}$

          $x > 5\frac{3}{4}$

You must remember to keep the inequality sign in your answer.

For example, if you leave (a) as $4 = y$ you will lose a mark because you have not included *all* possible values of *y*.

If you are asked for integer solutions to **(b)** the final answer will be $x \geqslant 6$.

Example 14

n is an integer.
List the values of *n* such that $-11 < 2(n - 3) < 1$.

$$-11 < 2(n-3) < 1$$
$$-11 < 2n - 6 < 1$$
$$-11 + 6 < 2n < 1 + 6$$
$$-5 < 2n < 7$$
$$-2.5 < n < 3.5$$
Values of n are $-2, -1, 0, 1, 2, 3$

This is a double inequality.
Expand the bracket.
Add 6 throughout.

Remember to list the integer solutions as you were asked to in the question.
Remember to include 0.

Exercise 11F

1 Solve these inequalities.

(a) $2(x - 7) \leq 8$

(b) $7 < 2(m + 5)$

(c) $4(3w - 1) > 20$

(d) $3(2y + 1) \leq -15$

(e) $2(p - 3) > 4 + 3p$

(f) $1 - 5k < 2(5 + 2k)$

(g) $5(x - 1) \geq 3(x + 2)$

(h) $2(n + 5) \leq 3(2n - 2)$

2 Solve these inequalities then list the integer solutions.

(a) $-4 \leq 2x \leq 8$

(b) $-6 < 3y < 15$

(c) $-8 \leq 4n < 17$

(d) $-12 < 6m \leq 30$

(e) $-5 < 2(t + 1) < 7$

(f) $-3 < 3(x - 4) < 6$

(g) $-6 \leq 5(y + 1) \leq 11$

(h) $-17 < 2(2x - 3) \leq 10$

11.4 Factorising by removing a common factor

Key words:
common factor

Factorising an algebraic expression is the opposite of expanding brackets. To factorise an expression, look for a **common factor** – that is, a number that divides into all the terms in the expression. To factorise completely, use the HCF of the terms.

HCF means highest common factor.

2 is a factor of 6x.
2 is also a factor of 10.
So 2 is a common factor of 6x and 10.

For example:

$6x + 10$ can be written as $2(3x + 5)$
because $6x = 2 \times 3x$
and $10 = 2 \times 5$

Notice that the common factor is the term outside the bracket.

Example 15

Copy and complete the following:

(a) $3t + 15 = 3(\square + 5)$ (b) $4n + 12 = \square(n + 3)$

(a) $3t + 15 = 3(t + 5)$ Because $3 \times t = 3t$ (and $3 \times 5 = 15$)

(b) $4n + 12 = 4(n + 3)$ $4 \times n = 4n$ and $4 \times 3 = 12$

Example 16

Factorise these expressions:

(a) $5a + 20$ (b) $4x - 12$ (c) $x^2 + 7x$ (d) $6p^2q^2 - 9pq^3$

(a) $5a + 20 = 5 \times a + 5 \times 4$

$\quad = 5(a + 4)$

$\quad = 5(a + 4)$

Check: $5(a + 4) = 5 \times a + 5 \times 4$

$\quad\quad\quad\quad = 5a + 20$ ✓

5 is a factor of $5a$: $5a = 5 \times a$
5 is also a factor of 20: $20 = 5 \times 4$
So 5 is a common factor of $5a$ and 20
and is the term outside the bracket.

Check your answer by removing the brackets.

(b) $4x - 12 = 4 \times x - 4 \times 3$

$\quad = 4(x - 3)$

$\quad = 4(x - 3)$

2 is a factor of $4x$ and 12.
4 is also a common factor of $4x$ and 12.
Use 4 because it is the highest common factor (HCF) of $4x$ and 12.
Always look for the HCF.

(c) $x^2 + 7x = x \times x + x \times 7$

$\quad = x(x + 7)$

$\quad = x(x + 7)$

x is a common factor of x^2 and $7x$.

(d) $6p^2q^2 - 9pq^3$

$\quad = 3 \times 2 \times p \times p \times q \times q - 3 \times 3 \times p \times q \times q \times q$

$\quad = 3pq^2(2p - 3q)$

$\quad = 3pq^2(2p - 3q)$

The HCF is $3pq^2$.

Exercise 11G

1 Copy and complete the following:

(a) $3x + 15 = 3(\square + 5)$ (b) $5a + 10 = 5(\square + 2)$

(c) $2x - 12 = 2(x - \square)$ (d) $4m - 16 = 4(m - \square)$

(e) $4t + 12 = \square(t + 3)$ (f) $3n + 18 = \square(n + 6)$

(g) $2b - 14 = \square(b - 7)$ (h) $4t - 20 = \square(t - 5)$

Use Example 15 to help.

Don't forget to check your answers by removing the brackets.

2 Factorise these expressions:

(a) $5p + 20$ (b) $2a + 12$ (c) $3y + 15$

(d) $7b + 21$ (e) $4q + 12p$ (f) $6k + 24l$

> Use Example 16(a) to help.
> Remember: $5 = 5 \times 1$

3 Factorise these expressions:

(a) $4t - 12$ (b) $3x - 9$ (c) $5n - 20$

(d) $2b - 8$ (e) $6a - 18b$ (f) $7k - 7$

4 Factorise these expressions:

(a) $y^2 + 7y$ (b) $x^2 + 5x$ (c) $n^2 + n$

(d) $x^2 - 7x$ (e) $p^2 - 8p$ (f) $a^2 - ab$

> Use Example 16(c) to help.
> Remember: $n = n \times 1$

5 Factorise these expressions:

(a) $6p + 4$ (b) $4a + 10$ (c) $6 - 4t$

(d) $12m - 8n$ (e) $25x + 15y$ (f) $12y - 9z$

> Remember: $6p = 2 \times 3p$

6 Factorise completely:

(a) $3x^2 - 6x$ (b) $8x^2 - xy$

(c) $8a + 4ab$ (d) $p^3 - 5p^2$

(e) $3t^3 + 6t^2$ (f) $10yz - 15y^2$

(g) $18a^2 + 12ab$ (h) $16p^2 - 12pq$

7 Factorise these expressions:

(a) $4ab^2 + 6ab^3$ (b) $10xy - 5x^2$

(c) $3p^2q - 6p^3q^2$ (d) $8mn^3 + 4n^2 - 6m^2n$

(e) $6h^2k - 12hk^3 - 18h^2k^2$

> Use Example 16(a) to help.
> Look for the common factors in the terms.

Expanding two brackets

You can use a grid method to multiply two numbers.

For example:

34×57

$\times$	50	7
30	1500	210
4	200	28

$34 \times 57 = (30 + 4) \times (50 + 7)$

$= 30 \times 50 + 30 \times 7 + 4 \times 50 + 4 \times 7$

$= 1500 + 210 + 200 + 28$

$= 1938$

You can also use a grid method when you multiply two brackets together. You have to **multiply each term in one bracket by each term in the other bracket**.

For example:

To expand and simplify $(x + 2)(x + 5)$

×	x	5
x	x^2	$5x$
2	$2x$	10

$$(x + 2)(x + 5) = x \times x + x \times 5 + 2 \times x + 2 \times 5$$
$$= x^2 + 5x + 2x + 10$$
$$= x^2 + 7x + 10$$

> You simplify the final expression by collecting the like terms:
> $5x + 2x = 7x$.

It is like working out the area of a rectangle of length $x + 5$ and width $x + 2$:

Total area $= (x + 2)(x + 5)$
$$= x^2 + 5x + 2x + 10$$
$$= x^2 + 7x + 10$$

	← x →	← 5 →
x	area $= x \times x$ $= x^2$	area $= x \times 5$ $= 5x$
2	area $= 2 \times x$ $= 2x$	area $= 2 \times 5$ $= 10$

Example 17

Expand and simplify:

(a) $(a + 4)(a + 10)$

(b) $(t + 6)(t - 2)$

(a)

×	a	10
a	a^2	$10a$
4	$4a$	40

> Remember: you can use a grid to help.

$$(a + 4)(a + 10) = a \times a + a \times 10 + 4 \times a + 4 \times 10$$
$$= a^2 + 10a + 4a + 40$$
$$= a^2 + 14a + 40$$

> Remember to multiply each term in the first bracket by each term in the second bracket.

(b)

×	t	-2
t	t^2	$-2t$
6	$6t$	-12

> Remember you are multiplying by -2.
> $^+$ve $\times$ $^-$ve $=$ $^-$ve.

$$(t + 6)(t - 2) = t \times t + t \times (-2) + 6 \times t + 6 \times (-2)$$
$$= t^2 - 2t + 6t - 12$$
$$= t^2 + 4t - 12$$

Look again at the last example.

$$(t + 6)\ (t - 2)\qquad = t^2 - 2t + 6t - 12$$
$$= t^2 + 4t - 12$$

The	First terms in each bracket	multiply to give t^2.
The	Outside pair of terms	multiply to give $-2t$.
The	Inside pair of terms	multiply to give $+6t$.
The	Last terms in each bracket	multiply to give -12.

This method is often known as **FOIL** and is another way of expanding brackets.

Exercise 11H

Expand and simplify:

1 $(a + 2)(a + 7)$ **2** $(x + 3)(x + 1)$ **3** $(x + 5)(x + 5)$

4 $(t + 5)(t - 2)$ **5** $(x + 7)(x - 4)$ **6** $(n - 5)(n + 8)$

7 $(x - 4)(x + 5)$ **8** $(p - 4)(p + 4)$ **9** $(x - 9)(x - 4)$

10 $(h - 3)(h - 8)$ **11** $(y - 3)(y - 3)$ **12** $(4 + a)(a + 7)$

13 $(m - 7)(8 + m)$ **14** $(6 + q)(7 + q)$ **15** $(d + 5)(4 - d)$

16 $(8 - x)(3 - x)$ **17** $(x - 12)(x - 7)$ **18** $(y - 16)(y + 6)$

Be careful when there are negative signs – this is where a lot of mistakes are made.

Squaring an expression

You can use the same method of expanding two brackets for examples involving the square of an expression.

To square an expression, write out the bracket twice and expand.

Example 18

Expand and simplify $(x + 4)^2$.

$$(x + 4)^2 = (x + 4)(x + 4)$$

$$(x + 4)(x + 4)$$

$$= x^2 + 4x + 4x + 16$$
$$= x^2 + 8x + 16$$

You need to multiply the expression $(x + 4)$ by itself so write down the bracket twice and expand as you did in Example 17 or use FOIL as in this example.

Notice that you do not just square the x and the 4, there are two other terms in the expansion.

Exercise 11I

1 Expand and simplify:

   (a) $(x + 5)^2$ **(b)** $(x + 6)^2$ **(c)** $(x - 3)^2$

   (d) $(x + 1)^2$ **(e)** $(x - 4)^2$ **(f)** $(x - 5)^2$

   (g) $(x + 7)^2$ **(h)** $(x - 8)^2$ **(i)** $(3 + x)^2$

   (j) $(2 + x)^2$ **(k)** $(5 - x)^2$ **(l)** $(x + a)^2$

> In question 1, see if you can spot the pattern between the terms in the brackets and the final expression.

2 Copy and complete the following by finding the correct number to go in each box:

   (a) $(x + \square)^2 = x^2 + \square x + 36$ **(b)** $(x - \square)^2 = x^2 - \square x + 49$

   (c) $(x + \square)^2 = x^2 + 18x + \square$ **(d)** $(x - \square)^2 = x^2 - 20x + \square$

3 Expand and simplify:

   (a) $(x + 4)(x - 4)$ **(b)** $(x + 5)(x - 5)$ **(c)** $(x + 2)(x - 2)$

   (d) $(x - 11)(x + 11)$ **(e)** $(x - 3)(x + 3)$ **(f)** $(x - 1)(x + 1)$

   (g) $(x + 9)(x - 9)$ **(h)** $(x + a)(x - a)$ **(i)** $(t + x)(t - x)$

> What happens to the x term when you multiply brackets of the form $(x + a)(x - a)$?

Example 19

Expand and simplify $(3x - y)(x - 2y)$

$\times$	x	$-2y$
$3x$	$3x^2$	$-6xy$
$-y$	$-xy$	$2y^2$

$$(3x - y)(x - 2y) = 3x^2 - 6xy - xy + 2y^2$$
$$= 3x^2 - 7xy + 2y^2$$

> Remember to multiply each term in the first bracket by each term in the second bracket.

> Be careful when there are negative signs. This is where a lot of mistakes are made.
> $^+$ve $\times$ $^-$ve $=$ $^-$ve.
> $^-$ve $\times$ $^-$ve $=$ $^+$ve.

Exercise 11J

Expand and simplify:

  1 $(3a + 2)(a + 4)$ **2** $(5x + 3)(x + 2)$ **3** $(2t + 3)(3t + 5)$

  4 $(4y + 1)(2y + 7)$ **5** $(6x + 5)(2x + 3)$ **6** $(4x + 3)(x - 1)$

  7 $(2z + 5)(3z - 2)$ **8** $(y + 1)(7y - 8)$ **9** $(3n - 5)(n + 8)$

  10 $(3b - 5)(2b + 1)$ **11** $(p - 4)(7p + 3)$ **12** $(2z - 3)(3z - 4)$

  13 $(5x - 9)(2x - 1)$ **14** $(2y - 3)(2y - 3)$ **15** $(2 + 3a)(4a + 5)$

16 $(3x + 4)^2$ **17** $(2x - 7)^2$ **18** $(5 - 4x)^2$

19 $(2x + 1)(2x - 1)$ **20** $(3y + 2)(3y - 2)$ **21** $(5n + 4)(5n - 4)$

22 $(3x + 5)(3x - 5)$ **23** $(1 + 2x)(1 - 2x)$ **24** $(3t + 2x)(3t - 2x)$

Can you see the connection between questions 19–24 and question 3 in Exercise 11I?

Examination questions

1 (a) Factorise $7x + 14$
  (b) Expand and simplify $4(m + 3) + 3(2m - 5)$

(3 marks)
AQA, Spec A, 1I, June 2003

2 Expand and simplify
  $5(2a - c) + 4(3a + 2c)$

(2 marks)
AQA, Spec A, 1I, November 2003

3 Solve $3(w - 2) = 9$

(3 marks)
AQA, Spec A, 2I, June 2003

4 (a) Factorise $4c + 12$
  (b) Factorise $x^2 + 5x$

(3 marks)
AQA, Spec A, 1I, November 2003

5 Factorise completely $2a^2 - a$

(2 marks)
AQA, Spec A, 2H, June 2003

6 Factorise completely $12y^2 - 8y$

(2 marks)
AQA, Spec A, 1H, June 2003

7 Factorise completely $3x^2 - 6xy$

(2 marks)
AQA, Spec B, 2I, June 2003

8 Expand and simplify $(x - 3)(x + 1)$

(2 marks)
AQA, Spec A, 1I, November 2003

9 Expand and simplify $(x + 4)^2$

(2 marks)
AQA, Spec A, 2H, June 2003

10 Expand and simplify $(4x - 3)(x + 5)$

(3 marks)
AQA, Spec A, 2I, June 2003

11 Expand and simplify
  (a) $5(2x + 1) - 3(x - 4)$
  (b) $(y - 4)(y - 2)$
  (c) $(2t + 5)(2t - 5)$

(6 marks)
AQA, Spec B, 1H, June 2003

Summary of key points

Expanding brackets (grade D to A)

When you expand brackets you must multiply each term inside the brackets by the term outside the bracket, for example $5(a + 6) = 5a + 30$ and $x(2x - y) = 2x^2 - xy$.

You often have to expand more than one bracket and simplify. For example

$$8(x + 1) - 3(2x - 5) = 8x + 8 - 6x + 15$$
$$= 8x - 6x + 8 + 15$$
$$= 2x + 23$$

Equations sometimes involve brackets.

When dealing with equations involving brackets, you usually expand the brackets first.

When an equation involves fractions, it can be transformed into an equation without fractions by multiplying all terms by the LCM of the numbers in the denominators.

You can use a grid method or FOIL when you multiply two brackets together.

You have to multiply each term in one bracket by each term in the other bracket.

So to expand $(x + 2)(x + 5)$

$$(x + 2)(x + 5) = x \times x + x \times 5 + 2 \times x + 2 \times 5$$
$$= x^2 + 5x + 2x + 10$$
$$= x^2 + 7x + 10$$

Squaring expressions (grade C/B)

When you square an expression, write the bracket next to itself and expand.

For example

$$(x - 7)^2 = (x - 7)(x - 7) = x^2 - 7x - 7x + 49 = x^2 - 14x + 49$$

Inequalities with brackets

Inequalities can involve brackets. Expand the brackets first, then follow the usual method for solving an inequality.

Factorising (grade D/B)

Factorising is the opposite operation to expanding.
A common factor is a factor that divides into all the terms. For example

$$6x + 10 = 2(3x + 5) \quad \text{and} \quad x^2 + 7x = x(x + 7)$$
$$6p^2q^2 - 9pq^3 = 3pq^2(2p - 3q)$$

To factorise completely, use the HCF of the terms.

Notice that the common factor is the term outside the bracket.

This chapter will show you how to:

✔ write a formula from a problem
✔ substitute numbers into expressions and formulae
✔ change the subject of a formula
✔ find approximate solutions to equations using trial and improvement

12.1 Formulae
You will need to know:

● the correct order of operations:

$$\text{Brackets} \rightarrow \text{Indices} \rightarrow \begin{array}{c} \text{Multiplication} \\ \text{Division} \end{array} \rightarrow \begin{array}{c} \text{Addition} \\ \text{Subtraction} \end{array}$$

> **Key words:**
> formula
> variable

A **formula** is a general rule that shows how quantities (or **variables**) are related to each other.

> Formulae is the plural of formula.

For example:

$$v = u + at$$

This is a formula that shows the relationship between an object's final velocity, v, its initial velocity, u, its acceleration, a, and the time it has been moving, t.

Deriving formulae

When solving a problem, it often helps to write a formula to express the problem. Start by deciding on a letter to represent an unknown value.

Example 1

Alex buys x melons.
Each melon costs 45 pence.
Alex pays with a £5 note.
Write a formula for the change, C, in pence, Alex should receive.

$x = $ number of melons

$C = 500 - 45x$

> £5 = 500p
> The melons cost 45p each so the cost, in pence, for x melons is 45x.

Exercise 12A

1 Nilesh buys y mangoes.
Each mango costs 48 pence.
Nilesh pays with a £5 note.
Write a formula for the change, C, in pence, Nilesh should receive.

2 Apples cost r pence each and bananas cost s pence each.
Sam buys 7 apples and 5 bananas.
Write a formula for the total cost, t, in pence, of these fruit.

3 To roast a chicken you allow 45 minutes per kg and then a further 20 minutes.
Write a formula for the time, t, in minutes, to roast a chicken that weighs w kg.

4 To roast lamb you allow 30 minutes plus a further 65 minutes per kg.
Write a formula for the time, t, in minutes, to roast a joint of lamb that weighs w kg.

5 A rectangle has a length of $3x + 1$ and a width of $x + 2$.
Write down a formula for the perimeter, p, of this rectangle.

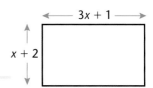

Substitution

This section shows you how to use substitution to find the values of different algebraic expressions.

> Use mathematical operations in the correct order when substituting values into an algebraic expression.

Example 2

If $a = 5$, $b = 4$ and $c = 3$ work out the value of these expressions:

(a) $\dfrac{a + 3}{2}$ **(b)** $3b^2 - 1$ **(c)** $\dfrac{5c + 1}{b}$

(a) $\dfrac{a + 3}{2} = \dfrac{5 + 3}{2}$

$\qquad = 8 \div 2$

$\qquad = 4$

The dividing line acts like a bracket. You must work out the numerator first.

Remember: $\dfrac{8}{2} = 8 \div 2$.

continued ▼

(b) $3b^2 - 1 = 3 \times 4^2 - 1$

$\qquad = 3 \times 16 - 1$

$\qquad = 48 - 1$

$\qquad = 47$

You must work out the indices first ($4^2 = 16$), then the multiplication (3×16), then the subtraction ($48 - 1$).

(c) $\dfrac{5c + 1}{b} = \dfrac{5 \times 3 + 1}{4}$

$\qquad = \dfrac{15 + 1}{4}$

$\qquad = \dfrac{16}{4}$

$\qquad = 4$

Exercise 12B

1 If $r = 5$, $s = 4$ and $t = 3$, work out the value of these expressions:

(a) $\dfrac{r + 3}{2}$ **(b)** $\dfrac{s + 5}{3}$ **(c)** $\dfrac{t + 7}{2}$

(d) $3r^2 + 1$ **(e)** $4t^2 - 6$ **(f)** $2s^2 + r$

$3r^2 = 3 \times r^2 = 3 \times r \times r$

(g) $4(5s + 1)$ **(h)** $t(r + s)$ **(i)** $5(2s - 3t)$

(j) $\dfrac{5t + 1}{s}$ **(k)** $\dfrac{4r - 2}{t}$ **(l)** $\dfrac{3s + t}{r}$

2 Copy and complete the following table:

x	1	2	3	4	5
$x^2 + 2x$			15		

$3^2 + 2 \times 3 = 9 + 6 = 15$

 3 Copy and complete the following table:

x	3	4	3.5	3.7	3.8
$x^3 - x$		60			

Remember: $x^3 = x \times x \times x$
$4^3 - 4 = 64 - 4 = 60$

4 If $A = 6$, $B = -4$, $C = 3$ and $D = 30$, work out the value of these expressions:

(a) $D(B + 7)$ **(b)** $A(B + 1)$ **(c)** $A^2 + 2B + C$

(d) $\dfrac{2A + 3}{C}$ **(e)** $\dfrac{4B + D}{2}$ **(f)** $\dfrac{A^2 + 3B}{C}$

Substituting into formulae

Example 3

A formula for working out acceleration is:

$$a = \frac{v - u}{t}$$

where v is the final velocity, u is the initial velocity and t is the time taken.

Work out the value of a when $v = 50$, $u = 10$, $t = 8$.

$$a = \frac{v - u}{t}$$

$$= \frac{50 - 10}{8}$$

$$= 40 \div 8$$

$$= 5$$

Remember: the line for division acts like a bracket. You must work out the numerator first.

Example 4

A formula for working out distance travelled is:

$$s = ut + \tfrac{1}{2}at^2$$

where u is the initial velocity, a is the acceleration and t is the time taken.

Work out the value of s when $u = 3$, $a = 8$, $t = 5$.

$$s = ut + \tfrac{1}{2}at^2$$

$$= 3 \times 5 + \tfrac{1}{2} \times 8 \times 5^2$$

$$= 15 + 4 \times 25$$

$$= 15 + 100$$

$$= 115$$

Do $5^2 = 25$ first;
then $3 \times 5 = 15$ and $\tfrac{1}{2} \times 8 = 4$;
then $4 \times 25 = 100$;
then $15 + 100 = 115$.

Exercise 12C

1 Use the formula $a = \dfrac{v - u}{t}$ to work out the value of a when:

(a) $v = 15$, $u = 3$, $t = 2$

(b) $v = 29$, $u = 5$, $t = 6$

(c) $v = 25$, $u = 7$, $t = 3$

(d) $v = 60$, $u = 10$, $t = 4$.

2 A person's Body Mass Index, b, is calculated using the formula

$$b = \frac{m}{h^2}$$

where m is their mass in kilograms and h is their height in metres.

Work out the value of b when:

Round each answer to the nearest whole number.

(a) $m = 70, h = 1.8$ **(b)** $m = 38, h = 1.4$

(c) $m = 85, h = 1.9$ **(d)** $m = 59, h = 1.7$.

3 The formula for the area of a trapezium is

$$A = \tfrac{1}{2}(a + b)h$$

Work out the value of A when:

(a) $a = 10, b = 6, h = 4$ **(b)** $a = 13, b = 9, h = 8$

(c) $a = 9, b = 6, h = 4$ **(d)** $a = 15, b = 10, h = 6$.

4 Use the formula $s = ut + \tfrac{1}{2}at^2$ to work out the value of s when:

(a) $u = 3, a = 10, t = 2$ **(b)** $u = 7, a = 6, t = 5$

(c) $u = 2.5, a = 5, t = 4$ **(d)** $u = -4, a = 8, t = 3$.

5 A formula for working out the velocity of a car is:

$$v = \sqrt{u^2 + 2as}$$

where u is the initial velocity, a is the acceleration and s is the distance travelled.

Work out the value of v when:

(a) $u = 3, a = 4, s = 5$ **(b)** $u = 6, a = 8, s = 4$

(c) $u = 9, a = 10, s = 2$ **(d)** $u = 7, a = 4, s = 15$.

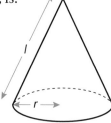

6 A formula for the surface area of a cone, including the base, is:

surface area of cone $= \pi r(r + l)$

where r is the radius and l is the slant height.

Work out the surface area of a cone with dimensions:

(a) $r = 2\,\text{cm}, l = 13\,\text{cm}$ **(b)** $r = 4\,\text{cm}, l = 10\,\text{cm}$.

Give your answers to 3 s.f.

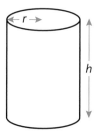

7 A formula for the total surface area of a cylinder is:

surface area of cylinder $= 2\pi r(r + h)$

where r is the radius and h is the height.

Work out the surface area of a cylinder with dimensions:

(a) $r = 3.7\,\text{cm}, h = 10.8\,\text{cm}$ **(b)** $r = 4.2\,\text{cm}, h = 4.1\,\text{cm}$.

Give your answers to 3 s.f.

Using formulae

Example 5

The perimeter of a rectangle is given by $P = 2l + 2w$, where l is the length and w is the width.

Work out the value of l when $P = 24$ and $w = 5$.

$$P = 2\ell + 2w$$
$$24 = 2\ell + 2 \times 5$$
$$24 = 2\ell + 10$$
$$24 - 10 = 2\ell$$
$$14 = 2\ell$$
$$7 = \ell$$

Substitute the values you know, of P and w, into the formula, then solve the equation to find l.

Exercise 12D

1 The formula for the area of a rectangle is $A = lw$, where l is the length and w is the width.
Work out the value of w when:

(a) $A = 12$ and $l = 4$

(b) $A = 36$ and $l = 9$

(c) $A = 42$ and $l = 7$

(d) $A = 60$ and $l = 15$.

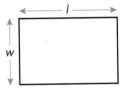

2 The formula for the voltage, V, in an electrical circuit is $V = IR$, where I is the current and R is the resistance.
Work out the following:

(a) The value of R when $V = 18$ and $I = 2$.

(b) The value of R when $V = 35$ and $I = 5$.

(c) The value of I when $V = 240$ and $R = 30$.

(d) The value of I when $V = 240$ and $R = 40$.

3 The perimeter of a rectangle is given by

$$P = 2l + 2w$$

where l is the length and w is the width.
Use the formula to:

(a) find l when $P = 18$ and $w = 4$.

(b) find l when $P = 32$ and $w = 7$.

(c) find w when $P = 60$ and $l = 17$.

(d) find w when $P = 50$ and $l = 13.5$.

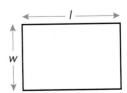

4 Use the formula $v = u + at$

    (a) to find u when $v = 30$, $a = 8$, $t = 3$

    (b) to find a when $v = 54$, $u = 19$, $t = 7$

    (c) to find t when $v = 60$, $u = 15$, $a = 5$

    (d) to find u when $v = 20$, $a = 7$, $t = 4$.

12.2 Changing the subject of a formula

Key words:
subject
rearrange

You will need to know that:
- addition and subtraction are inverse operations
- multiplication and division are inverse operations
- squaring and finding the square root are inverse operations

The **subject** of a formula appears only once, on its own, on one side of the formula.

In the formula $v = u + at$ the variable v is called the subject of the formula.

P is the subject of the formula $P = 2l + 2w$.

You can use the formula to find the value of P.

P is on its own on one side of the formula.

You can **rearrange** a formula to make a different variable the subject.

Example 6

Rearrange $d = a + 8$ to make a the subject.

$$d = a + 8$$
$$d - 8 = a$$

You need to have a on its own on one side.

Subtract 8 from both sides as you would when solving an equation.

Example 7

Make x the subject of the formula $y = 5x - 2$.

$$y = 5x - 2$$
$$y + 2 = 5x$$
$$\frac{y + 2}{5} = x$$

Add 2 to both sides to leave $5x$ on its own.

Then divide both sides by 5 to leave x on its own.

Example 8

Rearrange $P = 4g + 2h$ to make g the subject.

$P = 4g + 2h$

$P - 2h = 4g$

Subtract $2h$ from both sides.

$\dfrac{P - 2h}{4} = g$

Divide both sides by 4.

Example 9

Rearrange $V = \sqrt{(w + y)}$ to make w the subject.

$V = \sqrt{(w + y)}$

$V^2 = w + y$

Square both sides first.

$V^2 - y = w$

Then subtract y from both sides.

Example 10

Rearrange $p = \dfrac{q^2}{r} - s$ to make q the subject.

$p = \dfrac{q^2}{r} - s$

$p + s = \dfrac{q^2}{r}$

Add s to both sides to leave the term with q on its own.

$r(p + s) = q^2$

Multiply both sides by r.

$\pm \sqrt{r(p + s)} = q$

Square root both sides.

Exercise 12E

1 Rearrange each of these formulae to make a the subject:

(a) $d = a + 8$ (b) $t = a + 12$

(c) $k = a - 6$ (d) $w = a - 7$

2 Rearrange each of these formulae to make w the subject:

(a) $P = 4w$ (b) $a = 3w$

(c) $A = lw$ (d) $h = kw$

3 Make x the subject of each of these formulae:

(a) $y = 5x - 6$ (b) $y = 4x - 7$

(c) $y = 2x + 1$ (d) $y = 6x + 5$

Use Example 7 to help.

4 The formula $F = 1.8C + 32$ can be used to convert degrees Celsius, °C, to degrees Fahrenheit, °F.

(a) Convert 15°C to degrees Fahrenheit.

(b) Rearrange the formula to make C the subject.

(c) Use your new formula to convert 68°F to °C.

(d) What is 82°F in degrees Celsius to the nearest degree? You must show your working.

5 Make r the subject of each of these formulae:

(a) $p = 4r + 2t$ (b) $v = 7r + 4h$ (c) $w = 3r - 2s$ (d) $y = 6r - 5p$

> Use Example 8 to help.

6 A formula used to calculate velocity is:

$$v = u + at$$

where u is the initial velocity, a is the acceleration and t is the time taken.

(a) Rearrange the formula to make a the subject.

(b) Rearrange the formula to make t the subject.

7 Rearrange these formulae to make a the subject:

(a) $b = \frac{1}{2}a + 6$ (b) $b = \frac{1}{2}a + 7$

(c) $b = \frac{1}{3}a - 1$ (d) $b = \frac{1}{4}a - 3$

(e) $b = 2(a + 1)$ (f) $b = 3(a - 5)$

> Multiply $\frac{1}{2}a$ by 2 to get a.

8 Rearrange these formulae to make x the subject:

(a) $3(x + y) = 5y$ (b) $2(x - y) = y + 5$

(c) $z = \dfrac{x}{y} - 5$ (d) $3p = \dfrac{2x}{q} - s$

9 Rearrange these formulae to make w the subject:

(a) $K = \sqrt{w + t}$ (b) $A = \sqrt{w - a}$ (c) $h = 2\sqrt{w} + 1$ (d) $T = \sqrt{wr} + 5$

> Use Example 9 to help.

10 Rearrange these formulae to make r the subject:

(a) $t = \dfrac{r^2}{g} - m$ (b) $h = \dfrac{r^2}{4} + 3a$

(c) $V = \frac{1}{3}\pi r^2 h$ (d) $A = \pi(r^2 - s^2)$

> Use Example 10 to help.

11 A formula for the total surface area of a cylinder is:

$$A = 2\pi r(r + h)$$

where r is the radius and h is the height.
Rearrange the formula to make h the subject.

12 A formula for the period of a pendulum is:

$$T = 2\pi\sqrt{\frac{l}{g}}$$

Rearrange the formula to make l the subject.

> The period T is the time for one complete swing; l is the length and g is a constant.

12.3 Trial and improvement

Key words:
trial and improvement

You will need to know how to:
- substitute values into algebraic expressions
- use your calculator effectively

You cannot always solve equations using algebraic methods (like those you learned in Chapter 5).

Trial and improvement is a method for finding approximate solutions to complex equations.

1 Estimate what the answer might be.

↓

2 Work out the value of the expression using your estimate.

↓

3 Is your answer too big or too small?

↓

4 Use the answer to step 3 to improve your estimate.

↓

5 Work out the value of the expression using this improved estimate.

↓

6 Is this answer too big or too small?

↓

7 Improve your estimate again. Continue steps 2–6 until you have an answer to the required degree of accuracy.

Example 11

Use **trial and improvement** to find the solution to the equation

$$x^3 + x = 80$$

Give your answer to 1 decimal place.

x	$x^3 + x$	Comment
4	$4^3 + 4 = 68$	too small
5	$5^3 + 5 = 130$	too big
4.5	$4.5^3 + 4.5 = 95.625$	too big
4.2	$4.2^3 + 4.2 = 78.288$	too small
4.3	$4.3^3 + 4.3 = 83.807$	too big
4.25	$4.25^3 + 4.25 = 81.015...$	too big
	$x = 4.2$ to 1 decimal place	

First try a value for x which you think is close.
$4^3 = 64$ so try $x = 4$.

Try $x = 5$ next to find out which two whole numbers the solution is between.

As 5 is too big the solution must be between 4 and 5. Try $x = 4.5$ next.

Still too big so try between 4 and 4.5.

Too small, but close. Try $x = 4.3$ next.

The solution must be between 4.2 and 4.3.
You need to know which of these is closest so you **must** try halfway between, $x = 4.25$.

4.25 is too big so the solution must be closer to 4.2 than to 4.3.

Exercise 12F

1 Use trial and improvement to find the solution to the equation

$$x^3 + x = 100$$

Give your answer to 1 decimal place.
Copy and continue this table to help you.

x	$x^3 + x$	Comment
4	68	too small
5		

2 Use trial and improvement to find the solution to the equation

$$x^3 + x = 25$$

Give your answer to 1 decimal place.

Use a table like that in Example 11 and question 1.

3 Use trial and improvement to find the solution to the equation

$$x^3 - x = 30$$

Give your answer to 1 decimal place.

You must show your working in the exam – otherwise you will not be given any marks!

4 Use trial and improvement to find the solution to the equation

$$x^3 + 2x = 240$$

Give your answer to 1 decimal place.

5 Use trial and improvement to find the solution to the equation

$$x^3 - x^2 = 1000$$

Give your answer to 1 decimal place.

Examination questions

1 Sam buys x packets of sweets.
Each packet of sweets costs 22 pence.
Sam pays with a £5 note.
Write down an expression for the change, in pence, Sam should receive.   *(2 marks)*
AQA, Spec A, 1I, June 2003

2 Parveen is using trial and improvement to find a solution to the equation

$$x^3 + 7x = 30$$

This table shows her first two trials.
Continue the table to find a solution to the equation.
Give your answer to 1 decimal place.

x	$x^3 + 7x$	Comment
2	22	Too small
3	48	Too big

(3 marks)
AQA, Spec A, 2I, June 2003

3 Make x the subject of the formula

$$w = x^2 + y$$

(2 marks)
AQA, Spec A, 2I, June 2003

4 Make t the subject of the formula $u = \dfrac{t}{3} + 5$ *(2 marks)*
AQA, Spec A, 2I, November 2003

5 Ken has made y cakes.
He puts them into packs of 5.
He has n packs of cakes and 2 cakes left over.
Write down an equation which shows the relationship between y and n.   *(2 marks)*
AQA, Spec B, 2I, November 2003

Summary of key points

Formulae written using letters and symbols (up to grade D/C)

An algebraic formula uses letters to show the relationship between variable quantities. For example:

$$v = u + at$$ where v is the final velocity
u is the initial velocity
a is the acceleration
t is the time.

Substitution (grade D/C)

Use mathematical operations in the correct order when substituting values into an algebraic expression.

For example:
Work out the value of s when $u = 3$, $a = 8$, $t = 5$

$$s = ut + \tfrac{1}{2}at^2$$
$$= 3 \times 5 + \tfrac{1}{2} \times 8 \times 5^2$$
$$= 15 + 4 \times 25$$
$$= 15 + 100$$
$$= 115$$

Changing the subject of a formula (grade C/B)

The subject of a formula appears only once, on its own, on one side of the formula.

In the formula

$$v = u + at$$

the variable v is called the subject of the formula.

You can rearrange a formula to make a different variable the subject:

$v = u + at$ can be rearranged to $a = \dfrac{(v - u)}{t}$

Trial and improvement (grade C)

You can use a trial and improvement method to find an approximate solution to a complex equation.

13 Indices

This chapter will show you how to:

✔ calculate powers and roots
✔ revise, use and develop the laws of indices
✔ calculate using standard form

13.1 Calculating powers and roots

Key words:
index notation

You will need to know:

• how to use index (power) notation.

In Chapter 1 you revised how to calculate squares and cubes.
Expressions such as 4^2 and 5^3 are examples of **index notation** .
This can be used to simplify numerical or algebraic expressions
(for example $5 \times 5 \times 5 = 5^3$).
You can calculate any number raised to any power.
For example, $3^4 = 3 \times 3 \times 3 \times 3 = 81$
and $2^7 = 2 \times 2 \times 2 \times 2 \times 2 \times 2 \times 2 = 128$.
The above can be done without using a calculator, but a calculator
is useful when dealing with decimal numbers.
There is a key on your calculator which will work out powers of
any number.

Remember that calculators
vary and this key
sometimes has the x and
the y the other way round.
It may be a 'second
function' key.

To work out $(2.3)^6$ key in:

which gives an answer of 148.035889

Negative indices

Indices can be negative, for example, 3^{-4}. You work this out in the
same way as for positive indices. Key in:

There is more about
negative indices in Section
13.3.

This gives an answer of 0.01234568...

Roots

Calculators usually have a key for working out roots. It is often the
second function of the calculator key used for powers.

Your calculator has a key
for working out powers

... and a key to work out roots

So to work out $\sqrt[5]{360}$, you need to key:

5   3   6   0   =

which gives an answer of 3.2453...

Exercise 13A

1 Find the value of the following, giving your answer to 3 s.f.

(a) 2^{12} (b) 7.2^4 (c) -8^7 (d) $(-2.9)^6$

(e) $\sqrt[4]{81}$ (f) $\sqrt[6]{300}$ (g) $\sqrt[4]{100}$ (h) $\sqrt[5]{-36.25}$

(i) 3.7^{-3} (j) 6^{-2}

2 Which has the greater value of each pair?

(a) 4^5 or 5^4 (b) 3^9 or 9^3 (c) $\sqrt[4]{4.9}$ or $\sqrt[5]{9.4}$

3 Complete the following calculations, giving your answer to 3 s.f.

(a) $1.6^5 + 3.4^4$ (b) $2.6^7 - \sqrt[5]{16\,807}$ (c) $\dfrac{(4.9)^8}{\sqrt[6]{4096}}$

4 A student calculates a number as 3.6^{15}.
What is this value (to 3 s.f.)?

5 Calculate the following, giving your answers to 3 s.f.

(a) $5^5 \times 3^4$ (b) $5.6^{-6} \times \sqrt[4]{4.6}$ (c) $(-8.2)^5 \div -4.1$

(d) $\dfrac{-7^5 \times -6^6}{\sqrt[5]{9000}}$ (e) $\dfrac{2^9 + 1^{-12}}{\sqrt[6]{52 \times 14 + 1}}$ (f) $\dfrac{(-3.8)^4}{-2.9}$

6 Find the value of the following where $x = 4$ and $y = 5$:

(a) x^5 (b) y^6 (c) 4^x

(d) 7^y (e) x^y (f) $7x^y$

7 Sally is given a choice for her birthday. She can have
2 to the power of the number of days in April (in pence),
or £20 multiplied by the number of days in April.
Which one should she take?

13.2 Multiplying and dividing expressions involving indices

You will need to know:
- how to simplify fractions

Example 1

Simplify:

(a) $4^3 \times 4^4$ (b) $r^3 \times r^2$ (c) $4h^2 \times 5h^4$

(a) $4^3 \times 4^4 = 4 \times 4 \times 4 \times 4 \times 4 \times 4 \times 4$

       $= 4^7$

(b) $r^3 \times r^2 = r \times r \times r \times r \times r$

       $= r^5$

(c) $4h^2 \times 5h^4 = 4 \times h \times h \times 5 \times h \times h \times h \times h$

            $= 4 \times 5 \times h \times h \times h \times h \times h \times h$

            $= 20 \times h^6$

            $= 20h^6$

> Remember $r^3 = r \times r \times r$.

> The order in which you multiply does not matter, so you can rewrite this with the numbers first then the letters.

Example 2

Simplify:

(a) $6^5 \div 6^2$ (b) $c^3 \div c^4$ (c) $10b^2 \div 5b^4$

(a) $6^5 \div 6^2 = \dfrac{6 \times 6 \times 6 \times 6 \times 6}{6 \times 6}$

       $= \dfrac{6 \times 6 \times 6 \times \cancel{6} \times \cancel{6}}{\cancel{6} \times \cancel{6}}$

       $= 6 \times 6 \times 6$

       $= 6^3$

(b) $c^3 \div c^4 = \dfrac{c \times c \times c}{c \times c \times c \times c}$

       $= \dfrac{\cancel{c}^1 \times \cancel{c}^1 \times \cancel{c}^1}{c \times \cancel{c}^1 \times \cancel{c}^1 \times \cancel{c}^1}$

       $= \dfrac{1}{c}$

(c) $10b^2 \div 5b^4 = \dfrac{10 \times b \times b}{5 \times b \times b \times b \times b}$

       $= \dfrac{10 \times \cancel{b}^1 \times \cancel{b}^1}{5 \times b \times b \times \cancel{b}^1 \times \cancel{b}^1}$

> Write the division as a fraction.

> Because the top and bottom of the fraction are both just multiplications, you can cancel any terms which are on both the top and bottom.

continued ▼

$$= \frac{10}{5 \times b \times b}$$

$$= \frac{10}{5b^2}$$

> Remember to divide the 10 by the 5.

$$= \frac{2}{b^2}$$

Exercise 13B

1 Simplify each of the following:
   (a) $12^2 \times 12^7$ (b) $45^3 \times 45^4$ (c) $6^2 \times 6^{10}$
   (d) $5^2 \times 5^3 \times 5^6$ (e) $7^1 \times 7^2 \times 7^5 \times 7^8$

2 Simplify each of the following:
   (a) $a^3 \times a^2$ (b) $b^4 \times b^3$ (c) $c^2 \times c^4$ (d) $d^5 \times d^5$
   (e) $x^3 \times x$ (f) $y^2 \times y$ (g) $z \times z^4$ (h) $w \times w^3$

> Remember $x = x^1$.

3 Simplify each of the following:
   (a) $2a^3 \times 3a^4$ (b) $5b^4 \times 3b^5$
   (c) $4c^2 \times c^6$ (d) $3d^4 \times 3d^4$

> Use Example 1(c) to help.

4 Can you see a quicker way to answer questions 2 and 3? If you can, describe it using an example to help.

5 Simplify each of the following:
   (a) $p^5 \div p^2$ (b) $q^6 \div q^2$ (c) $r^4 \div r^3$ (d) $s^6 \div s^3$
   (e) $e^2 \div e^5$ (f) $f^2 \div f^6$ (g) $g^3 \div g^4$ (h) $h \div h^6$

> Use Example 2(b) to help.

6 Simplify each of the following:
   (a) $5x^5 \div x^2$ (b) $8y^6 \div 4y^2$
   (c) $12r^4 \div 3r^3$ (d) $2s^2 \div 10s^5$

> Use Example 2(c) to help.

7 Can you see a quicker way to answer questions 5 and 6? If you can, describe it using an example to help.

> Remember $\frac{2}{10}$ simplifies to $\frac{1}{5}$.

13.3 Laws of indices

> **Key words:**
> index laws
> laws of indices

You may have noticed by now some of the rules that make simplifying expressions involving indices easier to do. They are known as the **index laws** or **laws of indices** .

To multiply powers of the same number or variable, add the indices.

$$4^3 \times 4^2 = 4^{3+2}$$
$$= 4^5$$

$$x^3 \times x^2 = x^{3+2}$$
$$= x^5$$

$$4^3 \times 4^2 = 4 \times 4 \times 4 \times 4 \times 4$$
$$= 4^{3+2}$$
$$= 4^5$$

In general: $x^n \times x^m = x^{n+m}$

To divide powers of the same number or variable, subtract the indices.

$$6^5 \div 6^3 = 6^{5-3}$$
$$= 6^2$$

$$t^5 \div t^3 = t^{5-3}$$
$$= t^2$$

$$6^5 \div 6^3 = \frac{6 \times 6 \times 6^1 \times 6^1 \times 6^1}{6^1 \times 6^1 \times 6^1}$$
$$= 6^{5-3}$$
$$= 6^2$$

In general: $x^n \div x^m = x^{n-m}$

Remember: the multiplication and division laws only work when the powers are of the *same* number or variable.

Example 3

Simplify:

(a) $5^2 \times 5^3$

(b) $3a^4 \times 5a^3$

(c) $g^3 \times g \times g^2$

(a) $5^2 \times 5^3 = 5^{2+3}$
$$= 5^5$$

(b) $3a^4 \times 5a^3 = 3 \times a^4 \times 5 \times a^3$
$$= 3 \times 5 \times a^4 \times a^3$$
$$= 15 \times a^{4+3}$$
$$= 15a^7$$

(c) $g^3 \times g \times g^2 = g^3 \times g^1 \times g^2$
$$= g^{3+1+2}$$
$$= g^6$$

Multiply the numbers first then multiply the powers of *a*.

Remember that any value raised to the power 1 is itself. E.g. $7^1 = 7$; $5 = 5^1$; $x^1 = x$; $h = h^1$.
You don't normally write the power 1.

Example 4

Simplify:

(a) $3^6 \div 3^2$ (b) $x^3 \div x^5$ (c) $12b^5 \div 4b^3$ (d) $\dfrac{t^4 \times t}{t^3}$

(a) $3^6 \div 3^2 = 3^{6-2}$

$\qquad\qquad = 3^4$

(b) $x^3 \div x^5 = x^{3-5}$

$\qquad\qquad = x^{-2}$

You can have negative powers as well as positive ones.

(c) $12b^5 \div 4b^3 = \dfrac{12 \times b^5}{4 \times b^3}$

$\qquad\qquad = \dfrac{12}{4} \times \dfrac{b^5}{b^3}$

$\qquad\qquad = 3 \times b^{5-3}$

$\qquad\qquad = 3 \times b^2$

$\qquad\qquad = 3b^2$

As with the multiplication example, we can deal with the numbers first, $12 \div 4 = 3$; then deal with the letters, $b^5 \div b^3 = b^2$.

(d) $\dfrac{t^4 \times t}{t^3} = \dfrac{t^4 \times t^1}{t^3}$

$\qquad\quad = \dfrac{t^5}{t^3}$

$\qquad\quad = t^{5-3}$

$\qquad\quad = t^2$

Simplify the top first, $t^4 \times t = t^5$; then deal with the division, $t^5 \div t^3 = t^2$.

Exercise 13C

Simplify each of the following:

1 (a) $6^3 \times 6^4$ (b) $13^4 \times 13^8$ (c) $25^8 \times 25^9$
    (d) $5^4 \times 5^6 \times 5^7$ (e) $33 \times 33^6 \times 33^8$ (f) $12^3 \times 12^6 \times 12^9$

2 (a) $m^3 \times m^2$ (b) $a^4 \times a^3$ (c) $n^2 \times n^4$
    (d) $u^5 \times u^5$ (e) $t^3 \times t^6$

3 (a) $d^3 \times d$ (b) $a^2 \times a$ (c) $r \times r^4$
    (d) $e \times e^3$ (e) $t \times t^6$

4 (a) $2h^3 \times 3h^4$ (b) $5e^4 \times 3e^5$ (c) $4g^2 \times g^6$
    (d) $3r \times 3r^4$ (e) $6e^3 \times 4e$

5 (a) $a^5 \div a^2$ (b) $t^6 \div t^2$ (c) $e^4 \div e^3$
    (d) $s^6 \div s^3$ (e) $t^2 \div t^2$

6 (a) $e^2 \div e^5$ (b) $f^2 \div f^6$ (c) $g^3 \div g^4$

    (d) $h \div h^4$ (e) $w \div w^6$

7 (a) $5x^5 \div x^2$ (b) $6y^6 \div 2y^2$ (c) $12r^4 \div 4r^3$

    (d) $2s^2 \div 8s^5$ (e) $8t^4 \div 4t$

8 (a) $h^3 \times h^4 \times h^2$ (b) $e^4 \times e \times e^3$ (c) $4c^2 \times c^6 \times 6c$

9 (a) $\dfrac{d^4 \times d}{d^3}$ (b) $\dfrac{a^2 \times 3a^2}{a}$ (c) $\dfrac{4r \times 3r^3}{6r^2}$

Powers of 1 and 0

> **Any number or variable raised to the power 1 is equal to the number or variable itself.** Examples: $3^1 = 3$, $x^1 = x$.
>
> **Any number or variable raised to the power 0 is equal to 1.** Examples: $3^0 = 1$, $x^0 = 1$.

The following shows how the rule for a power of zero is obtained.

The division rule is 'subtract indices'. Look what happens when you apply it to this example:

$$x^7 \div x^7 = x^{7-7} = x^0$$

But when you divide anything by itself you always get an answer of 1, which means that $x^0 = 1$.

Example 5

Work out the value of:

(a) 4^0 (b) $8k^0$ (c) $(3wz)^0$

(a) $4^0 = 1$

(b) $8k^0 = 8 \times k^0 = 8 \times 1 = 8$

(c) $(3wz)^0 = 1$

In (b) it is only the k that is raised to power 0 so the answer is $8 \times 1 = 8$.

In (c) everything inside the bracket is raised to power 0 so the answer is simply 1. (There is no term outside the bracket.)

Key words:
reciprocal

Negative indices and reciprocals

In the expression $\dfrac{1}{x^7}$, you can replace 1 with x^0. Then follow the index law for dividing:

$$\frac{1}{x^7} = 1 \div x^7 = x^0 \div x^7 = x^{0-7} = x^{-7}$$

This shows you that the **reciprocal** of a power is the same as that quantity raised to the equivalent negative power.

Remember: the reciprocal of x^n means '1 divided by x^n'.

$$\frac{1}{x^n} = x^{-n}$$

For example

$$\frac{1}{3^4} = 3^{-4} \qquad \frac{1}{x^4} = x^{-4} \qquad \frac{1}{y} = \frac{1}{y^1} = y^{-1} \qquad a^{-5} = \frac{1}{a^5}$$

x^{-n} can be read as the reciprocal of x^n, or $\dfrac{1}{x^n}$.

Example 6

Work out the value of:

(a) 9^{-2} (b) $3^{-3} \times 6^2$ (c) $(0.2)^{-4}$

(a) $9^{-2} = \dfrac{1}{9^2} = \dfrac{1}{81}$

(b) $3^{-3} \times 6^2 = \dfrac{1}{3^3} \times 6^2 = \dfrac{1}{27} \times 36 = \dfrac{36}{27} = \dfrac{4}{3} = 1\frac{1}{3}$

(c) $(0.2)^{-4} = \dfrac{1}{(0.2)^4} = \dfrac{1}{0.0016} = \dfrac{10\,000}{16} = 625$

or

$(0.2)^{-4} = \left(\frac{1}{5}\right)^{-4} = \left(\frac{5}{1}\right)^4 = 5^4 = 625$

The fraction in (b) has been written in its simplest form then as a mixed number.

$0.0016 = \dfrac{16}{10\,000}$

Inverting the fraction and changing the negative power to a positive one is a very useful skill. It means you don't need a calculator.

Exercise 13D

1 Work out the value of the following:

(a) 12^1 (b) 7.6^0 (c) 1^{36} (d) $2.6^0 \times 9.7^1$

12 1 1 $9,7$

2 Write each of these in fraction form.

(a) 4^{-2} $\frac{1}{4^2}$ (b) 3^{-5} $\frac{1}{3^5}$ (c) 5^{-3} $\frac{1}{5^3}$ (d) 2^{-10} $\frac{1}{2^{10}}$

$\frac{1}{16}$ $\frac{1}{243}$ $\frac{1}{125}$ $\frac{1}{1024}$

3 Find the value of your answers to question 2, giving your answers in fraction form, e.g. $2^{-2} = \dfrac{1}{2^2} = \dfrac{1}{4}$.

4 Write each of these in negative index form.

(a) $\dfrac{1}{3^2}$ 3^{-2} (b) $\dfrac{1}{6^3}$ 6^{-3} (c) $\dfrac{1}{10^9}$ 10^{-9} (d) $\dfrac{1}{8^4}$ 8^{-4}

5 Work out:

(a) $4^{-2} \times 8^2$ (b) $7^3 \times 10^{-2}$ (c) $9^{-3} \div 3^{-2}$ (d) $(0.4)^4 \div 10^{-5}$

4 $3,43$ $\frac{1}{81}$ $\frac{2^4}{5^4} \cdot 10^5 =$

 $= 2^4 \cdot 5 \cdot 2^5 = 2^9 \cdot 5 = 2560$

6 Work out the value of the following and then write a sentence about what you notice.

$$\dfrac{1}{2^{-1}} \qquad \dfrac{1}{2^{-2}} \qquad \dfrac{1}{2^{-3}} \qquad \dfrac{1}{2^{-4}}$$

[handwritten: 2, 4, 8, 16]

Example 7

Write d^{-3} as a fraction and find its value when $d = 4$.

(a) $d^{-3} = \dfrac{1}{d^3}$

when $d = 4$ $d^{-3} = \dfrac{1}{d^3}$

$= \dfrac{1}{4^3}$

$= \dfrac{1}{64}$

Example 8

Write each of these using negative indices.

(a) $\dfrac{6}{a}$

(b) $\dfrac{12x^2}{3x^4}$

(a) $\dfrac{6}{a} = \dfrac{6}{a^1}$

$= 6a^{-1}$

(b) $\dfrac{12x^2}{3x^4} = \dfrac{12}{3} \times x^{2-4}$

$= 4x^{-2}$

Note that $a^{-1} = \dfrac{1}{a}$. This is very useful to remember.

Remember:
$x^n \div x^m = x^{n-m}$

Exercise 13E

1 Write each of these as a fraction:

(a) x^{-3} (b) x^{-4} (c) x^{-1} (d) x^{-5}

(e) $5x^{-3}$ (f) $3x^{-2}$ (g) $6x^{-4}$ (h) $4x^{-1}$

2 Work out the value of each of the expressions in question 1 when $x = 2$.

[handwritten notes in right margin:]

① $5^{-2} \times 5^4 = 25 = 5^2$

$(4^3)^{-2} = \dfrac{1}{1024} = \dfrac{1}{4^5}$

$6^3 / 6^5 = \dfrac{1}{36} = \dfrac{1}{6^2}$

$5^6 \times 5^9 = 5^{15}$

$6^{-3} / 6^{-5} = 6^2$

$(47)^0 = 1$

② $100^{\frac{1}{2}} = 10$

$121^{\frac{1}{2}} = 11$

$32^{\frac{1}{5}} = 2$ $8^{-\frac{1}{3}} = \dfrac{1}{2}$

$\left(\dfrac{81}{125}\right)^{\frac{1}{2}} = \dfrac{9}{5\sqrt{5}}$ $125^{\frac{2}{3}} = 25$

$36^{-\frac{3}{2}} = \dfrac{1}{216}$ $16^{-\frac{5}{4}} = \dfrac{1}{32}$

③ $5^{-2} = \dfrac{1}{5^2} = \dfrac{1}{25}$

$4^{-1} = \dfrac{1}{4}$

$7^4 : 7^5 = \dfrac{1}{7}$

3 Write each of these using negative indices:

(a) $\dfrac{1}{k^2}$ (b) $\dfrac{1}{e^5}$ (c) $\dfrac{1}{g}$

(d) $\dfrac{1}{x^n}$ (e) $\dfrac{6}{p}$ (f) $\dfrac{5}{a^3}$

(g) $\dfrac{4}{d^2}$ (h) $\dfrac{8}{h^m}$ (i) $\dfrac{3x^2}{x^4}$

(j) $\dfrac{12a^3}{3a^7}$ (k) $\dfrac{9b}{3b^2}$ (l) $\dfrac{4y^2}{6y^5}$

Use Example 8 to help.

4 Simplify each of these and write your answer:

(i) using negative indices

(ii) as a fraction.

(a) $e^2 \div e^5$ (b) $8s^3 \div s^4$ (c) $\dfrac{20f}{5f^3}$

(d) $\dfrac{12t^3}{15t^7}$ (e) $n^2 \times n^{-4}$ (f) $m^{-1} \times 3m^{-2}$

(g) $\dfrac{2a^2 \times 3a^2}{3a^6}$ (h) $\dfrac{5q \times 3q^2}{10q^3 \times q}$

Problems with more than one variable

Example 9

Simplify:

(a) $3xy^2 \times 5xy$

(b) $\dfrac{10a^3b^2}{2a^2b^2}$

(a) $3xy^2 \times 5xy = 3 \times x \times y^2 \times 5 \times x \times y$

$= 3 \times 5 \times x \times x \times y^2 \times y$

$= 15 \times x^2 \times y^3$

$= 15x^2y^3$

Deal with the numbers first, then each of the different letters in turn, first the xs then the ys.

Your answer does not need to show the intermediate lines of working.

(b) $\dfrac{10a^3b^2}{2a^2b^2} = \dfrac{10 \times a^3 \times b^2}{2 \times a^2 \times b^2}$

$= \dfrac{10}{2} \times \dfrac{a^3}{a^2} \times \dfrac{b^2}{b^2}$

$= 5 \times a^1 \times 1$

$= 5a$

Deal with the numbers first, then each of the different letters in turn, first the as then the bs.

$\dfrac{a^3}{a^2} = a^{3-2} = a^1$

$\dfrac{b^2}{b^2} = b^{2-2} = b^0 = 1$

Exercise 13F

1 Simplify these expressions:

(a) $4ab^2 \times 2ab$ (b) $3f^2g \times 7f^3g^2$

(c) $5xy^2 \times 4x^3y$ (d) $6p^4q^3 \times 6p^2q^2$

(e) $7r^2s \times 4s^5 \times 2r$ (f) $3c^3 \times 5cd \times 5d^4$

> Write the answers using negative indices *and* as algebraic fractions, where appropriate.

2 Simplify these expressions:

(a) $ab^2 \div a$ (b) $9f^2g \div 3f^2$

(c) $8xy^2 \div 4x$ (d) $6p^4q^3 \div 2p^2$

(e) $12r^2s^3 \div 4s^2$ (f) $7c^3d^4 \div 7d^3$

3 Simplify the following expressions:

(a) $\dfrac{6x^3y^2}{3x^2y^2}$ (b) $\dfrac{4j^2k^4}{jk^2}$ (c) $\dfrac{6a^2b^3}{6a^2b}$

(d) $\dfrac{9m^4n^3}{3mn^3}$ (e) $\dfrac{4d^3e^2}{6(de)^3}$ (f) $\dfrac{16p^3q^4}{4p^3q^2}$

4 Match the expression on the left with the correct one from the right:

A	$3a^3b \times 2ab^2$
B	$a^2 \times 6b \times ab$
C	$12a^4b^4 \div 2a^2b$
D	$6a^2b^3 \div 3ab \times 2a$
E	$2a^3b^2 \times 6a^4b^2 \div 3a^3b$
F	$8a^2b^3 \times 3a^4b^2 \div 4a^3b$

G	$4a^4b^3$
H	$6a^4b^3$
I	$4a^2b^2$
J	$6a^3b^2$
K	$6a^3b^4$
L	$6a^2b^3$

5 Simplify the following expressions:

(a) $\dfrac{6x^3y^2 \times 3x^2y}{9x^4y^2}$ (b) $\dfrac{4a^2b \times 6ab^3}{8a^3b^2}$

(c) $\dfrac{8c^3d^3}{2c^2d \times 2c^2d^3}$ (d) $\dfrac{9m^2n \times m^2n^2}{3m^2n^3 \times 2mn}$

6 Simplify the following expressions:

(a) $\dfrac{4r^2st \times 6rs^2t}{12r^2s^2t}$

(b) $\dfrac{2x^3y^2z \times 10x^2yz}{5x^2yz^3}$

(c) $\dfrac{5abc \times 2b^2c^3 \times a^3bc}{3ab^3 \times 10ab^2c^2}$

7 If $s = 3$ and $t = 0.5$ find the value of the following:

(a) s^{-2} (b) t^{-4} (c) $5s^3t^{-1}$ (d) $\dfrac{s^{-3}}{t}$

Power of a power

To raise a power of a number or variable to a further power multiply the indices.

$(5^2)^3 = 5^{2 \times 3}$
$\qquad = 5^6$

$(p^2)^3 = p^{2 \times 3}$
$\qquad = p^6$

$(5^2)^3 = 5^2 \times 5^2 \times 5^2$
$\qquad = (5 \times 5) \times (5 \times 5) \times (5 \times 5)$
$\qquad = 5^6$

In general: $(x^n)^m = x^{nm}$

Example 10

Simplify:

(a) $(r^3)^2$

(b) $(2k^2)^3$

(a) $(r^3)^2 = r^{3 \times 2}$

$\qquad = r^6$

(b) $(2k^2)^3 = 2^3 \times (k^2)^3$

$\qquad = 8 \times k^{2 \times 3}$

$\qquad = 8 \times k^6$

$\qquad = 8k^6$

Deal with the number first, $2^3 = 8$; then with the letter $(k^2)^3 = k^{2 \times 3} = k^6$.

Exercise 13G

1 Simplify each of the following:

(a) $(p^3)^2$ **(b)** $(q^2)^4$ **(c)** $(r^3)^4$ **(d)** $(f^5)^3$ **(e)** $(d^4)^3$

2 Simplify each of the following:

(a) $(2j^2)^3$ **(b)** $(2m^3)^4$ **(c)** $(3w^5)^3$ **(d)** $(5x^4)^3$ **(e)** $(7d^5)^2$

3 Write the following as a single power of 8 where possible.

(a) $8^5 \times 8^3$

(b) $8^3 \div 8^5$

(c) $8^4 + 8^6$

(d) $8^2 \times 8^9 \times 8^3$

(e) $(8^3)^5$

(f) $\dfrac{8^4 \times 8^2}{8^3}$

(g) $\dfrac{8 \times (8^2)^3}{8^6 \times 8^3}$

(h) $\dfrac{(8^3 \div 8^7) \times (8^3 \times 8^1)}{8}$

4 Write as a single power of 7 where possible

(a) $7^{-3} \times 7^{-2}$

(b) $7^{-3} \div 7^3$

(c) $(7^{-2})^3$

(d) $7^{-3} - 7^3$

(e) $7^4 \div 7^{-3}$

(f) $\dfrac{7^{-2} \times 7^{-5} \times (7^4)^{-2}}{(7^{-3})^6}$

(g) $\dfrac{1}{7^3} \times \dfrac{1}{7^7}$

(h) $\dfrac{1}{7^4} \times \dfrac{1}{7^2} \div \dfrac{1}{7^3}$

13.4 Complex expressions involving indices

In the rest of this section you will see how to develop the index laws so that you can handle quite complex expressions involving indices.

Suppose you are asked to answer this question:
Work out the value of $5^{\frac{1}{2}} \times 5^{\frac{1}{2}}$

Use the law for multiplying powers: $5^{\frac{1}{2}} \times 5^{\frac{1}{2}} = 5^{\frac{1}{2} + \frac{1}{2}} = 5^1 = 5$

You can see that you have multiplied something by itself to get 5 so this means you must have started with the square root of 5.

This follows directly from the definition of a square root.

This means that $\sqrt{5} = 5^{\frac{1}{2}}$
and the same is true for any square root, so that

$$\sqrt{7} = 7^{\frac{1}{2}} \quad \text{and} \quad \sqrt{20} = 20^{\frac{1}{2}} \quad \text{and} \quad \sqrt{x} = x^{\frac{1}{2}}.$$

In the same way can you see that since

$$x^{\frac{1}{3}} \times x^{\frac{1}{3}} \times x^{\frac{1}{3}} = x^{\frac{1}{3} + \frac{1}{3} + \frac{1}{3}} = x^1$$

then $x^{\frac{1}{3}}$ must be the cube root of x, i.e. $\sqrt[3]{x} = x^{\frac{1}{3}}$.

You can extend this idea to any fractional power.

In general: $\sqrt[n]{x} = x^{\frac{1}{n}}$

Example 11

Work out the value of

(a) $36^{\frac{1}{2}}$

b) $81^{\frac{1}{4}}$

(c) $64^{\frac{1}{3}} \times 8^{-1}$

(a) $36^{\frac{1}{2}} = \sqrt{36} = 6$

(b) $81^{\frac{1}{4}} = \sqrt[4]{81} = 3$

(c) $64^{\frac{1}{3}} \times 8^{-1} = \sqrt[3]{64} \times \dfrac{1}{8^1} = 4 \times \dfrac{1}{8} = \dfrac{4}{8} = \dfrac{1}{2}$

For $\sqrt{36}$ give the positive square root.

Part (c) uses the rules for negative and fractional indices.

Exercise 13H

1 Work out the value of:

(a) $25^{\frac{1}{2}}$ (b) $144^{\frac{1}{2}}$ (c) $27^{\frac{1}{3}}$

(d) $16^{\frac{1}{4}}$ (e) $216^{\frac{1}{3}}$ (f) $625^{\frac{1}{4}}$

(g) $100\,000^{\frac{1}{5}}$ (h) $(512/1000)^{\frac{1}{3}}$

2 Write the following using index notation.

(a) $\sqrt{875}$ (b) $\sqrt[4]{184}$ (c) $\sqrt[3]{89}$ (d) $\sqrt[8]{864}$

3 Work out the value of the following:

(a) $25^{-\frac{1}{2}}$ (b) $32^{-\frac{1}{5}}$ (c) $81^{-\frac{1}{4}}$ (d) $1024^{-\frac{1}{10}}$

Finding $x^{\frac{m}{n}}$ where $m \neq 1$

You also need to know how to find the value of an expression such as $x^{\frac{m}{n}}$ where the power is a fraction with a number other than 1 in the numerator.

This is a combination of the 'power to a power' rule and the rule for roots.

> Always remember that the number on the top of the fraction is the power and the number on the bottom is the root.

$$x^{\frac{m}{n}} = (\sqrt[n]{x})^m \quad \text{or} \quad x^{\frac{m}{n}} = \sqrt[n]{(x^m)}$$

These rules are best remembered as:

$x^{\frac{m}{n}} = (\sqrt[n]{x})^m$ 'take the root then raise to the power'

$x^{\frac{m}{n}} = \sqrt[n]{(x^m)}$ 'raise to the power then take the root'

You can see that these are the same in the following example:

$$(25^2)^{\frac{1}{2}} = 625^{\frac{1}{2}} = 25$$

and

$$(25^{\frac{1}{2}})^2 = 5^2 = 25$$

Using the rules you have learnt you can see why this works. The inverse of raising x to the power n (x^n) is raising x^n to the power $\frac{1}{n}$:

$$(x^n)^{\frac{1}{n}} = x^{n \times \frac{1}{n}}$$
$$= x^{\frac{n}{n}}$$
$$= x^1$$
$$= x$$

> As $(x^n)^m = x^{n \times m}$

> As $\frac{n}{n} = 1$

> As $x^1 = x$

You will find that it is usually easier to do the first of these (i.e. root first then power), because it keeps the numbers smaller and so easier to handle.

Example 12

Work out the value of

(a) $16^{\frac{3}{4}}$ (b) $125^{-\frac{2}{3}}$

(a) $16^{\frac{3}{4}} = (\sqrt[4]{16})^3$ *or* $\sqrt[4]{16^3}$

$(\sqrt[4]{16})^3 = (2)^3 = 8$

$\sqrt[4]{16^3} = \sqrt[4]{4096} = 8$

(b) $125^{-\frac{2}{3}} = \dfrac{1}{125^{\frac{2}{3}}} = \dfrac{1}{(\sqrt[3]{125})^2} = \dfrac{1}{(5)^2} = \dfrac{1}{25}$

$2^3 = 8$ is easy to do, $\sqrt[4]{4096} = 8$ is not one you would be expected to know!

You should know that $\sqrt[3]{125} = 5$ because you are expected to know the cubes of 2, 3, 4, 5 and 10.

Do not even try (b) the other way ... you would have to calculate 125^2!

Notice that (b) combines the rule for $x^{\frac{m}{n}}$ with the one for negative indices.

Example 13

Find the value of n in each of the following.

(a) $2^n = \frac{1}{32}$ (b) $9^n = 3$ (c) $4^{n+1} = 2^8$

(a) $2^n = \dfrac{1}{32}$ $32 = 2^5$ so $\dfrac{1}{32} = \dfrac{1}{2^5} = 2^{-5}$ giving $n = -5$

(b) $9^n = 3$ $\sqrt{9} = 3$ so $9^{\frac{1}{2}} = 3$ giving $n = \frac{1}{2}$

(c) $4^{n+1} = 2^8$ $2^8 = 256$ and $256 = 4^4$ so $4^{n+1} = 4^4$

giving $n + 1 = 4$ and so $n = 3$

In (c) it is easier to work out the value of 2^8 first, then think what power of 4 will be needed to give 256.

Exercise 13I

1 Work out the value of:

  (a) $125^{\frac{2}{3}}$ (b) $64^{\frac{2}{3}}$ (c) $81^{\frac{3}{4}}$ (d) $1296^{\frac{3}{4}}$

  (e) $729^{\frac{5}{6}}$ (f) $343^{-\frac{2}{3}}$ (g) $32^{\frac{7}{5}}$ (h) $625^{-1.25}$

Change the decimal into a fraction.

2 Write the following using index notation:

  (a) $\sqrt{52^3}$ (b) $\sqrt[3]{79^2}$ (c) $\sqrt[5]{143^3}$ (d) $\sqrt[4]{728^3}$

3 Which is greater?

  (a) $2401^{\frac{3}{4}}$ or $729^{\frac{2}{3}}$ (b) $216^{\frac{2}{3}}$ or $256^{\frac{3}{4}}$ (c) $3375^{\frac{2}{3}}$ or $3125^{\frac{4}{5}}$

4 Complete the following calculations:

  (a) $8^{\frac{2}{3}} \times 81^{\frac{3}{4}}$ (b) $512^{\frac{2}{3}} \div 128^{\frac{3}{7}}$ (c) $1728^{\frac{2}{3}} \times 729^{\frac{2}{3}} \div 10^{\frac{2}{3}}$

5 Find the value of n in each of the following:

(a) $2^n = 128$

(b) $5^n = \frac{1}{3125}$

(c) $81^n = 3$

(d) $25^{n-1} = 5^6$

(e) $64^n = 4$

(f) $8^{2n+1} = 2^{15}$

(g) $4^{3n-2} = 16^2$

(h) $8^n = 2^{-9}$

6 (a) Draw the graph of $y = 2^x$ for values of x from -2 to 3. (Use a scale of 2 cm to 1 unit.)

(b) Estimate the square root of 2 from your graph.

7 Find the value of a and b in the following:

(a) $8^{\frac{a}{b}} = 4$

(b) $1024^{\frac{a}{b}} = 8$

(c) $625^{\frac{a}{b}} = 125$

13.5 Standard form

> **Key words:**
> standard form

You will need to know:
- **how to handle powers of 10**
- **the rule for negative indices**

It is often convenient to write very large numbers or very small numbers in **standard form**.

> **All numbers can be expressed as $x \times 10^n$, where x = a number between 1 and 10 and n is an integer.**

Remember that
$$10 = 10^1$$
$$100 = 10 \times 10 = 10^2$$
$$1000 = 10 \times 10 \times 10 = 10^3$$

and that
$$0.1 = \tfrac{1}{10} = 10^{-1}$$
$$0.01 = \tfrac{1}{100} = 10^{-2}$$
$$0.001 = \tfrac{1}{1000} = 10^{-3}$$

> These are all examples of the rule for negative indices
> $$\frac{1}{x^n} = x^{-n}$$

Numbers greater than 1

These numbers are written in standard form:

400	=	4×100	=	4×10^2
6000	=	6×1000	=	6×10^3
81 200	=	$8.12 \times 10\,000$	=	8.12×10^4
370 900	=	$3.709 \times 100\,000$	=	3.709×10^5

$$3.709 \times 10^5$$

> This part is always a number between 1 and 10 ($1 \leqslant n < 10$).

> This part is always a power of 10.

> Don't forget to write the $\times$ in between the two parts.
>
> *Never* write answers as 3.709^{05} which is how your calculator may display numbers in standard form. (See page 234).

In the same way, $40\,827.59 = 4.082\,759 \times 10^4$
and $360.47 = 3.6047 \times 10^2$

Notice that the power of 10 is always one less than the number of digits before the decimal point in the original number.

Numbers less than 1

These numbers are written in standard form:

$0.6 \quad = \dfrac{6}{10} \quad = \dfrac{6}{10^1} \quad = \quad 6 \times 10^{-1}$

$0.03 \quad = \dfrac{3}{100} \quad = \dfrac{3}{10^2} \quad = \quad 3 \times 10^{-2}$

$0.0078 \quad = \dfrac{7.8}{1000} \quad = \dfrac{7.8}{10^3} \quad = \quad 7.8 \times 10^{-3}$

$0.000419 \quad = \dfrac{4.19}{10\,000} \quad = \dfrac{4.19}{10^4} \quad = \quad 4.19 \times 10^{-4}$

Note that the decimals are first of all written as fractions with a numerator between 1 and 10.

4.19×10^{-4}

This part is always a number between 1 and 10.
$(1 \leqslant n < 10)$

Note that the power of 10 is always **negative** for numbers less than 1.

Notice that the power of 10 is always negative and one more than the number of zeros between the decimal point and the first non-zero digit in the original number.

Exercise 13J

1 Write the following numbers in standard form:

    (a) 3 000 000   (b) 7400   (c) 32 000

    (d) 603 500   (e) 108   (f) 68

    (g) 650.5   (h) 99.9

2 Write in standard form:

    (a) 0.0005   (b) 0.006   (c) 0.4

    (d) 0.00012   (e) 0.0717   (f) 0.0001975

    (g) 0.9009   (h) 0.0010003

3 Match a number in the first column with a correct standard form number from the second column.

0.9	9×10^3
900 000	9×10^2
900	9×10^5
0.09	9×10^1
90	9×10^{-1}
9000	9×10^{-2}

4 Write these numbers in standard form.

   (a) The diameter of the Earth is approximately 12 735 kilometres.

   (b) The total mass of krill, a small sea shrimp, is estimated to be 650 million tonnes.

5 A single-celled organism is 2 tenths of a millimetre long. What is this in metres?

Changing between standard form and a decimal number

Look at these examples,

3.709×10^5 3 . 7 0 9 0 0 = 370 900

The power (5) is positive so count 5 places to the *right* from the decimal point, this gives the answer of 370 900.

4.19×10^{-4} 0 0 0 0 4 . 1 9

The power (-4) is negative so count 4 places to the *left* from the decimal point, this gives the answer of 0.000419.

Positive ↔ Right

Negative ↔ Left

Always put a zero in front of the decimal point when the number is small:
  0.000419
is much clearer than
  .000419

Exercise 13K

1 Write these as decimal numbers.

   (a) 5×10^4 **(b)** 3.8×10^3

   (c) 6×10^{-3} **(d)** 7.26×10^9

   (e) 8.492×10^{-2} **(f)** 4.37×10^6

   (g) 1.006×10^{-4} **(h)** 6.2387×10^3

2 Write these numbers in standard form as decimal numbers.

   (a) The distance from the Earth to the Sun is approximately 1.488×10^8 kilometres.

   (b) The average width of an iris of an eye is 1×10^{-2} metres.

   (c) A billion molecules of water has a mass of 3×10^{-11} grams.

3 These numbers are *not* in standard form. Rewrite them in standard form.

   (a) 123×10^2 **(b)** 0.8×10^7

   (c) 17×10^{-2} **(d)** 0.25×10^{-4}

   (e) 18 million   **(f)** $\frac{1}{8}$

   (g) $36 \times 10^4 \times 0.006$ **(h)** $\sqrt{40 \times 10}$

4 Complete these calculations, giving your answer in standard form.

(a) $(6.4 \times 10^5) - (8.34 \times 10^4)$

(b) $(4.2 \times 10^2) + (5.6 \times 10^3)$

(c) $(3.9 \times 10^{-2}) + (4.2 \times 10^{-3})$

(d) $(8.2 \times 10^{-3}) - (6.1 \times 10^{-4})$

(e) $(4 \times 10^2) - (7.2 \times 10^{-1})$

(f) $(6.1 \times 10^3) + (5.7 \times 10^2) - (4.8 \times 10^{-1})$

> Write these as decimal numbers before you do the calculation.
>
> Write your answers in standard form.
>
> Do not use a calculator.

5 On Jane's computer a picture of a dragonfly uses 7 832 450 bytes. Her picture of a bee uses 7.68×10^6 bytes. Which is larger and by how much?

Calculations involving numbers written in standard form

You will need to be able to use the index laws for multiplying and dividing powers.
Make sure that you leave your answer in standard form when asked to do so.

Example 14

$P = 6.6 \times 10^7$ and $Q = 3 \times 10^5$
Work out the value of

(a) $P \times Q$ (b) $P \div Q$, giving your answers in standard form.

(a) $(6.6 \times 10^7) \times (3 \times 10^5)$

$= (6.6 \times 3) \times (10^7 \times 10^5)$

$= 19.8 \times 10^{12}$

$= (1.98 \times 10) \times 10^{12}$

$= 1.98 \times 10^{13}$

(b) $(6.6 \times 10^7) \div (3 \times 10^5)$

$= \dfrac{6.6 \times 10^7}{3 \times 10^5}$

$= (6.6 \div 3) \times 10^2$

$= 2.2 \times 10^2$

> Rearrange to keep the powers of 10 together.
>
> Multiplying powers of 10 means add indices.
>
> Re-write 19.8 as 1.98×10 to get a number between 1 and 10, as required.
>
> Writing the question as a division in the form of a fraction helps you to focus on what needs to be done.
>
> Dividing powers of 10 means subtract indices.

Example 15

$G = 3.6 \times 10^3$ and $H = 4 \times 10^{-5}$
Work out the value of

(a) $G \times H$

(b) $G \div H$, giving your answers in standard form.

(a) $(3.6 \times 10^3) \times (4 \times 10^{-5})$

$= (3.6 \times 4) \times (10^3 \times 10^{-5})$

$= 14.4 \times 10^{-2}$

$= (1.44 \times 10) \times 10^{-2}$

$= 1.44 \times 10^{-1}$

(b) $(3.6 \times 10^3) \div (4 \times 10^{-5})$

$= \dfrac{3.6 \times 10^3}{4 \times 10^{-5}}$

$= (3.6 \div 4) \times 10^8$

$= 0.9 \times 10^8$

$= \tfrac{9}{10} \times 10^8$

$= 9 \times 10^7$

When the numbers involve negative powers you need to take more care. It is very easy to make a mistake!

Rearrange to keep the powers of 10 together.

Multiplying powers of 10 means add indices.

Re-write 14.4 as 1.44×10 to get a number between 1 and 10, as required.

Note that
$10 \times 10^{-2} = 10^1 \times 10^{-2}$
$= 10^{-1}$

Writing the question as a division in the form of a fraction helps you to focus on what needs to be done.

Dividing powers of 10 means subtract indices. Note that $3 - -5 = 3 + 5 = 8$

Re-write 0.9 as $\tfrac{9}{10} = 9 \times 10^{-1}$ to get a number between 1 and 10, as required.

Example 16

The distance of the Earth from the Sun is approximately 9.3×10^7 miles. Light travels at a speed of about 1.86×10^5 miles per second.
Use this information to calculate the time taken for light to travel to the Earth from the Sun.

$\text{Time} = \dfrac{\text{distance}}{\text{speed}}$

$= \dfrac{9.3 \times 10^7}{1.86 \times 10^5}$

$= 5 \times 10^2$

$\text{Time} = 500 \text{ seconds}$

$= 8 \text{ minutes } 20 \text{ seconds}$

Take care with the units.
$\dfrac{\text{miles}}{\text{miles per sec}}$
will give the time taken in seconds.

$9.3 \div 1.86$ is exactly 5.

Calculators and standard form

To enter a number given in standard form on your calculator you can use the button marked **EXP**

EXP stands for *exponent* which means 'a power of 10'.
So to enter 1.86×10^5 you press,

Never press $\times$ or **1** **0** when using **EXP**.

This is much quicker than pressing ...

... especially when you have quite a few numbers written in standard form in a calculation!

A word of warning!
When you use your calculator to obtain an answer that is either very big or very small, the number in the calculator display may or may not appear in standard form or be displayed to enough decimal places.

For example, to calculate the reciprocal of 2.3×10^8:

The reciprocal is $\dfrac{1}{2.3 \times 10^8}$

The reciprocal button is 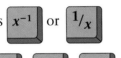 or

> The reciprocal button is the quickest way of doing this calculation.

Press

(making sure your calculator is in the correct mode).

> This may not give the answer to the required degree of accuracy ... it depends whether your calculator is in the correct mode.

The answer given is $4.347\,826\,087^{-09}$
Written in standard form this is 4.35×10^{-9} (3 s.f.)

The choice of 'NORM 1~2?' does not just write an answer in standard form when you select 1, it also gives you all the decimal places. When you select 2 you may not see enough decimal places, or the answer may disappear off the display screen.

> To check, press the mode button until the word 'NORM' appears (this is usually option 3).
> When you choose this you will see 'NORM 1 ~2 ?'
> If you select 2 your answer will **not** be in standard form, so select 1.

Exercise 13L

1 Work out the following, giving your answers in standard form.

> Remember that calculators vary. Check how yours works.

(a) $(2.2 \times 10^3) \times (4 \times 10^5)$ (b) $(3.3 \times 10^6) \times (3 \times 10^{-4})$

(c) $(8.4 \times 10^5) \div (2 \times 10^4)$ (d) $(5.6 \times 10^3) \div (4 \times 10^{-2})$

(e) $(4.2 \times 10^4) \times (5.3 \times 10^3)$ (f) $(1.6 \times 10^{-2}) \div (4 \times 10^{-4})$

2 This year it is estimated that, on average, each person in the United Kingdom will spend £480 making mobile phone calls. If the population is 5.4×10^7, what is the total amount spent on mobile phone calls? Give your answer in standard form.

3 The radius of the Earth is approximately 6.4×10^3 km. Estimate the volume of the Earth if we assume it is a sphere.

> Volume of a sphere $= \frac{4}{3}\pi r^3$

4 If $a = 1.32 \times 10^7$ and $b = 4.28 \times 10^6$, find the following giving your answers in standard form.

(a) $3b$ (b) ab (c) a^2 (d) $a + b$ (e) $b \div a$

5 If there are 6×10^9 people in the world and they eat 7×10^7 tonnes of fish per year, how much fish does each person eat?

6 The diameter of the Sun is approximately 1.392×10^6 km. Find its surface area using the formula $4\pi r^2$.

7 A rectangular picture measures 1.4×10^2 cm by 2.7×10^3 cm. What is   **(a)** the area   **(b)** the perimeter of the picture?

8 The number of krill (a type of plankton) is estimated to be 6×10^{15} and their mass is about 6.5×10^8 tonnes.

(a) What is the mass of 1 krill in grams?

(b) If the volume of 1 million krill is 100 cm³, what is the density of krill?

9 After the Sun, the next nearest star to the earth is Proxima Centauri. It takes about 4.24 years for light from this star to reach the Earth. If light travels at 1.86×10^5 miles per *second*, estimate the distance of Proxima Centauri from Earth.

10 A teaspoon of oil (5 ml) is spilled onto a flat area of water. It covers 0.4 hectares. What is the thickness of the oil?

1 hectare = 100 m × 100 m.

Examination questions

1 Simplify

(a) $m^2 \times m^5$ **(b)** $p^6 \div p^3$ **(c)** $(q^4)^2$

(3 marks)
AQA, Spec B, 1H, June 2003

2 (a) Work out the value of $2^4 \times 3^2$

(b) Simplify $(2x^3y) \times (3xy)$

(4 marks)
AQA, Spec B, 1I, November 2003

3 Simplify

(a) $x^5 \times x^{-2}$ **(b)** $y^5 \div y^{-2}$

(2 marks)
AQA, Spec A, 2I, June 2003

4 Simplify $(2xy^2)^3$

(2 marks)
AQA, Spec A, 1H, June 2003

5 (a) Write down the value of 11^0

(b) Find the value of $8^{2/3}$

(c) Simplify $6^{-2} \times 144^{0.5}$

(6 marks)
AQA, Spec A, 1H, June 2003

Summary of key points

The laws of indices (grade C)

To multiply powers of the same number or variable add the indices.
$$x^n \times x^m = x^{n+m}$$

To divide powers of the same number or variable subtract the indices.
$$x^n \div x^m = x^{n-m}$$

Powers of 1 and 0 (grade C/B)

Any number or variable raised to the power 1 is equal to the number or variable itself
For example: $3^1 = 3$, $x^1 = x$.

Any number or variable raised to the power 0 is equal to 1
For example: $3^0 = 1$, $x^0 = 1$.

A power to a power (grade A)

To raise a power of a number or variable to a further power you multiply the indices.
$$(x^n)^m = x^{nm}$$

Negative indices (grade B)

x^{-n} means the reciprocal of x^n. $x^{-n} = \dfrac{1}{x^n}$

Fractional powers (grade A/A*)

Fractional powers always give roots.

For example: $\sqrt{x} = x^{\frac{1}{2}}$, $\sqrt[3]{x} = x^{\frac{1}{3}}$ and in general $\sqrt[n]{x} = x^{\frac{1}{n}}$.

To work out $x^{\frac{m}{n}}$ either $x^{\frac{m}{n}} = (\sqrt[n]{x})^m$ 'take the root then raise to the power'

or $x^{\frac{m}{n}} = \sqrt[n]{(x^m)}$ 'raise to the power then take the root'

Standard form (grade B)

This is how you write a number in standard form:

370900   = 3.709×10^5

This part is always a number between 1 and 10 ($1 \le n < 10$).

This part is always a power of 10.

Notice that the power of 10 is always one less than the number of digits before the decimal point in the original number.

0.000419   = 4.19×10^{-4}

This part is always a number between 1 and 10 ($1 \le n < 10$).

Note that the power of 10 is always negative for numbers less than 1.

Notice that the power of 10 is always negative and one more than the number of zeros between the decimal point and the first non-zero digit in the original number.

This chapter will show you how to:

✔ use rules to generate sequences
✔ find rules for the *n*th term of a linear sequence
✔ find rules for the *n*th term of simple quadratic sequences

14.1 Rules for sequences

Key words:
sequence
term
general rule
general term
*n*th term

A number **sequence** is a list of numbers. There is often a connection between the numbers in the list.

Each number in a sequence is called a **term**.

A **general rule** is a rule you can use to work out any term in a sequence. You need to know the position of the term (the term number). The letter *n* is used for the term number.

The general rule is sometimes called the **general term**.

It is also often called the **nth term**.

Exam questions usually refer to the *n*th term.

Example 1

The *n*th term of a sequence is $2n + 5$.

(a) Write down the first four terms.
(b) What is the difference between consecutive terms?
(c) Which term has a value of 45?
(d) Explain why 36 cannot be a term in this sequence.

For the first term $n = 1$

(a) 1st term: $2 \times 1 + 5 = 2 + 5 = 7$
   2nd term: $2 \times 2 + 5 = 4 + 5 = 9$
   3rd term: $2 \times 3 + 5 = 6 + 5 = 11$
   4th term: $2 \times 4 + 5 = 8 + 5 = 13$

For the third term $n = 3$

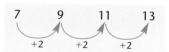

(b) The difference between consecutive terms is +2.

(c) $2n + 5 = 45$
   $2n = 40$
   $n = 20$

   The 20th term is 45.

continued ▶

Every term can be worked out using the rule $2n + 5$. So you need to find which value of *n* satisfies the equation $2n + 5 = 45$.

(d) If 36 is in the sequence then there is a whole number n for

which $2n + 5 = 36$

$2n + 5 = 36$

$2n = 31$

$n = 15\frac{1}{2}$

but $15\frac{1}{2}$ is not a whole number so 36 is not in the sequence.

Part **(d)** is an extension of the method used in part **(c)**.
This is a UAM question where you have to think of a method to solve the problem and use mathematical reasoning.

Example 2

The nth term of a sequence is $n^2 - 1$.
(a) Write down the first four terms of the sequence.
(b) Write down the 12th term.

(a) 1st term: $1^2 - 1 = 1 - 1 = 0$

2nd term: $2^2 - 1 = 4 - 1 = 3$

3rd term: $3^2 - 1 = 9 - 1 = 8$

4th term: $4^2 - 1 = 16 - 1 = 15$

(b) 12th term: $12^2 - 1 = 144 - 1 = 143$

Remember $n^2 = n \times n$
So $3^2 = 3 \times 3 = 9$

Exercise 14A

1 The nth term of a sequence is $2n + 3$.

(a) Write down the first four terms.

(b) What is the difference between consecutive terms?

(c) Write down the 20th term.

2 For each of the following sequences:
- find the first four terms
- write down the difference between consecutive terms
- find the 30th term.

(a) nth term: $3n + 5$

(b) nth term: $\dfrac{n + 2}{2}$

(c) nth term: $4 - 3n$

(d) nth term: $\dfrac{3 - n}{2}$

3 The nth term of a sequence is $2n + 4$.

(a) Work out the value of the 8th term.

(b) Which term has a value of 46?

(c) Explain why 35 is not a term in this sequence.

4   The *n*th term of a sequence is $3n - 1$.
   (a) Calculate the value of the 6th term.
   (b) Which term has a value of 59?
   (c) Explain why 90 is not a term in this sequence.

5   The *n*th term of a sequence is $n^2 + 1$.
   (a) Write down the first four terms of the sequence.
   (b) Write down the 12th term.

6   The *n*th term of a sequence is $n^2 - 3$.
   (a) Write down the first four terms of the sequence.
   (b) Write down the 13th term.

7   The *n*th term of a sequence is $\dfrac{n + 1}{n}$.
   (a) Write down the first four terms of the sequence.
   (b) Write down the 9th term.

8   The *n*th term of a sequence is $n^2 + 10$.
   (a) Which term has a value of 35?
   (b) Explain why 55 is not a term in this sequence.

14.2 Finding the *n*th term of a linear sequence

Key words:
linear sequence

If the difference between consecutive terms is the same, the sequence is a   linear sequence .

You can use the difference to help you find the rule for the *n*th term.

In the sequence: 7, 11, 15, 19, ... the terms go up in 4s.
This tells you that the rule will include **4*n***.

To find the rest of the rule you can compare the sequence to the 4× table:

4× table	4	8	12	16	...
Sequence	7	11	15	19	...

You have to **add 3** to the numbers in the 4× table to get the numbers in the sequence.
So the rule for the *n*th term is: $4n + 3$

Example 3

Find the nth term and the 50th term of the sequence:

2, 5, 8, 11, ...

Sequence: 2   5   8   11

Difference: +3   +3   +3

The difference is 3 so the rule will include $3n$.

3× table: 3, 6, 9, 12, ...

Sequence: 2, 5, 8, 11, ...

You have to subtract 1 from the numbers in the 3× table to get the numbers in the sequence.

The rule for the nth term is: $3n - 1$

50th term $= 3 \times 50 - 1 = 150 - 1 = 149$

Exercise 14B

Find the nth term and the 50th term of each of the following sequences:

1 4, 7, 10, 13, ... $3n+1$

2 5, 7, 9, 11, ... $2n+3$

3 7, 11, 15, 19, ... $4n+3$

4 3, 8, 13, 18, ... $5n-2$

5 6, 9, 12, 15, ... $3n+3$

6 6, 13, 20, 27, ... $7n-1$

7 2, 12, 22, 32, ... $10n-8$

8 16, 25, 34, 43, ... $9n+7$

9 5, 6, 7, 8, ... $n+4$

10 $-4, -1, 2, 5, ...$ $3n-1$

11 25, 21, 17, 13, ...
$-4n+29$

12 20, 15, 10, 5, ... $-5n+25$
5 10 15 20

14.3 Sequences of patterns

You can often use patterns to help find the rule for a sequence of numbers. For example, the square numbers can be shown as patterns of dots.

These are the first three square numbers:

Example 4

Look at this sequence of patterns:

(a) Draw the 4th pattern in the sequence.

(b) Write out the number of dots in the first five patterns.

(c) What is (i) the 6th term?   (ii) the nth term?

(a) ○ ○ ○ ○ ○
　　○ ○ ○ ○
　　○ ○ ○ ○
　　○ ○ ○ ○

(b) 2, 5, 10, 17, 26

(c) (i) 37　　　　　(ii) $n^2 + 1$

4th pattern = 5 × 5 square + 1 dot.

Compare the sequence to the sequence of square numbers to find the nth term.

○　　○ ○　　○ ○ ○
　　　○ ○　　○ ○ ○
　　　　　　 ○ ○ ○

6th pattern = 6 × 6 square + 1 dot.

Exercise 14C

1　○　　　　　○　　　　　　　○
　　○　　　　○ ○　　　　　○ ○ ○
　　○　　　　○ ○　　　　　○ ○ ○
　　　　　　　○　　　　　　 ○ ○ ○
　　　　　　　　　　　　　　　　○

(a) Draw the 4th pattern in the sequence.

(b) What is (i) the 5th term? (ii) the nth term?

(c) What do you notice about the differences between successive terms in the sequence?

2　Dots are used to make a sequence of triangles:

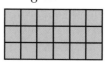

(a) The numbers in this sequence are called triangle numbers. One rule for finding the nth triangle number is $\frac{1}{2}n(n + 1)$
What is the 10th triangle number?

(b) What do you notice about the differences between each term for this sequence?

(c) The nth term is 210 and the $(n + 1)$th term is 231. Find the value of n.

Look at the link between the differences and term number.

3　The diagrams show a sequence of rectangles.

Diagram 1　　　　Diagram 2　　　　Diagram 3
Area = 1 × 4　　 Area = 2 × 5　　 Area = 3 × 6
　　 = 4　　　　　　 = 10　　　　　　 = 18

(a) (i)　Write down the area of diagram 7.
　　(ii)　Write down the area of diagram 20.
　　(iii)　Write down an expression for the area of diagram n.

(b) A rectangle in the sequence has an area of 130 cm². What are the dimensions of this rectangle? What is the perimeter of this rectangle?

(c) What is the perimeter of a rectangle in the sequence which has an area of 810 cm²?

14.4 Simple quadratic sequences

Key words:
second difference
quadratic sequence

If the difference between consecutive terms is not the same, it can help to look at the difference between the differences. These are called the **second differences** .

> If the second differences are the same, the sequence is a **quadratic sequence** .

> The nth term of a quadratic sequence involves n^2.

For example, for the sequence whose nth term is $n^2 + 4$:

sequence	5,	8,	13,	20,	29, ...
first difference		+3	+5	+7	+9 ...
second difference			+2	+2	+2 ...

Halving the second difference gives the coefficient of n^2.

You will be expected to find the nth term of quadratic sequences that are based on square numbers. In the example above, the terms in the sequence are all 4 more than the corresponding square number.

Example 5

Find the nth term and the 10th term of the sequence:

   0, 3, 8, 15, 24, ...

Sequence:	0,	3,	8,	15,	24, ...
First difference:		3,	5,	7,	9, ...
Second difference:			2,	2,	2, ...
Square numbers:	1, 4, 9, 16, 25, ...				
Sequence:	0, 3, 8, 15, 24, ...				

The rule for the nth term is: $n^2 - 1$

10th term $= 10^2 - 1 = 100 - 1 = 99$

The second differences are the same so it is a quadratic sequence and the nth term will involve n^2.
Compare the sequence to the sequence of square numbers.
You have to take 1 from the square numbers to get the numbers in the sequence, e.g. $1 - 1 = 0$, $4 - 1 = 3$.

Exercise 14D

Find the nth term and the 15th term of each of the following sequences:

1 $4, 7, 12, 19, ...$ **2** $-1, 2, 7, 14, ...$

3 $7, 10, 15, 22, ...$ **4** $-4, -1, 4, 11, ...$

5 $2, 8, 18, 32, ...$ **6** $3, 12, 27, 48, ...$

7 $0, -3, -8, -15, ...$ **8** $1, 2\frac{1}{2}, 5, 8\frac{1}{2}, ...$

> Try doubling the terms, then look at the second difference.

Examination questions

1 (a) The nth term of a sequence is $4n + 1$
    (i) Write down the first three terms of the sequence.
    (ii) Is 122 a term in this sequence?
    Explain your answer.

(b) Tom builds fencing from pieces of wood as shown below.

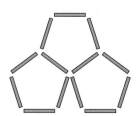

Diagram 1	Diagram 2	Diagram 3
4 pieces of wood	7 pieces of wood	10 pieces of wood

How many pieces of wood will be in Diagram n?

(5 marks)
AQA, Spec A, 1I, June 2003

2 A pattern using pentagons is made of sticks.

    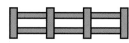

Diagram 1	Diagram 2	Diagram 3
5 sticks	9 sticks	13 sticks

(a) How many sticks are needed for Diagram 5?

(b) Write down an expression for the number of sticks in Diagram n.

(c) Which Diagram uses 201 sticks?

(7 marks)
AQA, Spec A, 1I, November 2003

Summary of key points

nth term (grade D)

The nth term is a general rule to work out any term in a sequence if you know its position.

For example:

nth term $= 5n + 3$

To work out the 40th term use $n = 40$,

40th term $= 5 \times 40 + 3 = 200 + 3 = 203$.

Finding the nth term of a linear sequence (grade C)

If the difference between consecutive terms is the same, the sequence is a linear sequence.

For example:

If the difference between consecutive terms is $+3$ then the nth term will include $3n$. Compare the sequence to the $3 \times$ table to help you find the rest of the rule for the nth term.

Sequence: 2,   5,   8,   11 ...

Difference: $+3$ $+3$ $+3$

The difference is 3 so the rule will include $3n$

$3\times$ table: 3, 6, 9, 12, ...

Sequence: 2, 5, 8, 11, ...

You have to subtract 1 from the numbers in the $3\times$ table to get the numbers in the sequence, so the rule for the nth term is $3n - 1$.

Finding the nth term of a simple quadratic sequence (grade B)

If the second differences are the same, the sequence is a quadratic sequence.

The nth term of a sequence involves n^2.

For example:

Sequence   5,   8,   13,   20,   29, ...

First difference $+3$ $+5$ $+7$ $+9$ ...

Second difference $+2$ $+2$ $+2$ ...

Compare with the sequence of square numbers

Square numbers   1, 4,   9, 16, 25, ...

Sequence   5, 8, 13, 20, 29, ...

You have to add 4 to the square numbers to get the numbers in the sequence so the rule for the nth term of the sequence is $n^2 + 4$.

This chapter will show you how to:

✔ define and use probability and relative frequency
✔ deal with mutually exclusive and independent events
✔ draw tree diagrams
✔ solve conditional probability problems

15.1 Probability and relative frequency

Key words:
event
outcome
probability scale

An **event** , such as throwing a dice, can have different **outcomes** . You can describe the chances of an outcome using words such as *certain, likely, even, unlikely* and *impossible*.

$0 \leqslant$ probability $\leqslant 1$.

> The **probability scale** uses numbers between 0 and 1.

The connection between the values on the probability scale and the descriptive words is shown below:

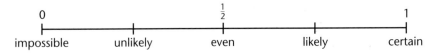

> You can write probability as a fraction, a decimal or a percentage.

Converting between fractions, decimals and percentages was covered in Chapter 3.

Example 1

What is the probability of obtaining a Head when you toss a fair coin?

A fair coin is one where Head and Tail are equally likely.

$$P(H) = P(T) = \tfrac{1}{2}$$
$$= 0.5$$
$$= 50\%$$

P(H) is shorthand for the probability of throwing a head, and P(T) for the probability of throwing a tail.

> The result which actually occurs is called an **outcome** .

Only two outcomes are possible when tossing a coin.

The probability of throwing a 3 on an ordinary six-sided dice must be one out of six, since there are six faces and six numbers (six possible outcomes). We write this as

$$P(3) = \tfrac{1}{6} = 0.167 = 16.7\% \text{ (3 s.f.)}$$

> You must write a probability as a fraction or a decimal or a percentage, not 1 out of 6 or 1 : 6.

$$\textbf{Probability} = \frac{\textbf{number of successful outcomes}}{\textbf{total number of possible outcomes}}$$

Example 2

An eight-sided (octahedral) fair dice is thrown.
Find the probability of obtaining
(a) a 3   **(b)** a prime number.

(a) The probability $P(3) = \tfrac{1}{8}$.

> The value 3 can only occur once so there is only 1 successful outcome. The total number of possible outcomes is 1, 2, 3, 4, 5, 6, 7 or 8 i.e. 8 possible outcomes altogether.

(b) The probability $P(\text{prime}) = \tfrac{4}{8} = \tfrac{1}{2}$.

> The only prime number values are 2, 3, 5 and 7, so there are 4 successful outcomes.

Theoretical probability and relative frequency

> **Key words:**
> theoretical probability
> relative frequency

The probability of obtaining a Head when a coin is tossed is $P(H) = \tfrac{1}{2}$. We call this the **theoretical probability**. If you toss the same coin ten times it does not mean that you will get a Head exactly five times. The number of successful trials (in this case, number of Heads) will approach the value of $\tfrac{1}{2}$ as you increase the total number of trials.

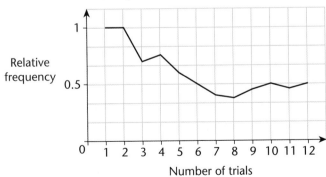

The **relative frequency** is given by the fraction

> Theoretical probability and relative frequency can be compared to each other.

$$\textbf{Relative frequency} = \frac{\textbf{number of successful trials}}{\textbf{total number of trials}}$$

Example 3

A coin was tossed 50 times and the results were:

	Frequency
Head (H)	28
Tail (T)	22
Total	50

Work out the relative frequency and estimate the probability of getting **(a)** Heads **(b)** Tails. Compare this answer with the theoretical probability for P(H) and P(T).

How could your relative frequency be made more accurate?

This is a typical exam question.

(a) Relative frequency $= \dfrac{\text{number of Heads obtained}}{\text{total number of throws}}$

$= \dfrac{28}{50} = 0.56$

When a larger number of outcomes is involved it is easier to extend the table to include a column for the relative frequencies.

(b) Relative frequency $= \dfrac{\text{number of Tails obtained}}{\text{total number of throws}}$

$= \dfrac{22}{50} = 0.44$

The theoretical probabilities are P(H) = P(T) = 0.5.

The relative frequency for a Head is greater than its theoretical probability.

The relative frequency for a Tail is less than its theoretical probability.

The relative frequency can be made more accurate by increasing the number of trials. Both relative frequencies will approach the theoretical probabilities as the number of trials increases.

You can also work back to calculate the frequency using the expression:

Frequency = total number of trials × relative frequency

In the example above, the frequency for obtaining Heads is $50 \times 0.56 = 28$.

Exercise 15A

1 Write the following as a decimal or percentage using the notation P(event) = ☐.

(a) the probability that you will be given homework is 40% (as a decimal)

(b) the probability that it will snow tomorrow is 0.01 (as a percentage)

(c) the probability of throwing two consecutive sixes with a dice is $\frac{1}{36}$ (as a decimal).

2 An ordinary six-sided dice is thrown. What is the probability of

(a) throwing a 2

(b) obtaining an even number

(c) throwing a number less than 4

(d) landing on a 3 or a 5

(e) not obtaining a six?

3 In a raffle one hundred tickets were sold. One ticket is picked as the winner. What is the probability of winning if you have bought

(a) 3 tickets   (b) 11 tickets?

4 In a bag of 16 marbles there are 11 blue and 5 red. What is the probability of picking out

(a) a blue marble   (b) a red marble

(c) a green marble   (d) a blue or red marble?

5 Copy the tally chart shown. Toss a coin 100 times and record your results in the chart.

	Tally	Frequency	Relative frequency
Head (H)			
Tail (T)			
	Total	100	

Now work out the relative frequencies for obtaining a Head and a Tail. Compare your results with the theoretical probabilities for P(H) and P(T).

How could your relative frequencies be made more accurate?

6 Put 7 red counters and 3 blue counters into a bag. Pick a counter from the bag, record its colour and replace in the bag. Repeat to give a total of 50 times.

> You could simulate this experiment using a random number generator on a calculator.

(a) Draw a table showing your results.

(b) Work out the relative frequencies for picking a red and blue counter.

(c) Compare your results with the theoretical values for P(R) and P(B).

15.2 Mutually exclusive outcomes

> **Key words:**
> mutually exclusive
> exhaustive

When you toss a coin you can get either a Head (H) *or* Tail (T) but not both at the same time. Obtaining a Head excludes a Tail. Such outcomes are called **mutually exclusive** outcomes. If you get one outcome you cannot get the other outcome.

> Notice it is the word 'or' that is important here.

In addition, there are no other possibilities, apart from Head or Tail. The outcomes Head and Tail are **exhaustive** . This means that the sum of the probabilities is 1.

If you know that outcomes are mutually exclusive and exhaustive, you can easily calculate the probabilities.

Example 4

Work out the probability of obtaining a Head (H) *or* a Tail (T) when tossing a coin.

> We know that $P(H) = P(T) = \frac{1}{2}$.
>
> The probability of obtaining a Head *or* a Tail is:
>
> $P(H \text{ or } T) = P(H) + P(T) = \frac{1}{2} + \frac{1}{2} = \frac{2}{2} = 1$.

> The key word used here is 'or'. As long as the outcomes are mutually exclusive, this means 'add' the probabilities.

For any two outcomes, say A and B, that are mutually exclusive and exhaustive,

$$P(A \text{ or } B) = P(A) + P(B) = 1$$

This is also true for more than two mutually exclusive exhaustive outcomes.

Example 5

A fair six-sided dice is thrown. What is the probability of getting a two or a four?

$P(2) = \frac{1}{6}$

$P(4) = \frac{1}{6}$

$P(2 \text{ or } 4) = \frac{1}{6} + \frac{1}{6}$

$\qquad = \frac{2}{6}$

$\qquad = \frac{1}{3}$

> Add the probabilities.

> The outcomes (2) and (4) are not exhaustive so the sum of the probabilities is less than 1.

Example 6

A spinner is divided into three equal sections coloured red, green and blue.

Find the probability that

(a) the spinner will land on red

(b) the spinner will not land on red.

> All sectors are equal so they are all equally likely. The events are also mutually exclusive.

(a) $\qquad P(R) = \frac{1}{3}$

$\qquad$ Also $\quad P(B) = P(G) = P(R) = \frac{1}{3}$.

> Also $P(B) = P(G) = P(R) = \frac{1}{3}$

(b) The probability that it will not land on red P(**not** R) is the same as landing on Green **or** Blue P(G **or** B).

$P(G \text{ or } B) = P(G) + P(B) = \frac{1}{3} + \frac{1}{3} = \frac{2}{3} = P(\text{not } R)$

> This is the same as
> $1 - P(R) = 1 - \frac{1}{3} = \frac{2}{3}$
> $P(\text{not } R) = 1 - P(R)$

For an event A, the probability of the event A *not* happening is given by

$$P(\text{not } A) = 1 - P(A)$$

> In probability you can use the word 'not' to indicate that the event will not happen.

Exercise 15B

1 A letter of the alphabet is chosen at random.

(a) What is the probability that it will be a vowel?

(b) What is the probability it will be a consonant?

2 One of the longest rivers in the world is the MISSISSIPPI. If one of these letters were chosen at random, what is the probability of

(a) choosing the letter I

(b) choosing the letter S

(c) choosing the letters M or P?

3 A fair six-sided dice is thrown. Work out the probability of

(a) throwing a 3 $\qquad\qquad$ (b) not throwing a 1

(c) throwing a 3 or a 4 $\qquad$ (d) not throwing a 2 or a 3.

4 A set of cards has the numbers 1 to 30 written on them.
What is the probability of choosing

(a) a square number

(b a prime number

(c) a number >10?

5 A bag contains 5 green counters, 4 blue counters, 2 red counters
and 3 yellow counters. What is the probability of

(a) picking a red counter   (b) picking a blue or green counter

(c) not picking a yellow counter   (d) not picking a green or red counter?

6 A bag contains 16 green, red and yellow marbles. The probability
of picking a green marble is $P(G) = \frac{1}{4}$ and the probability of picking
a red marble is $P(R) = \frac{3}{8}$.

(a) Work out the probability that

(i) the marble picked will be yellow

(ii) the marble will not be red or yellow.

(b) How many marbles of each colour are in the bag?

15.3 Listing outcomes

Key words:
sample space diagram

You should always record all the possible outcomes from a single
event or two successive events in a systematic way. Use simple lists
or **sample space diagrams** (two-way tables).

The result which actually
occurs is called an
outcome. Only two
outcomes are possible
when tossing a coin – a
Head or a Tail.

Example 7

Two coins (10p and 5p) are tossed at the same time. List all the
possible outcomes.

Let H mean Heads and T mean Tails.

The outcomes are: HH HT TH TT.

There are 4 possible outcomes in total.

Notice that HT and TH are
different because the H
could be on the 10p coin
and the T on the 5p coin or
vice versa.

Example 8

A four-sided spinner has 4 equal areas coloured red, blue,
green and yellow. The spinner is spun and a coin is tossed.
List all the possible outcomes in a sample space diagram.

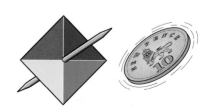

Let R, B, G and Y stand for Red, Blue, Green and Yellow.

		Spinner			
		R	B	G	Y
Coin	Heads (H)	R,H	B,H	G,H	Y,H
	Tails (T)	R,T	B,T	G,T	Y,T

There are 8 possible outcomes altogether.

A sample space diagram is simply a two-way table. It is useful if there are a large number of outcomes.

Exercise 15C

1 Three coins are tossed simultaneously.

(a) List all of the possible outcomes systematically.

(b) Complete a sample space diagram.

(c) Which method is better for recording the data?

2 Two spinners are spun at the same time and the two scores added.

(a) List all the possible outcomes.

(b) What is the probability that the final answer is negative?

3 In a game, you pick a ball from each box. To win a prize, you must select blue and a square number, or red and a prime. List all the outcomes and work out the probability of winning.

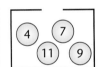

4 An ordinary six-sided dice is thrown. It is then tossed a second time. The scores are then added together. Construct a sample space diagram and from this work out

(a) the probability of throwing two 3s

(b) the probability of the sum being less than 5

(c) the probability that the score on the first throw is double the score on the second throw.

15.4 Independent events and tree diagrams

Key words:
independent events
tree diagram

When two or more events take place, they are **independent** if they have no effect on each other.

Examples are

● tossing a coin then tossing the coin again

● tossing a coin and rolling a dice

● rolling two dice together.

Independent events can happen at the same time or one after the other.

For two independent events A and B, then

$$P(A \text{ and } B) = P(A) \times P(B)$$

and

the total number of outcomes =
total number of outcomes from event A ×
total number of outcomes from event B

> For two independent events, you multiply the probabilities of each outcome.

> The key word used here is 'and'. As long as the outcomes are independent, this means 'multiply' the probabilities.

The results from these events can be put into a sample space diagram or **tree diagram** . In a tree diagram each 'branch' of the tree represents one of the possible outcomes.

Example 9

A six-sided dice and coin are thrown at the same time. One possible outcome is (3, Head).

(a) List all the possible outcomes using a sample space diagram.

(b) How many outcomes are possible?

(c) What is the probability of P(even, Tail)?

(d) Work out the probabilities of all events, using a tree diagram to show this information.

(a)

		Dice					
		1	2	3	4	5	6
Coin	Head (H)	1,H	2,H	3,H	4,H	5,H	6,H
	Tail (T)	1,T	2,T	3,T	4,T	5,T	6,T

> Total number of outcomes = number of outcomes from tossing the dice × number of outcomes from tossing the coin.

(b) There are 6 (from throwing the dice) × 2 (from tossing the coin) = 12 outcomes altogether.

(c) $P(\text{even,T}) = \dfrac{\text{number of successful outcomes}}{\text{total number of outcomes}}$

$$= \tfrac{3}{12} = \tfrac{1}{4}$$

> Those events with (even,T) are highlighted in red, i.e. there are 3 successful outcomes.

continued ▶

(d)

	1	$P(H, 1) = \frac{1}{2} \times \frac{1}{6} = \frac{1}{12}$
	2	$P(H, 2) = \frac{1}{2} \times \frac{1}{6} = \frac{1}{12}$
	3	$P(H, 3) = \frac{1}{2} \times \frac{1}{6} = \frac{1}{12}$
H	4	$P(H, 4) = \frac{1}{2} \times \frac{1}{6} = \frac{1}{12}$
	5	$P(H, 5) = \frac{1}{2} \times \frac{1}{6} = \frac{1}{12}$
	6	$P(H, 6) = \frac{1}{2} \times \frac{1}{6} = \frac{1}{12}$
	1	$P(T, 1) = \frac{1}{2} \times \frac{1}{6} = \frac{1}{12}$
	2	$P(T, 2) = \frac{1}{2} \times \frac{1}{6} = \frac{1}{12}$
	3	$P(T, 3) = \frac{1}{2} \times \frac{1}{6} = \frac{1}{12}$
T	4	$P(T, 4) = \frac{1}{2} \times \frac{1}{6} = \frac{1}{12}$
	5	$P(T, 5) = \frac{1}{2} \times \frac{1}{6} = \frac{1}{12}$
	6	$P(T, 6) = \frac{1}{2} \times \frac{1}{6} = \frac{1}{12}$

As the events are independent, the probability of even and a Tail is P(even *and* Tail)
= P(even) × P(Tail)
= $\frac{1}{2} \times \frac{1}{2} = \frac{1}{4}$.

The events P(even, Tail) are highlighted in red. Multiply the probabilities as you follow the branches.

The probability P(even, Tail) is the sum of three branches from the tree.

P(even, Tail)
= P(T,2) + P(T,4) + P(T,6)
= $\frac{1}{12} + \frac{1}{12} + \frac{1}{12} = \frac{3}{12} = \frac{1}{4}$
as expected.

The diagram could be drawn with the dice first followed by the coin. The probability on each branch would still be the same.

Three independent events

You can extend your tree diagram to find the probability of three or more independent events. For rolling two dice and tossing a coin, the probability of two even numbers and a tail is

$$P(\text{even, even, Tail}) = P(\text{even}) \times P(\text{even}) \times P(\text{Tail})$$
$$= \frac{1}{2} \times \frac{1}{2} \times \frac{1}{2}$$
$$= \frac{1}{8}$$

In general, $P(A, B, C) = P(A) \times P(B) \times P(C)$

Exercise 15D

1 A coin and a three-sided spinner are used at the same time.

(a) Draw a tree diagram to show all the possible outcomes.

(b) What is the probability of getting a red and a tail?

2 A four-sided dice with the numbers 1, 2, 3, 4 and a spinner with three equal sectors coloured red, white and blue are thrown at the same time.

(a) Draw a tree diagram to show all the possible outcomes.

(b) What is the probability of getting a red or a white and an even number?

3 Peter takes two exams. The probability that he passes mathematics is 0.8. The probability that he passes history is 0.6. The results of the two exams are independent.

(a) Draw a tree diagram to show all the possible outcomes.

(b) What is the probability that (i) he passes mathematics and fails history (ii) he fails both subjects?

4 A bag contains 5 red counters and 3 green counters. A counter is taken at random from the bag and then replaced. Another counter is then taken from the bag.

(a) Draw a tree diagram to show all the possible outcomes.

(b) What is the probability of picking:
(i) two red counters (ii) two counters of the same colour
(iii) at least one green counter?

5 An ordinary six-sided dice is tossed twice. Draw a tree diagram and show that the probability of obtaining just one six is $\frac{5}{18}$.

15.5 Conditional probability

Key words:
conditional
dependent events

The probability that one event occurs may depend upon another event having taken place. In this case the event is conditional . The second event is a dependent event .

In questions, look out for phrases such as 'without replacement' or 'not replaced', as these tell you that a second event is conditional.

Example 10

There are 6 red counters and 5 blue counters in a bag. A counter is taken at random from the bag and not replaced. A second counter is then taken out. What is the probability that they are both red?

First counter: $P(\text{red}) = \frac{6}{11}$

Second counter: $P(\text{red}) = \frac{5}{10}$

The probability that both counters are red is

$P(\text{red and red}) = \frac{6}{11} \times \frac{5}{10} = \frac{30}{110} = \frac{3}{11}$

The events are clearly dependent.

After the first counter has been removed it is not replaced. There are now only ten counters in the bag.

When more than one probability is required, draw a tree diagram. Work out the conditional probabilities for each of the branches.

Example 11

James has a drawer containing 4 pairs of white socks and 5 pairs of black socks. He takes two socks out of the drawer at random.
Draw a probability tree diagram to show all the possible outcomes.
Work out the probability that he takes out

(a) a pair of black socks   **(b)** a pair of white socks

(c) two socks of different colours.

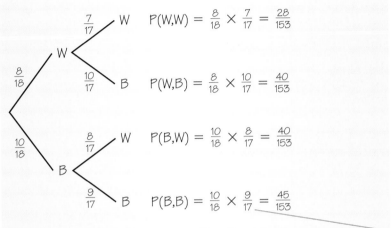

On the right of the tree diagram write out the conditional probabilities of each branch. Remember that the total probabilities from all the branches must add up to 1. This is always a good check.

The sum of all the conditional probabilities is

$\frac{28}{153} + \frac{40}{153} + \frac{40}{153} + \frac{45}{153} = 1$.

There are 17 socks left and 9 of them are black.

(a) Pair of black socks $P(B,B) = \frac{45}{153} = \frac{5}{17}$

(b) Pair of white socks $P(W,W) = \frac{28}{153}$

You need not cancel probability fractions unless the question asks you to do so.

(c) Pair of different coloured socks is

$$P(B,W) + P(W,B) = \frac{40}{153} + \frac{40}{153} = \frac{80}{153}$$

Exercise 15E

1 There are 5 black pens and 3 red pens in a drawer. A pen is taken and not replaced. A second pen is then taken.

   (a) Draw a tree diagram showing all the possible outcomes.

   (b) What is the probability that both pens taken will be red?

2 A bag contains 5 green balls and 4 blue balls. A ball is drawn at random and not replaced. A second ball is then picked from the bag. What is the probability that

   (a) both balls are blue   **(b)** both balls are a different colour?

3 There are ten white carnations and six red roses. Sophie picks two flowers at random. Find the probability that they are

   (a) different   **(b)** both roses.

4 A box contains 15 chocolates. Ten have hard centres and five have soft centres. Debbie selects two chocolates at random.

  (a) Draw a tree diagram to show all the possible outcomes.

  (b) What is the probability that she chooses
    (i)   two hard centres   **(ii)**   one of each kind
    (iii) two of the same kind?

5 A bag contains 10 counters, 2 are red, 3 are blue and 5 are yellow. Steve takes two counters out of the box, at random.

  (a) Draw a tree diagram to show all the possible outcomes.

  (b) What is the probability that he chooses
    (i)   two yellow counters   **(ii)**   two counters of the same colour
    (iii) two counters of different colours   **(iv)** at least one red counter?

Examination questions

1 Red, blue, white and green tickets are sold in a raffle.

Ticket colour	Probability of winning first prize
Red	0.4
Blue	0.3
White	0.1
Green	

  (a) Calculate the probability of a green ticket winning the first prize.

  (b) 2000 tickets were sold in this raffle. Calculate how many red and blue tickets were sold.

2 A spinner has a red sector (R) and a yellow sector (Y). The arrow is spun 1000 times.

  (a) The results for the first 20 spins are shown below.

     R R Y Y Y R Y Y R Y Y Y Y Y Y R Y R Y Y Y

    Work out the relative frequency of a red after 20 spins.

  (b) The table shows the relative frequency of a red after different number of spins.

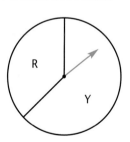

Number of spins	Relative frequency of a red
50	0.42
100	0.36
200	0.34
500	0.3
1000	0.32

  How many times was a red obtained after 200 spins?

                                       (4 marks)
                             AQA, Spec A, 1I, November 2004

3 The diagram shows a spinner. When the arrow is spun the probability of scoring 2 is 0.3. The arrow is spun twice and the scores are added.

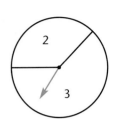

(a) Copy and complete the tree diagram

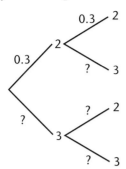

(b) What is the probability that the total score is 4? *(3 marks)*

AQA, Spec A, 1I, June 2003

4 (a) Matthew has a dice with 3 red faces, 2 blue faces and 1 green face.
He throws the dice 300 times. The results are shown in the table.

Red	Blue	Green
153	98	49

   (i) What is the relative frequency of throwing a red?

   (ii) Is the dice fair? Explain your answer.

(b) Emmie has a dice with 4 red faces and 2 blue faces.
She throws the dice 10 times and gets 2 reds.
Emmie says the dice is **not** fair. Explain why Emmie could be wrong. *(4 marks)*

AQA, Spec A, 1H, November 2003

5 A bag contains 3 green balls and 7 white balls.
A ball is taken from the bag at random and replaced.
Another ball is then taken from the bag at random.

(a) Draw a tree diagram to show all the possible outcomes.

(b) What is the probability that both balls are the same colour? *(3 marks)*

6 A bag contains 8 balls. 5 are black and 3 are white.
A ball is taken out of the bag at random and **not** replaced.
Another ball is taken out of the bag at random.
What is the probability that both of the balls are the same colour? *(3 marks)*

Summary of key points

Probability (grade D/C/B)

Probability can be shown on a probability scale between the values 0 and 1.

Probability can be expressed as a fraction, percentage or decimal.

An outcome is the result that actually occurs. The outcomes from tossing a coin are Head or Tail.

$$\text{Theoretical probability} = \frac{\text{number of successful outcomes}}{\text{total number of possible outcomes}}$$

Relative frequency (grade C)

$$\text{Relative frequency} = \frac{\text{number of successful trials}}{\text{total number of trials}}$$

Frequency = total number of trials × relative frequency

Mutually exclusive outcomes (grade D)

If there are n mutually exclusive outcomes, all equally likely, then the probability of just one outcome happening is $\frac{1}{n}$. If there are n mutually exclusive outcomes and m successful outcomes then the probability of a successful outcome is $\frac{m}{n}$.

For any two outcomes, A and B, that are mutually exclusive, and exhaustive, then
P(A or B) = P(A) + P(B) = 1.
The probability of the outcome A *not* happening is given by P(not A) = 1 − P(A).

Independent events (grades A to D)

For two independent events A and B

P(A and B) = P(A) × P(B)
and the total number of outcomes = total number of outcomes from event A × total number of outcomes from event B.

Recording outcomes (grades C/B)

Always list the collection of all possible outcomes systematically. Use lists, sample space diagrams and tree diagrams as appropriate.

Always work out the probabilities of each branch of a tree diagram. The total probability (adding all the branches together) should always be equal to 1.

Dependent or conditional events (grades A/A*)

Two or more events are dependent if one event is conditional on another. These problems often use the words 'without replacement'.

16 Linear graphs

Algebra 5

This chapter will show you how to:

- ✔ describe and use the equation $y = mx + c$
- ✔ find equations of parallel and perpendicular lines through given points
- ✔ solve simultaneous equations graphically
- ✔ solve graphical inequalities
- ✔ interpret conversion graphs, distance–time and velocity–time graphs

16.1 Linear graphs

Key words:
line segment
linear graph
variables
linear equation

Any two points $A(x_1, y_1)$ and $B(x_2, y_2)$ on an x–y coordinate grid can be joined together to form a straight line. The section of line between A and B is called a **line segment** .

The straight line can also be extended beyond A and B and the resulting graph is called a **linear graph** .

When points lie on a straight line there is a connection between the **variables** x and y.

A linear graph is simply the picture of a **linear equation** .

In a linear equation the highest power of x is 1. Examples of linear equations include $y = 2x$, $y = -3x + 2$, $2x + 3y = 5$, $y = -2$ and $x = 4$.

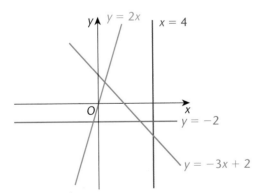

The two equations $y = -2$ and $x = 4$ do not contain a variable on the right-hand side. These equations represent straight line graphs parallel to the x-axis ($y = -2$) and parallel to the y-axis ($x = 4$).

To draw straight line graphs from a linear equation, first construct a table of results for values of x and y, then plot these on a graph.

You need a minimum of three points to draw a straight line graph. Two points to define a straight line and a third point as a check.

Example 1

Plot the graph of $y = x + 1$ for values $-2 \leqslant x \leqslant 2$.

When $x = -2$ $\quad y = -2 + 1 = -1$

When $x = 0$ $\quad\quad y = 0 + 1 = 1$

When $x = 2$ $\quad\quad y = 2 + 1 = 3$

x	-2	0	2
y	-1	1	3

Substitute the given x values into the equation to find the corresponding y values.

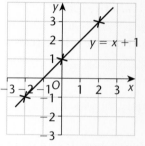

Plot these three points and join them up with a straight line. It is a good idea to extend your line just beyond the end points. Always label your axes and label the line with the equation.

Guidelines to help you plot straight lines

- Choose a minimum of three values for x (always include 0).
- Put these values into the equation and work out the corresponding values for y.
- Write the results in a table.
- From the results identify the extreme values for $+x$ and $-x$ and $+y$ and $-y$ and draw a coordinate grid using these values.
- Plot the points from the table.
- Draw a straight line through all the points.
- Label the line with the equation.

Exercise 16A

1 Draw a coordinate grid with both the x- and y-axes from -5 to 5. On the grid draw and label the lines

(a) $x = 4$ (b) $y = -1$ (c) $x = -5$ (d) $y = 4$.

2 In questions (a) to (f), use the same scale on both axes and draw x- and y-axes between -5 and $+5$.

Draw these straight line graphs. For each one

(i)   make a table of values, choosing at least three values of x

(ii)   work out the values of y using the equation of the line

(iii) plot the points and draw a straight line through them.

(a) $y = x - 3$ (b) $y = 2x + 3$ (c) $y = 4 - x$

(d) $y = 3x - 1$ (e) $y = 2 - 2x$ (f) $y = 1 - \frac{1}{2}x$

3 On squared paper draw x- and y-axes from -6 to 6.
Using x-values of -3, 0 and 3 find the corresponding values of y for the equation $y = x - 2$.
Put these results into a table and plot them on the graph.

4 On squared paper, draw x- and y-axes from -10 to 10.
Using the x values -5, 0 and 3, find the corresponding values for y for the equation $y = 2x + 3$.
Put these results into a table and plot the graph.

5 Using x values of -4, 0 and 2, find the corresponding values of y for the equation $y = -x + 4$.
Draw an appropriate coordinate grid and plot these results.

6 Draw the graph of $y = 3x - 1$.

7 Using the same scale on both axes, draw x- and y-axes from -3 to $+3$.
Draw the graphs of $y = x$ and $y = -x$ on the same coordinate grid.
What do you notice about these graphs?

8 Using the same scale on both axes, draw x- and y-axes from -5 to $+5$. Draw the graphs of $y = 2x - 1$ and $y = \frac{1}{2}x + 2$ on the same coordinate grid.
Write down the coordinates of the point where these graphs cross each other.

Finding the mid-point of a line

Key words:
mid-point

You can find the **mid-point** of a line segment AB, if you know the coordinates of A and B.
If A is at (x_1, y_1) and B is point (x_2, y_2), the mid-point
$$(x, y) = \left(\frac{x_1 + x_2}{2}, \frac{y_1 + y_2}{2}\right)$$

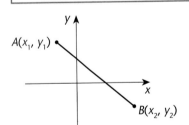

Example 2

Work out the coordinates of the mid-point of the line segment RS where R has coordinates $(1, 7)$ and S $(4, 6)$.

$$(x, y) = \left(\frac{1 + 4}{2}, \frac{7 + 6}{2}\right)$$

$$(x, y) = \left(\frac{5}{2}, \frac{13}{2}\right)$$

$$(x, y) = (2\tfrac{1}{2}, 6\tfrac{1}{2})$$

This is similar to finding the mean of two numbers.

Exercise 16B

1 Write down the coordinates of the end points for each of the
following line segments. For each line segment work out the
coordinates of the mid-point.

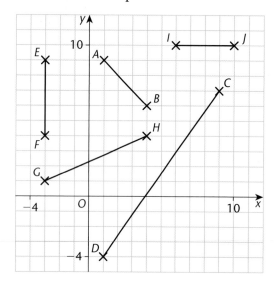

$$\frac{7+7}{2} , \frac{9+2}{2}$$

2 Without drawing these line segments, work out their mid-points.

  (a) $A(1, 1)\ B(8, 1)$ **(b)** $C(7, 9)\ D(7, 2)$

  (c) $E(2, 3)\ F(5, 9)$ **(d)** $G(-4, 5)\ H(2, 5)$

  (e) $I(-2, 2)\ J(3, -3)$

16.2 Finding equations of straight lines

Key words:
gradient (*m*)
positive gradient
negative gradient
y-intercept (*c*)

Here, the slope of the line, from left to right,
is upwards.
The ratio

    $\dfrac{\text{vertical distance}}{\text{horizontal distance}}$

gives a measure of the steepness of the line and is called the
gradient of the line. It is denoted by the letter *m*. This line has a
positive gradient .

If you have a line that slopes in the opposite direction, then one of
the values in the ratio will be negative. This line has a
negative gradient .

You can often find the gradient by looking at the line.

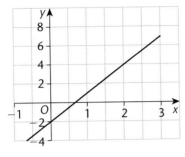

In simple terms, 3 up for every 1 across.

x	0	1	2	3
y	−2	1	4	7

Gradient = amount of increase in y-value for unit increase in x-value.

Look at the table of values for this line. You can see that each time the x-value increases by 1, the y-value increases by 3. This means that the gradient of the graph is 3.

For a line that passes through any two points $A(x_1, y_1)$ and $B(x_2, y_2)$ the gradient, m, is given by the expression

$$m = \frac{y_2 - y_1}{x_2 - x_1} = \frac{\text{difference in } y\text{-values}}{\text{difference in } x\text{-values}}$$

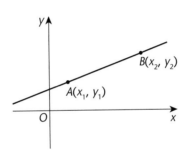

For example if A is $(-1, 1)$ and B is $(1, 5)$ then,

$$m = \frac{5 - 1}{1 - (-1)} = \frac{4}{2} = 2$$

It often helps to draw a sketch.

The place where the line crosses the y-axis is the **y-intercept** and is denoted by c.

It is the value of y when $x = 0$.
For this graph $c = 3$

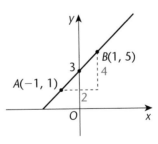

Straight lines that pass through the negative part of the y-axis will have negative intercept values. Lines that pass through the origin have an intercept value of zero.

The two values m and c are all that are needed to define a straight line.

All equations of straight lines have the same general form

$$y = mx + c$$

where m is the gradient of the line and c is the y-intercept.

For the line segment joining $A(-1, 1)$ and $B(1, 5)$ the equation is

$$y = 2x + 3$$

You can find the equation of any straight line if you know the coordinates of any two points on the line.

Example 3

A line passes through the points $A(-4, 5)$ and $B(2, -10)$. Find

(a) the gradient of the line

(b) the equation of the line.

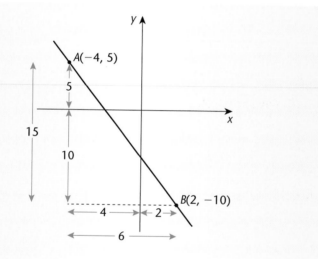

First draw a sketch.

(a) Difference in y-values $= 5 - (-10) = 5 + 10 = 15$

Difference in x-values $= -4 - 2 = -6$

To find the gradient, find the difference between the y values and the difference between the x values.

Gradient, $m = \dfrac{\text{Difference in y-values}}{\text{Difference in x-values}} = \dfrac{15}{-6} = -\dfrac{5}{2}$

Now find the ratio.

(b) $y = -\dfrac{5}{2}x + c$

This line passes through point A with x-value -4 and y-value 5.

$5 = -\dfrac{5}{2}(-4) + c$

$5 = 10 + c$

$c = -5$

To find the equation of the line choose one of the points, say $A(-4, 5)$. Using $y = mx + c$ with the m value given above you can find the value of c.

The same result for the value of c could have been obtained by using the other point $B(2, -10)$.

The equation of the straight line is

$$y = -\dfrac{5}{2}x - 5$$

Now combine the results of m and c into one equation. Don't forget the y and x!

Parallel lines

Key words:
parallel
perpendicular

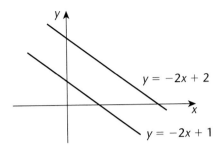

It is useful to know how the gradients of **parallel** lines and **perpendicular** lines are related.

Parallel lines have the same gradient.

The graph above shows two lines both with $m = -2$.

Perpendicular lines

Perpendicular lines are straight lines that cross over each other at right angles.

The gradient of a line that is perpendicular to a line with a gradient m is the reciprocal of m with a change in sign,
$$-\frac{1}{m}$$

This is always true whatever the value of m. If two lines are perpendicular then the *product* of their gradients is always equal to -1.
$$m \times -\frac{1}{m} = -1$$

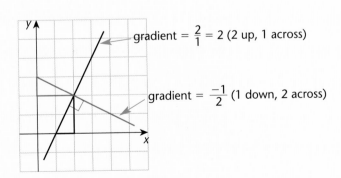

The product of the gradients of two perpendicular lines $= -1$.

Example 4

Find the equation of the line perpendicular to $y = -\frac{5}{2}x - 5$ that passes through the point $B(2, -10)$.

Gradient of line, $m = -\dfrac{5}{2}$

> Find the gradient of the perpendicular line by inverting the fraction and changing the sign.

Gradient of perpendicular line is $-\dfrac{1}{m} = \dfrac{2}{5}$

The equation of the perpendicular line is

$$y = \frac{2}{5}x + c$$

> Form the general equation of this line.

and this passes through point B with x-value 2 and y-value -10

$$-10 = \frac{2}{5}(2) + c$$

> Now work out the value of c knowing that the line passes through the point $B(2, -10)$.

$$-50 = 4 + 5c$$

$$5c = -54$$

$$c = \frac{54}{5}$$

The equation of the perpendicular line is

$$y = \frac{2}{5}x - \frac{54}{5}$$

$$5y = 2x - 54$$

> Now substitute the value of c back into the general equation. The fractional parts can be eliminated by multiplying throughout by 5.

or

$$2x - 5y = 54$$

If a line passes through two points $A(x_1, y_1)$ and $B(x_2, y_2)$ then the gradient of the line is given by

$$m = \frac{y_2 - y_1}{x_2 - x_1} = \frac{\text{difference in the } y \text{ values}}{\text{difference in the } x \text{ values}}$$

The value of the y-intercept is calculated using the equation given by $y_1 = mx_1 + c$ or $y_2 = mx_2 + c$.

The gradient of a line that is perpendicular to $y = mx + c$ is given by $-\dfrac{1}{m}$.

Exercise 16C

1 Find the gradient of the lines joining the points
   (a) $(1, 2)$ and $(5, 10)$
   (b) $(-1, 6)$ and $(2, -3)$
   (c) $(-4, -1)$ and $(0, -2)$

2 A line has a gradient of -2 and passes through the point with coordinate $(4, 6)$. Find the equation of the line.

3 A line has a gradient of $m = \frac{1}{3}$ and passes through the point $(-3, 5)$. Find the equation of the line.

4 Find **(a)** the gradients of the lines and **(b)** the equation of the lines that pass through the following points.
   (a) $A(5, 2)$ $B(4, 1)$ (b) $C(-6, -4)$ $D(-3, -7)$
   (c) $E(1, 4)$ $F(6, 3)$ (d) $G(-2, 6)$ $H(-1, -7)$

5 Write down the gradients of the lines that are perpendicular to the lines with the following gradients.
   (a) 6 (b) -4 (c) $\frac{2}{3}$ (d) $-\frac{1}{4}$
   (e) $1\frac{1}{2}$ (f) $-3\frac{1}{5}$ (g) 0.6 (h) -1.25

6 A line passes through the origin and is perpendicular to the line with equation $2x + y = 5$. Find the equation of the line.

7 A line is perpendicular to the line with equation $y = 4x - 3$ and crosses the y-axis at $y = 7$. Find the equation of the line.

8 **Without plotting** these straight lines identify which lines are parallel to the line $y = x - 1$.
   (a) $y = x + 1$ (b) $y = x - 7$ (c) $y = 2x - 1$
   (d) $2y = x - 1$ (e) $y = x + \frac{1}{2}$

9 **Without plotting** these straight lines identify which lines are parallel to $y = -3x - 2$.
   (a) $y = 3x - 2$ (b) $y = -3x + 2$ (c) $y = -2x - 3$
   (d) $y = -2 - 3x$ (e) $y = 2 + 3x$

16.3 Using the general equation $y = mx + c$

Key words:
coefficient

The equation of a straight line is $y = x + 2$. The gradient m of this line is 1 and the y-intercept value is 2. Both of these values can be obtained directly from the equation.

The gradient is given by the value of the number in front of the 'x' (the **coefficient** of x). It is important to include the sign. For the line $y = x + 2$, the gradient is 1. For an equation $y = -2x - 5$ the gradient is -2.

In algebra you write x to represent $1x$ and $-x$ to represent $-1x$.

The y-intercept, c, is given by the number term. In the equation $y = x + 1$ the intercept, c, is $+1$, and in the equation $y = -2x - 5$ the intercept, c, is -5.

It is important to include the sign in the values of both m and c.

Example 5

State the gradients and intercept values for the equations

(a) $y = x + 8$

(b) $y = -3x + 3$

(c) $y = -\frac{1}{2}x - 5$

(d) $y = 2.6x + 0.4$

(a) gradient = 1, intercept = 8

(b) gradient = -3, intercept = 3

(c) gradient = $-\frac{1}{2}$, intercept = -5

(d) gradient = 2.6, intercept = 0.4

Gradients and intercepts can also be fractions and decimals. Remember to include the sign in front of the gradient and intercept.

You can always apply this, provided the equation of the straight line is in the form

$$y = mx + c$$

This is not always the case. You may need to alter an equation to make y the subject of the equation. Then you can find the values of m and c directly. The values of m and c may also help you plot the graph.

You need to be aware that equations can be presented in a variety of ways. This is important for simultaneous equations, later on.

Example 6

Which of the following equations is equivalent to the equation $y = x + 1$?

(a) $y - x = 1$ (b) $-x + y = 1$ (c) $-x = 1 - y$ (d) $y - x - 1 = 0$ (e) $0 = x + 1 - y$

(a) $y - x = 1$

$y = 1 + x$

Add x to both sides.

(b) $-x + y = 1$

$y = 1 + x$

Add x to both sides.

(c) $-x = 1 - y$

$y + \cancel{x} - \cancel{x} = \cancel{y} + x + 1 - \cancel{y}$

$y = x + 1$

Add $x + y$ to both sides.

(d) $y - x - 1 = 0$

$y = x + 1$

Add $x + 1$ to both sides.

(e) $0 = x + 1 - y$

$y = x + 1$

Add y to both sides.

All the above equations are identical to $y = x + 1$.

Example 7

Write the following equations in the form $y = mx + c$.

(a) $2y = 6x - 12$ (b) $5y = -10x + 20$

(a) $2y = 6x - 12$

        $y = \dfrac{6x}{2} - \dfrac{12}{2}$ Divide both sides by 2.

        $y = 3x - 6$

(b) $5y = -10x + 20$ Divide both sides by 5.

        $y = -\dfrac{10x}{5} + \dfrac{20}{5}$

        $y = -2x + 4$

Example 8

Find the gradient and intercept of these graphs.

(a) $3x = y - 5$ (b) $\frac{1}{3}y + \frac{1}{4}x + \frac{1}{6} = 0$

(a) $3x = y - 5$ Add 5 to both sides and

      $3x + 5 = y$ turn the equation around

         $y = 3x + 5$ to make y the subject on the left hand side.

     Gradient $= 3$; intercept $= 5$

(b) $\frac{1}{3}y + \frac{1}{4}x + \frac{1}{6} = 0$ Remove fractional parts by

     $4y + 3x + 2 = 0$ multiplying throughout by the lowest common

         $4y = -3x - 2$ denominator, 12. Then

            $y = -\frac{3}{4}x - \frac{1}{2}$ make y the subject by dividing both sides by 4.

      Gradient $= -\frac{3}{4}$; intercept $= -\frac{1}{2}$

Exercise 16D

1 For the following equations write down the value of the gradient and intercept.

(a) $y = 3x + 2$ (b) $y = -x + 3$

(c) $y = 4x - 7$ (d) $y = -3x - 4$

(e) $y = \frac{2}{3}x + 8$ (f) $y = 0.8x - 0.3$

2 Re-write the following equations in the form $y = mx + c$. For each one write down the value of the gradient and intercept.

(a) $3y = 9x + 18$
(b) $2y = -8x - 4$
(c) $8y = -24x + 8$
(d) $y - 7 = -2x$
(e) $2x + 6 = 2y$
(f) $-3x = 4 + y$
(g) $2y + 3x = 2$
(h) $x = -3y - 2$
(i) $4y + 2 + 3x = 0$
(j) $0.3x = 2.4 + 0.6y$
(k) $\frac{1}{2}x - \frac{1}{3}y = \frac{1}{4}$
(l) $\frac{1}{6}y - \frac{1}{2} + \frac{1}{3}x = 0$

16.4 Using graphs to solve simultaneous equations

Key words:
graphical solutions
simultaneous equations

Approximate solutions to linear **simultaneous equations** can be found graphically. The point of intersection of the two straight lines gives the values of x and y that satisfy both equations simultaneously.

Solving simultaneous equations algebraically was covered in Chapter 5.

Graphical solutions to simultaneous equations are only approximate as they depend on the accuracy of your drawing.

Example 9

Solve the simultaneous equations graphically.

$2x - y = 3$
$x + 2y = 14$

Always read the question carefully. If you are asked to solve a pair of simultaneous equations graphically then you **must** draw the graphs. An algebraic solution will award you no marks.

$2x - y = 3$

x	0	2	4
y	-3	1	5

Construct tables of values of x and y for both linear equations.

$x + 2y = 14$

x	0	2	4
y	7	6	5

The range of x values has to be large enough for the two lines to cross over each other. You will only see this when the lines are plotted. If they do not cross you will need to plot more points or extend the line.

When plotting straight lines always use a minimum of three points.

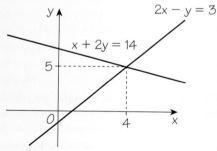

The point of intersection is $(4, 5)$

$x = 4$ and $y = 5$.

Check: $2x - y = 3$

$2 \times 4 - 5 = 8 - 5 = 3$ ✓

Always draw your lines as accurately as possible.

Always check your answer by substituting into one of the equations given.

Exercise 16E

1 Solve the following simultaneous equations graphically.

$$x - y = 1$$
$$x + 2y = 4$$

Tabulate the x- and y-values for x between 0 and 5. Draw the x-axis for the graph between −1 and 6 and the y-axis between 2 and 4.

2 Solve the following pair of simultaneous equations by drawing a graph with x-axis between −5 and 5 and y-axis between −5 and 5.

$$x + 3y = 8$$
$$3x + 2y = -4$$

3 Find a graphical solution to the simultaneous equations

$$3x - y = -3$$
$$x + y = -1$$

4 Solve the pair of simultaneous equations graphically

$$2x - y = 3$$
$$x + 2y = 4$$

What can you say about the two lines?

Look at the gradient of each line.

16.5 Graphical inequalities

Key words:
region
boundary region

Inequalities often describe a range of values that can be represented on a number line.

See Chapter 5, page 74.

Inequalities can also be represented by a region or area on a graph.

For the inequalities ⩽ and ⩾ the **boundary region** is indicated by a continuous line. For the inequalities < and > the boundary region is indicated by a dashed line.

Example 10

Show the region $x \geqslant 2$.

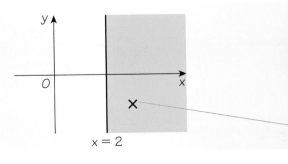

$x = 2$

Begin by drawing the line $x = 2$. Then shade in the region that satisfies the inequality. This may be found by substituting coordinate values into the inequality to see if it is true for a given point.

For example, for this point $(3, -1)$, the x-coordinate $\geqslant 2$.

Example 11

Show the region given by the inequality $y < 2x$.

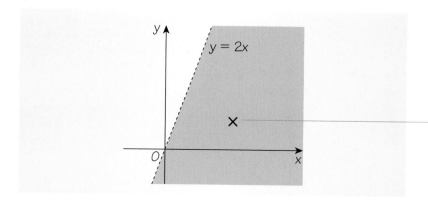

Remember:
$<$ or $>$ ---------------------------

$\leqslant$ or $\geqslant$ _____

For example, for the point $(3, 1)$, $y = 1$ and $2x = 6$, so $y < 2x$.

Example 12

Show the region that satisfies the inequality $3x + y \leqslant 3$.

When $y = 0$, $x = 1$ and when $x = 0$, $y = 3$.

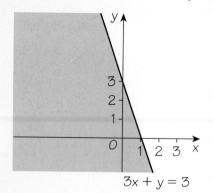

$3x + y = 3$

First draw the line $3x + y = 3$.

This line is easy to plot if you find the points where it crosses the x- and y-axes. These are found by putting $y = 0$ and working out x, then putting $x = 0$ and working out y.

It is not always obvious which side of the line should be shaded. Take a point e.g. the origin $(0, 0)$. Putting these values into the inequality gives $0 + 0 \leqslant 3$, which is true, so this point lies inside the required region.

In some cases more than one inequality is given. In these cases the region that satisfies all inequalities simultaneously is the required region.

Example 13

Draw a graph that shows the region satisfying the following inequalities.

$x < 7, y \geqslant 1, y < x - 3$

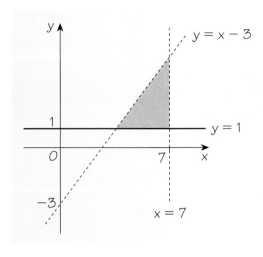

Begin by drawing the lines $x = 7$, $y = 1$ and $y = x - 3$.

The region required is

$x < 7$ i.e. to the left of the dashed line $x = 7$

$y \geqslant 1$ i.e. above the solid line $y = 1$

$y < x - 3$ i.e. below the dashed line $y = x - 3$

Only those points on the line $y = 1$ within the region shaded can be included.

Exercise 16F

1 For the following graphs, write down the inequalities which describe the region shaded.

(a)

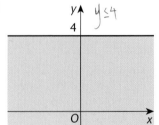

$y \leqslant 4$

(b)

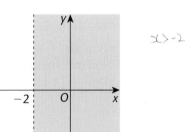

$x > -2$

$\frac{6}{3} = 2$

(c)

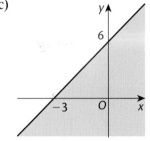

$(0, 6)$

$(-3, 0)$

(d)

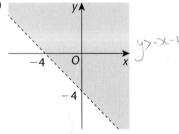

$y > -x - 4$

$y = xc$

$6 = 0 + c$

$c = 6$

$0 = 0 + 6$

$0 = 6$ $y < 2x + 6$

$0 < 6$

2 In the following questions draw graphs to show the region that satisfies the inequality.

   (a) $x \leqslant 2$ **(b)** $y > -2$ **(c)** $-1 < x \leqslant 4$

   (d) $0 \leqslant y \leqslant 3$ **(e)** $y \geqslant 3x$ **(f)** $y < 3x + 2$

   (g) $y \geqslant x - 5$ **(h)** $2x + y < 3$

3 Shade in the region which satisfies the following inequalities.

   (a) $x < 4$ and $x + y \geqslant 2$ **(b)** $y \leqslant 3$ and $2x + y \geqslant 1$

4 Sketch the region bounded by the inequalities $x \geqslant -2$, $y < 7$ and $x + y > 1$.

$x = -4$

$y > x - 4$

$y = x + -(x + 4)$

5 The region R is shown shaded. Write down the three inequalities which together describe the shaded region.

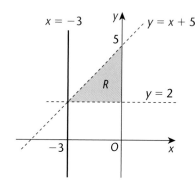

16.6 Conversion graphs

A **conversion graph** is used to convert one type of measurement into another type of measurement, usually with different units. Conversion graphs can be linear or curved. Only linear graphs are covered in this chapter.

Curved conversion graphs are covered in Chapter 25.

Example 14

(a) Construct a table to show the conversion between degrees Celsius (°C) and degrees Fahrenheit (°F) for values $-40 \leqslant C \leqslant 100$. Use the equation
$$F = 2C + 30$$

(b) Draw a conversion graph.

(c) What is the temperature in °F when it is 10°C?

(d) If the temperature is 160°F what is this in °C?

(a)

C	-40	-20	0	20	40	60	80	100
$2C$	-80	-40	0	40	80	120	160	200
$+30$	30	30	30	30	30	30	30	30
F	-50	-10	30	70	110	150	190	230

continued ▶

(b)

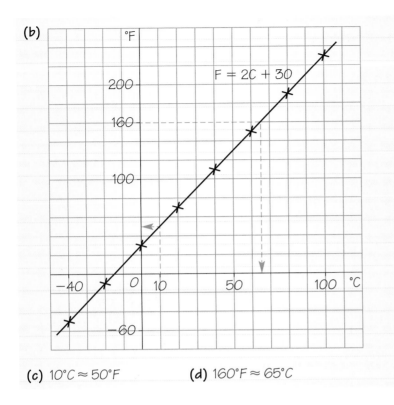

$F = 2C + 30$

Draw dashed lines on the graph at 10°C and 160°F to meet the conversion line. From these points draw dashed lines to meet the axes and read off the answers.

(c) $10°C \approx 50°F$ **(d)** $160°F \approx 65°C$

Exercise 16G

1 Copy and complete the conversion table between km and miles using the conversion 5 miles = 8 km.

miles (x)	0	5	10	20	50	100
km (y)		8			80	

Draw a conversion graph with x-values from 0 to 100 miles and y-values from 0 to 200 km.

(a) How many miles are there in 75 km?

(b) How many km are there in 75 miles?

2 Copy and complete the conversion table between centimetres (to 1 d.p.) and inches.

cm	0				12.5		25	
inches	0	1	2	3	5	6	10	12

Draw a conversion graph between cm and inches. From your graph work out

(a) how many cm there are in 4 inches and

(b) how many inches there are in 20 cm.

First work out how many cm there are in 1 inch.

16.7 Graphs describing journeys

Distance–time graphs

Always plot time on the horizontal axis (the *x*-axis) and distance on the vertical axis (the *y*-axis). You can use **distance–time graphs** to work out the **speed** for part of a journey.

$$\text{Speed} = \frac{\text{distance}}{\text{time}}$$

If distance is measured in metres and time in seconds then the units of speed are metres per second or m/s. Other units of speed are kilometres per hour (km/h) and miles per hour (mph).

m/s can also be represented by ms⁻¹ and km/h by kmh⁻¹ or kph. The short form mph usually means miles per hour.

You can also use a distance–time graph to work out an **average speed** for a journey.

Average speed for a journey means the same as the mean speed for the journey.

$$\textbf{Average speed} = \frac{\textbf{total distance}}{\textbf{total time}}$$

Units for average speed are identical to those used for speed.

Example 15

Part of Jen's cycle ride is shown in this distance–time graph.

(a) How long did Jen stop at (i) *B* and (ii) *C*?

(b) What was the speed between *A* and *B*?

(c) What was the average speed between *A* and *D*?

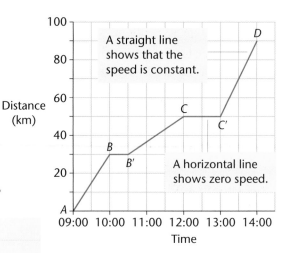

(a) (i) 30 mins **(ii)** 1 hour

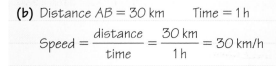
(b) Distance $AB = 30$ km Time $= 1$ h

$$\text{Speed} = \frac{\text{distance}}{\text{time}} = \frac{30 \text{ km}}{1 \text{ h}} = 30 \text{ km/h}$$

Always write down the data you need from the graph.

(c) Total distance $= 90$ km and total time $= 5$ hours

$$\text{Average speed} = \frac{\text{total distance}}{\text{total time}}$$

Average speed uses total distance and total time for the journey.

$$= \frac{90 \text{ km}}{5 \text{ h}} = 18 \text{ km/h}$$

Exercise 16H

1 Chloë walks to school each morning, stopping off at the shop on the way. The distance–time graph shows her journey to school.

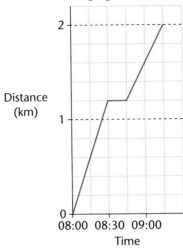

(a) What time does she get to the shop?

(b) How long does she spend in the shop?

(c) How long does it take Chloe to walk to school?

(d) What is her average walking speed?

2 The distance–time graph shows the cycle journey that Imran undertook from home. He made two stops at shops on his journey before returning home.

(a) How long did Imran's cycle ride take?

(b) Which part of his journey did he travel at the greatest speed?

(c) How can you tell?

(d) What was his average speed for the whole cycle ride?

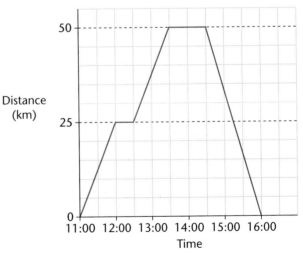

3 Glyn walked to the village shop, a distance of 800 m.
It took him 10 minutes to get there and he was in the shop for 5 minutes.
He then walked to the Post Office, another 200 m.
This took 3 minutes and he was in the Post Office for 6 minutes.
He then walked back home which took 16 minutes.

(a) Draw a distance–time graph for his journey.

(b) How fast did he walk to the shop?

(c) What was his average speed for the whole journey?

4 Marcus leaves home at 08.15 and walks 400 m to the bus stop in 6 minutes. He waits 5 minutes for the bus to arrive. The bus journey is 4000 m and the bus arrives near school at 08.38. He then walks another 100 m to school in 2 minutes.

(a) Draw a distance–time graph for his journey.

(b) What was the average speed of the bus in metres per minute?

(c) Convert this speed to km/h.

> Do the same as you did for question 3 in questions 4 and 5.
>
> Plan sensible scales for the axes.

5 Alison went to town in her car. She drove the 12 miles into town in 25 minutes. She was in town for $1\frac{1}{4}$ hours then left to drive home. After 20 minutes of her journey home, 9 miles from town, she stopped at her Aunt's house and stayed for $\frac{1}{2}$ hour. She then continued her journey home, taking another 10 minutes.

(a) Draw a distance–time graph for her journey.

(b) What was her average speed in mph on the way to town?

Velocity–time graphs

> **Key words:**
> velocity
> velocity–time graph
> acceleration
> deceleration

When the direction of travel is given, **velocity** is used instead of speed.

On a **velocity–time graph** plot time on the x-axis and plot velocity on the y-axis. The gradient gives the rate of change of velocity, or **acceleration**. Constant acceleration gives a straight line.

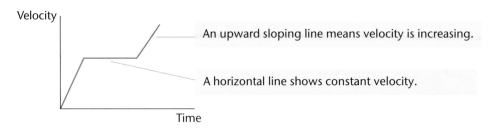

An upward sloping line means velocity is increasing.

A horizontal line shows constant velocity.

$$\text{Acceleration} = \frac{\text{velocity}}{\text{time}}$$

If the velocity is measured in metres per second (m/s) and time in seconds (s) the units of acceleration are metres per second per second or m/s^2 (ms^{-2}). A positive value for acceleration involves speeding up. A negative value involves slowing down (or **deceleration**).

The *area* under a velocity–time graph gives the

average velocity × time taken.

This is the same as the distance travelled.

> Working out areas under a velocity–time graph usually involves working out the areas of trapeziums.

Example 16

A car starts out from rest and travels, at constant acceleration, 75 m in 5 seconds before continuing its journey at a constant speed for a further 30 seconds. After this time it slows down uniformly, taking 10 seconds to come to a complete stop. Draw the velocity–time graph for this journey. From the graph find

(a) the acceleration in the first 5 seconds

(b) the deceleration

(c) the total distance travelled.

'Slows down uniformly' means that the deceleration is constant.

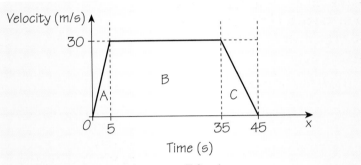

Average velocity $= \dfrac{\text{distance}}{\text{time}}$

$= \dfrac{75}{5}$ m/s

$= 15$ m/s

(a) Acceleration $= \dfrac{\text{velocity}}{\text{time}} = \dfrac{30 \text{ m/s}}{5 \text{ s}} = 6 \text{ m/s}^2$

(b) Deceleration $= \dfrac{30 \text{ m/s}}{10 \text{ s}} = 3 \text{ m/s}^2$

A deceleration is often written as an acceleration with a negative sign.

(c) Total distance travelled

$= $ area of A $+$ area of B $+$ area of C

$= \left(\dfrac{30 \times 5}{2}\right) + (30 \times 30) + \left(\dfrac{30 \times 10}{2}\right)$ m

$= 7.5 + 900 + 150$ m

$= 1125$ m

The area can be worked out by adding the area of two triangles and a rectangle or by working out the area of a trapezium.
Area of the trapezium

$= \tfrac{1}{2}(a + b)h$
$= \tfrac{1}{2}(30 + 45) \times 30$
$= 1125$ m

Exercise 16I

1 The diagram shows a velocity–time graph of the first 30 seconds of a car journey.

(a) Work out the acceleration of the car in the first 10 seconds.

(b) How far did the car travel altogether?

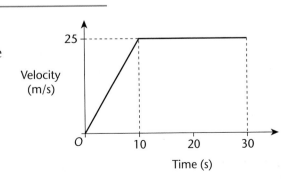

2 The diagram shows a velocity–time graph for a car journey. From the graph find

  (a) the total distance travelled

  (b) the average velocity for the whole journey

  (c) the distance travelled in the first 10 seconds

  (d) the deceleration of the car.

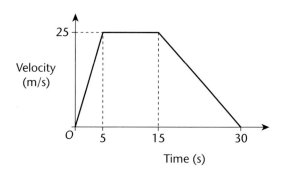

Examination questions

1 Complete this table of values for $y = 2x - 1$.

x	−1	0	1	2	3
y	−3		1		

On a grid with x-values between -2 and 4 and y-values between -4 and 6 draw the graph of $y = 2x - 1$ for values of x between -1 and $+3$. *(3 marks)*

AQA, Spec A, 1I, June 2003

2 The line $y = -3$ crosses the line $y = x - 2$ at the point P.
What are the coordinates of P?
The lines may be drawn on an x–y grid with x- and y-axes between -6 and 6. *(3 marks)*

AQA, Spec A, 1I, November 2004

3 Susan completes a journey in two stages. In stage 1 of her journey, she drives at an average speed of 80 km/h and takes 1 hour 45 minutes.

  (a) How far does Susan travel in stage 1 of her journey?

  (b) Altogether, Susan drives 190 km and takes a total time of 2 hours 15 minutes.
  What is her average speed, in km/h, in **stage 2** of her journey? *(4 marks)*

AQA, Spec A, 1H, November 2003

4 Solve this pair of simultaneous linear equations graphically.

$$5x + 3y = 21$$
$$3x + 5y = 19$$

(4 marks)

5 On an x–y grid with the x- and y-axes between -6 and $+6$ indicate clearly the region defined by the three inequalities

$$y \leqslant 4 \qquad x \geqslant -3 \qquad y \geqslant x + 2$$

Mark the region with an R. *(3 marks)*

AQA, Spec A, 2H, November 2004

6 (a) Find the equation of the line AB.

  (b) Give the y-coordinate of the point on the line with an x-coordinate of 6.

  (c) Write down the gradient of a line perpendicular to AB.

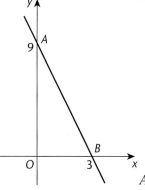

(6 marks)

AQA, Spec A, 2H, November 2004

Summary of key points

Linear equations (grade D/C/B)

A straight line graph is a graph of a **linear equation**.

The equation $y = mx + c$ is the general equation for any straight line, where m represents the **gradient** and c represents the **y-intercept**.

Mid-points (grade C)

The coordinates of the mid-point of a line segment are given $(x, y) = \left(\dfrac{x_1 + x_2}{2}, \dfrac{y_1 + y_2}{2} \right)$

Gradients (grade A/B)

Parallel lines have the same gradient. A line that is **perpendicular** to a line gradient m has gradient $-\dfrac{1}{m}$. The product of the gradients for two perpendicular lines is always -1.

The gradient of a line can be found from two known points $A(x_1, y_1)$ and $B(x_2, y_2)$ using the equation

$$m = \frac{\textbf{difference in the } y \textbf{ values}}{\textbf{difference in the } x \textbf{ values}} = \frac{y_2 - y_1}{x_2 - x_1}.$$

The value of the y-intercept is calculated using the equation given by $y_1 = mx_1 + c$ or $y_2 = mx_2 + c$

Simultaneous equations (grade B)

Approximate solutions to (linear) **simultaneous equations** can be found graphically. The point of intersection gives the values of x and y that satisfy both equations simultaneously.

Inequalities (grade B)

Linear inequalities can be represented graphically by an area or region in the x–y plane. Where more than one inequality is given, the region that satisfies all inequalities simultaneously is the required region.

Conversion graphs (grade D)

A **conversion graph** allows one form of measurement to be converted into another form of measurement.

Graphs of journeys (grade D/C/B)

From a **distance–time** graph the gradient represents speed.

$\text{Speed} = \dfrac{\text{distance}}{\text{time}}.$ $\text{Average speed} = \dfrac{\text{total distance}}{\text{total time}}.$

From a **velocity–time graph** the gradient represents acceleration.

$\text{Acceleration} = \dfrac{\text{velocity}}{\text{time}}.$

A negative value represents deceleration. The area under a velocity–time graph gives the distance travelled.

17 Ratio and proportion

This chapter will show you how to:

- ✔ write, use and simplify ratios
- ✔ divide quantities in a given ratio
- ✔ solve problems using direct and inverse proportion
- ✔ apply ratios to problems involving scales and maps

17.1 Ratio

Writing and using ratios

Key words:
ratio

A **ratio** compares two or more quantities.

A scale model of an Aston Martin Vanquish car is labelled 1 : 18. This means that the length of any part of the real car is 18 times longer than the corresponding part on the model.

Part	Length on model	Length on actual car
Diameter of wheels	3 cm	3 cm × 18 = 54 cm
Length of car	26 cm	26 cm × 18 = 468 cm

A ratio is written with a : symbol between the numbers.
For 1 : 18, you say '1 to 18'.

1 : 18 is an example of a ratio.

Example 1

A recipe for fudge pudding uses:
How much sugar and dates would you need if you used 9 ounces of butter?

> 6 ounces of butter
> 5 ounces of sugar
> 4 ounces of dates.

> 9 ounces of butter = 1.5 × 6 ounces
> Amount of sugar = 1.5 × 5 = 7.5 ounces
> Amount of dates = 1.5 × 4 = 6 ounces

The multiplier is $\frac{9}{6}$ = 1.5, so multiply the quantity of each ingredient by 1.5.

Exercise 17A

1 A cocktail uses cranberry juice, orange juice and tonic water in the ratio 5 : 3 : 1.
  (a) If I have 25 m*l* of tonic water, how much of the other ingredients will I need?
  (b) How much cocktail will I have altogether?

Find the total of all the ingredients.

2 A fruit stall sells 5 apples for every 3 oranges.

 (a) Apples come in boxes of 60. The stall has 4 boxes.
    How many oranges do they need? *144*

 (b) One day the stall sells 42 oranges.
    How many apples do they sell? *70*

3 Cupro-nickel is used to make key rings. It is made from mixing copper and nickel in the ratio 5 : 2.

 (a) How much copper would you need to mix with 3 kg
    of nickel? *7.5*

 (b) How much nickel would you need to mix with 2 kg
    of copper? *0.8*

4 Potting compost is made by mixing 8 kg of peat with 3 kg of sand.

 (a) How much sand would you need to mix with 20 kg of peat? *7.5*

 (b) If I buy a 33 kg bag of compost, how much peat and sand will
    be in the bag? *24, 9*

5 A recipe for 500 ml of spaghetti sauce includes 150 g of meat, 250 g tomatoes, 75 g mushrooms and 25 g onions.

 (a) If I have only 200 g of tomatoes, how much of each other
    ingredient will I need? *60 mushrooms, 120 meat, 3.25 onions*

 (b) How many millilitres of sauce would this make?

Simplifying ratios

You will need to know:
● how to simplify fractions

You can simplify a ratio in the same way as you simplify fractions.
In Example 1, the amounts of butter used were 9 ounces and
6 ounces.
These amounts can be written as a ratio 9 : 6
This means that you need 1.5 times as much of each of the
ingredients when you are using 9 ounces of butter, instead of 6.
The three ratios 9 : 6, 3 : 2 and 1.5 : 1 are equivalent.

A ratio is in its **simplest form** when the numbers in the
ratio are **integers** and they have no common factor other
than 1.

Key words:
simplest form
lowest terms

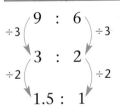

Here the simplest form of
the ratio is 3 : 2. You could
also say the ratio 3 : 2 is in
its lowest terms.

These rules are the same as
the ones used for finding
equivalent fractions.

Example 2

Write these ratios in their lowest terms:

(a) $18:24:6$ (b) $1\,\text{kg}:350\,\text{g}$ (c) $\frac{3}{4}:\frac{1}{3}$ (d) $2.7:3.6$

(a) $18:24:6$	
$\quad= 3:4:1$	Divide by 6.
(b) $1\,\text{kg}:350\,\text{g}$	
$\quad= 1000\,\text{g}:350\,\text{g}$	Units are not the same, change kg to g.
$\quad= 1000:350$	The units *must* be the same before you can simplify.
$\quad= 20:7$	
(c) $\frac{3}{4}:\frac{1}{3}$	Find the LCM of the denominators (in this case 12).
$\quad= (\frac{3}{4} \times 12):(\frac{1}{3} \times 12)$	Multiply both fractions by this number.
$\quad= 9:4$	
(d) $2.7:3.6$	
$\quad= 27:36$	Multiply by 10.
$\quad= 3:4$	Divide by 9.

1m=100cm

Exercise 17B

1 Write these ratios in their lowest terms.

    (a) $9:15$ *3:5*   (b) $40:25$ *8:5*   (c) $48:36$ *4:3*

    (d) $120:70$ *12:7*   (e) $28:49$ *4:7*   (f) $1000:250$ *4:1*

45:200
9:40

2 Write these ratios in their simplest form.

    (a) $£1:20\text{p}$ *5:1*   (b) $6\,\text{mm}:3\,\text{m}$ (c) $5l:250\,\text{ml}$

    (d) $300\,\text{m}:2\,\text{km}$ (e) $75\text{p}:£5$ (f) $350\,\text{g}:2\,\text{kg}$

    (g) $£3.50:£1.25$ (h) $45\,\text{cm}:2\,\text{m}$ (i) $50\,\text{mm}:1\,\text{m}$

Change each part into the same units first.

3 Change these ratios into their simplest form.

    (a) $\frac{1}{2}:\frac{1}{4}$ (b) $\frac{2}{3}:\frac{3}{4}$ (c) $\frac{3}{5}:\frac{7}{10}$

    (d) $3.2:1.6$ (e) $5.4:1.8$ (f) $4.8:1.2$

4 In an army corps there are 480 privates, 96 corporals, and 12 captains. Write the ratio of privates to corporals to captains in its simplest form.

5 A drink was made with $1\frac{1}{2}$ cups of lemonade and $\frac{1}{4}$ cup of blackcurrant cordial. Write the ratio of lemonade to blackcurrant in its lowest terms.

6 The cost of materials for a soft toy were: fabric £3.20, stuffing £1.20 and others 60p. Write the ratio of the cost of fabric to stuffing to others in its lowest terms.

Writing a ratio as a fraction

Any ratio can be written as a fraction.

The ratio $2:3$ can be written as the fraction $\frac{2}{3}$.
This gives you an alternative way of solving some simple
mathematical problems.

Example 3

In a school the ratio of boys to girls is $5:7$.
There are 265 boys. How many girls are there?

Method A (using 'multiplying')

Suppose there are x girls.

The ratio $5:7 = 265:x$

$265 \div 5 = 53$

$5:7 = (5 \times 53):(7 \times 53) = 265:371$

There are 371 girls in the school.

Multiply both sides of the
ratio by 53.
Multiplying both sides of a
ratio by the same value
gives an equivalent ratio.

Method B (using fractions)

Suppose there are x girls.

The ratio $5:7 = 265:x$

$\dfrac{5}{7} = \dfrac{265}{x}$

$\dfrac{7}{5} = \dfrac{x}{265}$

$\dfrac{7 \times 265}{5} = x$

$x = 371$

There are 371 girls in the school.

Inverting **both** sides of the
equation gets the unknown
(*x*) on the top, making it
easier to solve.

Multiply both sides by 265.
See Chapter 5 for how to
solve this kind of equation.

Exercise 17C

1 At a football match the ratio of boys to men is $3:8$. If there are
630 boys, how many men are there?

2 A piece of wood is cut into two pieces in the ratio $4:7$.
   (a) If the shortest piece is 140 cm, how long is the longer piece?
   (b) How long was the original piece of wood?

3 A recipe for jam uses 125 g of fruit and 100 g of sugar.

   (a) What is the ratio of fruit to sugar in its lowest terms?

   (b) If I had 300 g of fruit how much sugar would I need?

   (c) What would be the total mass of the jam in part **(b)**?

4 In a box of strawberries the ratio of good to damaged fruit is 16 : 3.

   (a) If I found 18 damaged fruits, how many good ones did I have?

   (b) How many strawberries are in the box?

5 A celebrity magazine has pictures to text in the ratio of 3 : 2. If a magazine has 75 sections of pictures, how many sections of text does it have?

Writing a ratio in the form 1 : *n* or *n* : 1

Some ratios can easily be written in a form where one side of the ratio is 1.

For example, a ratio of 20 : 4 simplifies to 5 : 1.

Other ratios do not naturally simplify in this way.

For example, 12 : 8 simplifies to 3 : 2.

> Remember: the numbers on each side must be integers with no common factors for a ratio to be in its simplest form.

If you are asked to write this ratio in the form *n* : 1 you cannot leave the answer as 3 : 2.

Look at which number has to be 1. Divide both sides of the ratio by that number.

12 : 8 = (12 ÷ 8) : (8 ÷ 8) = 1.5 : 1

All ratios can be written in the form 1 : *n* or *n* : 1.

> This is called unitary form.

Example 4

Write these ratios in the form *n* : 1.

(a) 27 : 4   **(b)** 6 cm : 25 mm   **(c)** £2.38 : 85p

(a) 27 : 4 = 6.75 : 1

> Divide both sides by 4.

(b) 6 cm : 25 mm = 60 mm : 25 mm

            = 60 : 25 = 2.4 : 1

> Divide both sides by 25.

(c) £2.38 : 85p = 238p : 85p = 238 : 85

             = 2.8 : 1

> Change of units needed in (b) and (c).

> Divide both sides by 85.

Example 5

Write these ratios in the form $1 : n$.

(a) $5 : 9$ **(b)** $400\,\text{g} : 1.3\,\text{kg}$ **(c)** $40\,\text{min} : 1\frac{1}{4}\,\text{h}$

(a) $5 : 9 = 1 : 1.8$

(b) $400\,g : 1.3\,kg = 400\,g : 1300\,g$

$\qquad = 400 : 1300 = 1 : 3.25$

(c) $40\,min : 1\frac{1}{4}\,h = 40\,min : 75\,min$

$\qquad = 40 : 75 = 1 : 1.875$

Divide both sides by 5.

Divide both sides by 400.

Change of units needed in (b) and (c).

Divide both sides by 40.

Exercise 17D

1 Write the following ratios in the form $n : 1$.

   (a) $15 : 3$ **(b)** $7 : 2$

   (c) $9 : 4$ **(d)** $12 : 5$

   (e) $3\,\text{kg} : 200\,\text{g}$ **(f)** $1\,\text{hour} : 40\,\text{min}$

   (g) $5\,\text{m} : 40\,\text{cm}$ **(h)** $1\,\text{day} : 10\,\text{hours}$

   (i) $2.5\,l : 500\,\text{m}l$

2 Write the following ratios in the form $1 : n$.

   (a) $16 : 24$ **(b)** $45 : 54$

   (c) $36 : 96$ **(d)** $15\,\text{hours} : 1\,\text{day}$

   (e) $5\,\text{days} : 4\,\text{weeks}$ **(f)** $6\,\text{mm} : 2.7\,\text{cm}$

   (g) $18\,l : 4500\,\text{m}l$ **(h)** $5\,\text{cm} : 15\,\text{mm}$

Dividing quantities in a given ratio

Key words:
part
share

Ratios can be used to share quantities.

Ian and Simon share £60 so that Ian receives twice as much as Simon.

You can think of this as a ratio Ian : Simon $= 2 : 1$

Ian will get this much: Simon will get this much:

Remember: a ratio compares quantities.

where these bags of money contain equal amounts.

To work out each man's share, you need to divide the money into 3 (2 + 1) equal **parts** (or **shares**).

$$1 \text{ part} = £60 \div 3 = £20$$

Ian receives	$2 \times £20 = £40$	2 parts
Simon receives	$1 \times £20 = £20$	1 part
Check:	$£20 + £40 = £60$ ✓	

To share quantities in a given ratio:

(1) **Work out the total number of equal parts.**
(2) **Work out the amount in 1 part.**
(3) **Work out the value of each share.**

You can use this method to share amounts between more than two people.

Example 6

The alloy *Alnico* is often used to make guitar pickups. It is made from five metals: cobalt, iron, nickel, aluminium and titanium in the ratio $6 : 5 : 3 : 1 : 1$.

How many grams of each metal are there in 544 g of this alloy?

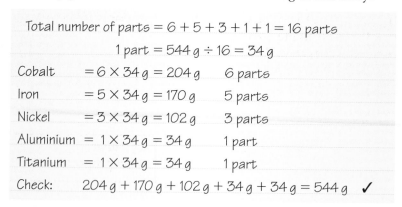

Total number of parts $= 6 + 5 + 3 + 1 + 1 = 16$ parts

$$1 \text{ part} = 544 \text{ g} \div 16 = 34 \text{ g}$$

Cobalt	$= 6 \times 34 \text{ g} = 204 \text{ g}$	6 parts
Iron	$= 5 \times 34 \text{ g} = 170 \text{ g}$	5 parts
Nickel	$= 3 \times 34 \text{ g} = 102 \text{ g}$	3 parts
Aluminium	$= 1 \times 34 \text{ g} = 34 \text{ g}$	1 part
Titanium	$= 1 \times 34 \text{ g} = 34 \text{ g}$	1 part
Check:	$204 \text{ g} + 170 \text{ g} + 102 \text{ g} + 34 \text{ g} + 34 \text{ g} = 544 \text{ g}$ ✓	

Check that your final answers add up to the original amount.

Exercise 17E

1 Divide these amounts in the ratio given:

(a) 50 in the ratio $3 : 2$
(b) 30 in the ratio $3 : 7$
(c) 45 in the ratio $5 : 4$
(d) 64 in the ratio $3 : 5$

2 Divide each amount in the ratio given:

(a) £200 in the ratio $9 : 1$
(b) 125 g in the ratio $2 : 3$
(c) 600 m in the ratio $7 : 5$
(d) 250 *l* in the ratio $16 : 9$

3 Divide £120 in the ratio:

(a) $7:3$ (b) $1:3$ (c) $3:2$ (d) $5:7$

4 The ratio of pupils to teachers for a school trip is $13:2$.
If 135 people go on the trip, how many of them are teachers?

5 At a farm the ratio of black
lambs to white lambs born
is $2:9$.
If 132 lambs are born one
year, how many of them
are black?

6 A cocktail is made from orange juice, sparkling wine and lime
juice in the ratio $5:6:1$. If I make 960 m*l* of cocktail, how much
of each ingredient will I need?

7 Concrete is made from cement, sand and gravel in the ratio
$1:3:4$. If I want to make 2 tonnes of concrete, how much of each
ingredient do I need?

8 Paul is 18 years old, John is 15 years old and Sarah is 12 years old.
Their grandfather leaves them £255 to be divided between them
in the ratio of their ages. How much will each of them receive?

> Simplify the ratio first.

17.2 Proportion

Direct proportion

> **Key words:**
> direct proportion

Two quantities are in direct proportion if
- **their ratio stays the same as they increase or decrease**
- **when one is zero, so is the other.**

For example, a mass of 10 g attached to a spring stretches it by
40 mm. A mass of 25 g stretches the same spring by 100 mm.
The ratio mass : extension is

    $10:40 \ = 1:4$ for the first mass
    $25:100 = 1:4$ for the second mass

The ratio of mass : extension is the same. For a weight of 0 g the
extension is 0 m. So the two quantities mass and extension are in
direct proportion. If we plot the quantities on a graph, we get a
straight line through the origin.

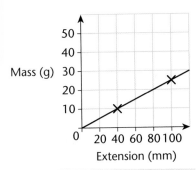

> You will find more about
> direct proportion in
> Section 17.4.

Example 7

The cost of 5 apples is £1.40. How much do 8 apples cost?

Suppose that 8 apples cost x pence.

$5 : 140 = 8 : x$

$$\frac{140}{5} = \frac{x}{8}$$

$$\frac{140 \times 8}{5} = x$$

$$x = 224$$

8 apples cost £2.24

> Cost and number of apples are in direct proportion. The ratio number of apples : cost must stay the same.

> Work in pence for both costs.
> Use the fractions method as in Example 3.

> Multiply both sides by 8.

Unitary method

> **Key words:**
> unitary method

Example 7 can be done by finding the cost of 1 apple.
This is called the **unitary method** .

5 apples cost £1.40 or 140p

1 apple costs $\frac{140}{5} = 28p$

8 apples cost $28 \times 8 = 224p$ or £2.24

> Notice how much easier this is and how it only takes three lines!

In the unitary method, you find the value of one unit of a quantity.

Example 8

Susan has a part-time job.
She works for 15 hours a week and is paid £93.
Her employer wants her to work for 25 hours a week.
How much will Susan be paid now?

For 15 hours she is paid £93

For 1 hour she is paid $\dfrac{£93}{15} = £6.20$

For 25 hours she will get $25 \times £6.20 = £155$

> First of all, find how much she is paid for **1** hour.

> To solve a *direct* proportion question you always do a *division* followed by a *multiplication*.

Exercise 17F

1 If 8 pens cost £2.96, how much will 3 pens cost?

2 40 litres of petrol will take me
200 miles.
How far can I go on 15 litres?

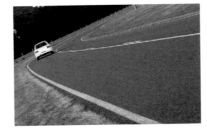

3 Five books weigh 450 g. How much would 18 books weigh?

4 Alice sews buttons onto trousers. She is paid £30 for sewing
200 buttons.
How much would she get for sewing 35 buttons?

5 6 packets of soap powder cost £7.38.
How much will 10 packets cost?

6 Kim earns £67.20 for 12 hours' work. One week she works
17 hours. How much does she earn for that week?

7 A chocolate cake for 8 people uses 120 g of sugar.
How much sugar would you need for a cake for 15 people?

8 Omar walks 500 m to work and it takes him 8 minutes. If he
continues to walk at the same speed,
  (a) how far will he walk in 1 hour?
  (b) how long will it take him to walk 2.25 km?

Inverse proportion

Key words:
inverse proportion

When two quantities are in direct proportion:

● as one increases, so does the other
● as one decreases, so does the other.

Suppose you travel from Leeds to Newcastle, a distance of 100 miles.

If you travel at an average speed of 50 mph it will take you 2 hours.

If you only average 40 mph it will take you $2\frac{1}{2}$ hours.

The slower you travel, the more time it takes:

● as the speed decreases the time increases
● as the speed increases the time decreases.

$$\text{Time} = \frac{\text{distance}}{\text{speed}}$$

$$\frac{100}{50} = 2$$

$$\frac{100}{40} = 2\frac{1}{2}$$

If you *halve* the speed to
25 mph you *double* the
time to 4 hours.

When two quantities are in **inverse proportion**, one quantity increases at the same rate as the other quantity decreases.

The best way to solve problems involving inverse proportion is to use the unitary method.

Example 9

Two people take 6 hours to paint a fence.

How long will it take 3 people?

> 2 people take 6 hours
>
> 1 person takes 6 × 2 = 12 hours
>
> 3 people will take 12 ÷ 3 = 4 hours

Read the problem first and decide whether it is *direct* or *inverse* proportion. Use your common sense! The more people painting, the less time it takes.

Work out how long it will take 1 person. 1 person takes twice as long as 2 people.

Example 10

Tarik is repaying a loan from a friend.

He agrees to pay £84 per month for 30 months.

If he can afford to pay £120 per month, how many months will it take to repay?

> Paying £84 per month takes 30 months
>
> Paying £1 per month takes (84 × 30) months
>
> Paying £120 per month will take $\dfrac{84 \times 30}{120}$ months = 21 months

Repaying *more* each month will take *less* time, so this is inverse proportion.

210 years !

To solve *inverse* proportion questions you always do a *multiplication* followed by a *division*.

Exercise 17G

1 It takes 3 people 4 days to paint a shop.
How long would it take 1 person?

2 It takes 2 people 6 hours to make a suit.
How long would it take 3 people?

3 Eight horses need a trailer of hay to feed them for a week.
How many horses could this feed for 4 days?

4 It takes 5 bricklayers 4 days to build a house.
How long would it take 2 bricklayers?

5 It takes 3 pumps 15 hours to fill a swimming pool.
How many pumps would be needed to fill the pool in 9 hours?

6 A refugee camp has enough food for 400 people for 25 days. If the food lasts only 20 days, how many people must be in the camp?

7 It takes 7 days for 6 people to dig a trench.
How long would it take 14 people?

8 In a library 48 paperback books 20 mm wide fit on a shelf.
How many books 24 mm wide would fit on the same shelf?

9 It takes 8 hours to fly to Chicago at a speed of 300 mph.
  (a) How long would it take to fly to Chicago at a speed of 400 mph?
  (b) If the journey took 12 hours, what speed was I travelling?

10 It takes 6 window cleaners 8 days to clean the windows at Rockingham Palace.
  (a) How long would it take 4 window cleaners?
  (b) The windows need to be cleaned in 3 days for a special event. How many window cleaners are needed?

Exercise 17H

1 If 10 metres of material cost £23.50, what will 7 metres cost?

2 A man cuts a hedge in 45 minutes using a cutter with a blade 36 cm wide. How long would it take if the blade was only 15 cm wide?

3 It takes 200 tiles to tile a room when each tile covers 36 cm². If I use tiles which cover 25 cm², how many tiles would I need?

4 Alan works for 9 hours a week in a shop. He is paid £45. He increases his hours to 15. How much will he be paid?

5 It takes $2\frac{1}{2}$ hours to travel to London by train at an average speed of 80 mph. A new train travels at an average speed of 100 mph. How long will the journey to London take on the new train?

6 If 14 kg of potatoes cost £2.38, how much will 6 kg cost?

7 One load of feed lasts 40 sheep 9 days.
  (a) How many sheep could this load feed for 5 days?
  (b) For how many days could it feed 60 sheep?

8 In a hotel it takes a maid 24 minutes to clean 3 rooms.
  (a) How long will it take her to clean 16 rooms?
  (b) How many rooms can she clean in 4 hours?

17.3 Map scales

You will need to know:
- how to convert between metric units of length

Map **scales** are written as ratios.

A scale of 1 : 50 000 means that 1 cm on the map represents 50 000 cm on the ground.

A scale of 1 : 25 000 means that 1 cm on the map represents 25 000 cm on the ground.

> Always look carefully to see what scale is being used.

When you answer questions involving map scales you need to:

- use the scale of the map
- convert between metric units of length so that your answer is in sensible units.

Example 11

The scale of a map is 1 : 50 000. The distance between Truro and St. Austell on the map is 40 cm. What is the actual distance between Truro and St. Austell? Give your answer in kilometres.

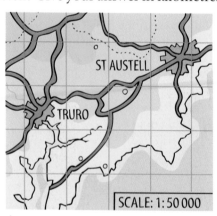

SCALE: 1 : 50 000

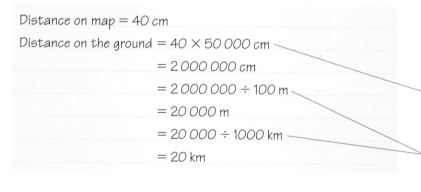

Distance on map = 40 cm

Distance on the ground = 40 × 50 000 cm

= 2 000 000 cm

= 2 000 000 ÷ 100 m

= 20 000 m

= 20 000 ÷ 1000 km

= 20 km

> Work out the real distance in cm then convert to km.

> Each 1 cm on the map is 50 000 on the ground.

> 1 m = 100 cm
> 1 km = 1000 m

Example 12

The distance between two towns is 24 km. How far apart will they be on a map of scale 1 : 180 000?

Distance on the ground = 24 km

$= 24 \times 1000$ m

$= 24\,000$ m

$= 24\,000 \times 100$ cm

$= 2\,400\,000$ cm

Distance on map $= \dfrac{2\,400\,000}{180\,000}$ cm $= 13.3$ cm (to 1 s.f.)

Convert the real distance to cm before you divide by the scale of the map.

Exercise 17I

1 The scale of a map is 1 : 25 000.
Find the actual distance represented by these measurements on the map.

(a) 4 cm   (b) 7 cm

(c) 8 mm   (d) 12.5 cm

2 A map has scale 1 : 200 000.
What measurement on the map will represent:

(a) 4 km   (b) 20 km

(c) 15 km   (d) 12.5 km?

3 This motorway map has a scale of 1 : 3 000 000.

(a) The distance on the map from Carlisle to Kendal is 2.5 cm. How far is the actual distance?

(b) Measure, in a straight line, the distance from Carlisle to Penrith. How far is the actual distance?

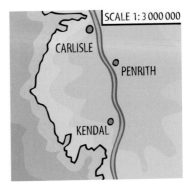

4 A map of Ireland has a scale of 1 : 500 000.
What measurement on the map would represent these distances:

(a) Waterford to Dundalk, 400 km

(b) Wicklow to Carrigart, 320 km

(c) Dublin to Tralee, 160 km?

5 On a map with a scale of 1 : 2 500 000 the distance from Leeds to Edinburgh is 13 cm.

(a) What is the actual distance from Leeds to Edinburgh?

(b) Newcastle is 150 km from Leeds.
How far is this on the map?

6 This map has a scale of 1 : 50 000. Use the map to work out the following distances.

   (a) Between the ends of the 2 piers at Tynemouth.

   (b) From Sharpness Point to Smuggler's Cave.

   (c) The Ferry crossing of the Tyne.

   (d) The length of both piers.

   (e) From the Coast Guard Station (CG Sta) to the Coast Guard Lookout (CG Lookout).

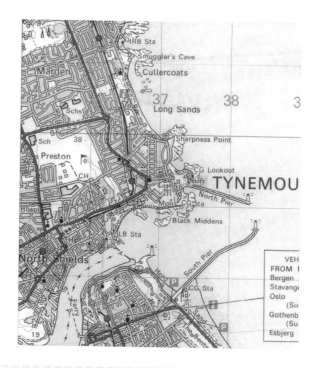

17.4 Variation

Direct variation

In Section 17.2 you saw that two quantities are in direct proportion if their ratio stays the same as they increase or decrease. The example used was stretching a spring.

If no mass is attached to the spring, the extension is 0 m. This is important because two quantities can only be in direct proportion **if they are both zero at the same time**.

> **So the two conditions for direct proportion are:**
> - **if one quantity is zero then the other is also zero,**
> - **their ratio stays the same as they increase or decrease.**

This can be interpreted graphically, by plotting one quantity (or **variable**) against the other.

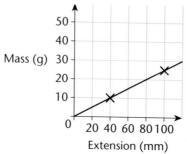

Key words:
variable
constant of proportionality

When two quantities are in direction proportion, their graph is a straight line graph passing through the origin.

In this case, the equation is $y = \dfrac{x}{4}$.

Suppose the gradient of the graph is k, then the equation of the graph will be $y = kx$

The symbol k is known as the **constant of proportionality**.

The gradient of the straight line and the constant of proportionality are always the same when two quantities are in direct proportion.

In this example, the relationship between these two quantities can be stated in several different ways, and you need to be familiar with all of them.

(1) y is directly proportional to x

(2) y varies directly as x

(3) $y \propto x$ where $\propto$ means 'is proportional to'

(4) $y = kx$ where k is the constant of proportionality

Look back at Chapter 16 ... this is a graph of the form $y = mx + c$ where the value of c is zero.
The value of the gradient k is $\frac{25}{100}$ or $\frac{1}{4}$, the same as the fractional equivalent of the ratio.

Notice that (3) and (4) are almost the same, except that the symbol for proportionality ($\propto$) has been replaced by $= k$. Always use the correct notation.

Example 13

The volume of a solid (V cm^3) is directly proportional to the length of one of its sides (l cm). When $l = 12$ cm, $V = 30$ cm^3.

(a) Sketch a graph of this relationship.

(b) Find an equation connecting V and l.

(c) Find the volume when $l = 26$ cm.

(d) Find the length of the solid when the volume is 37.5 cm^3.

(a)

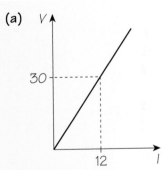

(b) Gradient $= \dfrac{30}{12} = 2.5 = k$

$V = 2.5\,l$

$k =$ constant of proportionality.

When $l = 0$, $V = 0$.

(c) When $l = 26$ cm

$V = 2.5 \times 26 = 65$ cm^3

(d) When $V = 37.5$ cm^3

$37.5 = 2.5 \times l$

$l = \dfrac{37.5}{2.5}$

$l = 15$ cm

The next example shows you how to handle questions when the information is given in the form of a table.

Example 14

In an experiment, the values of time (T min) and temperature (θ °C) were recorded in the following table.

Time (T min)	7.5	p	18
Temperature (θ°C)	16.35	25.07	q

If θ varies directly as T, find the missing values in the table.

$$7.5 : 16.35 = p : 25.07$$

$$\frac{7.5}{16.35} = \frac{p}{25.07}$$

> The ratio of time to temperature stays the same.

$$\frac{7.5 \times 25.07}{16.35} = p$$

$$p = 11.5 \text{ minutes}$$

$$16.35 : 7.5 = q : 18$$

$$\frac{16.35}{7.5} = \frac{q}{18}$$

> It is easier to use fractions.
>
> The ratio/fraction can be written either way round.

$$\frac{16.35 \times 18}{7.5} = q$$

> You will need simple equation solving skills.

$$q = 39.24°C$$

Exercise 17J

1 y varies directly as x. If $y = 6$ when $x = 2$ find
   (a) y when $x = 5.5$ (b) x when $y = 12$.

2 w is directly proportional to t. If $w = 20$ when $t = 5$ find
   (a) w when $t = 7$ (b) t when $w = 18$.

3 The table shows values of h and p.
   If $h \propto p$ find the missing values a and b.

h	1.4	5.2	a
p	b	18.2	39.55

4 The surface area of a solid (A cm²) is directly proportional to the length of one of its sides (l cm).
   When $l = 14$, $A = 33.6$ cm².

   (a) Sketch a graph of the relationship between l and A.

   (b) Find an equation connecting a and l.

   (c) Find the area, A, when $l = 19$ cm.

   (d) Find the length, l, when $A = 42$ cm².

5 The table shows values of d and m.
Show that m is proportional to d.

d	4	6.5	11.7
m	2.32	3.77	6.786

6 The table shows values of c and N.
Is N directly proportional to c?
Give reasons for your answer.

c	3	7.2	12.5	20
N	4.8	11.52	18.25	32

7 The extension, e, of a spring is directly proportional to w, the
weight attached to the end of the spring.
When $w = 2.5$ kg, $e = 80$ mm.
(a) Find an equation connecting e and w.
(b) Find e when $w = 3.5$ kg.
(c) Find w when $e = 182.4$ mm.

Inverse variation

In Section 17.2 you saw that when one quantity increases at the
same rate as another quantity decreases, the two quantities are in
inverse proportion.
Example 9 asked this question:
Two people take 6 hours to paint a fence.
How long will it take 3 people?

You solved this by first finding how long it would take 1 person
(12 hours).
It is then easy to see that times for varying numbers of people will
be:

Number of people (x)	1	2	3	4	6	8	9	12
Time (hours) (y)	12	6	4	3	2	$1\frac{1}{2}$	$1\frac{1}{3}$	1

> Inverse proportion means
> that more people take less
> time. For example, 9 people
> will do the job three times
> faster than 3 people.

Notice that the product of any x–y pair is always 12 and the
equation of this relationship can be written as

$$xy = 12 \qquad y = \frac{12}{x}$$

A graph of these results looks like this:

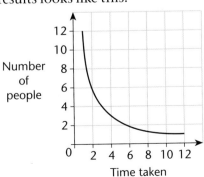

You can see that if the example stated that ...
Two *people take 5 hours to paint a fence*

... the equation would be $y = \dfrac{10}{x}$.

One person will take 10 hours.

or if it is
Two *people take 8 hours to paint a fence*

then the equation would be $y = \dfrac{16}{x}$.

One person will take 16 hours.

In this example, whatever the time is, the equation will always be of the form $y = \dfrac{k}{x}$

and the graph will be the same shape as the one above – it is known as a rectangular hyperbola.

You will meet this again in Chapter 25.

For two quantities in inverse proportion the graph will be a rectangular hyperbola.

When two quantities behave in this way:

(1) *y* is inversely proportional to *x*

(2) *y* varies inversely as *x*

(3) $y \propto \dfrac{1}{x}$ where $\propto$ means 'is proportional to'

(4) $y = \dfrac{k}{x}$ where *k* is the constant of proportionality

Notice that (3) and (4) are almost the same, except that the symbol for proportionality ($\propto$) has been replaced by $= k$. Always use the correct notation.

Example 15

In this table the two quantities *x* and *y* vary inversely.

Find the missing values, *d* and *e*.

x	d	7.5	16
y	20	8	e

$y = \dfrac{k}{x}$

$8 = \dfrac{k}{7.5}$

$8 \times 7.5 = k$ or $k = 60$

$y = \dfrac{60}{x}$

Using this equation, $20 = \dfrac{60}{d}$ which gives $d = 3$

$e = \dfrac{60}{16}$ which gives $e = 3.75$

The two quantities *x* and *y* vary inversely so the equation connecting them must be $y = \dfrac{k}{x}$

Use the pair of *x* and *y* values that you know, to find *k*.

Notice that as *x* gets bigger, *y* gets smaller (and vice versa), which is what you would expect when two quantities vary inversely.

Exercise 17K

1 y varies inversely as x.
If $y = 3$ when $x = 4$, find

(a) y when $x = 24$

(b) x when $y = \frac{1}{3}$.

2 The table shows values of x and y, where x and y vary inversely.
Find the missing values a and b.

x	0.5	2.5	a
y	b	6	1.2

3 If $h \propto \dfrac{1}{l}$ copy and complete this table of values.

l	0.2		0.6	1.5	3.6	
h		48	40			3.2

4 z is inversely proportional to w.
$z = 54$ when $w = \frac{2}{3}$

(a) Find an equation connecting z and w,

(b) Find z when $w = 1.5$,

(c) Find the possible values of w when $z = w$.

5 The table shows values of p and Q.
Is Q inversely proportional to p?
Give reasons for your answer.

p	0.6	2.4	9	30
Q	18	4.5	1.2	0.36

6 In a physics experiment it is found that the volume (V cm³) of a
gas varies inversely as its pressure (P bar).
When the pressure is 1.6 bar the volume of the gas is 180 cm³.

(a) Find an equation connecting V and P.

(b) Find the volume when the pressure is 2.5 bar.

(c) Find the pressure when the volume is 150 cm³.

(d) Sketch a graph of the relationship between P and V.

Other kinds of variation

When two quantities are in direct proportion, their ratio stays the
same as they increase or decrease. So, for example, when one
doubles the other also doubles.
Sometimes two quantities increase together **but not at the same
rate**.

Example 16

y is proportional to x^2. If $y = 12$ when $x = 3$, find

(a) the value of y when $x = 9$

(b) the value of x when $y = 3$.

$y \propto x^2$ or $\qquad\qquad\qquad y = kx^2$

Using $y = 12$ when $x = 3 \qquad 12 = k \times 3^2$

$$12 = k \times 9$$

$$\frac{12}{9} = k$$

$k = \frac{4}{3}$

$y = \frac{4}{3}x^2$

(a) when $x = 9 \quad y = \frac{4}{3} \times 9^2 = \frac{4}{3} \times 81 = 108$

(b) when $y = 3 \quad 3 = \frac{4}{3} \times x^2$

$$\frac{9}{4} = x^2$$

$$x = \pm\frac{3}{2} = \pm 1\frac{1}{2}$$

> Use the $\propto$ notation to mean 'is proportional to' just as in the previous examples. Then replace $\propto$ by $= k$, where k is the constant of proportionality.
>
> Note that this time the equation contains x^2 because y is proportional to x^2.

> Notice that there are **two** answers for x in part (b) since a square root has been taken.

Always start with a statement of proportionality.
(1) Use the given values of x and y to find the value of k, the constant of proportionality.
(2) Write down the equation connecting x and y using this value of k.
(3) Use the equation to calculate any answers required.

> Notice that, when x multiplies by 3, y multiplies by 9 (part a) and when y divides by 4, x divides by 2 (part b).
> This is entirely consistent with the 'squared' relationship between the two variables.

The most common ways that two quantities increase together, but not at the same rate, are shown in this table.

Variation		Equation
y is proportional to x^2 or y varies as x^2	$y \propto x^2$	$y = kx^2$
y is proportional to $\sqrt{x}$ or y varies as $\sqrt{x}$	$y \propto \sqrt{x}$	$y = k\sqrt{x}$
y is proportional to x^3 or y varies as x^3	$y \propto x^3$	$y = kx^3$

> The graphs of these relationships can be seen in Chapter 25.

Exercise 17L

1 y is directly proportional to x^2. If $y = 20$ when $x = 2$, find

    (a) y when $x = 3$ **(b)** x when $y = 245$.

2 Q varies directly as m^2. If $Q = 48$ when $m = 4$,

    (a) find Q when $m = 1.5$ **(b)** m when $Q = 75$.

3 Given that $F \propto d^2$ copy and complete the table to find the missing values a and b.

d	6	8	a
F	126	b	423.5

4 y varies directly as $\sqrt{x}$.
  If $y = 20$ when $x = 25$, find

    (a) y when $x = 81$ **(b)** x when $y = 12$.

5 Given that $y \propto x^3$ and $y = 162$ when $x = 3$.

    (a) Find an expression for y in terms of x.

    (b) Find y when $x = 4$.

    (c) Find x when $y = 750$.

6 The volume of a sphere is directly proportional to the cube of its radius.
  A sphere of radius 3 cm has a volume of 113.1 cm³. Find

    (a) the volume of a sphere of radius 2 cm

    (b) the radius of a sphere of volume 565.5 cm³.

7 In an experiment, measurements of Q and h were taken, as shown in this table.

h	3	6	7
Q	10.8	86.4	137.2

  Which of these laws fits the results?
  (A) $Q \propto \sqrt{h}$ (B) $Q \propto h^2$ (C) $Q \propto h^3$
  You must explain your answer.

When two quantities are in inverse proportion, one quantity increases at the same rate as the other quantity decreases.
So, for example, if one quantity doubles the other halves.
Sometimes two quantities are such that one increases and the other decreases **but not at the same rate**.

Example 17

y is inversely proportional to the square root of x.
If $y = 8$ when $x = 9$, find

(a) the value of y when $x = 144$ **(b)** the value of x when $y = 6$.

$$y \propto \frac{1}{\sqrt{x}} \quad \text{or} \quad y = \frac{k}{\sqrt{x}}$$

$$y = 8 \text{ when } x = 9 \qquad 8 = \frac{k}{\sqrt{9}}$$

$$8 = \frac{k}{3}$$

$$8 \times 3 = k$$

So $k = 24$ and the equation is $y = \dfrac{24}{\sqrt{x}}$

(a) when $x = 144$ $y = \dfrac{24}{\sqrt{144}} = \dfrac{24}{12} = 2$

(b) when $y = 6$ $6 = \dfrac{24}{\sqrt{x}}$

$$\sqrt{x} = \frac{24}{6} = 4$$

$$x = 4^2 = 16$$ ——————————— Square both sides.

> y is inversely proportional to the square root of x.
> Use the $\propto$ notation to mean 'is proportional to' just as in the previous examples. Then replace $\propto$ by $= k$, where k is the constant of proportionality.
>
> Note that this time the equation contains $\sqrt{x}$ and this term is in the **denominator** because y is **inversely** proportional to $\sqrt{x}$.

The most common ways that one quantity increases and the other decreases, but not at the same rate, are shown in this table.

Variation		Equation
y is inversely proportional to x^2 or y varies inversely as x^2	$y \propto \dfrac{1}{x^2}$	$y = \dfrac{k}{x^2}$
y is inversely proportional to $\sqrt{x}$ or y varies inversely as $\sqrt{x}$	$y \propto \dfrac{1}{\sqrt{x}}$	$y = \dfrac{k}{\sqrt{x}}$
y is inversely proportional to x^3 or y varies inversely as x^3	$y \propto \dfrac{1}{x^3}$	$y = \dfrac{k}{x^3}$

The graphs of these relationships can be seen in Chapter 25.

Exercise 17M

1 y is inversely proportional to $\sqrt{x}$.
If $y = 1.8$ when $x = 100$, find

    (a) y when $x = 9$ **(b)** x when $y = 4.5$.

2 y varies inversely as x^2.
If $y = 5$ when $x = 4$, find

    (a) y when $x = 10$ **(b)** x when $y = 20$.

3 If $y \propto \dfrac{1}{x^3}$ and $y = 6$ when $x = 2$, find

    (a) y when $x = \frac{1}{2}$ **(b)** x when $y = \frac{3}{4}$

4 Given that V varies inversely as t^2, copy and complete the table to find the missing values.

t	2	4		
V	9		36	144

5 W is inversely proportional to the square of m.
If $W = 2.4$ when $m = 5$, find

    (a) an equation connecting W and m

    (b) W when $m = 6$

    (c) m when $W = 15$.

6 When a cylinder has a fixed volume, the radius, r cm, is inversely proportional to the square root of the height, h cm.
When the height is 16 cm the radius is 5 cm.

    (a) Find the radius when the height is 4 cm.

    (b) Find the height when the radius is 10 cm.

7 The force of attraction, F newtons, between two magnets varies inversely as the square of the distance, d cm, between them.
When the magnets are 5 cm apart the force of attraction is 3 newtons.

    (a) Find the force of attraction when the magnets are 2 cm apart.

    (b) How far apart are the magnets when the force of attraction is $\frac{1}{3}$ newton?

8 If $y \propto \dfrac{1}{\sqrt{x}}$ find, in its simplest form, the ratio of the values of y when $x = 2.25$ and $x = 36$.

Examination questions

1 Mrs Jones inherits £12 000.
   She divides the £12 000 between her three children Laura, Mark and Nancy in the ratio 7 : 8 : 9, respectively.
   How much does Laura receive?

   (2 marks)
   AQA, Spec A, 2H, June 2003

2 The sizes of the interior angles of a quadrilateral are in the ratio 3 : 4 : 6 : 7
   Calculate the size of the largest angle.

   (2 marks)
   AQA, Spec A, 2H, November 2003

3 In an experiment measurements of t and h were taken.
   These are the results.

t	2	5	6
h	10	62.5	90

   Which of these rules fits the results?
   (a) $h \propto t$ **(b)** $h \propto t^2$ **(c)** $h \propto t^3$
   You **must** show all your working.

   (4 marks)
   AQA, Spec A, 1H, November 2003

4 M and G are positive quantities. M is inversely proportional to G. When $M = 90$, $G = 40$.
   Find the value of M when $G = M$.

   (4 marks)
   AQA, Spec A, 1H, November 2004

Summary of key points

Ratio (grade D/C)

A ratio is a way of comparing two or more quantities. For example, a ratio of 3 : 1 means that one quantity is 3 times larger than the other.

Ratios can be simplified by dividing by a common factor, or by multiplication. For example, 12 : 3 = 4 : 1 (dividing by 3) and 2.5 : 6 = 5 : 12 (multiplying by 2).

Ratios such as 4 : 1 and 5 : 12 are said to be in their simplest form or lowest terms. The numbers are integers and have no common factors other than 1.

A ratio can be written as a fraction. For example, $4 : 5 = \frac{4}{5}$, which gives you an alternative way of solving simple problems.

Ratios can be written in the form $1 : n$ or $n : 1$.
For example, $4 : 5 = 1 : 1\frac{1}{4}$ (dividing by 4) and $3 : 2 = 1\frac{1}{2} : 1$ (dividing by 2).

Ratios can be used to share or divide quantities.

To share quantities in a given ratio:
(1) Work out the total number of equal parts.
(2) Work out the amount in 1 part.
(3) Work out the value of each share.

Proportion (grade D/C)

Two quantities are in direct proportion if
- their ratio stays the same as they increase or decrease
- when one is zero, so is the other.

Two quantities are in inverse proportion if one increases at the same rate as the other decreases.

Unitary method (grade D)

The unitary method is the best way of solving simple direct and inverse proportion problems. It involves first finding the value of 1 unit of a quantity.

Map scales (grade C)

Ratios are used to describe the scale of a map. For example a ratio of $1 : 200\,000$ means that 1 cm on the map is equivalent to $200\,000$ cm on the ground. Always use sensible units for your answer, for example, $200\,000$ cm $= 2000$ m $= 2$ km.

Direct and inverse proportion (grade A/A*)

The two conditions for direct proportion are:
- if one quantity is zero then the other is also zero
- their ratio stays the same as they increase or decrease.

When two quantities are in direct proportion, their graph is a straight line passing through the origin.

For direct proportion:
(1) y is directly proportional to x
(2) y varies directly as x
(3) $y \propto x$ where $\propto$ means 'is proportional to'
(4) $y = kx$ where k is the constant of proportionality

When one quantity increases at the same rate as another quantity decreases, the two quantities are said to be in inverse proportion.
For two quantities in inverse proportion the graph will be a rectangular hyperbola.

It will have an equation of the form $y = \dfrac{k}{x}$

For inverse proportion:
(1) y is inversely proportional to x
(2) y varies inversely as x
(3) $y \propto \dfrac{1}{x}$ where $\propto$ means 'is proportional to'
(4) $y = \dfrac{k}{x}$ where k is the constant of proportionality

Always start with a statement of proportionality.
(1) Use the given values of x and y to find the value of k, the constant of proportionality.
(2) Write down the equation connecting x and y using this value of k.
(3) Use the equation to calculate any answers required.

Sometimes two quantities increase together **but not at the same rate**.

y is proportional to x^2 or y varies as x^2	$y \propto x^2$	$y = kx^2$
y is proportional to $\sqrt{x}$ or y varies as $\sqrt{x}$	$y \propto \sqrt{x}$	$y = k\sqrt{x}$
y is proportional to x^3 or y varies as x^3	$y \propto x^3$	$y = kx^3$

Sometimes two quantities are such that one increases and the other decreases **but not at the same rate**.

y is inversely proportional to x^2 or y varies inversely as x^2	$y \propto \dfrac{1}{x^2}$	$y = \dfrac{k}{x^2}$
y is inversely proportional to $\sqrt{x}$ or y varies inversely as $\sqrt{x}$	$y \propto \dfrac{1}{\sqrt{x}}$	$y = \dfrac{k}{\sqrt{x}}$
y is inversely proportional to x^3 or y varies inversely as x^3	$y \propto \dfrac{1}{x^3}$	$y = \dfrac{k}{x^3}$

This chapter will show you how to:

✔ recognise and use the four types of transformation – reflection, rotation, translation and enlargement
✔ enlarge shapes with fractional and negative scale factors
✔ find the centre and scale factor of an enlargement
✔ use a combination of transformations

18.1 Transformations

Key words:
transformation
image
reflection
rotation
translation
congruent
enlargement
similar

A **transformation** changes the size or position of an object. The original shape is called the object and the transformed shape is called the **image**. You need to know about four types of transformation:

- Reflection
- Rotation
- Translation
- Enlargement

The first three of these – **reflection** , **rotation** and **translation** – only alter the *position* of an object. The size of the image and the object are identical – the image and object are **congruent** .

You will find more about congruent and similar shapes in Chapter 19.

Enlargement alters not only the position of an object but also its size. The object and image are mathematically **similar** .

18.2 Reflection

You will need to know:
- what is meant by a line of symmetry

Key words:
reflection
mirror image
mirror line

Reflections take place along a mirror line to produce a **mirror image** .

This image is exactly the same size and shape as the original shape (congruent) and points on the image are the same distance behind the **mirror line** as they are on the object in front of the mirror line.

A line connecting a point *P* on the object with the corresponding point *P'* on the image will always cross the mirror line at right-angles.

In a reflection

$$PM = P'M$$

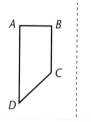

Every point on an object is reflected in this way. When answering questions on reflecting 2-dimensional shapes, just consider the corners (vertices) of the object.

Example 1

The object *ABCD* is reflected in a mirror line as shown. Draw the image of the object and label this *A'B'C'D'*.

The shape of an object can be drawn on tracing paper. Turning the paper over and lining up the mirror line will allow you to draw the reflected image. This method will work in all questions on reflections.

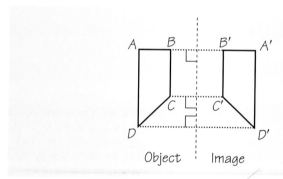

Draw a line from *C* at right-angles to the mirror line and continue the line to a point the same distance the other side and label it *C'*. Repeat this for points *A*, *B* and *D*. Now join the points in order with straight lines.

The object shape is labelled *A*, *B*, *C* and *D* in a clockwise direction. After reflection, the image shape is labelled *A'*, *B'*, *C'* and *D'* in an anticlockwise direction.

Object shapes can also cross a mirror line but the points are reflected in the same way.

Example 2

The triangle *ABC* is reflected in the mirror line as shown.
Draw the reflection of this triangle and label it *A'B'C'*.

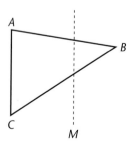

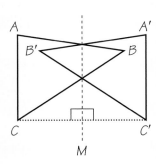

Draw a line from C at right angles to the mirror line and continue to the point C' at the same distance on the other side. Note that lines AB and A'B', and lines BC and B'C' cross exactly on the mirror line. This is always true for shapes that cross the mirror line.

Mirror lines can be vertical, horizontal or diagonal.

Mirror lines can be described with an equation, such as $x = 0$, $y = 4$, $y = x$.

Diagonal mirror lines (mirror lines at 45°) need some care when constructing the image.

Example 3

The trapezium *ABCD* is to be reflected in the diagonal mirror line as shown. Draw the image and label it *A'B'C'D'*.

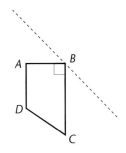

When the mirror line is sloping it is useful to rotate the object and mirror line until the mirror line is vertical.

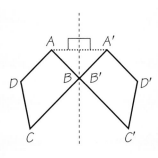

The mirror line is now vertical and the line *AA'* crosses the mirror line at right angles. Point *B* lies on the mirror line, so the image point *B'* also lies on the mirror line in the same position as *B*.

When you reflect an object on squared paper and in a diagonal mirror line, you can count squares diagonally to find the position of an image point.

Do this for each point separately, then join up the points to produce the final image.

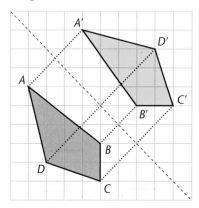

From A, count $1\frac{1}{2}$ squares diagonally to the mirror line. Count another $1\frac{1}{2}$ and mark the point A'. For B, count 1 square diagonally to the mirror line, then 1 more to B'. Find C' and D' in the same way.

Sometimes you are given the object and image and asked to draw the mirror line.

To describe a reflection fully you need to give the equation of the mirror line.

You need to be able to reflect in the lines $y = x$ and $y = -x$. They are the lines at 45° to the coordinate axes.

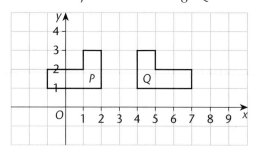

Example 4

The diagram shows an object P and its image Q after reflection.

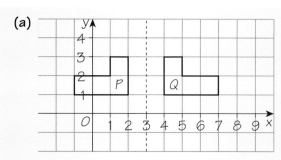

(a) Draw in the mirror line as a dashed line.

(b) What is the equation of this mirror line?

(a)

The mirror line must be the same distance from P as from Q.

The mirror line is parallel to the y-axis. The x-values of all points on it are 3. So the equation is $x = 3$.

(b) The equation of the mirror line is $x = 3$.

Exercise 18A

You will need squared paper for each question.

1 Each diagram shows an object with its coloured image. Copy these diagrams onto squared paper and draw in the mirror line in each case.

(a)

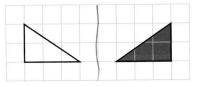

(b)

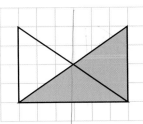

(c)

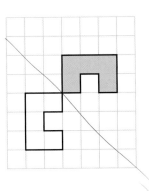

(d)

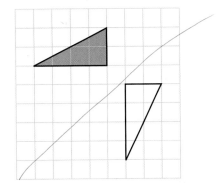

2 Copy these shapes and the dashed mirror line onto squared paper. Draw the reflected image in each case.

(a)

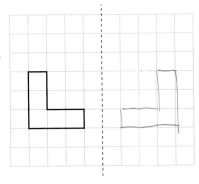

(b)

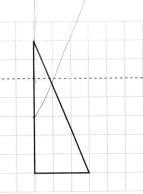

(c)

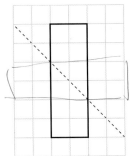

3 Copy the axes and triangle **A** onto squared paper.
Reflect triangle **A**

(a) in the *y*-axis and label the image **B**.

(b) in the *x*-axis and label it **C**.

(c) in the line $y = -x$ and label the image **D**.

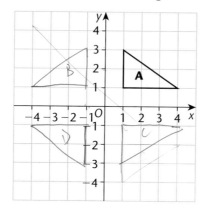

4 Copy the axes and the shape **P** onto squared paper. Reflect the shape

(a) in the line $x = 1$ (b) in the line $y = x$.

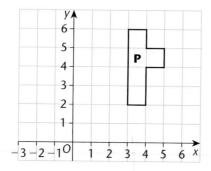

5 Copy the axes and polygon **A** onto squared paper.

(a) Reflect polygon **A** in the line $y = 1$. Label it **B**.

(b) Reflect polygon **A** in the line $y = x$. Label it **C**.

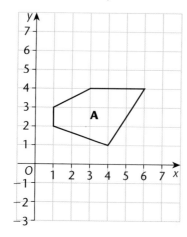

18.3 Rotation

Key words:
rotation
centre of rotation
quarter turn
half turn
three-quarter turn

A **rotation** turns an object either clockwise or anticlockwise through a given angle about a fixed point.

The fixed point is called the **centre of rotation**. It can be either inside or outside the object.

An object can be rotated through any angle (in degrees). Common rotations are **quarter turn** (90°), **half turn** (180°) and **three-quarter turn** (270°).

> Always look carefully to see whether the rotation is clockwise or anticlockwise.

Example 5

Draw the image of this shape after it has been rotated through 90° anticlockwise about the centre of rotation at

(a) *A*   (b) *B*   (c) *C*.

The image after each rotation is shown shaded.

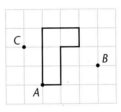

> Tracing paper is very useful in all questions on rotation.

Rotation about A

(a)
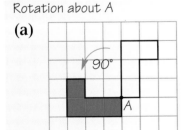

Rotation about B

(b)
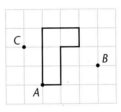

> Trace the shape and the centre of rotation on to tracing paper. Hold the centre of rotation fixed with your pencil point. Turn the tracing paper through the required angle.

> Draw the image you see on your diagram.

Rotation about C

(c)
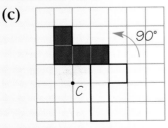

> The image formed after rotation is the same size and shape as the original object.

An object and its image after reflection are congruent.

Example 6

Draw the image of the shape after rotation about $P(1, -1)$:

(a) a quarter of a turn anticlockwise

(b) rotation through 180° clockwise.

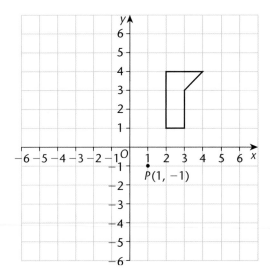

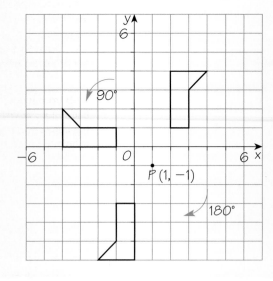

Use tracing paper to help you.

An image formed by a rotation through 180° clockwise is in the same position as an image formed by a rotation through 180° anticlockwise.
They are both 'half turns' and it doesn't matter which direction you turn.

To describe a rotation fully you need to give:

- **the centre of rotation**
- **the angle of turn**
- **the direction of turn**

Usually 90° or 180°.

Clockwise or anticlockwise.

If you want to describe fully the rotation that takes shape P on to shape Q, the easiest way is to use tracing paper.

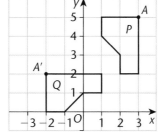

Remember to ask for some tracing paper when you attempt any transformation question on the exam paper. It is particularly useful for rotation questions.

Place the tracing paper over the whole of the diagram and draw shape P on it.

The longest side of the diagram in shape P is vertical. The corresponding side in shape Q is horizontal. So the angle of rotation must be 90° (or 270° if you rotate in the opposite direction).

Look for clues like this.

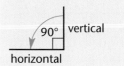

1 Try a point somewhere near P and Q. Put your pencil point on the point and hold it fixed.

2 Turn through 90° (or 270°) and see if shape P lands exactly on top of shape Q.

Once you have drawn the diagram, you can try several different centres of rotation quite quickly.

3 If it does not, try another centre of rotation. Keep trying different centres of rotation until shape P fits exactly on top of shape Q.

4 Describe the rotation, giving

- the angle

- the direction

- the centre of rotation (as coordinates).

Notice that the centre of rotation lies on the perpendicular bisector of AA'. This is true for any pair of corresponding points.

The rotation that takes P to Q is 90° anticlockwise about point (2, 1).

With experience, you will find it a lot easier to identify the correct centre of rotation.

Exercise 18B

1 Copy these shapes onto squared paper.

(a)
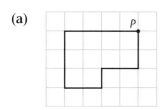

(b)
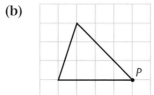

(c)
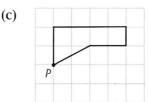

Rotate each shape about the point P

(i) a half turn clockwise

(ii) a quarter turn anticlockwise.

2 Copy this shape onto squared paper.

Draw the image of the shape after it has been rotated about the point *P*:

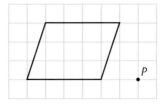

(a) 90° clockwise

(b) 180° anticlockwise

(c) three quarters of a turn clockwise.

3 On squared paper draw *x*- and *y*-axes going from −6 to +6. Copy this shape onto your axes.

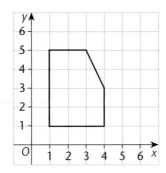

Draw the image of the shape after:

(a) a quarter turn clockwise about the origin (0, 0). Label the image *A*.

(b) a quarter turn anticlockwise about the origin (0, 0). Label the image *B*.

(c) 180° rotation anticlockwise about the origin (0, 0). Label the image *C*.

4 Describe fully the transformation which maps shape *P* onto shape *Q*.

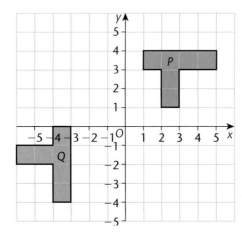

You need to give three pieces of information:
- the amount of turn
- the direction of turn
- the centre of rotation.

Use tracing paper to help you.

Key words:
translation
column vector

18.4 Translation

A **translation** slides a shape from one position to another.

In a translation every point on the shape moves the same distance in the same direction.

To describe a translation you need to give the distance and the direction of the movement.

Example 7

Translate this shape 4 squares to the right and 2 squares up.

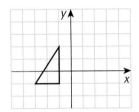

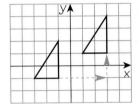

Choose any corner (vertex) of the shape and move this point 4 squares to the right and 2 squares up. Repeat this for the other vertices. Join up the vertices with straight lines.

An object and its image after a translation are congruent.

You can describe a translation by a **column vector** .

The translation in Example 7 has column vector $\begin{pmatrix} 4 \\ 2 \end{pmatrix}$.

Column vectors always have tall brackets round them.

The top number represents the movement in the *x*-direction and the bottom number represents the movement in the *y*-direction.

They are not fractions so do not draw a line between the numbers.

Notice that the translation to take the triangle back to its original position is $\begin{pmatrix} -4 \\ -2 \end{pmatrix}$.

Movements right and up are positive but left and down are negative.

Example 8

Translate this shape by

the vector $\begin{pmatrix} -3 \\ -3 \end{pmatrix}$.

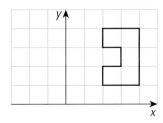

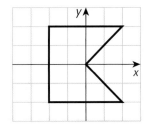

$\begin{pmatrix} -3 \\ -3 \end{pmatrix}$ means 3 squares *left* and 3 squares *down*.

Exercise 18C

1 Copy these shapes onto squared paper and translate them by the amounts shown.

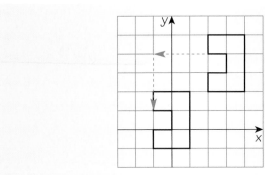

(a) $\begin{pmatrix} 3 \\ -2 \end{pmatrix}$ (b) $\begin{pmatrix} -4 \\ 2 \end{pmatrix}$ (c) $\begin{pmatrix} -5 \\ 0 \end{pmatrix}$ (d) $\begin{pmatrix} -1 \\ -6 \end{pmatrix}$ (e) $\begin{pmatrix} 0 \\ 3 \end{pmatrix}$

2 Copy this shape onto squared paper.
A translation of the shape moves the point P to the point P' on the image. Draw the complete image.
What is the column vector that describes the translation?

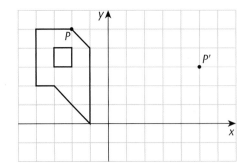

3 The triangle *A* is translated to new positions at *B*, *C*, *D* and *E*. Describe each transformation by giving the column vector.

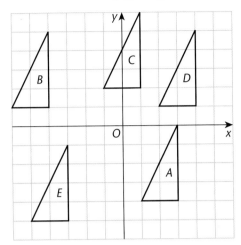

4 The letter *N* has vertices at positions *A*(0, 0), *B*(2, 4), *C*(4, 0) and *D*(4, 4). On an *x–y* grid with *x* and *y* between −6 and +6 plot these points and join the lines in order *A* to *D*.

(a) Translate the letter *N* by the translation vector $\begin{pmatrix} -5 \\ 2 \end{pmatrix}$ followed by a second translation using the vector $\begin{pmatrix} 3 \\ -7 \end{pmatrix}$.

Draw these new shapes on your diagram and label them *A′B′C′D′* and *A″B″C″D″*.

(b) What are the coordinates of the new vertex *B″*?

(c) What single transformation vector could be used to describe the movement from *ABCD* to *A″B″C″D″*?

18.5 Enlargement

An **enlargement** changes the size of an object but not its shape.

The number of times the shape is enlarged is called the **scale factor** or **multiplier**. This can be a whole number or a fraction.

In an enlargement, all the angles stay the same but all the lengths are changed in the same **proportion**. The image is **similar** to the object.

Key words:
enlargement
scale factor
multiplier
proportion
similar
centre of enlargement

An enlargement by a scale factor of a fraction less than 1 makes the image smaller.

For help with proportionality see Chapter 17 .

Example 9

Enlarge the rectangle *ABCD* by a scale factor 2.

Similar shapes:
- have equal angles
- have lengths in the same proportion.

Every length on the object is multiplied by 2 (doubled) on the image.

$AB = 3$, so $A'B' = 2 \times 3 = 6$
$BC = 2$, so $B'C' = 2 \times 2 = 4$

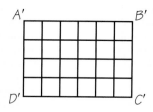

Example 10

What is the scale factor of the enlargement that takes shape *A* to shape *B*?

Shape A is only 1 square wide.

Shape B is 3 squares wide.

> Compare the lengths of corresponding sides.

Shape A is 3 squares long. Shape B is 9 squares long.

Shape B is 3 times longer and 3 times wider than shape A.

The scale factor is 3.

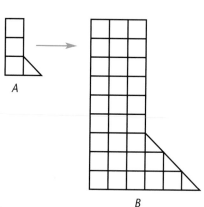

The final position of an enlargement is determined by the position of the **centre of enlargement**. In Examples 9 and 10 there is no centre of enlargement so the image can be drawn anywhere.

When you enlarge from a centre of enlargement, the distances from the centre to each point are multiplied by the scale factor.

If no centre of enlargement is given you can then draw the image close to the original shape.

Example 11

Copy the triangle *ABC*. Enlarge the triangle by scale factor 2 using the point *O* as the centre of enlargement.

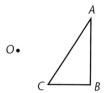

Multiply the distance *OA* by **2** to get *OA'*, and similarly for the other vertices.
$OA' = 2 \times OA$
$OB' = 2 \times OB$
$OC' = 2 \times OC$

Always draw your diagram as accurately as possible using a pencil and ruler. Leave the construction lines on your diagram.

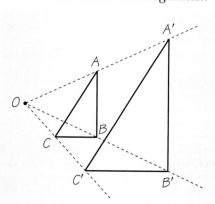

The centre of enlargement can be a point on the shape.

Example 12

Enlarge the triangle by scale factor 3 using point P as the centre of enlargement.

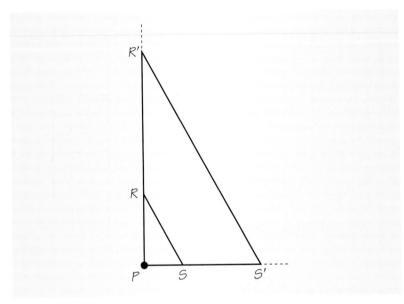

For an enlargement by a scale factor 3
$PR' = 3 \times PR$
$PS' = 3 \times PS$
The enlargement overlaps the original shape.

Sometimes the centre of enlargement is inside the shape.
Here, $ABCD$ has been enlarged about centre O, by scale factor $\frac{1}{3}$.

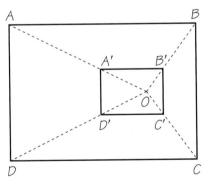

$OA' = \frac{1}{3}OA$
$OB' = \frac{1}{3}OB$
$OC' = \frac{1}{3}OC$
$OD' = \frac{1}{3}OD$

The enlargement is smaller than the original.

To describe an enlargement fully you need to give the scale factor and the centre of enlargement.

To find the position of the centre of enlargement, join vertices in the enlargement to the corresponding vertices in the original and continue these lines until they meet at a point.

This point is the centre of enlargement.

You draw in the construction lines for the enlargement.

Example 13

Triangle *ABC* has been enlarged to produce the shaded triangle *A'B'C'*.

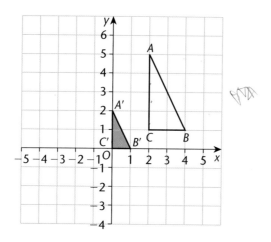

(a) What is the scale factor of the enlargement from *ABC* to *A'B'C'*?

(b) Mark on the grid the position of the centre of enlargement *P*, and give its coordinates.

> This enlargement has made a smaller image so the scale factor must be less than 1.

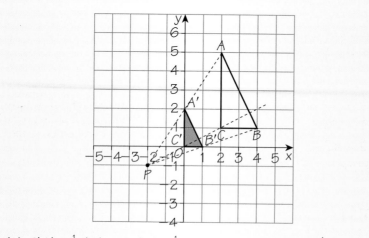

> All lengths on triangle *A'B'C'* are half the lengths on triangle *ABC*. So the scale factor is $\frac{1}{2}$.

> Draw in the construction lines joining *A* to *A'*, *B* to *B'* etc. These lines meet at point *P*(−2, −1), the centre of enlargement.

(a) $C'A' = \frac{1}{2}C'A'$ and $C'B' = \frac{1}{2}C'B'$. The scale factor is $\frac{1}{2}$.

(b) The centre of enlargement is at *P*(−2, −1).

Negative scale factor

An enlargement by a negative scale factor produces an image that is in the opposite direction to the object. The size of this image is given by the size of the scale factor.

Example 14

Enlarge triangle *PQR* with a scale factor −1 using the centre of enlargement *C*(−1, 1).

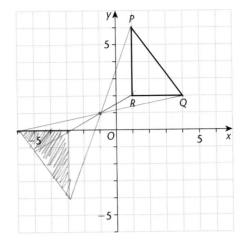

Mark the centre C at position (−1, 1). Draw extended lines from each vertex through to the opposite side of C. For a scale factor −1 the length PC = P'C, QC = Q'C and RC = R'C.

Label the image *P'Q'R'*. Leave your construction lines on the diagram.

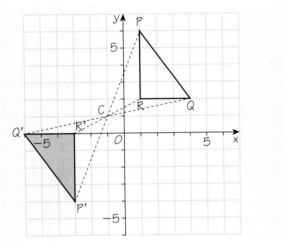

For a scale factor of −2, $P'C = 2 \times PC$

$Q'C = 2 \times QC$

$R'C = 2 \times RC$

... and so on for any other scale factor.

For a negative scale factor, the image is always 'upside down' in relation to the object.

Scale factors and similarity

- A scale factor of 1 produces an image shape that is the same size as the object shape. The object and image shapes are *congruent*.
- A scale factor > 1 produces an image shape that is larger than the object shape and the two shapes are *similar*.
- A scale factor < 1 produces an image shape that is smaller than the object shape and the two shapes are *similar*.
- A negative scale factor produces an image 'upside down' in relation to the object and on the opposite side of the centre of enlargement. The size of the image is given by the size of the scale factor.

Exercise 18D

1 For each of the following shapes work out the scale factor of the enlargement.

(a)

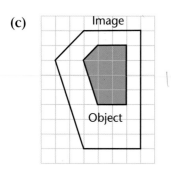

(b)

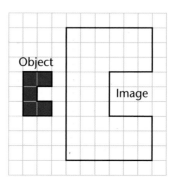

Use Example 10 to help you.

(c)

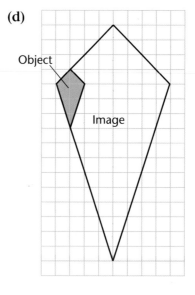

(d)

2 Copy each of the following shapes onto squared paper. Enlarge each one by scale factor 2.

(a)   **(b)**   **(c)**

No centre of enlargement is given, so draw the enlargement close to the original shape.

3 Copy the following shapes onto squared paper. Enlarge each one by scale factor 2 from the centre of enlargement C.

(a)

(b)

4 The vertices of triangle K are (1, 1), (1, 2) and (4, 1).

Enlarge the triangle K by scale factor 3 with (0, 0) as the centre of enlargement.

What are the coordinates of the image triangle K'?

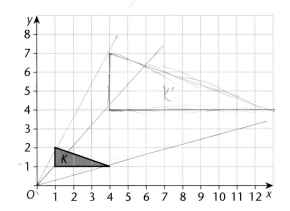

5 This right-angled triangle is to be enlarged by scale factor 4.

The centre of enlargement is at a point P(1, 2).

Copy the diagram and draw the enlargement.

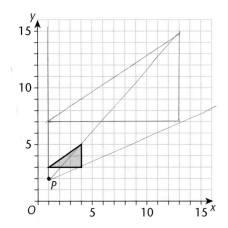

6 The rectangle X is enlarged to produce the image Y.

Use Example 13 to help you.

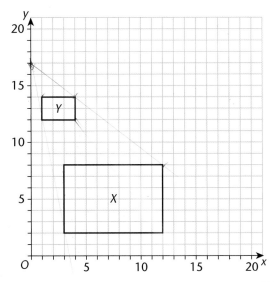

(a) What is the scale factor of this enlargement?

(b) Construct lines to show the position of the centre of enlargement.

(c) What are the coordinates of the centre of enlargement?

7 Copy the shape below onto squared paper and draw an
   enlargement using a scale factor of $\frac{1}{3}$ about the point (4, 4).

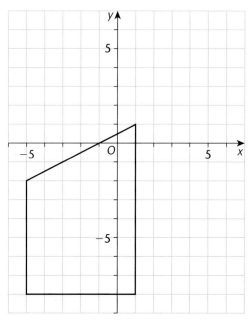

8 Copy the shape onto squared paper and draw an enlargement
   with a scale factor −2 about the origin.

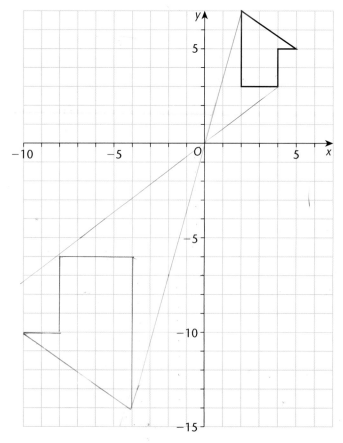

18.6 Combined transformations

Transformations can be performed one after the other.

Example 15

Reflect the object *P* about the *y*-axis (*x* = 0) and label it *P'*.
Now reflect *P'* about the line *x* = 4 and label this *P''*.

What single transformation takes *P* to *P''*?

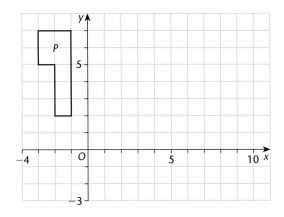

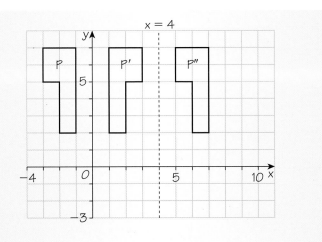

Draw on the mirror line
x = 4. Reflect the object in
the *y*-axis to give *P'*. Then
reflect *P'* in the line *x* = 4 to
give *P''*.

P'' is a translation of *P* 8 units to the right.

It is a translation of $\begin{pmatrix} 8 \\ 0 \end{pmatrix}$.

Example 16

Using the object shape *P* in Example 15, reflect *P* in the
y-axis followed by a reflection in the line *y* = *x*.

What single transformation takes *P* to *P''*?

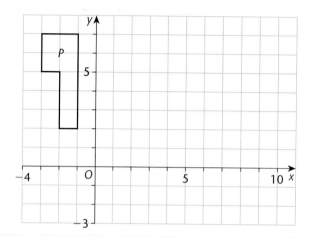

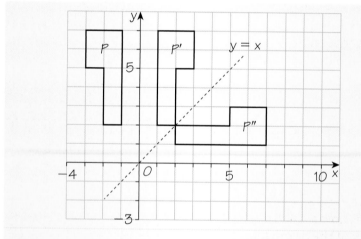

Draw on the mirror line
y = *x*.

P'' is obtained by a rotation of *P* through 90° clockwise about
the origin (0, 0).

The centre of rotation is the
intersection of the two
mirror lines *x* = 0 and *y* = *x*.
The point of intersection
is (0, 0).

A reflection followed by a second reflection along mirror lines that
are parallel (Example 15) can be replaced by a single transformation
involving a translation.

A reflection followed by a second reflection along mirror lines that
are not parallel to each other (Example 16) can be replaced by a
single transformation involving a rotation about the point of
intersection of the two mirror lines.

Exercise 18E

1 Copy shape *D* onto squared paper. Reflect the shape in the *y*-axis (label *D'*) followed by a reflection in the *x*-axis (label *D"*).
Describe the single transformation that takes *D* to *D"*.

Rotation in a anti-clockwise on a

point (0,0).

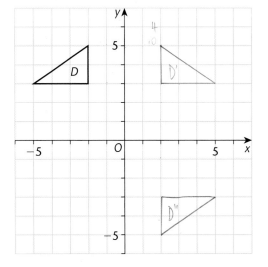

2 Construct a grid with *x*-axis as shown and *y*-axis from 0 to 10. Copy the object *E* and reflect the shape in the line *x* = −1. Label this *E'*. Now reflect shape *E'* in the *y*-axis and label this *E"*.
What is the single transformation that takes *E* to *E"*?

Translate 2 squares to the right.

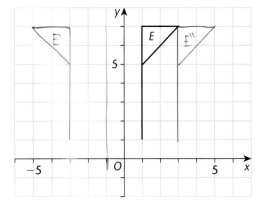

3 The object *T* is reflected in the line *y* = −*x* to produce an image *T'*. This image is now rotated a quarter of a turn anticlockwise about the origin to produce a second image *T"*.
What single transformation takes *T* to *T"*?

Reflect in x

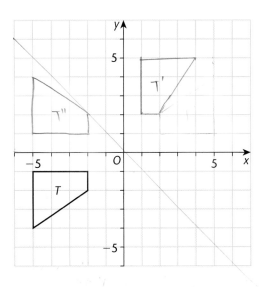

4 Draw an x–y grid with x and y between -4 and $+4$. On the grid draw the triangle ABC which has vertices at $A(2, 3)$, $B(3.5, 1.5)$ and $C(2.5, 1)$. On the same diagram

(a) Rotate ABC 90° anticlockwise about the point $(0, 0)$. $A(-3,2)$ $B(-1.5,3.5)$ $C(-1,2.5)$
Label this shape L.

(b) Reflect this new shape about the x-axis and label this $A(-3,-2)$ $B(-1.5,-3.5)$ $C(-1,-2.5)$
triangle M.

(c) Rotate this new shape 270° clockwise about the origin and $A(3,-2)$ $B(1.5,-3.5)$ $C(1,-2.5)$
label this N.

(c) Describe the single transformation that takes ABC to N.

$\xrightarrow{clockwise} (y,-x)$

$(x,y) \xrightarrow{antclock} (-y,x) \xrightarrow{180} (y,x)$

Examination questions

1 Triangle A is drawn on the grid below.

(a) Reflect triangle A along the line $y = -1$.
Label the triangle B.

(b) Rotate triangle A a quarter of a turn clockwise about the origin O.
Label the triangle C.

$2\cdot1$

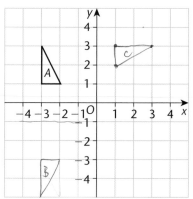

(5 marks)

2 The grid shows several transformations of the shaded triangle.

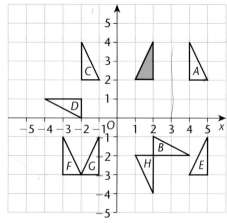

(a) Write down the letter of the triangle
(i) after the shaded triangle is reflected in the line $x = 3$ A
(ii) after the shaded triangle is translated by 3 squares to the right and 5 squares down E
(iii) after the shaded triangle is rotated 90° clockwise about O B

(b) Describe fully the single transformation which takes triangle F onto triangle G.

Reflection on m $x = -2$

(5 marks)

AQA, Spec A, 2I, November 2003

3 The diagram shows a right-angled triangle *A*.

Draw the new position of triangle *A* after
a rotation of
90° clockwise about the point $(-2, -1)$

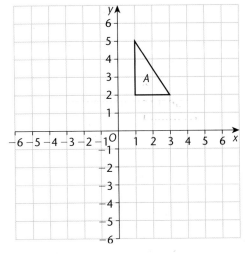

(3 marks)

4 Enlarge the shaded shape by scale factor $-\frac{1}{2}$ with centre of enlargement $(-1, 0)$.

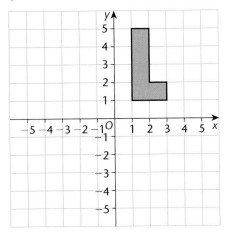

(2 marks)
AQA, Spec A, 2H, June 2004

5 On the grid opposite there are two shapes,
A and *B*.

(a) Describe fully the single transformation
that takes shape *A* to shape *B*.

(b) Copy the grid. Draw the enlargement of
shape *A* with scale factor $\frac{1}{3}$ and centre of
enlargement $(0, 0)$.

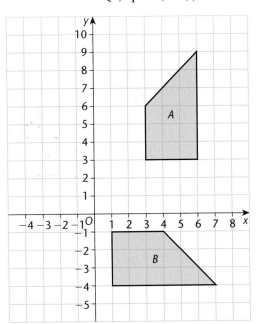

(5 marks)
AQA, Spec A, 1H, November 2004

Summary of key points

Reflections (grade D/C)

Reflections take place along a mirror line to produce a mirror image.

Mirror lines can be vertical, horizontal or at 45°. Mirror lines can be given as simple equations, such as $x = 0$, $y = 0$ or $y = x$.

To describe a reflection fully you need to give the equation of the mirror line.

Rotation (grade D/C)

A rotation turns an object either clockwise or anticlockwise through a given angle about a fixed point, called the *centre of rotation*.

To describe a rotation fully you need to give:

- the centre of rotation
- the angle of turn
- the direction of turn

Translation (grade D)

A translation slides a shape from one position to another.

Translations are described by *column vectors*, for example $\begin{pmatrix} 3 \\ -2 \end{pmatrix}$ the top number represents the movement in the x-direction and the bottom number represents the movement in the y-direction.

Enlargements (grades D to A)

An enlargement changes the size of an object but not its shape. In an enlargement, all the angles stay the same but all the lengths are changed in the same *proportion*. The image is *similar* to the object.

To describe an enlargement fully you need to give the scale factor and centre of enlargement.

- A scale factor of 1 produces an image that is the same size as the object. The object and image are *congruent*.
- A scale factor > 1 produces an image shape that is larger than the object shape and the two shapes are *similar*.
- A scale factor < 1 produces an image shape that is smaller than the object shape and the two shapes are *similar*.
- A negative scale factor produces an image 'upside down' in relation to the object and on the opposite side of the centre of enlargement.

Combined transformations (grade C/B/A)

Combined transformations produce more than one image, and can often be described in terms of a single transformation.

19 Congruency and similarity

This chapter will show you how to:

✔ define congruency in triangles using the four definitions of SSS, SAS, ASA and RHS
✔ define similarity and use scale factors
✔ find areas and volumes of similar figures using scale factors

19.1 Congruency and similarity

You will need to know:

● about ratio and proportion, right-angled triangles and enlargement

Key words:
congruent
similar

Congruent shapes are identical. Reflection, rotation and translation transformations all produce images that are congruent to the original objects.

Similar shapes have exactly the same shape but are not the same size.

All angles in the object and image are equal but all lengths are not. In similar shapes, corresponding lengths are in the same ratio or proportion.

This means they have exactly the same shape and are exactly the same size. All lengths and angles in the object and image are equal.

You may need to turn shapes over before they fit exactly. A reflection in a mirror line produces congruent shapes.

Example 1

Which pair of shapes are congruent to shape *A* and which shapes are similar to shape *A*?

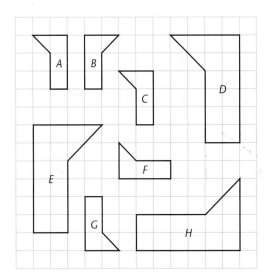

Some images need to be translated, rotated or reflected.

B, C, F and G are all congruent to A.

D, E and H are all similar to A.

The four conditions for congruent triangles are:

(1) Three sides are equal (known as Side, Side, Side – **SSS**).

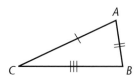

(2) Two sides are equal and the included angle is the same (known as Side, Angle, Side – **SAS**).

The *included* angle is the angle made by the two known sides.

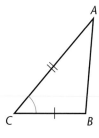

(3) Two angles are the same and a corresponding side is equal (known as Angle, Side, Angle – **ASA**).

Also called **S**ide, **A**ngle, **A**ngle – **SAA**.

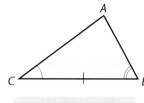

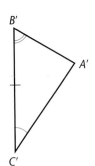

or

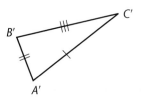

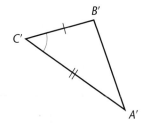

BC and *B'C'* are the corresponding sides.

AB and *A'B'* are the corresponding sides.

(4) A right angle, a hypotenuse and a corresponding side are equal (right angle, hypotenuse, side – **RHS**).

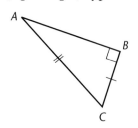

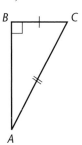

The hypotenuse is the longest side of a right-angled triangle. It is always opposite the right angle.

BC and *B'C'* are the corresponding sides.

There is one case where triangles are sometimes thought to be congruent, but they are not.
Look at these diagrams.

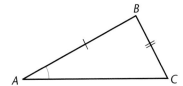

These triangles have two sides which are the same length and an angle which is the same size.
You might think that SSA is a condition for congruency, but one look at these triangles should convince you that it is not!

Try constructing a triangle with $AC = 8$ cm, $BC = 5$ cm and $B\hat{A}C = 30°$. Draw a line AB as your starting point.

Exercise 19A

1 Look at the shapes in the diagram.

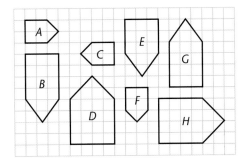

   (a) Write down the shapes that are congruent to shape A.

   (b) Write down the shapes that are similar to shape A.

2 Copy this shape onto squared paper.
On the same paper draw one shape that is similar and one shape that is congruent.

3 Look at the following figures. For each one write down the letters of the shapes that are similar to each other.

(a)

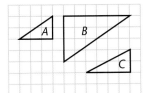

(b)

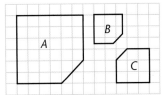

(c)

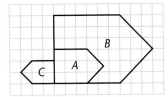

(d)

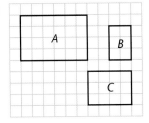

4 Look at the following pairs of triangles. State which pairs are congruent and give the appropriate reason.

(a)
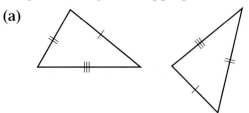

(b)
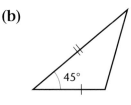
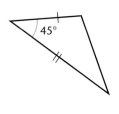

(c)
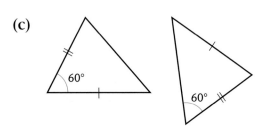

(d)
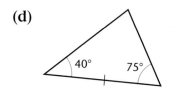
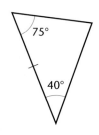

(e)
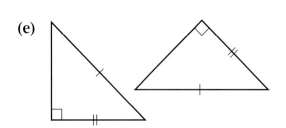

(f)
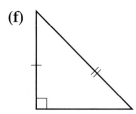
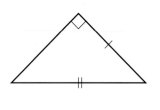

19.2 Similarity

Key words:
ratio
proportion
scale factor

Similar shapes have exactly the same shape but are not exactly the same size. All angles in the object and image are equal but all lengths are not.

Enlargement always produces an image that is similar to the object.

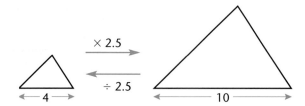

In similar shapes corresponding lengths are in the same **ratio** or **proportion**. The ratio is the **scale factor** of the enlargement.

Ratio and proportion were covered in Chapter 17.

The scale factor may sometimes be called the linear scale factor (k).

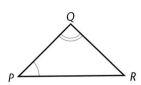

For two shapes to be similar

- the corresponding angles must be equal.
- the ratios of corresponding sides must be the same.

In the triangles *ABC* and *PQR*:

  angle *BAC* = angle *QPR*
  angle *CBA* = angle *RQP*
  angle *ACB* = angle *PRQ*.

Always write the letters of the triangles above one another, to help you pick out the corresponding lengths and ratios.

So in this case $\dfrac{ABC}{PQR}$

from which you can pick out the three ratios

$$\frac{AB}{PQ} = \frac{BC}{QR} = \frac{CA}{RP}$$

These can also be written as

  AB : *PQ*,   *BC* : *QR*,   *CA* : *RP*.

Either method defines the scale factor, *k*, in going from triangle *PQR* to triangle *ABC*.

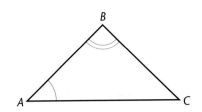

It is important to obtain the correct scale factor by identifying in which direction you are going e.g.

$PQR \rightarrow ABC$ is $k = \dfrac{AB}{PQ}$

whereas

$ABC \rightarrow PQR$ is $\dfrac{1}{k} = \dfrac{PQ}{AB}$.

Example 2

The two triangles opposite are similar with equal angles as shown.
Find the unknown lengths *x* and *y*.

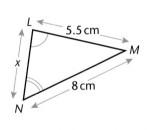

Method A

The triangles BAC and LMN are similar

$$\frac{LMN}{BAC}$$

The three equivalent ratios are:

$$\frac{LN}{BC} = \frac{LM}{BA} = \frac{MN}{AC}$$

$$\frac{x}{9} = \frac{5.5}{y} = \frac{8}{12}$$

So $x = \dfrac{72}{12} = 6 \text{ cm}$ and $y = \dfrac{5.5 \times 12}{8} = 8.25 \text{ cm}$

Notice how important it is to label the triangles accurately.
  angle *L* = angle *B*
  angle *M* = angle *A*
  angle *N* = angle *C*

Using two equivalent fractions allows either *x* or *y* to be determined.

continued ▼

Method B

The triangles BAC and LMN are similar

$$\frac{BAC}{LMN}$$

The scale factor of the enlargement $= \dfrac{AC}{MN} = \dfrac{12}{8} = 1.5$

So $AB = 1.5 \times LM = 1.5 \times 5.5 = 8.25\,cm$

and $LN = BC \div 1.5 = 9 \div 1.5 = 6\,cm.$

> Since $\triangle ABC$ is larger than $\triangle LMN$ the scale factor is worked out this way.
>
> Lengths in $\triangle LMN \times 1.5$ = Lengths in $\triangle ABC$.
>
> Lengths in $\triangle ABC \div 1.5$ = Lengths in $\triangle LMN$.

Example 3

Triangle ABC has $AB = 9$ cm.
A line PQ is drawn parallel to BC
so that $PQ = 5$ cm, $AP = 6$ cm
and $PB = 3$ cm.
Show that triangle APQ is
similar to triangle ABC.
Find the length BC.

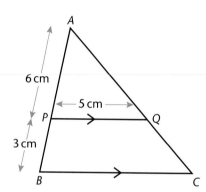

> Questions asking you to show a statement is true are becoming more popular in examinations. You must give reasons for your statements.

In triangles APQ and ABC

      angle BAC = angle PAQ   (a common angle)

      angle APQ = angle ABC   (PQ is parallel to BC so these

                            angles are corresponding).

So the third angles of both triangles must also be equal.

Method A

All angles in triangle ABC and triangle APQ are equal, which

means that triangles ABC and APQ are similar so

$$\frac{ABC}{APQ}$$

$$\frac{BC}{PQ} = \frac{AB}{AP}$$

$$\frac{BC}{5} = \frac{9}{6}$$

$$BC = \frac{9 \times 5}{6} = 7.5\,cm$$

continued ▶

Method B

$\triangle ABC$ is an enlargement of $\triangle APQ$ $\dfrac{ABC}{APQ}$

The scale factor $= \dfrac{AB}{AP} = \dfrac{9}{6} = 1.5$

So $BC = 1.5 \times PQ = 1.5 \times 5 = 7.5$ cm

> Always use corresponding sides to work out the scale factor.

Exercise 19B

1 The diagram shows two similar shapes. Find the values of the unknown letters.

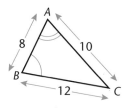
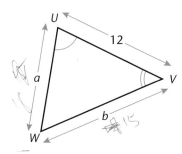

2 The drawing shows a picture frame containing a mount.

Are the two rectangles similar?

NO

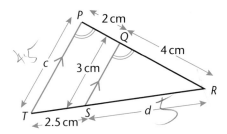

24 cm
20 cm
16 cm
12 cm

3 In these triangles find the values of *c* and *d*.

P   2 cm
Q
4 cm
c   3 cm
R
T   2.5 cm   S   d

4 Which of the following are True (T) and which are False (F)?

(a) Any two squares are similar to each other. T

(b) Any two isosceles triangles are similar to each other.

(c) Any two rhombuses are similar to each other.

(d) Any two rectangles are similar to each other.

(e) Any two circles are similar to each other.

(f) Any two regular hexagons are similar to each other.

> See Section 4.4 for a reminder about rhombuses.

5 A mobile phone mast is 22.5 m high and casts a shadow of length 25 m at midday. A tree next to the mast casts a shadow of length 15 m at midday. Find the height of the tree.

6 In the diagram, *ABC* is a right-angled triangle. Find the values of *x* and *y*.

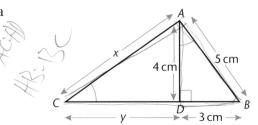

> Remember that angle *CAB* = 90°. There are 3 similar triangles.

19.3 Areas and volumes of similar shapes

> **Key words:**
> linear scale factor
> area scale factor
> volume scale factor

You can work out the area and volume of similar shapes if you know the linear scale factor.

The linear scale factor is the scale factor that applies to *lengths* of similar shapes.

Look at the rectangle below, size 1 × 3.
Enlarge it by a linear scale factor of 3.

> Enlargement was covered in Chapter 18.

The original shape becomes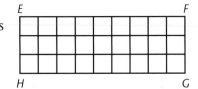

Length *EF* = 3 × length *AB* = 9
Length *FG* = 3 × length *BC* = 3

Area *ABCD* = 1 × 3 = 3
Area *EFGH* = 9 × 3 = 27
Enlarged area EFGH = 9 × area *ABCD*
           EFGH = 3^2 × area *ABCD*

This is the (linear scale factor)² and leads to the more general result:

When a shape is enlarged by a linear scale factor *k* the area of the enlarged shape is k^2 × the area of the original shape.

Enlarged area = k^2 × original area

> The k^2 factor is called the
> **area scale factor** .

Now look at a three-dimensional object, size $1 \times 1 \times 3$.
Enlarge this by a linear scale factor 3.

Original shape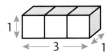

Enlarged shape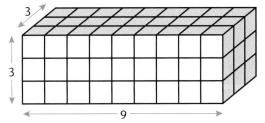

The original volume $= 1 \times 1 \times 3 = 3$.
The enlarged volume $= 3 \times 3 \times 9 = 81$.

The enlarged volume $= 27 \times$ the original volume.
$\qquad\qquad\qquad = 3^3 \times$ the original volume.

This is the (linear scale factor)^3 and leads to the more general result:

> **When a shape is enlarged by a linear scale factor k the**
> **volume of the enlarged shape is $k^3 \times$ the volume of the**
> **original shape.**
>
> **Enlarged volume $= k^3 \times$ original volume**

The k^3 factor is called the
volume scale factor .

> **Area and volume factors can be applied to any 2- or**
> **3-dimensional figures that are mathematically similar.**
> **All squares and circles are mathematically similar.**

Example 4

Two similar cylinders have heights which are in the ratio of $2 : 3$
What is the ratio of:

(a) their surface areas **(b)** their volumes?

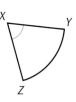

(a) Ratio of surface areas $= 2^2 : 3^2 = 4 : 9$

(b) Ratio of volumes $= 2^3 : 3^3 = 8 : 27$

This means that the large
cylinder holds over three
times more than the small
one $(27 \div 8 = 3.375)$.

Example 5

The diagram shows a sector of a circle that has been
enlarged by a linear scale factor of $\frac{3}{4}$.

(a) What is: **(i)** the length of XY **(ii)** the angle ZXY?

(b) If the area of sector $CAB = 16\pi\,\text{cm}^2$ what is the area of
  sector ZXY?

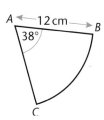

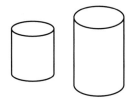

(a) (i) $XY = AB \times$ scale factor

$XY = 12 \times \frac{3}{4} = 9$ cm

> The scale factor of $\frac{3}{4}$ applies to the radius AB because it is a linear scale factor.

(ii) For similar shapes corresponding angles are equal, so angle $ZXY = 38°$.

(b) Area $ZXY = ($linear scale factor$)^2 \times$ Area CAB

Area $ZXY = \frac{9}{16} \times 16\pi = 9\pi$ cm^2

Example 6

The diagram shows two shampoo bottles. The shapes are mathematically similar. The smaller bottle contains 400 ml of shampoo.

Calculate how much shampoo the larger bottle holds. Give your answer to 2 s.f.

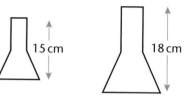

The linear scale factor is $\dfrac{\text{height of large bottle}}{\text{height of small bottle}} = \frac{18}{15} = 1.2$

Volume of large bottle $= (1.2)^3 \times 400$

$\qquad = 691.2$ ml

$\qquad = 690$ ml (to 2 s.f.)

> Volume of large bottle = (linear scale factor)^3 × volume of small bottle

Example 7

A supermarket sells two sizes of jars of coffee. The jars are in the shape of cylinders which are mathematically similar.

(a) If the diameter of the large jar is 8.2 cm, calculate the diameter of the small jar.

(b) Calculate the ratio of the areas of circular bases of the two jars.

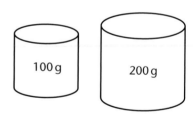

(a) Volume scale factor $= \dfrac{\text{amount in large jar}}{\text{amount in small jar}}$

$\qquad = \frac{200}{100} = 2$ (this is k^3)

So linear scale factor $(k) = \sqrt[3]{2} = 1.259\,92\ldots$

Diameter of base of small jar $\times k$

$\quad =$ Diameter of base of large jar

Diameter of base of small jar $\times 1.25992\ldots = 8.2$ cm

Diameter of base of small jar $= \dfrac{8.2}{1.25992}\ldots = 6.5083\ldots$

$\qquad = 6.5$ cm (1 d.p.)

> The amount of coffee in the jars represents the **volume** of the jars.

> Do not round 1.25992... Work with this value and only round the final answer to a sensible degree of accuracy.

continued ▶

(b) The ratio of the diameters is $1 : 1.25992\ldots$

The ratio of the areas of the bases of the jars

$= 1^2 : (1.25992\ldots)^2$

$= 1 : 1.58740\ldots \quad (k^2 = 1.58740\ldots)$

$= 1 : 1.6 \quad$ (approximately)

Exercise 19C

1 A photograph is being enlarged by a scale factor 3. If the original area of the print was 24 cm², what is the new area of the photograph?

2 Two similar triangles have areas of 16 cm² and 40 cm². If the base of the smaller triangle is 4 cm, find the base of the larger triangle.

3 In the shape opposite, $BX = 4$ cm, $AB = 6$ cm and the area of triangle $BXY = 10$ cm².

Find **(a)** the area of triangle ABC and
(b) the area of the trapezium $XYCA$.

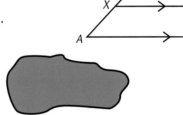

4 The following shape with an area of 12.2 cm² is being enlarged by a scale factor of 3.4.

What is the area of the enlarged shape?

5 The surface area of a cone is 176 cm². If it is enlarged by a scale factor of $\frac{2}{3}$, what is the surface area of the smaller cone?

6 Garden peas are sold in mathematically similar tins of two different sizes, 300 g and 400 g. If the height of the small tin is 10 cm, calculate the height of the large tin.

7 The radii of two spheres are in the ratio $2 : 5$.
   (a) If the volume of the smaller sphere is 8 cm³ calculate the volume of the larger sphere.
   (b) What is the ratio of the surface areas of the two spheres?

Examination questions

1 The two triangles shown are congruent.

Write down the values of p and q.

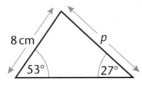
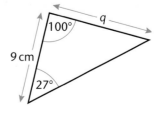

(2 marks)

2 Gary, *G*, can just see the top of a radio mast, *R*, over a wall, *W*. Gary is 15 m from the wall. The wall is 45 m from the radio mast. The wall is 2.7 m high.

Calculate the height of the radio mast, marked *h* on the diagram.

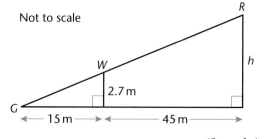

Not to scale

(3 marks)
AQA, Spec B, 2H, June 2004

3 Triangles *ADE* and *ABC* are similar. *DE* is parallel to *BC*, *AD* = 4 cm, *DE* = 6 cm and *BC* = 9 cm.

Calculate the length of *BD*.

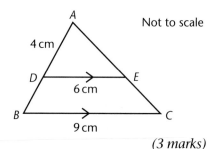

Not to scale

(3 marks)
AQA, Spec A, 2I, June 2003

4 *PQRS* is an enlargement of *ABCD* with scale factor $\frac{2}{3}$.
QR = 3.6 cm and angle *BAD* = 45°.

(a) Calculate the length of *BC*.

(b) Find the size of angle *QPS*.

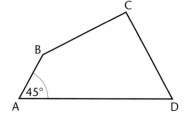

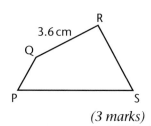

(3 marks)

5 Two similar bottles are shown. The smaller bottle is 20 cm tall and holds 480 ml of water. The larger bottle is 30 cm tall.

How much water does the larger bottle hold?

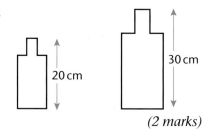

(2 marks)
AQA, Spec A, 2H, June 2003

6 A square-based pyramid with a base of side 2 cm has a volume of 2.75 cm³.

What is the volume of a similar square-based pyramid with a base of side 6 cm?

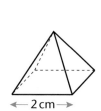

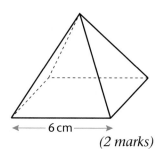

(2 marks)
AQA, Spec A, 2H, November 2004

Summary of key points

Congruency (grades C/B/A)

Congruent shapes have identical size and shape. Reflection, rotation and translation transformations all produce images that are congruent with the original objects.

The four conditions for congruent triangles are:

Side, Side, Side	SSS
Side, Angle, Side	SAS
Angle, Side, Angle **or** Side, Angle, Angle	ASA, SAA
Right angle, Hypotenuse, Side	RHS

SSA is *not* a condition for congruence.

Similarity (grade B)

Similar shapes have identical shapes but are different in size.

Enlargement always produces objects and images that are similar.

In similar shapes corresponding lengths are in the same *ratio* or *proportion*. The ratio is the *scale factor* of the enlargement.

For two shapes to be similar

- the corresponding angles must be equal.
- the ratios of corresponding sides must be the same.

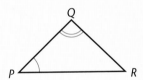
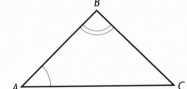

Area and volume of similar shapes (grade A)

You can work out the area and volume of similar shapes if you know the *linear scale factor*. The linear scale factor can be worked out using ratios of corresponding lengths $a : b$ or as equivalent fractions $\dfrac{a}{b}$.

When a shape is enlarged by a linear scale factor k the area of the enlarged shape is $k^2 \times$ the area of the original shape.

> Enlarged area = $k^2 \times$ original area

k^2 is called the *area scale factor*.

When a shape is enlarged by a linear scale factor k the volume of the enlarged shape is $k^3 \times$ the volume of the original shape.

> Enlarged volume = $k^3 \times$ original volume

k^3 is called the *volume scale factor*.

Area and volume factors can be applied to any 2- or 3-dimensional figures that are mathematically similar. All squares and circles are mathematically similar.

This chapter will show you how to:
- ✔ find the 'averages' – mean, median and mode for discrete data
- ✔ find the range and understand what is meant by the spread of the data
- ✔ find the mean, median and mode from frequency distributions and grouped frequency distributions for both discrete and continuous data
- ✔ find moving averages

20.1 Averages – mean, median, mode

You will need to know:
- how to interpret a stem-and-leaf diagram

Key words:
average
mean
mode
median
range
sigma

An **average** is a value that is representative of a set of data.

Three averages that you need to know are the **mean**, the **mode** and the **median**.

You can also describe how spread out the data is, using the **range**.

The range is found by subtracting the lowest value from the highest value.

Example 1

The list shows the hourly rate received by nine employees.

£12   £7.20   £8.60   £7.50   £10   £7.50   £13.40   £10   £7.50

Find the

(a) mean   **(b)** mode   **(c)** median   **(d)** range for this data.

(a) mean $= \dfrac{12 + 7.20 + 8.60 + 7.50 + 10 + 7.50 + 13.40 + 10 + 7.50}{9}$

$= \dfrac{83.70}{9} = 9.30$

The mean is £9.30.

Find the mean hourly rate by adding together all the values and dividing this by the total number of values.

There are 9 values in this example.

This is the mean hourly rate.

continued ▶

(b) The mode is £7.50.

> The mode is the value that occurs most often.

(c) 7.20 , 7.50 , 7.50 , 7.50 , **8.60** , 10 , 10 , 12 , 13.40

The median value is £8.60.

> The median is the middle value when the values are arranged in order of size.

d) The range = £13.40 − £7.20 = £6.20.

> The lowest hourly rate is £7.20 and the highest is £13.40. The range is the difference between these two amounts. It gives you an idea of how the data is spread out.

Finding the median for an even number of values

If you have an *odd* number of data values then there is *only one middle value* as in Example 1.

If you have an *even* number of data values then there are *two middle values*. The median is the number mid-way between these two values. For example, if a tenth person has an hourly rate of £6.50, then the list in ascending order is

6.50, 7.20, 7.50, 7.50, **7.50, 8.60** , 10, 10, 12, 13.40

and the median value is

$$\text{median value} = \frac{7.50 + 8.60}{2} = £8.05$$

The addition of this extra value could also affect the mean, range and mode.

These four quantities – mean, mode, median and range – tell you the main features of the data and allow you to make comparisons with other data sets.

Special notation for finding the mean

If there are n items of data in a list you could label them as $x_1, x_2, x_3, \dots , x_n$.

The sum of all these values can then be written as

$$\Sigma x = x_1 + x_2 + x_3 + \dots + x_n.$$

The symbol Σ (pronounced **sigma**) means '*the sum of*'. Σx means '*the sum of all the x values*'.

Using this notation,

$$\text{mean} = \frac{x_1 + x_2 + x_3 + \dots + x_n}{n}$$

$$= \frac{\Sigma x}{n}$$

$$\bar{x} = \frac{\Sigma x}{n}$$

> The symbol $\bar{x}$ (pronounced *x* bar) is shorthand for the mean.

Example 2

Jen obtained these scores for the first eight modules of her course.

63   49   51   52   70   67   52   76

(a) Find her mean score.

(b) She needs a mean score of 62 over 9 modules to pass her course. What does she need to score in her ninth module?

(a) $\bar{x} = \dfrac{\Sigma x}{8}$

$= \dfrac{63 + 49 + 51 + 52 + 70 + 67 + 52 + 76}{8} = \dfrac{480}{8}$

$= 60$

(b)   New mean $= 62$

$= \dfrac{\Sigma x}{9}$

So $\Sigma x = 9 \times 62 = 558$ ——— Total of 9 scores $= 9 \times 62 = 558$.

Score needed $= 558 - 480 = 78$ ——— Subtract sum of 8 scores from sum of 9 scores.

Exercise 20A

1 Find the mean, median, mode and range of the following.
   (a) 6   3   9   12   1   9   8   8   6   1   3   6
   (b) 20.1   20.7   21.4   22.7   29.6   22.6
   (c) $\frac{1}{4}$ $\frac{1}{2}$ $\frac{3}{4}$ $\frac{1}{4}$ $\frac{5}{4}$ $\frac{3}{2}$ $\frac{5}{2}$ $\frac{1}{4}$
   (d) 151   154   161   179   180   124   162   180   134

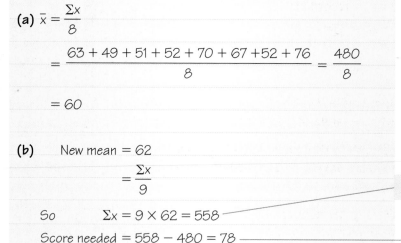

2 In a driving theory test the following percentage marks were recorded.

   75   61   52   82   64   71   90   46   55   57   64   63   67

   Find the mean, median and mode for the marks.

3 During a ten-pin bowling game the following scores were recorded.

   7   8   4   1   10   8   3   6   5   9

   Find the mean score (to the nearest whole number) and the mode.

4 The prices of a particular new car in a local paper were

   £9900, £10 200, £9625, £9865, £10 150, £9950

   What is the mean price for this new car?

5 Work out the mean of the following six amounts:

£150, £75, £62, £87, £46, £102

When a seventh amount is added, the new mean value is £83.
What is the amount that has been added?

> Use Example 2 to help you.

6 The mean height of the 13 boys in a class is 162 cm and the mean height for the 12 girls is 153 cm. What is the mean height for the whole class?

> The answer is not 157.5 cm!

7 In a test the following results were recorded.

21, 25, 18, 27, 22, 23, 19, 16, 21, 27
24, 24, 16, 18, 23, 24, 25, 20, 28

Find:

(a) the mean score   (b) the median score

(c) the mode   (d) the range of scores for this class.

8 Construct a list of six whole numbers that have range 4, a mode value of 6 and also a mean value of 6. What is the median value for this set of numbers?

9 This stem-and-leaf diagram shows the temperatures recorded in different cities around the world at noon one day.
Find the mean, mode, median and range of the temperatures.

> Look back to Section 10.4 for a reminder about stem-and-leaf diagrams.

$1|2$ represents 12°C.

```
0 | 1 4 5 8 8 9
1 | 0 0 3 5 6 6 9 9
2 | 0 1 1 1. 2 5 8
3 | 0 0 1
```

10 A class of students had their heights measured to the nearest centimetre.
The results for the girls and the boys are shown in this back-to-back stem-and-leaf diagram.

$140|7$ represents 147 cm.

Boys		Girls
8	120	4 9
0	130	2 7 7 7
7	140	2
3 5 5 8 9	150	1 3 5 5 6
0 1 1 1	160	0 2 8
5 7	170	0

Find the mean, mode, median and range of these heights

(a) for the girls only   (b) for the boys only   (c) for the whole class.

Which is the best average to use?

Key words:
extreme value

Sometimes when you calculate an average, the answer might not be very representative of the data set.

You must be able to select an appropriate average to use, and give your reasons.

Example 3

A factory's staff consists of a managing director, the works' supervisor and twelve employees.

Their net weekly wages (in £) are shown below.

135   135   135   135   150   150   150   164

164   178   193   193   276   957

(a) Find the mean, mode and median wage.

(b) Discuss how representative each of these averages are.

(a)   Mean $= \dfrac{135 + 135 + \dots + 276 + 957}{14}$

$\bar{x} = \dfrac{3115}{14} = £222.50$

Mode $= £135$

Median $= \dfrac{150 + 164}{2} = £157$

(b) The mean wage is £222.50.

The only reason why the mean is so high is that the managing director's £957 is included in the total of £3115. Re-calculating the mean without the £957 gives

Mean $= \dfrac{2158}{13}$

$\bar{x} = £166$ which is a much fairer answer.

The modal wage is £135.

The mode is not a good average to use because no-one earns less than this and eight of the employees earn quite a lot more.

continued ▶

All of the twelve employees earn less than this. Some of them earn nearly £90 less!

The value 957 is called an **extreme value** . A measure of average that leaves out extreme values will be more representative of the data.

It is important to remember that neither the median nor mode are affected by extreme values, but the mean value will be.

The median wage is £157.

All of the twelve employees earn a wage that is not too far away from this figure. The median sits in the middle of the list and is the most sensible answer to use.

Exercise 20B

1 In a school test, the marks out of 40 for ten students were as follows:

41 38 38 35 34 31 29 28 26 9   9 26 28 29

Find the mode, median and mean marks and the range for this set of data. Comment on your results.
Which average would best represent the data?

2 The number of cars parked during the day in a city centre car park during a two-week period was recorded as follows:

52 45 61 67 48 70 12 56 41 57 53 62 70 9

Find the mode, median and mean number of cars parked.
Comment on your results.
Which would be the best average to use?

3 A six-sided dice has numbers 1, 2, 3, 4, 4, 6. What are the mean, mode and median values of these numbers? Why would the mean value be an inappropriate measure of the typical value thrown?

4 'The average score for a batsman during one season was 17 runs.'
Explain why this statement could be misleading.

20.2 Frequency distributions

Key words:
frequency table
frequency

When you have a large amount of data you can place the results into a frequency table .

Example 4

In a soccer tournament the number of goals scored in each game was as follows.

0 2 5 4 3 2 0 0 3 5 3 4 2 2 2 3 1 1 0 1
6 1 0 0 3 1 0 1 2 2 1 4 3 0 0 1 2 1 2 1
5 0 1 3 2 0 6 3 1 1 1 0 2 0 2 1 0 2 2 0
1 2 3 0 4 2 3 1 5 1 6 1 1 3 2 2 0 0 3 0

Put this data into a frequency table. Find:
(a) the mean   **(b)** the median   **(c)** the mode   **(d)** the range
for the goals scored in the tournament.

Number of goals scored (x)	Number of games (f)	Total number of goals (fx)
0	19	0
1	20	20
2	18	36
3	12	36
4	4	16
5	4	20
6	3	18
Totals	$\Sigma f = 80$	$\Sigma fx = 146$

Use x for the data and f for the **frequency** (the number of times each value occurs).

The column fx records the total number of goals e.g. 12 games resulted in 3 goals, so $12 \times 3 = 36$ goals.

Σ means 'the sum of'.
Σfx is the total number of goals scored.
Σf is the total number of games played.

(a) Mean $= \dfrac{\text{total goals scored}}{\text{total games played}}$

$$\bar{x} = \frac{\Sigma fx}{\Sigma f} = \frac{146}{80}$$

$\bar{x}$ is the symbol for the mean.

$$\bar{x} = 1.825 \text{ goals per game}$$

There are 80 items of data. When they are arranged in order of size, the median will be mid-way between the 40th and 41st values.

Number of goals scored (x)	Number of games (f)	Total number of goals (fx)
0	19	0
1	20	20
2	18	36

$20 + 19 = 39$ pieces of data in first 2 rows.

40th and 41st pieces must be in here.

(b) Median = 2 goals per game

(c) Mode = 1 goal per game

The mode is the number of goals with the highest frequency. There were 20 games in which 1 goal was scored.

(d) Range $= 6 - 0 = 6$ goals

The range is the difference between the largest and the smallest values.

For grouped data, the mean is $\bar{x} = \dfrac{\Sigma fx}{\Sigma f}$.

How to find the median

- See whether the number of data values, n, is odd or even.
- If n is an *odd* number, the median is the $\frac{1}{2}(n+1)$th value in the ordered list.
- If n is an *even* number, there will be two middle values and the median is the mean of the two middle values in the ordered list.

The two middle values are in the positions given by the integers below and above $\frac{1}{2}(n+1)$.

Exercise 20C

1 The police recorded the speeds of cars to the nearest 10 mph on a country road.

Speed mph (x)	Frequency (f)	(fx)
10	0	
20	23	
30	48	
40	16	
50	2	
60	1	
Totals	$\Sigma f =$	$\Sigma fx =$

$fx = f \times x$

(a) How many cars were recorded?

(b) Copy and complete the table and find the mean, median and mode speeds for this road.

2 In a hockey tournament the following numbers of goals were scored throughout the competition.

Goals scored	x	0	1	2	3	4	5	6	7
Frequency	f	12	14	12	9	4	6	0	3

Use Example 4 to help you.

Find the mean, median and mode of the goals scored in the tournament.

3 The following scores out of ten for a test were recorded for 75 students.

5 7 8 2 1 9 3 8 7 4 2 8 4 9 8
2 4 9 5 5 6 6 5 8 9 10 6 10 7 3
8 3 4 2 5 7 3 7 7 8 9 7 8 4 10
6 4 3 8 2 9 9 2 3 8 5 6 6 8 10
5 6 6 2 8 10 5 5 6 7 1 8 7 4 6

Put this data into a frequency distribution table and find the mean, median, mode and range of the marks.

20.3 Dealing with large data sets

Key words:
estimated mean
class
class intervals
grouped frequency table
class limits
mid-interval value

When there are a large number of data items it is best to group the data together. This makes it easier to record the information, but individual data values are lost. This means that you can find only estimated values for the mean.

An **estimated mean** is not a guess because it comes from a calculation. It is called an estimate because you do not know the individual values.

When you group your data, the groups are called **classes** or **class intervals** and the frequency table is referred to as a **grouped frequency table**. The class intervals do not have to be the same size but they usually are. The numbers at the start and the end of each class interval are called the **class limits**.

Continuous data that is grouped can also be displayed as a histogram. This was covered in Chapter 10.

In order to calculate the mean from a frequency table, as you learned in Section 20.2, you need to use a single number instead of a class interval.

This single number is called the **mid-interval value** of each of the class intervals. It is found by working out the mean of the two class limits.

So for a class interval 11–15, the mid-interval value is

$$\frac{11 + 15}{2} = \frac{26}{2} = 13$$

Extend your grouped frequency table to include the mid-interval values. These mid-interval values replace the class intervals and so become the 'x values' in your calculation.

For continuous data rounded to the nearest integer, a class interval of 11–15 is strictly 10.5–15.5.

Example 5

60 students measure the time it takes them to travel to school (correct to the nearest minute).

Calculate estimates of:

(a) the mean

(b) the median class

(c) the modal class.

Time (minutes)	Number of students (f)
1–5	6
6–10	12
11–15	8
16–20	11
21–25	14
26–30	4
31–35	3
36–40	2
TOTAL	60

Note that the class intervals must not overlap or have any gaps between them.

Time (minutes)	Number of students (f)	Mid-interval value (x)	Total time (minutes) (fx)
1–5	6	3	18
6–10	12	8	96
11–15	8	13	104
16–20	11	18	198
21–25	14	23	322
26–30	4	28	112
31–35	3	33	99
36–40	2	38	76
	$\Sigma f = 60$		$\Sigma fx = 1025$

First calculate the mid-interval values, e.g.

$$x = \frac{(1+5)}{2} = \frac{6}{2} = 3$$

Calculate total time by multiplying frequency f by mid-interval value x.

Find the totals of each column.

(a) Estimated mean $= \dfrac{\text{total time taken}}{\text{number of students}}$

$$\bar{x} = \frac{\Sigma fx}{\Sigma f} = \frac{1025}{60}$$

$$\bar{x} = 17.1 \ (1 \text{ d.p.})$$

The estimated mean journey time is 17.1 minutes.

continued ▼

(b) The median is halfway between the 30th and 31st values.

$6 + 12 + 8 = 26$

$6 + 12 + 8 + 11 = 37$

30th and 31st items are in the 16–20 minutes class interval.

The median is between 16 and 20 minutes.

(c) The highest frequency is 14.

The class interval with this frequency is 21–25 minutes.

The modal class is 21–25 minutes.

> You cannot give an exact answer for the median, but you can work out which class interval contains the median.

> You also cannot give an exact value for the mode. The best you can do is to give the modal class interval.

The class with the highest frequency is the modal class and the class containing the median value is the median class.

Example 6

The frequency table below shows the heights of students in a Year 11 class.
What is the estimated mean height for boys and for girls?

Height (cm)	Frequency (boys)	Frequency (girls)
$155 \leqslant h < 160$	2	5
$160 \leqslant h < 165$	6	9
$165 \leqslant h < 170$	14	22
$170 \leqslant h < 175$	19	11
$175 \leqslant h < 180$	8	3
$180 \leqslant h < 185$	1	0

Height (h) (cm)	Frequency (f) (boys)	Mid-point values (x)	(fx)
$155 \leqslant h < 160$	2	157.5	315
$160 \leqslant h < 165$	6	162.5	975
$165 \leqslant h < 170$	14	167.5	2345
$170 \leqslant h < 175$	19	172.5	3277.5
$175 \leqslant h < 180$	8	177.5	1420
$180 \leqslant h < 185$	1	182.5	182.5
	$\Sigma f = 50$		$\Sigma fx = 8515$

Boys' heights

To work out the mid-point value for each class interval add together the boundary values and divide by 2 e.g. the mid-point value of $160 \leqslant h < 165$ is given by

$\dfrac{160 + 165}{2} = 162.5.$

Extend the table to include the mid-point values. These replace the 'x values' in your calculation.

$$\text{Estimated mean} = \frac{\text{total height}}{\text{total frequency}}$$

> Now work out the estimated mean using the formula.

$$\bar{x} = \frac{\Sigma fx}{\Sigma f} = \frac{8515}{50}$$

$$\bar{x} = 170.3 \text{ cm}$$

continued ▶

Girls' heights

Height (h) (cm)	Frequency (f) (girls)	Mid-point values (x)	fx
$155 \leqslant h < 160$	5	157.5	787.5
$160 \leqslant h < 165$	9	162.5	1462.5
$165 \leqslant h < 170$	22	167.5	3685
$170 \leqslant h < 175$	11	172.5	1897.5
$175 \leqslant h < 180$	3	177.5	532.5
$180 \leqslant h < 185$	0	182.5	0
	$\Sigma f = 50$		$\Sigma fx = 8365$

$$\text{Estimated mean} = \frac{\text{total height}}{\text{total frequency}}$$

$$\bar{x} = \frac{\Sigma fx}{\Sigma f} = \frac{8365}{50}$$

$$\bar{x} = 167.3 \text{ cm}$$

Remember to include the units. Check that your answer is sensible.

The estimated mean height is 3 cm greater for boys than for girls.

Exercise 20D

1 The number of telephone calls made from a particular house over a 72-day period was recorded. The results were as follows.

Copy and complete the table and find an estimate of the mean and the class intervals for the mode and median.

Calls made	Frequency (f)	Mid-interval value (x)	fx
0–2	12		
3–5	18		
6–8	31		
9–11	11		
Totals	$\Sigma f =$		$\Sigma fx =$

2 The shoe size of students in a class was recorded to the nearest whole size.

Complete the table and find an estimate of the mean and the class intervals for the mode and median.

Shoe size	Frequency (f)	Mid-interval value (x)	fx
3–4	5		
5–6	8		
7–8	12		
9–10	5		
11–12	2		
Totals	$\Sigma f =$		$\Sigma fx =$

3 Some students were asked how many CDs they had in their collection. The results are shown in this grouped frequency table.

Copy the table and extend it to include the mid-interval values.

Use this information to find an estimate for the mean number of CDs and work out the median and modal class intervals.

Number of CDs	Frequency (f)
50–64	12
65–79	0
80–94	11
95–109	9
110–124	15
125–139	33
140–154	28
155–169	41
170–184	16

4 The number of students in a particular Year 8 set using the school library was recorded over a term.

Using this data, work out an estimate for the mean number of visits to the library for this class. How does this compare with the median and modal class intervals?

Number of visits	Frequency (f)
0–4	84
5–9	46
10–14	38
15–19	51
20–24	22
25–29	13

5 The frequency table shows the distribution of the mass of fish caught during a local fishing competition. Work out an estimate for the mean mass.

Catch (kg)	Frequency (f)
0.1–0.5	6
0.6–1.0	6
1.1–1.5	9
1.6–2.0	4
2.1–2.5	3
2.6–3.0	4

20.4 Moving averages

> Key words:
> time series
> moving average

A graph showing how a particular value changes over a period of time is called a **time series** . Such graphs are useful when looking for patterns or trends in the data.

> You met time series graphs in Chapter 10.

Example 7

The table shows the amount of gas units used in each quarter of a year during a three-year period.

Year	1st Quarter Jan–Mar	2nd Quarter Apr–June	3rd Quarter July–Sept	4th Quarter Oct–Dec
2002	1000	395	233	536
2003	834	510	188	703
2004	976	470	215	608

Plot the information **(a)** as a time series for the quarterly figures
(b) as a time series for the yearly totals.
Comment on your results.

(a)

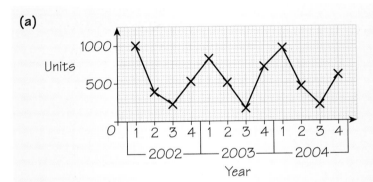

Plot the periods of time along the *x*-axis and the number of gas units used up the *y*-axis. Choose sensible starting values for the axes.

Although these are a collection of individual points, it is usual to join the points up with a straight line. This helps to find any trends in the data.

The time series for the quarterly figures indicates

● that the heating costs are at their lowest during the summer months (3rd quarter)

● that the heating costs are at their highest during the winter months (1st quarter).

Plotting the quarterly figures in a time series shows the seasonal patterns.

(b)

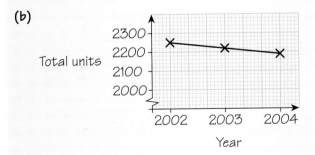

Plotting the annual totals (only three points here) shows that the trend appears to be that less gas is being used in successive years.

The time series for the yearly figures shows that over these three years the amount of gas used decreased each year.

A **moving average** is a set of mean values calculated from a time series using 2, 3 or more points taken from a section of the data set. Each new successive mean value is found by moving the data on to the next group of 2, 3 or more values. Moving averages are used to smooth out either seasonal variations or data containing extreme values, in order to look at long-term trends. For quarterly data use a 4-point moving average; for twice yearly data use a 2-point moving average, and so on.

Example 8

Using the data for the number of gas units used over a three-year period given in Example 7:

Year	1st Quarter Jan–Mar	2nd Quarter Apr–June	3rd Quarter July–Sept	4th Quarter Oct–Dec
2002	1000	395	233	536
2003	834	510	188	703
2004	976	470	215	608

(a) calculate the 4-point moving averages

(b) plot a graph of the moving averages

(c) comment on the results.

(a) $\dfrac{1000 + 395 + 233 + 536}{4} = 541$

The calculations show only the first three moving averages based on a 4-point data set.

$\dfrac{395 + 233 + 536 + 834}{4} = 499.5$

$\dfrac{233 + 536 + 834 + 510}{4} = 528.25$

The full set of moving averages is shown here.

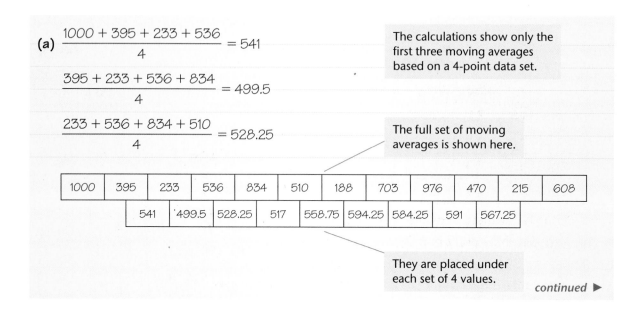

1000	395	233	536	834	510	188	703	976	470	215	608
	541	499.5	528.25	517	558.75	594.25	584.25	591	567.25		

They are placed under each set of 4 values.

continued ▶

(b)

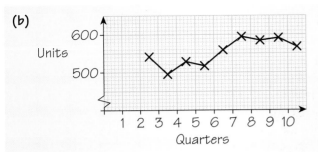

Join the points up with line segments (straight lines).

(c) The moving average suggests that the amount of gas used is fairly steady during this period.

Exercise 20E

1 The table shows the number of telephone calls made from a household in each quarter of a year over a three-year period.

Year	1st Quarter Jan–Mar	2nd Quarter Apr–June	3rd Quarter July–Sept	4th Quarter Oct–Dec
2002	469	758	591	446
2003	517	822	648	531
2004	496	729	604	463

(a) Plot this information as a time series.

(b) Work out the 4-point moving averages for the data.

(c) Plot the 4-point moving averages.

(d) Comment on the trend (if any) in the number of telephone calls made from this house.

2 The number of passengers (in millions) on a low-cost airline at quarterly intervals for the last three years is shown in the table.

	2002	2003	2004
1st quarter	2.1	2.8	5.2
2nd quarter	2.8	4.5	7.1
3rd quarter	3.3	4.1	6.2
4th quarter	2.9	3.5	5.8

(a) Plot this information as a time series graph.

(b) Work out the 4-point moving averages for the data.

(c) Plot the 4-point moving averages on a graph.

(d) Comment on the trend (if any) in the number of passengers carried.

Examination questions

1 The stem-and-leaf diagram shows the ages, in years, of 15 members of a badminton club.

Key: 2 | 7 means an age of 27 years.

```
2 | 7 8
3 | 0 2 4 8
4 | 1 2 3 3 4 6
5 | 3 6
6 | 2
```

(a) What is the median age of the members?

(b) What is the range of the ages?

(2 marks)

AQA, Spec A, 1I, June 2003

2 Phil counts the number of people in 50 cars that enter a car park. His results are shown in the table.

Calculate the mean number of people per car.

Number of people	Frequency
1	25
2	17
3	6
4	2
More than 4	0

(3 marks)

AQA November 2002

3 The following test scores were recorded.

```
40  28   9  54  61  93  53  39  23  14  43
35  52  64  27  70  23  14  58  38  19  61
26  51  59  63  73  24  38  63  27  51  72
```

(a) Construct a grouped frequency diagram for this data using the intervals 0–9, 10–19 etc.

(b) What is the modal class interval?

(c) Work out an estimate of the mean score for this set of data.

4 The lengths of 10 boxes are measured. The results are summarised in the table.

Length x (cm)	Frequency	Mid-point value
$1 < x \leqslant 3$	5	
$3 < x \leqslant 5$	2	
$5 < x \leqslant 7$	2	
$7 < x \leqslant 9$	1	

Complete the mid-point value column and use it to calculate an estimate of the mean length.

(3 marks)

AQA, Spec B, 1I, June 2002

5 A police officer records the speeds of 60 cars on a dual carriageway.

Speed (mph)	Frequency	Mid-point
40 to less than 50	9	
50 to less than 60	27	
60 to less than 70	21	
70 to less than 80	3	

(a) Write down the modal class.

(b) Use the class mid-points to calculate an estimate of the mean
speed of these cars.

(4 marks)

AQA, Spec B, 1I, June 2004

6 Jane records the times taken by 30 pupils to complete a number puzzle.

Time, t (minutes)	Number of pupils
$2 < t \leqslant 4$	3
$4 < t \leqslant 6$	6
$6 < t \leqslant 8$	7
$8 < t \leqslant 10$	8
$10 < t \leqslant 12$	5
$12 < t \leqslant 14$	1

(a) Calculate an estimate of the mean time taken to complete the puzzle.

(b) Which time interval contains the median time taken to complete
the puzzle?

(5 marks)

AQA, Spec A, 2I, June 2003

7 The table shows the amount of Bindi's gas bills from September 2004
to December 2005.

Date	Sept 2004	Dec 2004	Mar 2005	June 2005	Sept 2005	Dec 2005
Bill (£)	24.70	32.40	29.10	7.80	30.30	38.60

(a) Explain why a four-point moving average is appropriate for
these data.

(b) Show that the first value of the four-point moving average is £23.50.

(c) Calculate the second value of the four-point moving average for
these data.

(5 marks)

Summary of key points

Averages and range (grade D/C to B/A)

An *average* represents a set of data. There are three averages that you need to know: mean, median and mode .

The *mode* of a set of data is the value that occurs most often; the *median* of a set of data is the middle value when the data has been arranged in order of size, and the *mean* of a set of data is the sum of the values divided by the number of values.

- For individual data the mean is given by $\bar{x} = \dfrac{\Sigma x}{n}$ where n is the number of data values.

- For grouped data the mean is given by $\bar{x} = \dfrac{\Sigma fx}{\Sigma f}$

The *frequency* (f) is the number of times an answer or result has occurred and the symbol Σ represents 'the sum of'.

It is important to be able to select the appropriate average to use, and give reasons for your choice.

The range of a set of data is the difference between the highest and the lowest values. It gives a measure of spread of the data.

Frequency distributions (grades D/C)

Data sets can be put into a table called a *frequency table* (or frequency distribution table).

Large amounts of data, either discrete or continuous, are best collected and placed in groups with no gaps. These are called *class intervals* and the usual notation is 0–9 or $0 < x \leqslant 9$.

The values at the start and end of each class interval are called the *class limits*. A frequency table showing data in groups is called a *grouped frequency table*.

For grouped data, only an *estimate of the mean* is possible as individual information is lost.

The *mid-interval value* is the middle value in the class interval. It is calculated as the sum of the two class limits divided by 2.

The class with the highest frequency is the *modal class* and the class containing the median value is the *median class*.

Moving averages (grade B/A)

A *moving average* is a set of mean values calculated from a time series using 2, 3 or more points taken from a section of the data set. Each new successive mean value is found by moving the data on to the next group of 2, 3 or more values.

Moving averages are used to smooth out seasonal variations or data containing extreme values, to find long-term trends.

This chapter will show you how to:

✔ construct cumulative frequency tables
✔ find quartiles and interquartile ranges
✔ draw box plots
✔ compare and interpret distributions

You will need to know

● how to construct frequency tables

21.1 Cumulative frequency

Key words:
cumulative frequency table
grouped frequency table
cumulative frequency graph
upper class boundary

You can construct a **cumulative frequency table** from a **grouped frequency table** by calculating the running total of the frequency up to the end of each class interval.

You can use the table to plot a **cumulative frequency graph** (or cumulative frequency diagram). You can use these graphs to compare data sets.
You can draw cumulative frequency graphs for both discrete and continuous data sets.

Example 1

The following grouped frequency distribution shows the time taken for students to solve a puzzle. Using this data, draw a cumulative frequency diagram.

Class intervals (such as $0 < t \leqslant 5$) were discussed in Chapter 20.

Time taken (seconds)	Frequency
$0 < t \leqslant 5$	2
$5 < t \leqslant 10$	9
$10 < t \leqslant 15$	9
$15 < t \leqslant 20$	8
$20 < t \leqslant 25$	3
$25 < t \leqslant 30$	1

Time taken (t) (seconds)	Cumulative frequency
$t \leqslant 5$	2
$t \leqslant 10$	$2 + 9 = 11$
$t \leqslant 15$	$2 + 9 + 9 = 20$
$t \leqslant 20$	$2 + 9 + 9 + 8 = 28$
$t \leqslant 25$	$2 + 9 + 9 + 8 + 3 = 31$
$t \leqslant 30$	$2 + 9 + 9 + 8 + 3 + 1 = 32$

Begin by creating a cumulative frequency table. This cumulative frequency column is a running total of the frequency so far.

The figures in bold are the values of the cumulative frequency. The class intervals ($t \leqslant$) are also changed to show the values 'up to and including' i.e. the **upper class boundary** value.

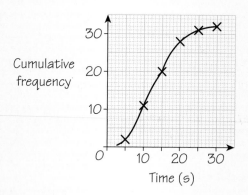

You can now use this data to draw a cumulative frequency diagram by plotting the cumulative frequency against the upper class boundary of each class interval. Cumulative frequency is always plotted on the vertical axis. The maximum value on the y-axis (the cumulative frequency axis) should be the total number of values in the data set, in this case 32.

Draw a smooth curve through the points (to give a cumulative frequency curve), as shown, or join the points using straight lines.

The shape of the cumulative frequency diagram reflects the characteristics of the data and how this data is spread or distributed within the range. This characteristic S shape (called an ogive) appears in nearly all cumulative frequency diagrams.

Cumulative frequency graphs

- Choose a suitable scale for each axis.

- Always plot the cumulative frequencies on the vertical axis (the y-axis).

- Plot the points using the upper class boundary e.g. (**5**, 2), (**10**, 11), (**15**, 20) and so on.

- Draw a smooth curve through the points (points can also be joined with straight lines) and include the origin.

- Check that you have plotted your points correctly.

- Check that your graph is an S-shape.

Estimating values using a cumulative frequency graph

Once you have drawn your cumulative frequency diagram you can use it to estimate values for the data.

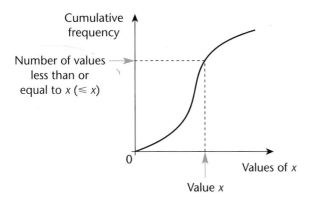

For example, given a value of x you can estimate the number of values less than or greater than x by drawing a straight line that meets the cumulative frequency curve and then drawing a corresponding line to meet the other axis.

Example 2

Use the cumulative frequency curve from Example 1 to estimate

(a) how many students solved the puzzle
   (i) within 12 seconds   (ii) in more than 18 seconds
(b) the time by which 28 students had solved the puzzle.

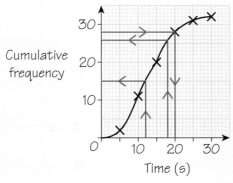

(a) (i) 15 students solved the puzzle within 12 seconds.

   (ii) $32 - 26 = 6$

       6 students took longer than 18 seconds.

(b) $x = 20$

   28 students solved the puzzle within 20 seconds.

Always draw lines on your graph to show how you obtained your answers. Do not rub them out.

(a) Draw a vertical line for $x = 12$ to intersect the cumulative frequency curve. Now draw a line to meet the cumulative frequency axis. The value obtained is 15.

(b) The line at $x = 18$ seconds gives a value of 26. This means that $32 - 26 = 6$ students took longer than 18 seconds to solve the puzzle.

(c) Draw a line from the cumulative frequency value of 28 to intersect the curve. Now draw a vertical line to meet the x-axis at the point $x = 20$. This is the time by which 28 students solved the puzzle.

Exercise 21A

1 (a) Copy and complete the cumulative frequency table showing the distribution of test marks for a group of 32 students.

Mark (%)	Number of students
1–10	1
11–20	2
21–30	4
31–40	7
41–50	5
51–60	8
61–70	2
71–80	2
81–90	1
91–100	0

Mark (%)	Cumulative frequency
⩽10	1
⩽20	3

Your x-axis (mark) should run from 0 to 100, and your y-axis from 0 to 32.

(b) Plot the results in a cumulative frequency graph.

2 The table below shows the frequency distribution of test marks for 120 students.

Marks	Number of students
1–10	1
11–20	6
21–30	8
31–40	15
41–50	17
51–60	24
61–70	22
71–80	15
81–90	9
91–100	3

Construct a cumulative frequency table (taking the first class interval to be ⩽10 and the last interval to be ⩽100). Draw the corresponding cumulative frequency graph for this distribution.

3 The results for the long jump at a school sports day are shown in the table.
Draw a cumulative frequency diagram for this distribution.
Estimate how many students jumped over 2.35 m.

Distance x (m)	Frequency
$1.70 < x \leqslant 1.80$	2
$1.80 < x \leqslant 1.90$	6
$1.90 < x \leqslant 2.00$	9
$2.00 < x \leqslant 2.10$	7
$2.10 < x \leqslant 2.20$	15
$2.20 < x \leqslant 2.30$	8
$2.30 < x \leqslant 2.40$	8
$2.40 < x \leqslant 2.50$	2

4 The temperature in °C recorded over a 66-day period is shown in the table.
Draw a cumulative frequency diagram and estimate the number of days that the temperature was above 18°C.

Temperature t (°C)	Number of days
$0 < t \leqslant 3$	1
$3 < t \leqslant 7$	7
$7 < t \leqslant 11$	18
$11 < t \leqslant 15$	20
$15 < t \leqslant 19$	17
$19 < t \leqslant 23$	2
$23 < t \leqslant 27$	1

5 The cumulative frequency curve for the amount of time spent on a homework task is shown below.

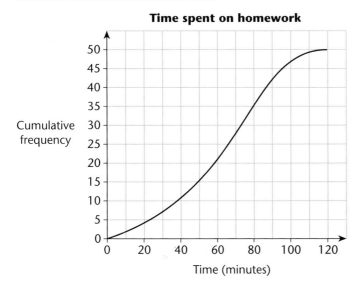

Time spent on homework

Cumulative frequency

Time (minutes)

Construct a cumulative frequency table and use this to estimate the number of students who spent more than 1 hour on the task.

21.2 The median, quartiles and interquartile range

Key words:
median
lower quartile
upper quartile
interquartile range

You can obtain important statistical measures from a cumulative frequency graph. These are useful when you want to compare two or more data sets.

- The **median** is the middle value of the distribution. This occurs half-way up the cumulative frequency axis – at 50% of the maximum value.
- The **lower quartile (LQ)** appears one-quarter of the way up the cumulative frequency axis at 25% of the maximum value.
- The **upper quartile (UQ)** appears three-quarters of the way up the cumulative frequency axis at 75% of the maximum value.
- The **interquartile range (IQR)** gives an improved measure of the spread of the data and is given by

  Interquartile range = upper quartile − lower quartile

The answers for the median, the lower quartile and the upper quartile are read off on the *horizontal* axis.
The first step in this process is to locate the appropriate points on the cumulative frequency axis.

The IQR is a measure of spread of the middle 50% of the data, so it excludes extreme values.

Example 3

Use the cumulative frequency graph of Example 1 to find an estimate for

(a) the median

(b) the lower quartile

(c) the upper quartile

(d) the interquartile range.

(a) To find an estimate of the median value, first find the point half-way up the cumulative frequency axis. Here the maximum value is 32 so the half-way value is 16. Now draw a horizontal line from this point to meet the curve. The median value is where this vertical line meets the *x*-axis, as shown.

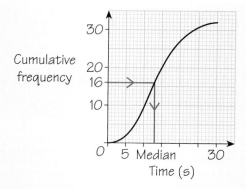

(a) The median = 13 seconds.

The middle value is really $\frac{1}{2}(32 + 1) = 16\frac{1}{2}$th value, but using $32 \div 2 = 16$th value for the median is accurate enough.

continued ▶

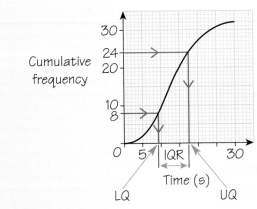

(b) Total = 32

$\frac{1}{4}$ of 32 = 8

LQ = 9.5 seconds

(c) $\frac{3}{4}$ of 32 = 24

UQ = 17.5 seconds

(d) IQR = 17.5 − 9.5 = 8 seconds

(b) To find the lower quartile (LQ), first locate a point a quarter of the way up the cumulative frequency axis and draw a horizontal line to intersect the curve. Draw a vertical line to meet the *x*-axis. This gives the lower quartile.

(c) Find the upper quartile (UQ) in a similar way, but this time use a point three-quarters of the way up the cumulative frequency axis.

(d) Remember:
IQR = UQ − LQ

Exercise 21B

1 The cumulative frequency graph for the distances travelled to school by 50 students is shown below.

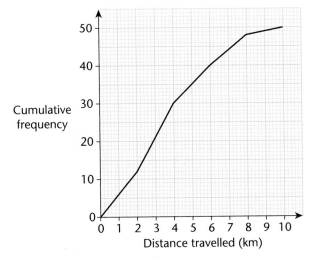

From the graph, estimate
(a) the median distance
(b) the lower quartile distance
(c) the upper quartile distance
(d) the interquartile range.

2 A survey was conducted to find the amount of time students spent on homework each week. The results are shown in the cumulative frequency graph.

From the graph, estimate

(a) the median time

(b) the lower quartile time

(c) the upper quartile time

(d) the IQR.

Give your answers in hours and minutes.

(e) Estimate what percentage of students spent 5 hours or less on homework each week.

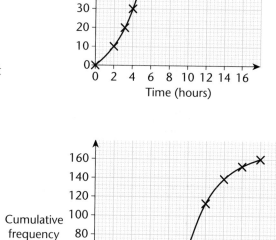

3 The cumulative frequency diagram shows the marks gained in a geography exam.

Use the graph to find mark estimates for

(a) the median

(b) LQ

(c) UQ

(d) IQR.

4 A tyre company carried out a survey to find how far cars travel before they need new tyres. The results for 100 cars is shown in the cumulative frequency diagram.

Use the graph to estimate

(a) the median distance travelled

(b) the lower quartile distance travelled

(c) the upper quartile distance travelled

(d) the interquartile range.

(e) What percentage of cars travelled a distance of more than 30 000 miles before the tyres needed to be changed?

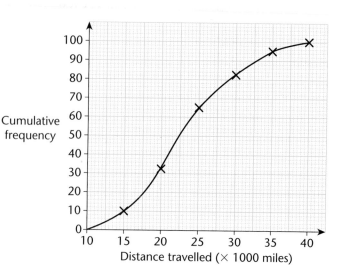

21.3 Box plots

Key words:
box plot

A **box plot** (or a box-and-whisker diagram) shows the important statistical information from a frequency distribution as a diagram.

At a glance, you can see the range (from the minimum and maximum values) and the interquartile range (from the lower and upper quartiles).

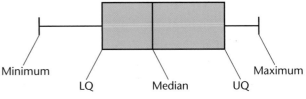

You can obtain a box plot from a cumulative frequency graph. Box plots are extremely useful when comparing two or more distributions that have similar cumulative frequency diagrams.

You will use box plots for comparing data sets in Section 21.4.

It is easier to draw the box plot directly under the cumulative frequency graph, as the scale for the plot is already given by the scale of the *x*-axis.

If you do not draw it under the cumulative frequency graph you must put a scale underneath the box-plot diagram.

Example 4

The times taken in seconds by eleven athletes to run a 400 m race are given in order below.

    58, 61, 62, 65, 67, 69, 70, 75, 78, 86, 97

Draw a box plot for this data. What is the range?

Minimum = 58

Maximum = 97

$$\text{Median} = \frac{1 + 11}{2} = \text{6th value} = 69$$

$$\text{LQ} = \frac{1 + 11}{4} = \text{3rd value} = 62$$

$$\text{UQ} = \frac{1 + 11}{4} \times 3 = \text{9th value} = 78$$

Begin by summarising the key statistical information.

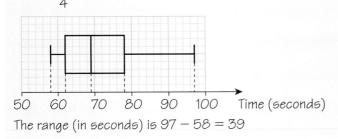

The range (in seconds) is $97 - 58 = 39$

Mark the positions of the maximum and minimum points. Draw a box showing the LQ, UQ and median positions. The width of the box reflects the IQR.

Example 5

Use the data from Example 3 to construct a box plot for the times
for solving the puzzle.

- Median = 13 s
- Lower quartile = 9.5 s
- Upper quartile = 17.5 s
- Interquartile range = 17.5 − 9.5 = 8 s
- Minimum = 0 s
- Maximum = 32 s
- Range = 32 − 0 = 32 s.

The corresponding box plot is:

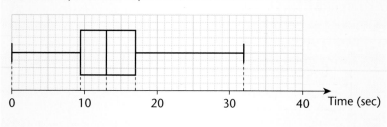

Use the cumulative
frequency graph on page
375 to obtain your
summary.

Exercise 21C

1 The total time (in days) spent in space by the top 11 astronauts is
listed below:

678, 651, 541, 489, 483, 430, 391, 387, 381, 375, 362

Draw a box plot for this data. What is the range for this data?

2 The number of words of between 1 and 11 letters long in a
newspaper article are given below.

Number of letters per word	1	2	3	4	5	6	7	8	9	10	11
Frequency	5	15	31	12	7	6	14	5	3	0	2

Draw a box plot for this data.

3 The box plot shown represents the results from
a history examination where 100 marks
were available. 64 students sat the exam.

Sketch the cumulative frequency curve
for the distribution of the marks.

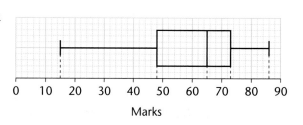

4 During a busy day at a doctor's surgery the amount of waiting time for patients was recorded as shown in the table.
The shortest waiting time was $1\frac{1}{2}$ mins. and the longest time was $9\frac{1}{2}$ mins.

(a) Construct a cumulative frequency table.

(b) Draw a cumulative frequency curve.

(c) Give an estimate of the median waiting time.

(d) Draw a box plot to show the key statistical data.

Waiting time, t (minutes)	Frequency (f)
$0 < t \le 1$	0
$1 < t \le 2$	4
$2 < t \le 3$	23
$3 < t \le 4$	43
$4 < t \le 5$	58
$5 < t \le 6$	37
$6 < t \le 7$	11
$7 < t \le 8$	3
$8 < t \le 9$	0
$9 < t \le 10$	1

5 The local leisure centre conducted a survey on the age distribution of its 800 members and the results are shown in the table.
The youngest person was 6 years old and the oldest was 78.

Age, a (years)	Frequency (f)
$0 < a \le 10$	41
$10 < a \le 20$	138
$20 < a \le 30$	168
$30 < a \le 40$	192
$40 < a \le 50$	126
$50 < a \le 60$	85
$60 < a \le 70$	39
$70 < a \le 80$	11

(a) Construct a cumulative frequency table for this data.

(b) Draw a cumulative frequency graph.

(c) Use the curve to estimate the median age of the leisure centre members.

(d) Find an estimate of the lower and upper quartiles and hence draw a box plot for this distribution.

Key words:
spread

21.4 Comparing and interpreting data

Box plots are useful when comparing two (or more) sets of data.

You can draw two or more cumulative frequency curves on the same set of axes, and draw your box plots one beneath the other, using the same scale. This lets you see at a glance how the data is spread for each set.

It's easy to draw the wrong conclusion unless you have all the available information!

Example 6

Four hundred batteries of two types, A and B, were tested to find how long they lasted, and the results are presented in the frequency table.

The minimum life of both types of battery was 19 hours. The maximum life of Type A was 32 hours and the maximum life of Type B was 35 hours.

(a) Construct
    (i) cumulative frequency diagrams
    (ii) box plots for both types of battery.

(b) Estimate the median life of each battery and comment on the results.

Battery life, t (hours)	Battery A	Battery B
$18 < t \leq 20$	5	15
$20 < t \leq 22$	10	30
$22 < t \leq 24$	20	55
$24 < t \leq 26$	25	100
$26 < t \leq 28$	110	75
$28 < t \leq 30$	190	55
$30 < t \leq 32$	40	40
$32 < t \leq 34$	0	25
$34 < t \leq 36$	0	5

(a) (i)

Battery life t (hours)	Battery A	Cumulative frequency A	Battery B	Cumulative frequency B
$18 < t \leq 20$	5	5	15	15
$20 < t \leq 22$	10	15	30	45
$22 < t \leq 24$	20	35	55	100
$24 < t \leq 26$	25	60	100	200
$26 < t \leq 28$	110	170	75	275
$28 < t \leq 30$	190	360	55	330
$30 < t \leq 32$	40	400	40	370
$32 < t \leq 34$	0	400	25	395
$34 < t \leq 36$	0	400	5	400

First work out the cumulative frequencies for both types of battery.

continued ▶

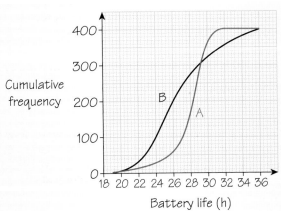

Plot the cumulative frequencies for A and B against the battery lifetimes.

Find the LQ, median and UQ from the graphs. Use these values to draw the box plots.

(ii)

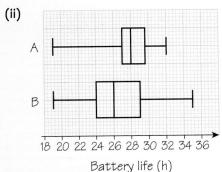

A

B

Battery life (h)

(b)

	Battery A	Battery B
Minimum	19	19
Maximum	32	35
Median	28	26
LQ	27	24
UQ	29.6	29
IQR	2.6	5.0

The IQR for A is almost half of that for B. This means that battery A is more reliable. (There are fewer extreme values.)

Although the maximum for type B lies in a higher class, type A is likely to last longer than type B.

Remember: the IQR is the spread of the middle 50% of the data.

Exercise 21D

1 Two swimmers A and B train together and record their practice sessions of lengths completed over several weeks. The table summarises the results.

Using the same scale, draw box plots for A and B. Compare and contrast the two box plots.

	Swimmer A	Swimmer B
Minimum	32	23
Maximum	56	49
Median	45	41
Lower quartile	41	36
Upper quartile	52	45

2 The frequency table shows the results of tests in Maths and English for a particular class. Draw a cumulative frequency curve and box plot for both distributions and comment on their differences.

Mark	Maths	English
0–2	1	0
3–5	2	0
6–8	1	2
9–11	5	4
12–14	7	11
15–17	9	11
18–20	5	2

3 The frequency table shows the male and female population (×10 000) of a city by age group. The youngest age was 0 for both males and females. The oldest male was 98 and the oldest female was 104.

(a) Draw a cumulative frequency diagram for both males and females on the same set of axes.

(b) What are the median ages for males and females?

(c) Construct a box plot for males and females and comment on the results obtained.

Age range	Males	Females
0–14	4.2	4.0
15–29	4.8	5.2
30–44	4.6	4.9
45–59	3.6	3.8
60–74	2.8	3.1
75–89	1.0	1.7
90 and over	0.5	1.8

Examination questions

1 A manager recorded the number of customers that entered his supermarket each hour over five days in June. The table shows a summary of his results.

Draw a box plot to show these results.

	Number of customers
Minimum	8
Lower quartile	23
Median	25
Upper quartile	33
Maximum	42

(3 marks)
AQA, Spec B, 1I, March 2004

2 Altogether 56 boys and 52 girls take two English tests. The box plots show the distributions of their total marks.

(a) Give two differences between the boys' marks and the girls' marks.

(b) A certificate is given to each student who scores a total of more than 55 marks. How many students are given a certificate?

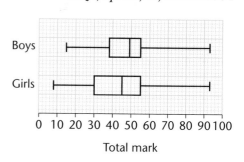

(4 marks)

3 The length of time, in minutes, of
40 telephone calls was recorded.
A cumulative frequency diagram of
this data is shown on the grid.

Use the diagram to find the limits
between which the middle 50% of
the times lie.

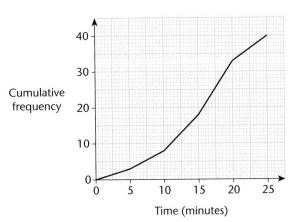

(2 marks)

4 The weights of 80 bags of rice are measured. The table
summarises the results.

(a) Draw a box plot to show this information.

(b) Write down the interquartile range for these data.

(c) How many bags weigh (i) less than 480 g
(ii) less than 500 g?

(d) Draw a cumulative frequency diagram to show the
information.

Minimum	480 g
Lower quartile	500 g
Median	540 g
Upper quartile	620 g
Maximum	720 g

(9 marks)
AQA, Spec B, 1I, March 2002

5 Brian and George played
40 games of golf.
The cumulative frequency
diagram shows information
about Brian's scores.
The box plot shows
information about
George's scores.

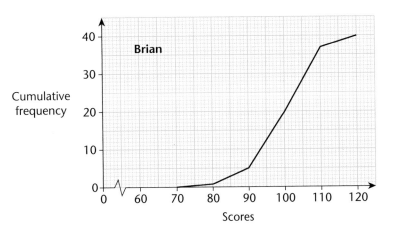

(a) Showing your method
clearly, find
(i) Brian's median score
(ii) Brian's interquartile
range.

(b) Use the cumulative
frequency diagram and the
box plot to answer the
following.

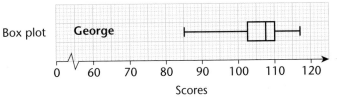

(i) Which player is more
consistent in his
scoring?

(ii) The winner of a game of golf is the player who has the lowest score.
Who do you think is the better player? Give a reason for your choice. *(5 marks)*

Summary of key points

Cumulative frequency (grade B)

You can construct a cumulative frequency table from a grouped frequency distribution by calculating the running total of the frequency up to the upper class boundary.

From a graph of the cumulative frequency, the following statistical measures can be obtained:

- Median – this is the value one half of the way into a distribution.
- Lower quartile (LQ) – this is the value one quarter of the way into a distribution.
- Upper quartile (UQ) – this is the value three quarters of the way into a distribution.
- Interquartile range (IQR) is the difference between the upper and lower quartiles. It gives a measure of the spread of the middle 50% of the data. IQR = UQ − LQ

Box plots (grade B)

A box plot shows the important statistical information from a frequency distribution as a diagram.

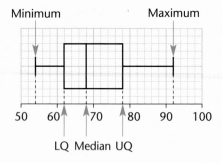

Box plots are useful when comparing two (or more) sets of data.

This chapter will show you how to:

✔ apply Pythagoras' theorem to right-angled triangles
✔ find the distance between two coordinate points
✔ use the basic trigonometric functions sine, cosine and tangent
✔ solve problems in three dimensions

The theorem of Pythagoras and trigonometry are connected as they both involve right-angled triangles. This chapter shows how they can be used to solve a number of problems in both two and three dimensions.

There is more on trigonometry in Chapter 27.

22.1 Pythagoras' theorem

Pythagoras' theorem involves only

right-angled triangles – those containing a 90° angle.

Key words:
Pythagoras' theorem
right-angled triangle
hypotenuse
Pythagorean triple
surd

You can apply Pythagoras' theorem to find the longest side of a triangle (called the **hypotenuse**) or to find the length of one of the shorter sides.

The hypotenuse is always opposite the right angle.

Look at the right-angled triangle ABC with $AB = 4$ cm, $BC = 3$ cm and $AC = 5$ cm.

Side AC is the hypotenuse.

This is the right angle.

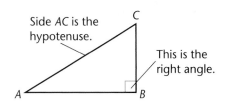

You can construct a square on each side of the triangle.

Area A + Area B
$\quad = 9 \text{ cm}^2 + 16 \text{ cm}^2$
$\quad = 25 \text{ cm}^2 = \text{Area } C$

$$
\begin{array}{ccccc}
\text{Area } A & + & \text{Area } B & = & \text{Area } C \\
3^2 & + & 4^2 & = & 5^2
\end{array}
$$

In other words

$$
\begin{array}{ccccc}
a^2 & + & b^2 & = & c^2
\end{array}
$$

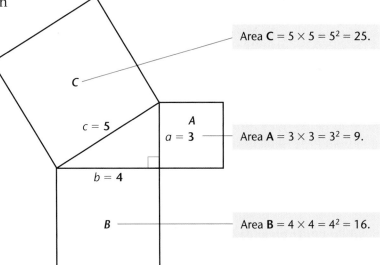

Area **C** = $5 \times 5 = 5^2 = 25$.

Area **A** = $3 \times 3 = 3^2 = 9$.

Area **B** = $4 \times 4 = 4^2 = 16$.

This leads to Pythagoras' theorem.

In any right-angled triangle the square of the hypotenuse (c^2) is equal to the sum of the squares on the other two sides ($a^2 + b^2$).

'Sum' means 'add'.

For a right-angled triangle with sides of lengths a, b and c, where c is the hypotenuse, Pythagoras' theorem states that $\boldsymbol{a^2 + b^2 = c^2}$

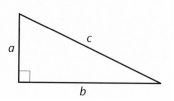

The above equation allows you to work out the longest side. It can also be used to find the length of one of the shorter sides by making side a or side b the subject of the equation.

For more on changing the subject of a formula, see Chapter 12.

$$a^2 + b^2 = c^2$$
$$a^2 + b^2 - b^2 = c^2 - b^2$$
$$a^2 = c^2 - b^2$$

Subtract b^2 from both sides.

Similarly you can find the length b using $b^2 = c^2 - a^2$.

Applying Pythagoras' theorem allows you to find the length of any side of a right-angled triangle.

Example 1

Work out the length of the hypotenuse (the side marked x) in this right-angled triangle.

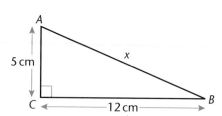

$$c^2 = a^2 + b^2$$
$$x^2 = 5^2 + 12^2$$
$$x^2 = 25 + 144$$
$$x^2 = 169$$
$$x = \sqrt{169}$$
$$x = 13 \text{ cm}$$

Write out Pythagoras' theorem and then substitute in the values given.

Any set of three positive whole numbers a, b, c that satisfies Pythagoras' theorem is called a **Pythagorean triple** .

A right-angled triangle with sides 3, 4 and 5 is one example (the easiest) of a Pythagorean triple. Another triple is 5, 12 and 13.

All triangles whose sides are multiples of these sets of lengths must be mathematically similar, and so have identical angles. For example, the triangle (6, 8, 10) is mathematically similar to triangle (3, 4, 5).

Most Pythagoras questions do not involve integer quantities, particularly those in 'real' situations, so the answers must be rounded to a sensible degree of accuracy.

> These two basic Pythagorean triples are worth remembering.

> You met proportionality in Chapter 17. Similar shapes were covered in Chapter 19.

Example 2

In the triangle PQR, angle $Q = 90°$, $QP = 4$ cm and $QR = 7$ cm. Calculate y, the length of PR, to 1 decimal place.

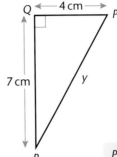

$y^2 = 4^2 + 7^2$

$y^2 = 16 + 49$

$y^2 = 65$

$y = \sqrt{65}$

$y = 8.06\,cm \approx 8.1\,cm$ (to 1 d.p.)

> PR is the hypotenuse, because it is opposite the right angle.

> Use your calculator to find the square root.

> Always put the units in your answer.

The answer $\sqrt{65}$ is called a **surd**. This is an exact answer. Forming the decimal only gives an approximate answer in most cases. Here the answer is given to 1 d.p.

> There is more about surds in Chapter 26.

Example 3

Calculate the length a (to 1 d.p.) in this triangle, where $b = 4$ cm and $c = 7$ cm.

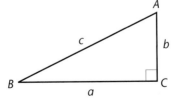

> c is the longest side (the hypotenuse) since it is opposite the right angle.

$a^2 = c^2 - b^2$

$a^2 = 7^2 - 4^2$

$a^2 = 49 - 16 = 33$

$a = \sqrt{33} = 5.744$

$a \approx 5.7\,cm$ (1 d.p.)

> Always subtract the squares when you are finding one of the shorter sides. If you add by mistake you get
> $$a^2 = 49 + 16$$
> $$a^2 = 65$$
> $$a = \sqrt{65} = 8.06...$$
> This is impossible because it is longer than the hypotenuse which is, by definition, the longest side.

Exercise 22A

1 Calculate the lengths marked with letters in each of the following triangles.

(i)

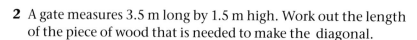

(ii)

(iii)

(iv)

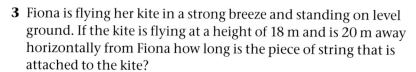

2 A gate measures 3.5 m long by 1.5 m high. Work out the length of the piece of wood that is needed to make the diagonal.

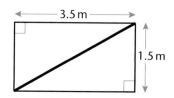

3 Fiona is flying her kite in a strong breeze and standing on level ground. If the kite is flying at a height of 18 m and is 20 m away horizontally from Fiona how long is the piece of string that is attached to the kite?

For questions 3, 4 and 6 it is useful to sketch a diagram and label it.

4 A boat sails due east for 24 km. However, the current takes the boat 10 km due south. How far does the boat actually travel?

5 Calculate the lengths
(a) *XY* and **(b)** *QR*.
Give your answers correct to 1 d.p.

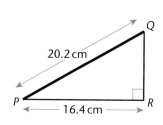

6 A children's slide is 3.6 m long. The vertical height of the slide above the ground is 2.1 m. Work out the horizontal distance between each end of the slide.

7 Work out the lengths of *c* and *d*.

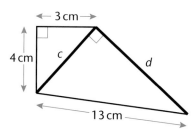

8 A section through a bicycle shed is shown opposite. Work out the width of the bicycle shed, giving your answer correct to 2 d.p.

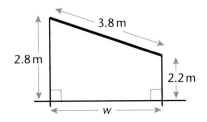

9 Find the perimeter of this kite:

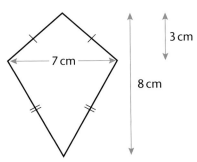

> Look back at Chapter 4 for the properties of the diagonals of a kite.

22.2 Finding the distance between two coordinate points

> **Key words:**
> coordinate points
> quadrant

Pythagoras' theorem can also be used to find the straight line distance between two **coordinate points** on a graph. Example 4 has positive coordinate values (points contained within the first **quadrant**) and Example 5 shows how to calculate the distance if one or more of the coordinates are negative.

Example 4

Two points on a graph have coordinates $P(6, 4)$ and $Q(2, 1)$. Find the length of PQ.

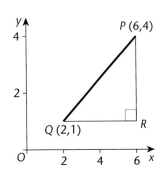

> A sketch will help you. Complete the diagram by drawing the triangle PQR where the angle at R is 90°.

> Use the x-coordinates to calculate horizontal distances and the y-coordinates to calculate vertical distances.

The length QR = 6 − 2 = 4 and the length PR = 4 − 1 = 3.

$$PQ^2 = QR^2 + PR^2$$
$$PQ^2 = 4^2 + 3^2$$
$$PQ^2 = 16 + 9 = 25$$
$$PQ = \sqrt{25}$$
$$PQ = 5$$

> R has the same y–coordinate as Q and the same x-coordinate as P, so the coordinate of R is (6, 1).

> Apply Pythagoras' theorem to triangle PQR.

Example 5

Find the length of line *MN* where *M* is point $(-2, 3)$ and *N* is point $(3, -5)$.

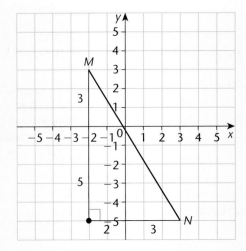

Draw a sketch, labelling the ends of the line, and construct a right-angled triangle.

$$MN^2 = (3 + 5)^2 + (2 + 3)^2 \qquad\qquad MN = \sqrt{8^2 + 5^2}$$
$$= 8^2 + 5^2 \qquad\qquad\qquad = \sqrt{89}$$

Apply Pythagoras.

You can leave your answer in surd form unless the question asks for a different format.

Exercise 22B

1 Two points on an *x–y* graph have coordinates $A(15, 9)$ and $B(3, 4)$. Find the length of *AB*.

2 The coordinates of two points on a graph have coordinates $X(9, 12)$ and $Y(5, 6)$. Find the length of *XY* to 1 d.p.

If in doubt always sketch a diagram and label the points. Draw a right-angled triangle and use Pythagoras' theorem.

3 Find the lengths (to 1 d.p.) between the following two points that have coordinates:

(a) $G(1, 8)\, H(7, 5)$
(b) $C(-3, 7)\, D(5, 4)$
(c) $W(9, 4)\, V(1, -6)$
(d) $F(2, -2)\, E(-4, -6)$
(e) $X(-4, -7)\, Y(2, 3)$
(f) $A(4, 0)\, B(0, 4)$
(g) $R(-1, 3)\, S(4, -1)$
(h) $M(-7, -6)\, N(-8, -9)$

22.3 Trigonometry – the ratios of sine, cosine and tangent

Key words:
trigonometry
sine
cosine
tangent
opposite
adjacent

Trigonometry is concerned with calculating sides and angles in triangles and involves three ratios called **sine**, **cosine** and **tangent**.

These ratios are usually abbreviated to **sin, cos** and **tan**.

Consider the right-angled triangle *ABC* shown below.

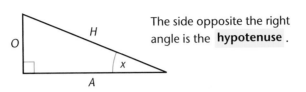

The side opposite the angle *x* is called the **opposite** .

The side opposite the right angle is the **hypotenuse** .

The side next to the angle *x* is called the **adjacent** .

Questions on trigonometry are often on the exam paper, so it is well worth learning the basics.

The hypotenuse is always the same side but the adjacent and opposite sides switch depending on which angle you are using.

The ratios for sine, cosine and tangent are defined as follows.

$$\sin x = \frac{\text{opposite}}{\text{hypotenuse}} = \frac{O}{H}$$

$$\cos x = \frac{\text{adjacent}}{\text{hypotenuse}} = \frac{A}{H}$$

$$\tan x = \frac{\text{opposite}}{\text{adjacent}} = \frac{O}{A}$$

$\sin x = \dfrac{O}{H}$

$\cos x = \dfrac{A}{H}$

$\tan x = \dfrac{O}{A}$

A useful memory aid is

SOHCAHTOA

Using a calculator for sin, cos and tan

The abbreviated forms (sin, cos and tan) are used on your calculator keys.

When using your calculator, make sure it is working in *degrees*. You will be able to tell because D or DEG will appear in the display. If it is not in degree mode, press the MODE key on your calculator until you see a choice of DEG, RAD or GRA appear in the display, then press whichever number corresponds to DEG (it is usually number 1). Remember that different makes of calculator have different keys... you may have to read your instruction booklet.

To find the sin, cos or tan of an angle press the appropriate key followed by the angle.

So for sin 34° press

You will see that sin 34° = 0.559192...

The inverse function

You will sometimes have to work backwards or use the inverse function.

For example, if $\cos x = 0.6418$ and you want to find the value of angle x, it is no use just pressing the cos key.

This is the *inverse operation* of finding the cosine, since you are given the value of the cosine and want to find the angle that goes with it. The inverse function on your calculator may be labelled

SHIFT   or perhaps   2nd F   or perhaps   INV

So to find angle x when $\cos x = 0.6418$, press

SHIFT   cos   0   .   6   4   1   8   =

You will see that $x = 50.0738... = 50.1°$ (3 s.f.)

This calculation is sometimes written as

$$\cos^{-1} 0.6148 = 50.1°.$$

When you press the shift key followed by the cos key, you may see $\cos^{-1}$ appear in the calculator display.

Exercise 22C

1 Sketch these triangles and label the sides A, O and H, where x is the angle to be found.

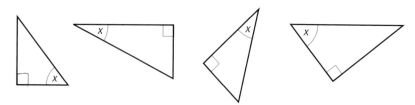

2 Using a calculator, work out the value of the following trigonometric functions giving your answer to 4 d.p.

  (a) $\sin 27°$ **(b)** $\tan 76°$ **(c)** $\cos 18°$ **(d)** $\tan 4°$

  (e) $\cos 85°$ **(f)** $\sin 90°$ **(g)** $\cos 0°$ **(h)** $\tan 20°$

  (i) $\cos 165°$ **(j)** $\tan 180°$ **(k)** $\sin 100°$ **(l)** $\cos 126.5°$

3 Use your calculator to find the value of each angle correct to 0.1°.

  (a) $\cos x = 0.6482$ **(b)** $\sin y = 0.8224$

  (c) $\tan \theta = 2.3385$ **(d)** $\sin A = 0.3264$

  (e) $\cos B = 0.6756$ **(f)** $\tan \phi = 21.2$

If the lengths of two sides are given, you will be able to work out an angle using one of the trigonometric ratios. To do this you will need to take the inverse of the function using your calculator.

Example 6

In the following triangle work out the size of angle x.

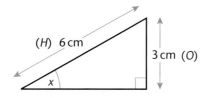

$$\sin x = \frac{O}{H} = \frac{3}{6} = 0.5$$

$$x = \sin^{-1}(0.5) = 30°$$

Use sine to work out the value of $\sin x$. Now use your calculator to work out the value of x by using the inverse of sin, which is $\sin^{-1}$.

Example 7

In the following triangle work out the size of angle x, to 2 d.p.

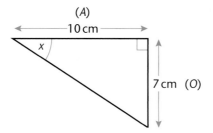

$$\tan x = \frac{O}{A} = \frac{7}{10} = 0.7$$

$$x = \tan^{-1}(0.7) = 34.99° \ (2 \ d.p.)$$

Use $\tan^{-1}$ to work out the value of x from $\tan x$.

Exercise 22D

1 Work out the size of the unknown angles, to the nearest degree, in the following triangles. (All lengths are given in cm.)

(i)

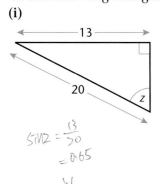

$\sin z = \frac{13}{20}$

$= 0.65$

$z = 41$

(ii)

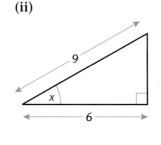

(iii)

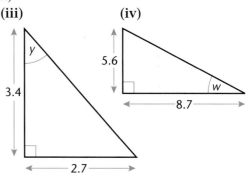

(iv)

2 Find all the missing angles in these triangles:

(i)

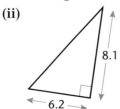

6.7

3.4

(ii)

8.1

6.2

(iii)

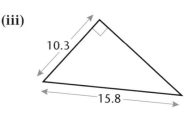

10.3

15.8

If the length of one side and the size of one angle are given then you can work out the lengths of the remaining sides using trigonometry. The information that is given in the question will always direct you to the correct trigonometric function to be used.

Example 8

In triangle PQR, length $QR = 15$ cm, angle $RPQ = 90°$ and angle $PRQ = 60°$. Work out the length of PR.

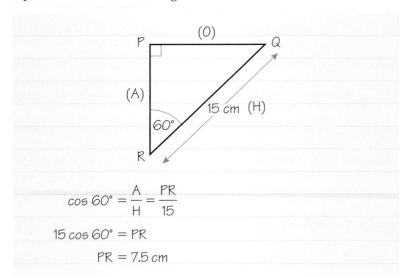

$$\cos 60° = \frac{A}{H} = \frac{PR}{15}$$

$$15 \cos 60° = PR$$

$$PR = 7.5 \, cm$$

Always draw and label a diagram with the given information. It is also useful to label the sides: opposite (*O*), adjacent (*A*) and hypotenuse (*H*).

Name the side you are given and the side you want to find and it is then obvious which ratio (sin, cos or tan) you need to use. QR is the hypotenuse and PR is adjacent (*A*) to the angle given so use the cosine ratio.

Example 9

Work out the length of **(a)** XZ and **(b)** YZ in the triangle below.

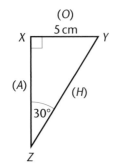

(*O*)
5 cm

(*A*)

(*H*)

30°

(a)   Let $XZ = y$

$$\tan 30° = \frac{O}{A} = \frac{5}{y}$$

$$y = \frac{5}{\tan 30°} = \frac{5}{0.5774}$$

$$y = 8.66\,cm \approx 8.7\,cm\ (1\ d.p.)$$

> XY is opposite to the angle given ($O = 5$ cm) and XZ is adjacent (A) to the angle given so use the tangent ratio.

> You need to rearrange the equation to make y the subject.

(b) Let $YZ = x$

Method A: sine ratio

$$\sin 30° = \frac{O}{H} = \frac{5}{x}$$

$$x = \frac{5}{\sin 30°} = \frac{5}{0.5}$$

$$x = 10.0\,cm$$

> Always use the original value in subsequent calculations and not the rounded value as this will lead to more errors.

Method B: Pythagoras

$$x^2 = y^2 + 5^2$$
$$= 8.66^2 + 5^2$$
$$= 99.99...$$
so $$x = \sqrt{99.99...}$$
$$= 10.0\,cm$$

> You can use either method for calculating the remaining side. Always read the question carefully just in case it asks you to use trigonometry.

Exercise 22E

1 Calculate the unknown length marked on each diagram to 1 d.p. (all lengths are in cm).

(i)
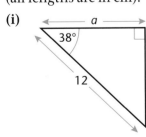

(ii)
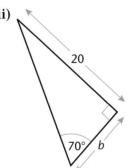

(iii)
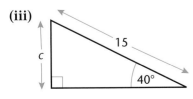

(iv)
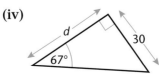

2 Find all the missing lengths in these triangles (to 1 d.p.).

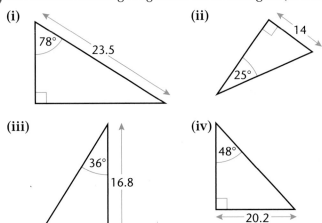

(i)   78°   23.5

(ii)   14   25°

(iii)   36°   16.8

(iv)   48°   20.2

Often you need to use trigonometry more than once in a question to find missing lengths or angles.

Example 10

ABCD is a parallelogram, length $AB = 4$ cm, length $BC = 8$ cm and angle $BAD = 65°$.

A triangle DEF is joined to the side AD of the parallelogram as shown. Work out the angle marked a.

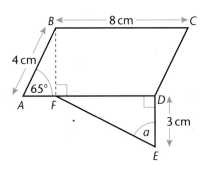

Let $AF = x$

$$\cos 65° = \frac{A}{H} = \frac{x}{4}$$

$$x = 4\cos 65° \approx 1.69 \text{ cm} \approx 1.7 \text{ cm (1 d.p.)}$$

Let $FD = y$

$$= 8 - x = 6.309\ldots \text{ cm}$$

$$\tan a = \frac{O}{A} = \frac{y}{3} = \frac{6.309\ldots}{3}$$

$$a = \tan^{-1}\left(\frac{6.309\ldots}{3}\right)$$

$$a = 64.57° \approx 64.6°$$

$x = 1.69047\ldots$
Use this value to work out angle FD(y).

$y = 6.30952\ldots$
Use this value to work out angle a.

Exercise 22F

1 In the following diagram, find the length marked x. Give your answer to the nearest cm.

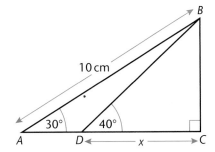

2 Triangle ABC is a right-angled triangle. The line BD meets the side AC at right angles.

If $AD = 5$ cm calculate the length DC (marked as y). Give your answer to the nearest cm.

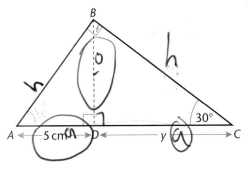

3 In the following shape work out the length marked b. Give your answer to 1 d.p.

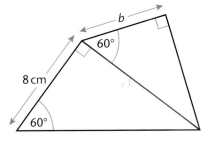

4 The diagram shows two right-angled triangles.

   (a) Given that $\cos x° = \frac{2}{3}$, calculate the length BD.

   (b) Find the value of $\sin y$.

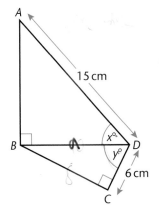

22.4 Applications of trigonometry and Pythagoras

In Chapter 7, you met the formulae for surface area and volume of a cone, given the radius of the base (r) and the slant height (l).

The *total* surface area = area of base + curved surface area

$$= \pi r^2 + \pi r l$$
$$= \pi r (r + l)$$

Volume of cone $= \frac{1}{3} \pi r^2 h$

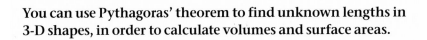

Notice that the radius of the base, the perpendicular height and the slant height form a right-angled triangle. So if you are given the radius and *one* of the heights, you can use Pythagoras to find the missing height.

> **(radius)^2 + (perpendicular height)^2 = (slant height)^2**
>
> $r^2 + h^2 = l^2$

> You can use Pythagoras' theorem to find unknown lengths in 3-D shapes, in order to calculate volumes and surface areas.

Example 11

A cone has a base radius of 5 cm and a slant height of 12 cm.

(a) Calculate the curved surface area of the cone.
 Give your answer in terms of π.

(b) Calculate the volume of the cone, giving your answer to 3 s.f.

(a) Curved surface area

$= \pi r l = \pi \times 5 \times 12 \text{ cm}^2 = 60\pi \text{ cm}^2$

(b) Using Pythagoras' theorem,

$r^2 + h^2 = l^2$

$5^2 + h^2 = 12^2$

$h^2 = 12^2 - 5^2$

$= 144 - 25$

$h = \sqrt{119}$

$= 10.9087...$

Volume of cone $= \frac{1}{3} \pi r^2 h$

$= \frac{1}{3} \times \pi \times 5^2 \times 10.9087...$

$= 285.58...$

Volume $= 286 \text{ cm}^2$ (3 s.f.)

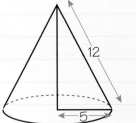

Draw a sketch first.

You need to find h before you can find the volume of the cone so use Pythagoras' theorem.

Do **not** round the answer for h or you will lose accuracy. Keep the value in the calculator display and only correct to 3 s.f. at the end.

Exercise 22G

1 A cone has a radius of 7 cm and a
perpendicular height of 10 cm.

   (a) What is the slant height of the cone?

   (b) What is its curved surface area?
      Give your answer to 3 s.f.

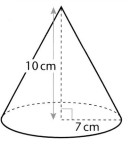

UAM **2** The curved surface of this ice cream cone is 75.4 cm².

   (a) What is the radius of the base?

   (b) What volume of ice cream can it hold?

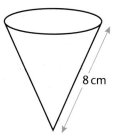

UAM **3** A square-based pyramid has a volume
of 351 cm³ and a base length of 9 cm.
Find its surface area.
Give your answer to 2 d.p.

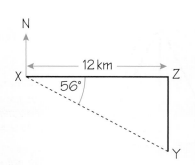

Find the perpendicular
height first.

Example 12

A yacht sets sail from a harbour X at 11.00 am.
The current makes the yacht drift south so that after 2 hours it is
12 km east of the harbour and on a bearing of 146°. How far south
has the yacht travelled? If it maintains its initial speed and
direction, how far south will the yacht be at 5 pm?

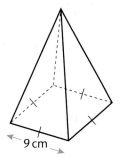

Look back at Chapter 4 for
a reminder of bearings.

This is a sketch. It is not
drawn to scale. Note that a
bearing of 146° means an
angle of 56° between XZ
and XY since bearings are
measured from the North.

$$\tan 56° = \frac{O}{A} = \frac{ZY}{12}$$

$$ZY = 12 \tan 56°$$

$$ZY = 12 \times 1.4826$$

$$ZY = 17.79 \text{ km (2 d.p.)}$$

After 6 hours it will be 3× further

$$3 \times 17.79 \text{ km} = 53.37 \text{ km.}$$

Combining topics in this
way is usual when asking
questions on trigonometry
and Pythagoras.

Exercise 22H

1 A flag pole is kept vertical by two guy ropes attached to the top of the pole and the ground which is level.

If $FM = 6$ m, $AM = 4.5$ m and angle $MCF = 42°$, calculate

(a) the length FC

(b) the length FA

(c) the distance between A and C.

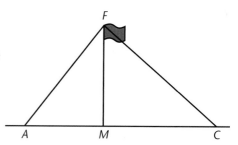

2 A rocket lifts off vertically and travels to a height of 5000 m. The second stage cuts in and takes the rocket at an angle of 25° to the vertical, covering a further distance of 2000 m. Calculate the height of the rocket above the ground.

3 A kite flying in a strong wind makes an angle of 35° with the ground. If the horizontal distance between the kite and the end of the string is 28 m, find the height of the kite above the ground.

4 An aeroplane is approaching an airport at a height of 1000 m. If the aeroplane is 8 km from the airport, work out the angle of approach.

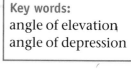

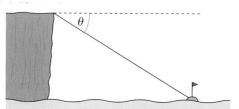

Angles of elevation and depression

When solving problems using Pythagoras or trigonometry, you might meet the following terms.

The **angle of elevation** is the angle measured upward from the horizontal.

The **angle of depression** is the angle measured downward from the horizontal.

Exercise 22I

UAM **1** The angle of elevation of a church steeple, measured at a distance of 45 m from the base of the steeple, is 65°. If it is measured from a point 2 m above ground level, what is the height of the steeple? Another measurement is taken giving an angle of 25°. How far from the base of the church was this measurement taken?

UAM **2** A surveyor stands 150 m away from the centre of a church tower. The angle of elevation to the top of the tower is 38°. The angle of elevation to the tip of the flagpole on top of the tower is 41°. Work out the height of the flagpole to 1 d.p.

UAM **3** An aeroplane is flying at a constant altitude of 10 000 m. An observer measures the angle of elevation to the plane as 25° and 15 seconds later as 22.6°.

   (a) Work out the horizontal distance the plane has covered in this time.

   (b) What is the speed of the aeroplane in km/h?

UAM **4** From the top of a sea cliff 35 m above the water the angles of depression of two boats A and B at sea are 19° and 4° respectively.

   Calculate the distance between the two boats to 1 s.f.

22.5 Problems in three dimensions

Key words:
three dimensions (3-D)

You can use Pythagoras' theorem and trigonometry to solve problems in **three dimensions** **(3-D)** .

A reminder about 3-D shapes can be found in Chapter 7.

You need to be able to identify and draw the correct right-angled triangles that contain the length or angle to be found.

Example 13

This cuboid has $a = 8$ cm, $b = 4$ cm and $c = 5$ cm. Work out

(a) the length of the longest diagonal HD (marked x) to 1 d.p.

(b) the angle between HD and DE (marked θ).

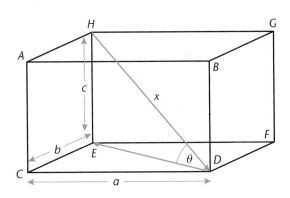

(a)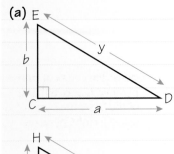

$$y^2 = a^2 + b^2$$
$$y^2 = 8^2 + 4^2$$
$$y^2 = 64 + 16 = 80$$
$$y = \sqrt{80}$$

You need to find the length *ED* first (label this with the letter *y*). Draw the right-angled triangle that contains the length *ED*.

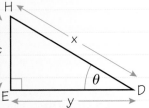

$$x^2 = c^2 + y^2$$
$$x^2 = 5^2 + 80$$
$$x^2 = 25 + 80 = 105$$
$$x = \sqrt{105}$$
$$x = 10.24 \text{ cm} \approx 10.2 \text{ cm (1 d.p.)}$$

Apply Pythagoras' theorem to the right-angled triangle *ECD*.

To find *x*, use Pythagoras' theorem a second time on the right-angled triangle *HED*.

(b) $$\tan \theta = \frac{O}{A} = \frac{5}{8.94} = 0.5590\ldots$$
$$\theta = \tan^{-1}(0.5590\ldots)$$
$$\theta = 29.21° \text{ (2 d.p.)}$$

You must use the value of *y* you found in your previous calculation, not the rounded value. In fact it is best to use y^2 directly in the calculation.

To find the angle θ, use the triangle *HED* shown above. As all three lengths are known you can use any one of the trigonometric functions.

Diagonal of a cuboid

The length of the diagonal of a cuboid is found by applying Pythagoras' theorem to two triangles.

For $\triangle AGE$ $\quad x^2 = c^2 + y^2$,

      where *x* is the longest diagonal in the cuboid.

      $x^2 = c^2 + (a^2 + b^2)$,

      since $y^2 = a^2 + b^2$ ($\triangle EGH$).

So $\quad\quad\quad x^2 = a^2 + b^2 + c^2$

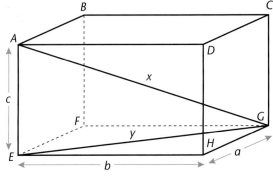

The length, *x*, of the longest diagonal in a cuboid with dimensions $a \times b \times c$ is given by

$$x^2 = a^2 + b^2 + c^2$$

This is a 3-D version of Pythagoras' theorem. It is the same method that you used in Example 13.

Angle between a line and a plane

Example 14

ABCDE is a pyramid with a rectangular base, *ABCD*. *AB* = 10 cm and *BC* = 5 cm.

The vertex of the pyramid, *E*, is vertically above the centre of the base, *O*. *EA* = 8 cm.

Find the angle between *EA* and the plane *ABCD*.

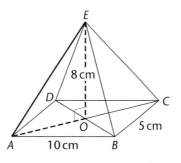

When the vertex of the pyramid is above the centre of the base, the pyramid is called a **right pyramid**.

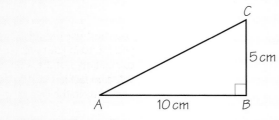

Drop a perpendicular from *E* onto the base. This line meets the base at *O*. Join *O* to A and use triangle *EAO* to calculate the angle *EAO*, the angle between *EA* and the base, *ABCD*.

$AC^2 = AB^2 + BC^2$

$AC^2 = 10^2 + 5^2$

$AC^2 = 125$

$AC^2 = \sqrt{125} = 11.1803\ldots$

$AO = \dfrac{11.1803\ldots}{2} = 5.59016\ldots$

First find *AC* by using Pythagoras' theorem on the base triangle *ABC*, then divide it by 2 to find *AO*.

ABCD is a rectangle so angle *B* = 90°

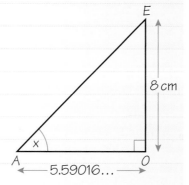

$\tan x = \dfrac{O}{A} = \dfrac{8}{5.59016\ldots}$

$x = 55.055\ldots$

$x = 55.1 \ (1\ d.p.)$

The angle between *EA* and the plane *ABCD* is 55.1° (1 d.p.)

Draw the right-angled triangle *EAO*. Label the angle *x*.

It is important to draw the correct right-angled triangle that contains the length or angle you require.

Exercise 22J

1 A cuboid has side lengths *a* = 3 cm, *b* = 4 cm and *c* = 12 cm. Work out the length of the longest diagonal of the cuboid.

First sketch the cuboid.

2 A square-based pyramid *ABCDE* has a base of side 4 cm and vertical height *EO* = 6 cm.

Find **(a)** the length of *EC*
(b) the angle between *EC* and *AC*.

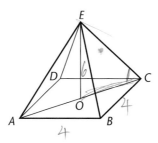

UAM **3** A vertical pole, *CP*, stands at one corner of a horizontal rectangular field *ABCD*.

AB = 25 m and *BC* = 18 m

The angle of elevation of *P* from *D* is 27°

Calculate

(a) the height of the pole

(b) the angle of elevation of *P* from *B*,

(c) the length *AC*,

(d) the angle of elevation of *P* from *A*.

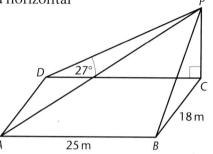

Pythagoras and 3-D coordinates

Just as a point on a plane (or flat surface) can be described by 2 coordinates, a point in 3-dimensional space can be described by 3 coordinates.

For example, the point, *A*, at one corner of the cuboid below, is at (4, 3, 2).

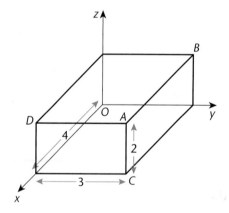

A is 4 units in the *x*-direction, 3 units in the *y*-direction and 2 units in the *z*-direction

O is the origin and has the coordinates (0, 0, 0)

Example 15

Using the diagram of the cuboid above, write down the coordinates of the points *B*, *C* and *D*.

B is 0 units in the *x*-direction, 3 units in the *y*-direction and 2 units in the *z*-direction so *B* is the point (0, 3, 2).

C is 4 units in the *x*-direction, 3 units in the *y*-direction and 0 units in the *z*-direction so *C* is the point (4, 3, 0).

D is 4 units in the *x*-direction, 0 units in the *y*-direction and 2 units in the *z*-direction so *D* is the point (4, 0, 2).

Exercise 22K

1 (a) Write down the coordinates
of *A, B, C, D, E, F, G*.

(b) Write down the coordinates of
the mid-points of
(i) *OC* **(ii)** *CF* **(iii)** *BC* **(iv)** *BE*

(c) Write down the coordinates of
the centre of the face
(i) *OABC* **(ii)** *GDEF* **(iii)** *EBCF*

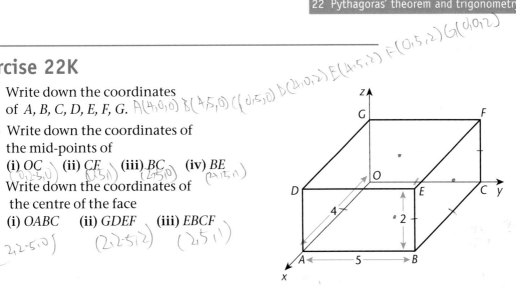

2 (a) Write down the coordinates of
P, S, V.

(b) Write down the coordinates of the
mid-points of
(i) *OP* **(ii)** *OS* **(iii)** *PQ*

(c) Write down the coordinates of the centre
of the face
(i) *OPQR* **(ii)** *OSTP* **(iii)** *PQUT*

(d) Write down the coordinates of the centre
of the box.

(e) Calculate the lengths of the lines joining
these pairs of points:
(i) *O* and *T* **(ii)** *O* and *V* **(iii)** *O* and *Q*
(iv) *O* and *U* **(v)** *P* and *V*.

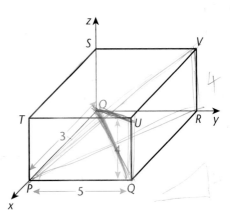

3 On the axes below, *O* is the origin.

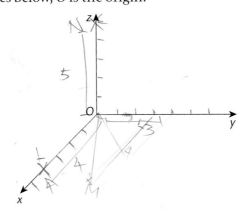

(a) Copy the axes and plot the points $L(4, 0, 0)$, $M(4, 3, 0)$ and
$N(0, 0, 5)$.

(b) Calculate the lengths of these lines **(i)** *OM* **(ii)** *MN*

Examination questions

1 A rectangular field *ABCD* is shown.
The length of the field, *AB* = 160 m.
The width of the field, *BC* = 75 m.

Calculate the length of the diagonal *BD*.
Give your answer to a suitable degree of accuracy.

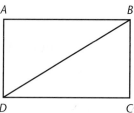

(4 marks)
AQA, Spec A, 2I, November 2004

2 ABC is a right-angled triangle. BC = 125 m.
Angle *CAB* = 33°.

Find the length of *AC* (marked *x* in the diagram).
Give your answer to an appropriate degree
of accuracy.

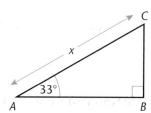

(4 marks)
AQA, Spec A, 2H, June 2003

3 In the diagram *PQ* = 14 cm and *QR* = 8.6 cm.
Angle *PSQ* = angle *SQR* = 90°.
Angle *PQS* = 25°.

Calculate angle *R*.

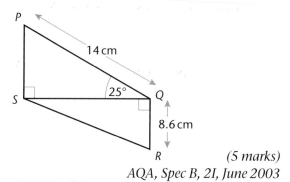

(5 marks)
AQA, Spec B, 2I, June 2003

4 *VABCD* is a right pyramid on a rectangular base.
VA = *VB* = *VC* = *VD* = 16 cm.
AB = 20 cm and *BC* = 14 cm.

Calculate the angle between the edge *VC* and
the base plane *ABCD*.

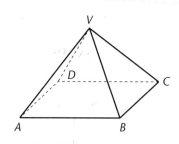

(5 marks)

5 *ABCDEFGH* is a cuboid with sides of 5 cm,
5 cm and 12 cm as shown.

Calculate angle *DFH*.

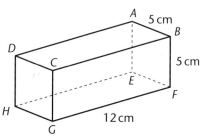

(5 marks)
AQA, Spec A, 2H, June 2003

Summary of key points

Pythagoras' theorem (grade C/B/A)

Pythagoras' theorem states that in a right-angled triangle the square of the hypotenuse is equal to the sum of the squares on the other two sides.

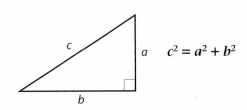

$$c^2 = a^2 + b^2$$

The lengths of the shorter sides can be calculated using

$$a^2 = c^2 - b^2 \quad \text{and} \quad b^2 = c^2 - a^2$$

Trigonometry (grade B)

Trigonometric functions are used to find lengths or angles in right-angled triangles.

The basic trigonometric functions are: sin x, cos x and tan x.
The three trigonometric ratios are given by

$$\sin x = \frac{\text{opposite}}{\text{hypotenuse}} = \frac{O}{H}$$

$$\cos x = \frac{\text{adjacent}}{\text{hypotenuse}} = \frac{A}{H}$$

$$\tan x = \frac{\text{opposite}}{\text{adjacent}} = \frac{O}{A}$$

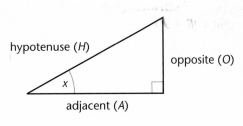

Applications of Pythagoras and trigonometry (grade A/A*)

The longest diagonal (x) in a cuboid with dimensions $a \times b \times c$ is found by applying Pythagoras' theorem to two triangles.

$$x^2 = a^2 + b^2 + c^2$$

You can use trigonometry and Pythagoras' theorem to find the angle between a line and a plane.

For a cone, Pythagoras' theorem connecting r, h and l is $r^2 + h^2 = l^2$

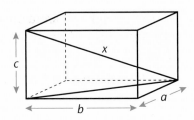

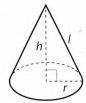

3-D coordinates

A point in 3-dimensional space can be described by 3 coordinates.

This chapter will show you how to:
✔ apply your knowledge of circle properties to solving problems
✔ prove important circle theorems
✔ apply the circle theorems to a variety of problems
✔ set out your solutions in a clear and logical way
✔ write out reasons for your deductions

23.1 Circle properties

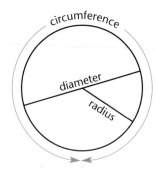

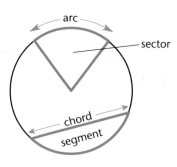

Key words:
chord
segment
tangent
arc
sector

The diagrams above show some of the terms you will use when solving problems involving circles. A **chord** is a straight line joining two points on the circumference. A **segment** is the region between a chord and the circumference.

Arcs were introduced in Chapter 7.

A **tangent** to a circle is a line that just touches the circumference at only one point of contact.

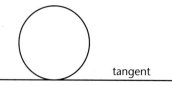

In this section we start with three important circle properties.

1 The perpendicular from the centre of a circle to a chord bisects the chord.

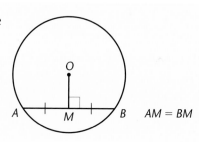

$AM = BM$

2 Tangents drawn to a circle from an external point are equal in length.

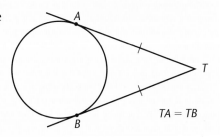

$TA = TB$

Proofs of these two results can be found in Chapter 32.

3 The angle between a tangent and a radius is 90°.

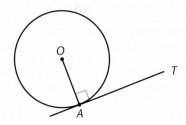

A proof of this result is quite difficult and is well beyond the scope of this book.

You need to remember these three results because they are often needed for exam questions.

Example 1

In the diagram, O is the centre of the circle and AB is a chord.
$AO = 10$ cm and $AB = 12$ cm.
M is the mid-point of AB. Find: the length of OM.

Using Pythagoras' theorem in $\triangle AOM$

$OM^2 = 10^2 - 6^2$

$\quad = 64$

$OM = 8$

M is the mid-point of chord AB so angle $OMA = 90°$

$\triangle AOM$ is similar to a 3, 4, 5 triangle.
3, 4, 5 is a Pythagorean triple (see Chapter 22).

Example 2

In the diagram, O is the centre of the circle and AB is a chord.
T is an external point and tangents from T to the circle touch the circle at points A and B.
If angle $ATB = 40°$, find the size of angles **(a)** TAB and **(b)** AOB.

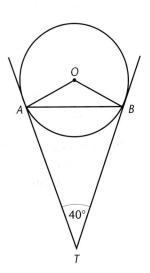

(a) Tangents from an external point are equal in length,

so $TA = TB$ which means that $\triangle TAB$ is isosceles.

angle $TAB = 70° =$ angle TBA (base angles of isosceles $\triangle$ TAB)

(b) TA is perpendicular to AO (tangent & radius meet at 90°)

so, angle $OAB = 90° - 70° = 20°$

In the same way, angle $OBA = 20°$

so, angle $AOB = 180° - 20° - 20° = 140°$

Exercise 23A

1

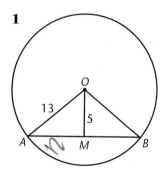

Find the length of *AM*.

2

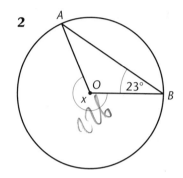

Find angle *x*.

3

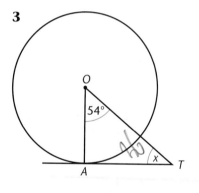

Find angle *x*.

4

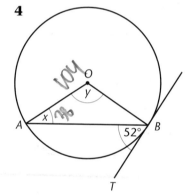

Find angles *x* and *y*.

5

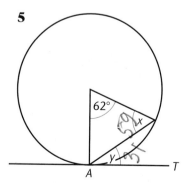

Find angles *x* and *y*.

6

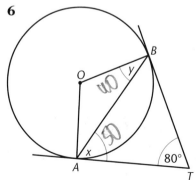

Find angles *x* and *y*.

23.2 Circle theorems

In this section you are going to see some important results about the relationship between angles of circles.

Theorem 1

You could be asked to prove any of these results in your GCSE exam.

The angle **subtended** by an arc at the centre of a circle is twice the angle that it subtends at the circumference.

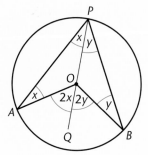

'Subtended' means 'held up by'. So the angle starts and finishes at the ends of the arc (or chord) of the circle.

To prove: **angle AOB = 2 × angle APB**

Proof: Let angle $APO = x$ and angle $BPO = y$
              Join P to O and extend the line to Q

              angle $PAO = x$ (base angles isosceles $\triangle OAP$)
              angle $PBO = y$ (base angles isosceles $\triangle OBP$)
              angle $AOQ = 2x$ (exterior angle property $\triangle APO$)
              angle $BOQ = 2y$ (exterior angle property $\triangle BPO$)
              so angle $AOB = 2x + 2y$
              but angle $APB = x + y$
              angle $AOB = 2 \times$ angle APB

Always state any basic information such as this at the start of your proof.

Exterior angle of triangle = sum of interior opposite angles (see Chapter 4).

Notice how each statement is supported by a reason why it must be true. This is an essential feature of any proof, as you will see in all the other proofs in this chapter and those in Chapter 31. You *must* set out your work in this way.

The diagram used above for this proof can appear in many different forms.
The result is true no matter which of these forms it takes.
Here are the ones you are likely to meet.

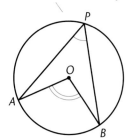
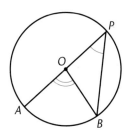
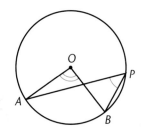
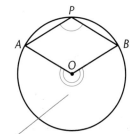

In all of these diagrams, angle AOB = 2 × angle APB

Notice that it is reflex angle AOB here.

Theorem 2

The angle in a semi-circle is a right angle.

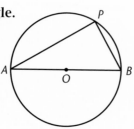

Note that *AB* is a diameter.

To prove: **angle *APB* = 90°**

Proof: angle *AOB* = 2 × angle *APB*
(angle at centre = twice angle at circumference)

but angle *AOB* = 180°
(*AOB* is a diameter so it is a straight line)

angle *APB* = 90°

Notice how convenient it is to use the result you proved in the first circle theorem!

This is standard practice in proofs ... once you have proved something you can then use the result to prove other results.

Example 3

In the diagram, *O* is the centre of the circle.
Calculate the size of the angle marked *x*, giving reasons for your answer.

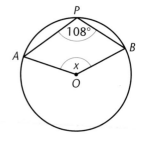

reflex angle *AOB* = 216°   (angle at centre = twice angle
                                  at circumference)

x = 360° − 216° (angles round a point)

x = 144°

Example 4

In the diagram, *O* is the centre of the circle and *AB* is a diameter.
Calculate the value of *x*, giving reasons for your answer.

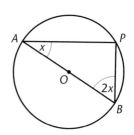

angle *APB* = 90°   (angle in a semi-circle)

x + 2*x* + 90° = 180°   (angle sum of triangle = 180°)

3*x* = 90°

x = 30°

Example 5

In the diagram, *O* is the centre of the circle and *TA* and *TB* are
tangents to the circle from *T*. Angle *ATB* = 46°.
Calculate the size of

(a) angle *AOB*   **(b)** angle *ACB*

giving reasons for your answers.

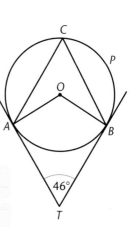

(a)   angle TAO = 90°   (tangent & radius meet at 90°)

angle TBO = 90°   (tangent & radius meet at 90°)

angle AOB = 360° − 90° − 90° − 46°

(angle sum of quadrilateral = 360°)

so   angle AOB = 134°

(b)   angle AOB = 2 × angle ACB

(angle at centre = twice angle at circumference)

so   angle ACB = 134° ÷ 2

= 67°

Exercise 23B

1

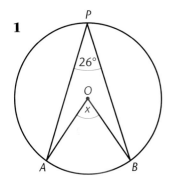

Find angle *x*.

2

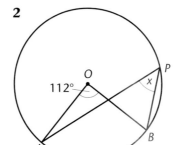

Find angle *x*.

3

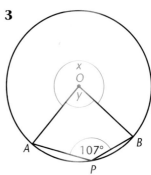

Find angles *x* and *y*.

4

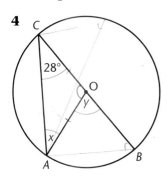

Find angles *x* and *y*.

5

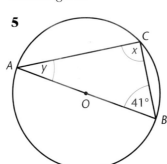

Find angles *x* and *y*.

6

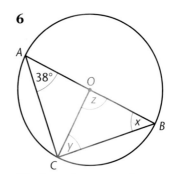

Find angles *x*, *y* and *z*.

7

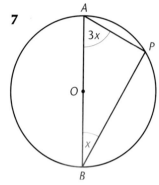

Find angle x.

8

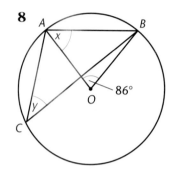

Find angles x and y.

9

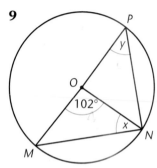

Find angles x and y.

10

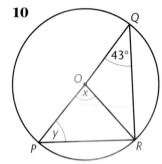

Find angles x and y.

11

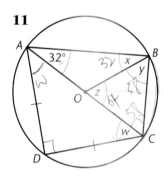

Find angles x, y, z and w.

12

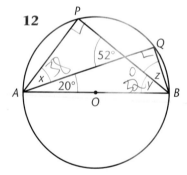

Find angles x, y and z.

In the two circle theorems you have met so far, the centre was important. The first result involved the angle at the centre of the circle and the second one involved a diameter.

The next two circle theorems do not depend on the centre of the circle (although you will see it is needed if you are *proving* either of the results).

Remember that a segment is a part of a circle cut off by a chord. There is always a major segment and a minor segment.

This next theorem concerns angles in the *same* segment, which means that they must both lie on the same side of the chord.

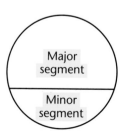

Theorem 3

Angles in the same segment are equal.

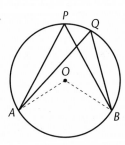

To prove: **angle APB = angle AQB**

Proof: Join A and B to O

angle $AOB = 2 \times$ angle APB
(angle at centre = twice angle at circumference)

angle $AOB = 2 \times$ angle AQB
(angle at centre = twice angle at circumference)

angle APB = angle AQB

The next theorem involves two words you may not know.
A **cyclic** quadrilateral is a quadrilateral whose four vertices
(points) lie on the circumference of a circle.
Angles that are **supplementary** add up to 180°.

> Notice that the chord does
> not need to be drawn. You
> can imagine where it
> would be and you can see
> that both angles APB and
> AQB are in the same
> segment.

> A similar word to this is
> **complementary**, which
> means that angles add up
> to 90°.

Theorem 4

Opposite angles of a cyclic
quadrilateral are supplementary.

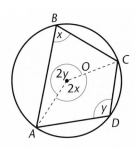

To prove: **angle ABC + angle ADC = 180°**

Proof: Let angle $ABC = x$ and angle $ADC = y$

angle $AOC = 2 \times$ angle ABC
(angle at centre = twice angle at circumference)

so angle $AOC = 2x$

reflex angle $AOC = 2 \times$ angle ADC
(angle at centre = twice angle at circumference)

so reflex angle $AOC = 2y$

but angle AOC and reflex angle AOC together make up a complete circle (360°)   so $2x + 2y = 360°$
which means that $x + y = 180°$

angle ABC + angle $ADC = 180°$

> It is also true that the other opposite pair are supplementary, namely angle BAD + angle $BCD = 180°$.

Notice that, as before, you are using results already proved to deduce new results. You will see this many times.

Theorem 4 has a useful 'extra' result. Look at this diagram.

Extend side AD to point E.

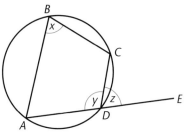

angle ABC + angle $ADC = 180°$ (proved above)
angle ADC + angle $CDE = 180°$ (angles on a straight line)
so   angle ABC = angle CDE

$$x + y = 180°$$
and $y + z = 180°$
∴ $x = z$

Angle CDE is an *exterior* angle of the cyclic quadrilateral because it is formed by extending one of the sides. Angle ABC is the interior angle *opposite* to angle CDE, so ...

An exterior angle of a cyclic quadrilateral is equal to the opposite interior angle.

Examples 6 and 7 use the four theorems that you have proved so far.

Example 6

In the diagram angle $QRT = 78°$ and angle $PTS = 25°$.
Calculate the size of
(a) angle PQS **(b)** angle QPT,
giving reasons for your answers.

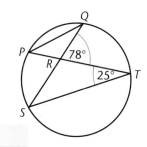

(a)   angle $PQS = 25°$ (angles in the same segment)

(b) angle QPT + angle $PQS = 78°$ (exterior angle property △QPR)
    so   angle $QPT = 53°$

Example 7

$ABCD$ is a cyclic quadrilateral. Angle $ADC = 100°$ and angle $ACD = 33°$.
Calculate the size of
(a) angle ABD **(b)** angle DBC,
giving reasons for your answers.

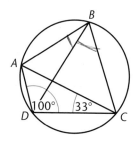

(a)   angle ABD = angle ACD   (angles in the same segment)

so   angle ABD = 33°

(b) angle ABC + angle ADC = 180° (opposite angles of cyclic quadrilateral)

so   angle ABC = 80°

       angle DBC = angle ABC − angle ABD = 80° − 33°

so   angle DBC = 47°

Exercise 23C

1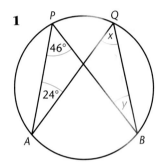

Find angles x and y.

2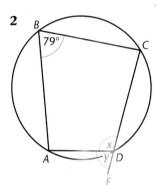

Find angles x and y.

3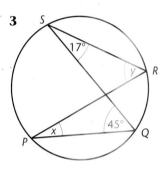

Find angles x and y.

4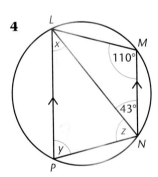

Find angles x, y and z.

5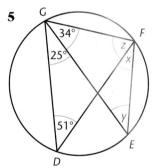

Find angles x, y and z.

6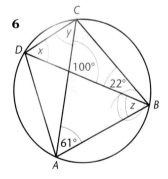

Find angles x, y and z.

7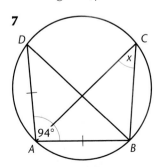

Find angle x.

8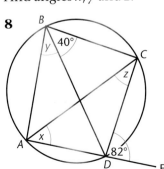

Find angles x, y and z.

9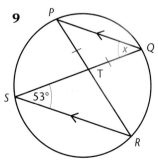

Find angle x.

10

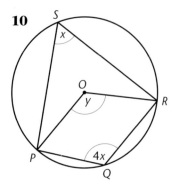

Find angles x and y.

11

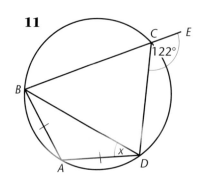

Find angle x.

12

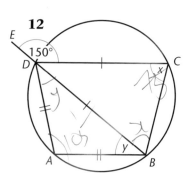

Find angles x and y.

Examples 8, 9 and 10 use all the theorems that you have learned so far.

Example 8

ABCD is a cyclic quadrilateral and O is the centre of the circle.
Angle $AOC = 82°$.

Calculate the size of

(a) angle ABC   **(b)** angle ADC,

giving reasons for your answers.

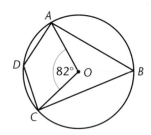

(a)   angle ABC = 41°   (angle at centre = twice angle at circumference)

(b) angle ADC + angle ABC = 180°   (opposite angles of cyclic quadrilateral)

so   angle ADC = 139°

Example 9

PQRS is a cyclic quadrilateral with exterior angle $QPM = 112°$.
The diagonals of the cyclic quadrilateral intersect at L.
Angle $PSQ = 44°$ and angle $QSR = 37°$.

Calculate the size of

(a) angle QRS   **(b)** angle SQR   **(c)** angle PLS,

giving reasons for your answers.

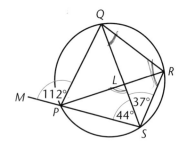

(a)   angle QRS = 112°   (exterior angle of cyclic quadrilateral = interior opposite angle)

(b)   angle SQR = 180° − 112° − 37°   (angle sum of △QRS)

so angle SQR = 31°

continued ▶

(c)   angle LPS = angle SQR   (angles in the same segment)
    so   angle LPS = 31°
         angle PLS = 180° − 31° − 44° (angle sum of △PLS)
    so   angle PLS = 105°

Example 10

In the diagram, O is the centre of the circle, triangle DEG is isosceles with $DE = EG$ and angle $GEF = 18°$.

Calculate the sizes of the angles marked x, y and z, giving reasons for your answers.

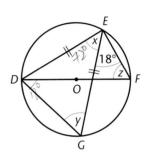

O is the centre of the circle so DF is a diameter,

so   angle DEF = 90°   (angle in a semi-circle)

which means that $x = 90° − 18° = 72°$

y is one of the base angles of isosceles triangle DEG

since $x = 72°$ then $y = (180° − 72°) ÷ 2 = 54°$

                 $z = y$ (angles in the same segment)

so $z = 54°$

Notice how, in all of these examples, each step follows logically from the previous step(s).

You can see that it is sometimes necessary to find angles other than those you are asked for ... they often form an important link in the thought process.

Always write a reason for each statement you make.

The next exercise uses all of the circle theorems you have learnt so far.

Exercise 23D

1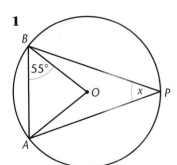

Find angle x.

2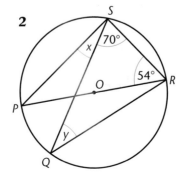

Find angles x and y.

3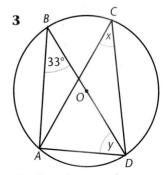

Find angles x and y.

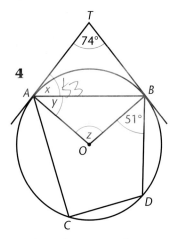

4

Find angles x, y and z.

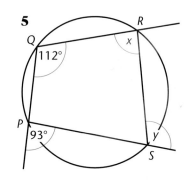

5

Find angles x and y.

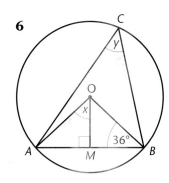

6

Find angles x and y.

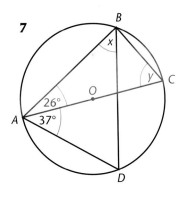

7

Find angles x and y.

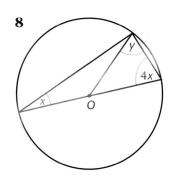

8

Find angles x and y.

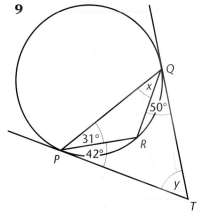

9

Find angles x and y.

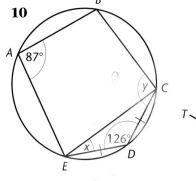

10

Find angles x and y.

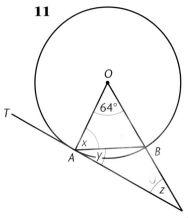

11

Find angles x, y and z.

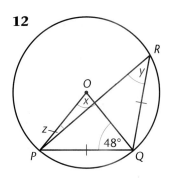

12

Find angles x, y and z.

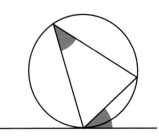

The last of the circle theorems involves the angle between a tangent and a chord.

As you can see in this diagram, the angle lies on one side of the chord and the angle it is equal to lies on the other side of the chord ... subtended by the chord.

It is called the angle in the alternate segment.

Theorem 5

The angle between a tangent and a chord is equal to the angle in the alternate segment.

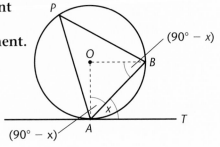

To prove: **angle TAB = angle APB**

Proof: Let angle $TAB = x$
angle $OAB = (90° - x)$ (tangent and radius meet at 90°)
but $\triangle OAB$ is isosceles since OA and OB are both radii
so angle $OBA = (90° - x)$
so, in $\triangle OAB$, using the fact that all angles add up to 180°,
angle $AOB = 180° - (90° - x) - (90° - x)$
$= 180° - 90° + x - 90° + x$
$= 2x$

but angle $AOB = 2 \times$ angle APB
(angle at centre = twice angle at circumference)

so angle $APB = x$

which means that   angle TAB = angle APB

> This theorem is called the Alternate segment theorem.

(angle between tangent and chord)   (angle in the alternate segment)

Example 11

In the diagram, TA is a tangent to the circle at A, $PA = PB$ and angle $BAT = 48°$.

Calculate the sizes of the angles marked x and y, giving reasons for your answers.

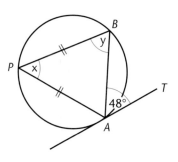

$x = 48°$ (alternate segment theorem)

y is one of the base angles of isosceles triangle PAB

so $y = (180° - 48°) \div 2 = 66°$

Example 12

In the diagram, *XY* is a tangent to the circle at *F*. *DE* is parallel to *XY* and angle *XFD* = 65°.

Calculate the sizes of the angles marked *x* and *y*, giving reasons for your answers.

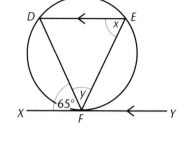

$x = 65°$ (alternate segment theorem)

angle EDF = 65° (alternate)

so $y = 180° - 65° - 65°$ (angle sum of △DEF)

       $y = 50°$

Exercise 23E

1

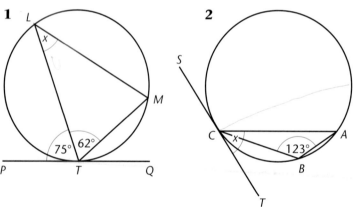

Find angle *x*.

2

Find angle *x*.

3

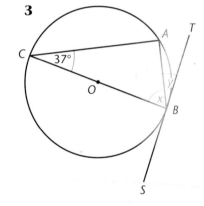

Find angles *x* and *y*.

4

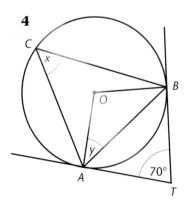

Find angles *x* and *y*.

5

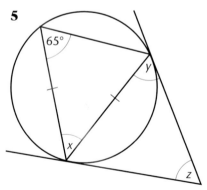

Find angles *x*, *y* and *z*.

6

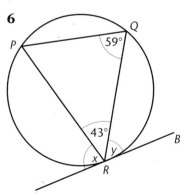

Find angles *x* and *y*.

7

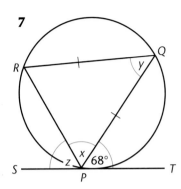

Find angles x, y and z.

8

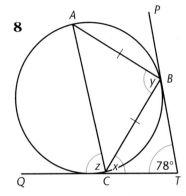

Find angles x, y and z.

9

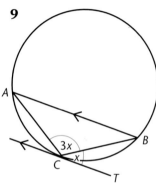

Find angle x.

10

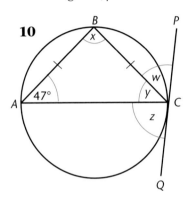

Find angles w, x, y and z.

23.3 Using circle properties and theorems

The next four examples recap all of the work in this chapter. The exercise that follows them tests your knowledge of all the circle properties and theorems.

Example 13

In the diagram, O is the centre of the circle, TB is a tangent to the circle at B and angle $TBA = 58°$.

Calculate the size of angle AOB, giving reasons for your answer.

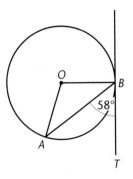

angle $OBA = 90° - 58° = 32°$

       (tangent & radius meet at $90°$)

$OA = OB$ (radii) so $\triangle OAB$ is isosceles

so   angle $OAB = 32°$ (base angles of isosceles $\triangle OAB$)

angle $AOB = 180° - 32° - 32° = 116°$

     (angle sum of $\triangle OAB$)

Example 14

In the diagram, O is the centre of the circle, reflex angle
ROQ = 206° and angle PRO = 45°.

Calculate the size of

(a) angle RPQ   (b) angle OQP,

giving reasons for your answers.

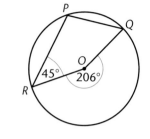

(a)   angle RPQ = 103°   (angle at centre = twice angle at circumference)

(b)   angle ROQ = 360° − 206° = 154°   (angles round a point)
         angle OQP = 360° − 154° − 45° − 103°   (angles sum of quadrilateral = 360°)
    so angle OQP = 58°

Example 15

In the diagram, LMNP is a cyclic quadrilateral and QR is a tangent to
the circle at P. Angle MNP = 93° and angle LPR = 38°.

Calculate the size of

(a) angle MLP   (b) angle MPL,

giving reasons for your answers.

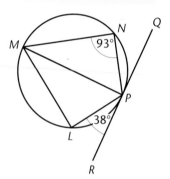

(a)   angle MLP = 180° − 93° = 87°   (opposite angles of cyclic quadrilateral)

(b)   angle LMP = angle LPR = 38°   (alternate segment theorem)
         angle MPL = 180° − 87° − 38°   (angle sum of △MLP)
    so   angle MPL = 55°

Example 16

In the diagram, O is the centre of the circle and points A, B, C and D
lie on the circumference. AC is a diameter and angle CAD = 71°.

Calculate the size of angle ABD, giving reasons for your answer.

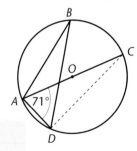

Join D to C

AC is a diameter so angle ADC = 90° (angle in a semi-circle)

                    angle ACD = 180° − 90° − 71° (angle sum of △ACD)
so   angle ACD = 19°

                    angle ABD = angle ACD = 19° (angles in the same segment)

Exercise 23F

1

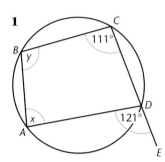

Find angles x and y.

2

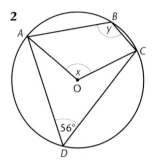

Find angles x and y.

3

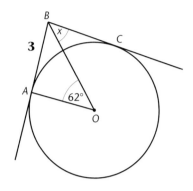

Find angle x.

4

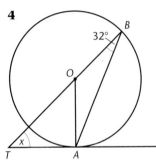

Find angle x.

5

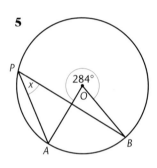

Find angle x.

6

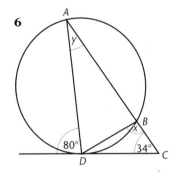

Find angles x and y.

7

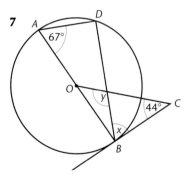

Find angles x and y.

8

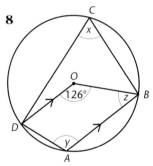

Find angles x, y and z.

9

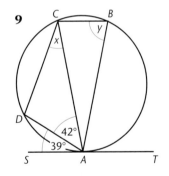

Find angles x and y.

10

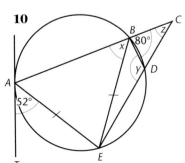

Find angles x, y and z.

Examination questions

1 In the diagram, the sides of triangle *ABC* are tangents to the circle.
D, *E* and *F* are the points of contact.
AE = 5 cm and *EC* = 4 cm.

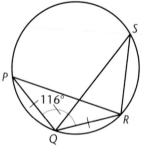

Not to scale

(a) Write down the length of *CD*.

(b) The perimeter of the triangle is 32 cm.
Calculate the length of *DB*.

(3 marks)

AQA, Spec A, 1H, November 2004

2 Points *P*, *Q*, *R* and *S* lie on a circle.
PQ = *QR*
Angle *PQR* = 116°

Not drawn accurately

(a) Explain why angle *QSR* = 32°.

(b) The diagram shows a circle, centre *O*.
TA is a tangent to the circle at *A*.
Angle *BAC* = 58° and angle *BAT* = 74°.

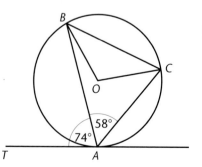

Not drawn accurately

(i) Calculate angle *BOC*.
(ii) Calculate angle *OCA*.

(6 marks)

AQA, Spec A, 1H, June 2004

Summary of key points

Proof of all circle theorems is grade A

Circle properties (grade B)

The perpendicular from the centre of a circle to a chord bisects the chord.

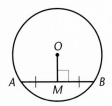

Tangents drawn to a circle from an external point are equal in length.

The angle between a tangent and a radius is 90°.

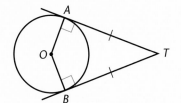

Circle theorems (grade B)

The angle subtended by an arc at the centre of a circle is twice the angle that it subtends at the circumference.

angle *LON* = 2 × angle *LMN*

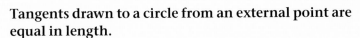

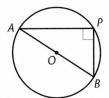

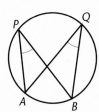

The angle in a semi-circle is a right angle.

angle *APB* = 90°

Angles in the same segment are equal.

angle *APB* = angle *AQB*

Opposite angles of a cyclic quadrilateral are supplementary.

angle *ABC* + angle *ADC* = 180°

An exterior angle of a cyclic quadrilateral is equal to the opposite interior angle.

angle *ABC* = angle *CDE*

Alternate segment theorem (grade A)

The angle between a tangent and a chord is equal to the angle in the alternate segment.

angle *TAB* = angle *APB*

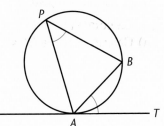

This chapter will show you how to:

✔ factorise quadratic expressions
✔ solve quadratic equations
✔ use the quadratic formula
✔ complete the square
✔ solve problems using quadratic equations

24.1 Factorising quadratic expressions

Key words:
quadratic

An expression of the form $ax^2 + bx + c$ where a, b and c are numbers, and $a \neq 0$, is called a **quadratic** expression in x.

Remember that $\neq$ means 'is not equal to'.

For example:

$$x^2 + 3x - 4, \qquad 3x^2 - 12, \qquad 2x^2 + 10x,$$

are all quadratics in x.

Difference of two squares

Key words:
difference of two squares

Example 1

Expand **(a)** $(x + 9)(x - 9)$ **(b)** $(5n + 4)(5n - 4)$

(a) $(x + 9)(x - 9) = x^2 - 81 + 9x - 9x$

$\qquad\qquad\qquad\quad = x^2 - 81 \qquad\qquad = x^2 - 9^2$

(b) $(5n + 4)(5n - 4) = 25n^2 - 16 + 20n - 20n$

$\qquad\qquad\qquad\quad = 25n^2 - 16 \qquad = (5n)^2 - 4^2$

You could use F first
                       O outside
                       I inside
                       L last

You first met this in Chapter 11.

In general: $(x + a)(x - a) = x^2 - a^2$

This result is known as 'the **difference of two squares**'.

This means that any expression of the form $x^2 - a^2$ can be factorised into two factors that differ only in that the sign in one bracket is + and in the other is −

Multiply out the brackets $(x + a)(x - a)$ to check.

Example 1

Factorise:

(a) $x^2 - 49$ **(b)** $9t^2 - y^2$

(a) $x^2 - 49 = x^2 - 7^2$

$$= (x + 7)(x - 7)$$

(b) $9t^2 - y^2 = (3t)^2 - y^2$

$$= (3t + y)(3t - y)$$

Exercise 24A

Factorise:

(a) $x^2 - 4$ (x-2)(x-2)   **(b)** $x^2 - 25$ (x+5)(x-5)   **(c)** $x^2 - 64$ (x-8)(x+8)

(d) $y^2 - 36$ (y-6)(y+6)   **(e)** $n^2 - 100$ (n+10)(n+10)   **(f)** $t^2 - 1$ (t-1)(t+1)

(g) $a^2 - 121$ (a-11)(a+11)   **(h)** $49 - x^2$ (7-x)(7+x)   **(i)** $81 - x^2$ (9-x)(9+x)

(j) $x^2 - y^2$ (x+y)(x-y)   **(k)** $a^2 - b^2$ (a+b)(a-b)   **(l)** $4x^2 - 81$ (2x-9)(2x+9)

(m) $9x^2 - 16$ (3x+4)(3x-4)   **(n)** $25x^2 - 4$ (5x-2)(5x+2)   **(o)** $9x^2 - y^2$ (3x-y)(3x+y)

(p) $16a^2 - b^2$ (4a-b)(4a+b)   **(q)** $9x^2 - 25y^2$ (3x-5y)(3x+5y)   **(r)** $36n^2 - 4m^2$ (6n-2m)(6n+2m)

(s) $20p^2 - 45q^2$ 5(2p-3q)(2p+3q)   **(t)** $75t^2 - 48w^2$ 3(5t-4w)(5t+4w)

> Questions **(r)**, **(s)** and **(t)** need an extra step ... can you see what it is?

Factorising quadratics of the form $x^2 + bx + c$

To factorise a quadratic you need to write it as the product of two expressions.

> Factorising is the opposite process to expanding two brackets.

For example:

To factorise $x^2 + 6x + 8$ you are trying to find the two linear expressions which multiply together to give $x^2 + 6x + 8$.

$x \times x = x^2$ so the first term in each bracket is x

So $x^2 + 6x + 8 = (x + ?)(x + ?)$

×	x	?
x	x^2	?
?	?	8

Now you need to find the two numbers which multiply together to give 8 and which add together to give 6 (the two numbers with a product of 8 and a sum of 6).

$1 \times 8 = 8$, but $1 + 8 = 9$, so the two numbers are *not* 1 and 8;
$2 \times 4 = 8$, *and* $2 + 4 = 6$, so the two numbers are 2 and 4.

> Check the factor pairs of 8.

Check:

×	x	2
x	x^2	$2x$
4	$4x$	8

$(x + 4)(x + 2) = x^2 + 2x + 4x + 8$

$$= x^2 + 6x + 8$$

So $x^2 + 6x + 8$ factorises to $(x + 4)(x + 2)$

Example 2

Factorise:

(a) $x^2 + 10x + 9$ (b) $a^2 + 4a - 5$ (c) $t^2 - 8t + 12$

(a) $x^2 + 10x + 9$

$x \times x = x^2$ so the first term in each bracket is x

$(x + ?)(x + ?)$

×	x	?
x	x^2	?
?	?	9

$1 \times 9 = 9$ and $1 + 9 = 10$

Check: $(x + 1)(x + 9) = x^2 + 9x + x + 9$

$= x^2 + 10x + 9$ ✓

$x^2 + 10x + 9 = (x + 1)(x + 9)$

(b) $a^2 + 4a - 5 = (a + ?)(a + ?)$

×	a	?
a	a^2	?
?	?	-5

$-1 \times 5 = -5$ and $-1 + 5 = 4$

$a^2 + 4a - 5 = (a - 1)(a + 5)$

(c) $t^2 - 8t + 12$

×	t	?
t	t^2	?
?	?	12

$-2 \times -6 = 12$ and $-2 + -6 = -8$

so $t^2 - 8t + 12 = (t - 2)(t - 6)$

Remember: you can use a grid to help.

Find two numbers that multiply to give 9 and add together to give 10. In this example all the terms in the quadratic are positive so both numbers you are looking for will be positive.

So the two numbers are 1 and 9.

Find two numbers that multiply to give -5 and add together to give 4. In this example the numbers you are looking for must multiply to give -5, so one will be negative and the other positive. They could be -1 and 5 or they could be 1 and -5. $-1 + 5 = 4$, so the two numbers are -1 and 5.

Check by multiplying out the brackets.

$^-$ve $\times$ $^-$ve $=$ $^+$ve so in this example the two numbers are both negative because they must multiply to give $+12$ but add together to give -8.

Check by multiplying out the brackets.

Exercise 24B

Factorise:

1 (a) $a^2 + 10a + 9$ **(b)** $x^2 + 4x + 3$ **(c)** $n^2 + 8n + 7$
  $(a+1)(a+9)$ $(x+1)(x+3)$ $(n+1)(n+7)$

  (d) $x^2 + 8x + 15$ **(e)** $y^2 + 6y + 8$ **(f)** $p^2 + 7p + 12$
  $(x+5)(x+3)$ $(y+2)(y+4)$ $(p+4)(p+3)$

2 (a) $x^2 + 4x - 5$ **(b)** $k^2 + 2k - 3$ **(c)** $r^2 + 7r - 8$
  $(x-1)(x+5)$ $(k+3)(k-1)$ $(r-1)(r+8)$

  (d) $z^2 + 4z - 12$ **(e)** $d^2 + 3d - 10$ **(f)** $x^2 + x - 6$
  $(z+6)(z-2)$ $(d+5)(d-2)$ $(x+3)(x-2)$

$x^2 + mx + c$

$m+n$ mn -1×5

3 (a) $b^2 - 4b - 5$ (b) $x^2 - 6x - 7$ (c) $g^2 - 9g - 10$

   (d) $m^2 - 3m - 10$ (e) $t^2 - 2t - 8$ (f) $f^2 - f - 12$

4 (a) $q^2 - 7q + 12$ (b) $y^2 - 8y + 15$ (c) $x^2 - 11x + 24$

   (d) $a^2 - 12a + 11$ (e) $x^2 - 10x + 16$ (f) $h^2 - 6h + 9$

5 (a) $t^2 + 5t - 14$ (b) $z^2 - 7z - 18$ (c) $p^2 + 9p + 20$

   (d) $d^2 - 5d - 36$ (e) $n^2 + 23n - 24$ (f) $x^2 - x - 20$

Key words:
coefficient

Factorising quadratics of the form $ax^2 + bx + c$

In the expression $ax^2 + bx + c$, a is known as the **coefficient** of x^2, b is the coefficient of x, and c is the constant term.

To factorise an expression like this, first consider the factors of the coefficient of x^2.

For example:
Factorise $2x^2 + 7x + 6$.
The coefficient of x^2 is 2 and the factors of 2 are 2 and 1, so the first terms in the brackets are $2x$ and x, because $2x \times x = 2x^2$.

2 is a prime number so only has factors of 2 and 1.

Remember $x = 1x$.

×	x	?
$2x$	$2x^2$	?
?	?	6

Now you need to find the two numbers with a product of 6 which you can combine with $2x$ and $1x$ to give $7x$.

$1 \times 6 = 6$, but $1 \times 2x + 6 \times 1x = 10x$, and $6 \times 2x + 1x \times 1 = 13x$
so the numbers are *not* 1 and 6.

$2 \times 3 = 6$, and $2 \times 2x + 3 \times x = 7x$ which is correct.

List the factors of 6 first and try them in different ways to see which combination gives $+7x$.

Check:

×	x	2
$2x$	$2x^2$	$4x$
3	$3x$	6

$(2x + 3)(x + 2) = 2x^2 + 4x + 3x + 6$
$= 2x^2 + 7x + 6$

So $2x^2 + 7x + 6$ factorises to $(2x + 3)(x + 2)$.

The questions in the next example show you what to do when the number of possible combinations increases. The method is the same as in the example above but there is more work to do to arrive at the correct combination.

Example 3

Factorise

(a) $3x^2 - 2x - 8$ **(b)** $6x^2 + 11x + 4$ **(c)** $10x^2 + 34x - 24$

(a) $3x^2 - 2x - 8$

The coefficient of x^2 is 3 (which is prime).

×	x	?
$3x$	$3x^2$	?
?	?	-8

$3x^2 - 2x - 8 = (3x + ?)(x + ?)$

$1 \times 8 = 8$ or $2 \times 4 = 8$

$3 \times 1 = 3$

The number term is negative so one of the numbers you are looking for will be positive and the other will be negative.

$(3x \times 1) + (1x \times -8) = -5x$ $(3x \times -8) + (1x \times 1) = -23x$

$(3x \times -1) + (1x \times 8) = +5x$ $(3x \times 8) + (1x \times -1) = +23x$

$(3x \times 2) + (1x \times -4) = +2x$ $(3x \times -4) + (1x \times 2) = -10x$

$(3x \times -2) + (1x \times 4) = -2x$ $(3x \times 4) + (1x \times -2) = +10x$

Try different combinations until you find which will give $-2x$.

You can see that there are quite a lot of combinations but notice that swapping the signs round in any one only changes the sign of the answer, not the amount of x's.

×	x	-2
$3x$	$3x^2$	$-6x$
$+4$	$+4x$	-8

So $3x^2 - 2x - 8 = (3x + 4)(x - 2)$

Check by expanding the brackets.

(b) $6x^2 + 11x + 4$

The coefficient of x^2 is 6. $6 \times 1 = 6$ and $3 \times 2 = 6$

In this example all the terms in the quadratic are positive so both numbers you are looking for will be positive.

×	x	?
$6x$	$6x^2$	?
?	?	4

or

×	$2x$	?
$3x$	$6x^2$	?
?	?	4

At this stage you cannot tell which of these is correct.

$(6x + ?)(x + ?)$ **or** $(3x + ?)(2x + ?)$

$1 \times 4 = 4$ and $2 \times 2 = 4$

$3x \times 1 + 2x \times 4 = 11x$ ✓

You need to find which of 1 and 4 or 2 and 2 combines with either $6x$ and $1x$, or $3x$ and $2x$, to give $11x$.
You will get better at this the more you practise!

×	$2x$	1
$3x$	$6x^2$	$3x$
4	$8x$	4

So $6x^2 + 11x + 4 = (3x + 4)(2x + 1)$

You should always check by multiplying out the brackets.

continued ▶

(c) $10x^2 + 34x - 24 = 2(5x^2 + 17x - 12)$

The coefficient of x^2 is 5, which is prime,

so $5x^2 + 17x - 12 = (5x + ?)(x + ?)$

The number term is 12 so possible products are

$1 \times 12 = 12 \quad 2 \times 6 = 12 \quad 3 \times 4 = 12$

Some possibilities are:

$(5x \times -1) + (1x \times 12) = 7x \qquad$ not correct

$(5x \times 2) + (1x \times -6) = 4x \qquad$ not correct

$(5x \times -3) + (1x \times 4) = -11x \quad$ not correct

$\dots$ and so on $\dots$

$(5x \times 4) + (1x \times -3) = 17x \quad$ ✓

×	x	4
$5x$	$5x^2$	$20x$
-3	$-3x$	-12

So $\quad 5x^2 + 17x - 12 = (5x - 3)(x + 4)$

So $\quad 10x^2 + 34x - 24 = 2(5x^2 + 17x - 12) = 2(5x - 3)(x + 4)$

> Notice that the **first** step is to remove a common factor. Always look for this, it makes factorising easier. Now factorise $5x^2 + 17x - 12$

> You need to find which of 1 and 12, 2 and 6, or 3 and 4, combine with 5 and 1 to give +17, the x coefficient.

> Remember that the number term is negative so one of the numbers you are looking for will be positive and the other will be negative.

> Remember to check by expanding the brackets.

> Don't forget the factor of 2 at the beginning of the final answer.

Look at Example 3(a) again:

$$3x^2 - 2x - 8 = (3x + 4)(x - 2)$$

The   First terms in each bracket multiply to give $3x^2$,
the   Outside pair of terms   multiply to give $-6x$,
the   Inside pair of terms   multiply to give $+4x$,
the   Last terms in each bracket multiply to give -8

You can use FOIL to check that your factorising is correct. It acts in the same way as putting the terms into a grid because it generates all four terms of the expansion.

Exercise 24C

Factorise:

(a) $2x^2 + 5x + 3$

(b) $3x^2 + 16x + 5$

(c) $2x^2 + 3x + 1$

(d) $3t^2 + 5t + 2$

(e) $4y^2 + 8y + 3$

(f) $4x^2 + 23x + 15$

(g) $4x^2 + 17x + 15$

(h) $6a^2 + 25a + 4$

(i) $2x^2 + 5x - 3$

(j) $3x^2 + x - 2$

(k) $2x^2 - 9x - 5$

(l) $4x^2 - 4x - 3$

(m) $2x^2 - 3x + 1$

(n) $3x^2 - 11x + 10$

(o) $8x^2 - 18x + 9$

(p) $4x^2 - 26x - 14$

(q) $6x^2 + 21x - 90$

(r) $6x^2 + 32x - 24$

(s) $12x^2 - 34x + 24$

(t) $24x^2 - 30x - 21$

24.2 Solving quadratic equations

Key words:
quadratic equation
root

When a quadratic expression forms part of an equation, the equation is called a **quadratic equation**.

Here are three examples of quadratic equations:

$$x^2 + 3x - 4 = 0 \qquad 3x^2 = 12 \qquad x^2 + 5x = 0$$

You can use different methods to solve, or find the **roots** of, quadratic equations.

Solving quadratic equations by rearranging

A quadratic equation may have 0, 1 or 2 solutions.

Some quadratic equations can be solved by rearranging the terms.

Example 4

Solve these equations:

(a) $3x^2 = 12$

(b) $2t^2 - 72 = 0$

(c) $2(x - 3)^2 - 50 = 0$

(a) $3x^2 = 12$

$x^2 = 4$

$x = \pm\sqrt{4}$

$x = \pm 2$

You must remember the negative square root as well as the positive one. Read ± 2 as 'positive or negative 2', or 'plus or minus 2'.

(b) $2t^2 - 72 = 0$

$2t^2 = 72$

$t^2 = 36$

$t = \pm 6$

Or you could divide throughout by the common factor of 2 as a first step.

(c) $2(x - 3)^2 - 50 = 0$

$2(x - 3)^2 = 50$

$(x - 3)^2 = 25$

$x - 3 = \pm 5$

$x = 3 \pm 5$

$x = 3 + 5 \quad$ or $\quad x = 3 - 5$

$= 8 \qquad\qquad = -2$

so the two solutions are $x = 8$ and $x = -2$

Solutions are sometimes referred to as roots ... you can use either word.

Exercise 24D

1 Solve these equations:

(a) $t^2 = 9$ (b) $3x^2 = 75$

(c) $5y^2 = 80$ (d) $z^2 - 49 = 0$

(e) $2a^2 - 32 = 0$ (f) $4x^2 - 16 = 0$

(g) $(x - 2)^2 = 25$ (h) $(y - 1)^2 = 36$

(i) $2(x + 2)^2 = 18$ (j) $3(x - 5)^2 - 48 = 0$

(k) $2(t + 3)^2 - 50 = 0$

Handwritten annotations:
$3(x-5)^2 = 48$
$(x-5)^2 = 16$
$x - 5 = 4$
$x = 9$

2 Solve these quadratic equations, giving your answers to 2 d.p.

(a) $2(x - 1)^2 = 6$ (b) $3(x - 4)^2 = 15$

(c) $5(x + 1)^2 - 10 = 0$ (d) $4(x + 3)^2 - 32 = 0$

3 What problem arises if you try to solve $3(x - 2)^2 + 27 = 0$?

> This is an example of a quadratic equation that has no real solutions. You will meet these again later.

Solving quadratic equations by factorising

Some quadratic equations can be solved by factorising. First rearrange the equation so that all the terms are on one side and are equated to zero.

Then, factorise the quadratic expression using the methods in Section 24.1

Before looking at an example, consider this:

If $A \times B = 0$ then either $A = 0$ or $B = 0$

This is a key statement in the solution of quadratic equations. Examples 5, 6 and 7 all use this fact and show you how to set out your work when solving quadratic equations.

> The product of two terms can only be zero if either one or the other (or both) of the terms is zero.

Example 5

Solve these equations:

(a) $x^2 + 5x = 0$ (b) $x^2 = 7x$

> (a) $x^2 + 5x = 0$
> $x(x + 5) = 0$
> $x = 0$ or $x + 5 = 0$
> $x = -5$
>
> The equation has two solutions:
> $x = 0$ and $x = -5$ *continued* ▶

> Don't divide both terms by x because x could be 0 and you would lose the $x = 0$ solution.

> If the product of two numbers is 0 then one of the numbers must be 0. So if $x \times (x + 5) = 0$ then either $x = 0$ or $x + 5 = 0$.

(b) $\quad x^2 = 7x$

$x^2 - 7x = 0$ ——————————————— First rearrange the equation so that all the terms are on one side.

$x(x - 7) = 0$ ———

Now factorise.

$\quad x = 0 \quad$ or $\quad x - 7 = 0$

$\qquad\qquad\qquad x = 7$

The two solutions are $x = 0$ and $x = 7$

Example 6

Solve these equations:

(a) $x^2 + 10x + 9 = 0$ **(b)** $x^2 - 6x + 9 = 0$ **(c)** $x^2 = 4x + 21$

(a) $\quad x^2 + 10x + 9 = 0$ ——————————————— All the terms are on one side of the equation already, so factorise this side.

$(x + 1)(x + 9) = 0$

$\quad x + 1 = 0 \quad$ or $\quad x + 9 = 0$

$\qquad x = -1 \qquad\qquad x = -9$

Remember: if the product of two numbers is 0 then one of the numbers must be 0.

The two solutions are $x = -1$ and $x = -9$

(b) $\quad x^2 - 6x + 9 = 0$

$(x - 3)(x - 3) = 0$ ——————————————— This equation has only one solution, $x = 3$.

$\quad x - 3 = 0 \quad$ or $\quad x - 3 = 0$

$\qquad x = 3 \qquad\qquad x = 3$

This is sometimes called a repeated solution or a repeated root.

(c) $\qquad\qquad x^2 = 4x + 21$

$x^2 - 4x - 21 = 0$ ——————————————— First rearrange the equation so that all the terms are on one side. Then factorise.

$(x + 3)(x - 7) = 0$

$\quad x + 3 = 0 \quad$ or $\quad x - 7 = 0$

$\qquad x = -3 \qquad\qquad x = 7$

The two solutions are $x = -3$ and $x = 7$

Remember: always set out your work as in these examples.

The steps are: **1** Rearrange (equation = 0).

2 Factorise the quadratic expression.

3 Use 'either/or' to find the two solutions.

Exercise 24E

Solve these quadratic equations:

1 (a) $t^2 + 5t = 0$ **(b)** $x^2 + 6x = 0$ **(c)** $n^2 - 9n = 0$

(d) $x^2 - 8x = 0$ **(e)** $z^2 = 2z$ **(f)** $a^2 - 10a = 0$

(g) $b^2 = 7b$ **(h)** $y^2 + y = 0$ **(i)** $x^2 = x$

2 (a) $b^2 + 10b + 9 = 0$ **(b)** $x^2 + 5x + 4 = 0$

(c) $m^2 + m - 2 = 0$ **(d)** $t^2 - 8t + 12 = 0$

(e) $x^2 + 2x - 15 = 0$ **(f)** $y^2 - 12y + 36 = 0$

(g) $z^2 - 2z - 24 = 0$ **(h)** $a^2 - 5a - 14 = 0$

(i) $x^2 + 11x + 30 = 0$ **(j)** $x^2 + 14x + 49 = 0$

3 (a) $x^2 = 5x + 6$ **(b)** $x^2 - 12 = x$ **(c)** $x^2 + 4 = 4x$

(d) $x^2 + 3x = 10$ **(e)** $x^2 = 5x - 6$ **(f)** $18x = x^2 + 81$

(g) $30 - x = x^2$ **(h)** $x^2 = 24 - 5x$ **(i)** $15x - 54 = x^2$

> Remember to rearrange the equations so that all the terms are on one side.

All of the questions in Exercise 24E have a squared term whose coefficient is 1. This means that factorising the quadratic expression is usually straightforward.

The next examples (and almost all of the questions in Exercise 24F) have a squared term whose coefficient is not 1. This means that the factorising is more difficult.

> Look back at Example 3 if you need help with factorising expressions of the form $ax^2 + bx + c$.

Example 7

Solve the equations:

(a) $3x^2 - x = 0$ **(b)** $2x^2 + x - 6 = 0$

(a) $3x^2 - x = 0$

$x(3x - 1) = 0$

$x = 0$ or $3x - 1 = 0$

$3x = 1$

$x = \frac{1}{3}$

The two solutions are $x = 0$ and $x = \frac{1}{3}$

> These two terms have a common factor of x.

(b) $2x^2 + x - 6 = 0$

$(2x - 3)(x + 2) = 0$

$2x - 3 = 0$ or $x + 2 = 0$

$2x = 3$ $x = -2$

$x = \frac{3}{2}$

The two solutions are $x = \frac{3}{2}$ $(1\frac{1}{2})$ and $x = -2$

Exercise 24F

Solve these quadratic equations:

1 (a) $2t^2 - t = 0$ **(b)** $3x^2 + 2x = 0$ **(c)** $2y^2 - 5y = 0$

(d) $3n^2 = n$ **(e)** $3z^2 = 5z$ **(f)** $5a^2 + 3a = 0$

2 (a) $2x^2 + 7x + 3 = 0$ **(b)** $2x^2 + 3x - 2 = 0$

(c) $3x^2 - 8x - 3 = 0$ **(d)** $3a^2 + 10a - 8 = 0$

(e) $2y^2 - y - 10 = 0$ **(f)** $4k^2 + 8k + 3 = 0$

(g) $4z^2 + 7z - 2 = 0$ **(h)** $3d^2 - 14d + 8 = 0$

(i) $6r^2 + 7r - 3 = 0$ **(j)** $4p^2 + 4p + 1 = 0$

3 (a) $3x^2 + 5x = 2$ **(b)** $3x^2 = 10x - 3$

(c) $2x^2 = 3 - 5x$ **(d)** $(x + 1)(x - 2) = 10$

(e) $(2x - 3)^2 = 8 - 8x$ **(f)** $2x = 35 - x^2$

(g) $2 - x - 6x^2 = 0$ **(h)** $12x^2 = 16x + 3$

> Remember to rearrange the equations so that all the terms are on one side.

Using the quadratic formula

You can solve a quadratic equation by factorisation if it has simple factors. For many quadratics this is not the case.

For example:
$x^2 + 2x - 5$ does not have simple factors (try factorising it!).
So you cannot solve the equation $x^2 + 2x - 5 = 0$ by factorisation.

One way to solve a quadratic equation that does have solutions is to use the **quadratic formula** :

> **Key words:**
> quadratic formula
> discriminant

> Use a graph plotter to draw $y = x^2 + 2x - 5$ and you can see there are 2 values of x which satisfy $x^2 + 2x - 5 = 0$. So the equation does have solutions.

> The solutions of the quadratic equation $ax^2 + bx + c = 0$, where $a \neq 0$, are given by the formula:
>
> $$x = \frac{-b \pm \sqrt{b^2 - 4ac}}{2a}$$

> This formula can always be used to solve a quadratic equation, even when the quadratic factorises. You do not need to remember the formula (it will be given on your exam paper), but you must know how to use it.

Example 8

Solve the equation $x^2 + 2x - 5 = 0$
Give your solutions to 2 d.p.

$x^2 + 2x - 5 = 0$

$a = 1, \quad b = 2, \quad c = -5$

$x = \dfrac{-b \pm \sqrt{b^2 - 4ac}}{2a}$

$= \dfrac{-2 \pm \sqrt{2^2 - 4 \times 1 \times (-5)}}{2 \times 1}$

$= \dfrac{-2 \pm \sqrt{4 - -20}}{2}$

> It is a good idea to start by writing down the values of a, b, and c.

> Substitute the values of a, b, and c into the quadratic formula.

continued ▶

$$= \frac{-2 \pm \sqrt{4 + 20}}{2}$$

$$= \frac{-2 \pm \sqrt{24}}{2}$$

$$x = \frac{-2 + \sqrt{24}}{2} \quad \text{or} \quad x = \frac{-2 - \sqrt{24}}{2}$$

$$x = 1.45 \quad \text{or} \quad x = -3.45 \quad (2 \text{ d.p.})$$

> You can simplify $\sqrt{24}$ to $2\sqrt{6}$ and leave your answer in surd form. This gives the exact values of the answers without any rounding. Simplifying surds is covered in Chapter 26.

Exercise 24G

1 Use the quadratic formula to solve the following equations. Give your answers to 2 d.p.

(a) $x^2 + 3x + 1 = 0$

(b) $x^2 + 5x + 2 = 0$

(c) $x^2 + 4x - 3 = 0$

(d) $d^2 - 2d - 5 = 0$

(e) $y^2 - y - 8 = 0$

(f) $k^2 - 8k + 6 = 0$

(g) $2x^2 + 7x - 3 = 0$

(h) $2x^2 - 9x + 8 = 0$

(i) $3x^2 + 5x - 1 = 0$

(j) $4p^2 + 7p - 5 = 0$

(k) $5y^2 - 2y - 5 = 0$

(l) $4t^2 + 9t + 1 = 0$

(m) $3z^2 + 5z - 1 = 0$

(n) $6p^2 - 7p - 2 = 0$

> If an exam question asks for the solutions to a quadratic equation to a certain number of decimal places or significant figures it is no use trying to factorise … it will be impossible to do so.

2 (a) Solve the equation $6p^2 + 7p - 3 = 0$, using the quadratic formula.

(b) How can you tell from your working and answers that you did not need to use the quadratic formula?

3 (a) Try to solve the equation $3x^2 + 5x + 4 = 0$, using the quadratic formula.

(b) What prevents you from being able to find the solutions?

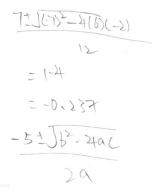

The discriminant

The part of the quadratic formula underneath the square root sign is called the **discriminant** because it discriminates between the following three possible situations.

- If $b^2 - 4ac > 0$ there are two distinct solutions of the quadratic equation. (If $b^2 - 4ac$ is a perfect square, then the quadratic will factorise.)

- If $b^2 - 4ac = 0$ there is one solution (it is a repeated root as in Example 6(b)).

- If $b^2 - 4ac < 0$ there are no real solutions because you cannot find the square root of a negative number.

Graphically, these three cases can be shown as:

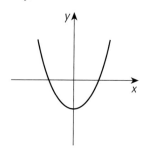

$b^2 - 4ac > 0$

Two solutions, so two points where the quadratic graph crosses the x-axis.

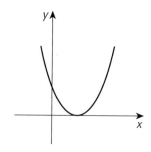

$b^2 - 4ac = 0$

One solution, so the quadratic graph touches the x-axis.

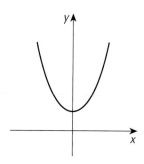

$b^2 - 4ac < 0$

No solutions, so the quadratic graph never crosses the x-axis.

$6x^2 - 10x + 5 = 0$

Chapter 28 has more on the intersection of quadratic and linear graphs. It builds on these ideas.

Exercise 24H

$b^2 - 4ac$

State whether these quadratic equations will have two, one, or zero solutions.

Consider the discriminant.

(a) $x^2 - 5x + 6 = 0$

(b) $x^2 + 3x - 7 = 0$

(c) $x^2 + 2x + 4 = 0$

(d) $x^2 - 6x + 9 = 0$

(e) $2x^2 + x - 3 = 0$

(f) $3x^2 - 2x + 1 = 0$

(g) $4x^2 - 3x - 4 = 0$

(h) $6x^2 + 5 = 10x$

(i) $2x^2 = 7x - 5$

(j) $4x = 12 - \dfrac{9}{x}$

You need to rewrite questions **(h)**, **(i)** and **(j)**.

$2x^2 - 7x + 5$

Completing the square

$4x^2 - 12x + 9$

Key words:
completing the square

There is another way to solve quadratic equations that cannot be factorised easily. You rewrite the equation in a way that is known as **completing the square** .

Expanding expressions of the form $(x + a)^2$ produces a term in x^2, a term in x and a square number. For example:

$$(x + 7)^2 = x^2 + 14x + 49$$

$x^2 = 20 \times x^2 + 2 \times 6 + 20$

Subtracting 49 from both sides gives

$$(x + 7)^2 - 49 = x^2 + 14x$$

You can see this using a diagram:

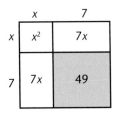

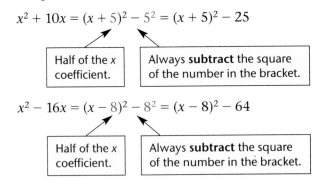

$$(x + 7)^2 = x^2 + 14x + 49 \qquad x^2 + 14x = (x + 7)^2 - 49$$

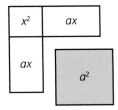

In general:
$$(x + a)^2 = x^2 + 2ax + a^2$$
$$(x + a)^2 - a^2 = x^2 + 2ax$$

To write a quadratic expression in completed square form you use this result the other way round:

$$x^2 + 2ax = (x + a)^2 - a^2$$

The number inside the brackets is half of the coefficient of x.
This is always the first step in completing the square.

For example:

$$x^2 + 10x = (x + 5)^2 - 5^2 = (x + 5)^2 - 25$$

| Half of the x coefficient. | Always **subtract** the square of the number in the bracket. |

$$x^2 - 16x = (x - 8)^2 - 8^2 = (x - 8)^2 - 64$$

| Half of the x coefficient. | Always **subtract** the square of the number in the bracket. |

Always make sure that the sign in the squared bracket is the same as the sign of the x coefficient.

When completing the square,

$x^2 + bx + c$ can be rewritten in the form $(x + p)^2 + q$

The values of b, c, p and q can be positive or negative.

Examples 9 and 10 show you how this result works in practice.

Example 9

Write the expression $x^2 + 8x + 10$ in completed square form.

$$x^2 + 8x + 10 = (x + 4)^2 - 4^2 + 10$$
$$= (x + 4)^2 - 16 + 10$$
$$= (x + 4)^2 - 6$$

Half the coefficient of x

Subtract 4^2

Put in the original number term.

Example 10

Find the values of p and q such that $x^2 - 6x + 2 = (x - p)^2 + q$

$$x^2 - 6x + 2 = (x - 3)^2 - 3^2 + 2$$
$$= (x - 3)^2 - 9 + 2$$
$$= (x - 3)^2 - 7$$

So $p = 3, q = -7$

Half the coefficient of x

Subtract 3^2

Put in the original number term.

Example 11

Solve the equation $x^2 + 10x + 11 = 0$ by completing the square.
Give your answers correct to 2 decimal places.

(a) $x^2 + 10x + 11 = 0$

$(x + 5)^2 - 25 + 11 = 0$

$(x + 5)^2 - 14 = 0$

$(x + 5)^2 = 14$

$x + 5 = \pm\sqrt{14}$

$x = -5 \pm \sqrt{14}$

so $x = -5 + \sqrt{14}$ or $x = -5 - \sqrt{14}$

$x = -1.26$ $x = -8.74$ (to 2 d.p.)

Start by completing the square on the left hand side.

Now solve the equation by rearranging.

Don't forget both the positive and negative square root.

Example 12

Solve the equation $x^2 - 6x + 7 = 0$ by completing the square.
Leave your answers in the form $a \pm \sqrt{b}$.

(a) $x^2 - 6x + 7 = 0$

$(x - 3)^2 - 9 + 7 = 0$

$(x - 3)^2 - 2 = 0$

$(x - 3)^2 = 2$

$x - 3 = \pm\sqrt{2}$

$x = 3 \pm \sqrt{2}$

Notice that these answers contain $\sqrt{2}$ and have not been written as decimals. $3 \pm \sqrt{2}$ is an example of a solution left in surd form; it gives the exact values of the answer without any rounding. Surds are covered in Chapter 26.

Exercise 24I

1 Write these expressions in completed square form.

(a) $x^2 + 10x + 5$ (b) $x^2 + 4x - 2$

(c) $x^2 - 12x + 20$ (d) $y^2 + 2y + 9$

(e) $x^2 - 3x - 1$ (f) $y^2 + 5y + 7$

2 Write each of the following expressions in the form $(x - p)^2 + q$.
State clearly the values of p and q.

(a) $x^2 - 6x + 11$ (b) $x^2 - 4x - 1$

(c) $x^2 + 14x + 8$ (d) $x^2 - 2x + 5$

(e) $x^2 - 7x + 4$ (f) $x^2 + x - 6$

3 Solve these equations by completing the square.
Give your answers correct to 2 decimal places.

(a) $x^2 + 8x + 5 = 0$ (b) $x^2 - 10x + 3 = 0$

(c) $y^2 + 12y - 3 = 0$ (d) $x^2 - 4x - 2 = 0$

(e) $x^2 - 16x - 4 = 0$ (f) $z^2 + 6z + 7 = 0$

(g) $y^2 - 5y + 2 = 0$ (h) $x^2 + 3x - 5 = 0$

4 Solve these equations by completing the square.
Leave your answers in the form $a \pm \sqrt{b}$.

(a) $x^2 + 4x + 1 = 0$ (b) $x^2 - 12x + 16 = 0$

(c) $x^2 - 8x - 6 = 0$ (d) $x^2 + 6x - 1 = 0$

(e) $x^2 - 2x - 4 = 0$ (f) $x^2 + 8x + 2 = 0$

5 Solve these equations by completing the square.
Give your answers correct to 2 decimal places.

(a) $x^2 + 4 = 6x$ (b) $x^2 = 10x - 5$

(c) $x^2 + 10x = 2x - 6$ (d) $8x = 2 - x^2$

(e) $2x^2 - 3 = x^2 + 6x$ (f) $2x^2 + 4x = x^2 + x + 1$

> Remember to rearrange the equations so that all the terms are on one side.

6 Try to solve $x^2 + 6x + 10 = 0$ by completing the square.
Explain why you cannot complete the solution.

7 How can you solve $2x^2 + 8x - 5 = 0$ by completing the square?
Complete the solution.

> You can complete the square when you have an expression in the form $x^2 + bx + c = 0$.

8 Start with the equation $x^2 + bx + c = 0$ and by completing the square, find a formula for the solution.

24.3 Applications of quadratic equations

The theory of quadratic equations can be applied to problems.

These questions are typical of the UAM questions you can expect on your exam paper.

Remember that UAM stands for Using and Applying Mathematics.

Example 13

I think of a number, square it then subtract four times the number. The answer is 77. Find the two possible values of the original number, showing all your working.

Let the original number be x.

Squaring gives x^2,

then subtract four times the number to give a final expression of $x^2 - 4x$

So $\qquad x^2 - 4x = 77$

$\qquad x^2 - 4x - 77 = 0$

$\qquad (x + 7)(x - 11) = 0$

Either $\qquad x + 7 = 0$, giving $x = -7$

or $\qquad x - 11 = 0$, giving $x = 11$

So the two possible values of the original number are -7 and 11.

Set up the quadratic equation, reading the question carefully.

Solve the quadratic equation by rearranging and factorising.

Remember to state the answer to the question you were asked.

Example 14

The diagram shows a right-angled triangle with sides of lengths $(2x - 1)$ cm, $(x + 4)$ cm and 15 cm.

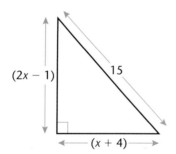

Calculate the lengths of the sides of the triangle.
Give your answers correct to 2 decimal places.

Using Pythagoras' theorem:

$$(x + 4)^2 + (2x - 1)^2 = 15^2$$

$$x^2 + 8x + 16 + 4x^2 - 4x + 1 = 225$$

$$5x^2 + 4x - 208 = 0$$

$$x = \frac{-b \pm \sqrt{(b^2 - 4ac)}}{2a}$$

$a = 5$, $b = 4$ and $c = -208$

$$x = \frac{-4 \pm \sqrt{(4^2 - 4 \times 5 \times -208)}}{2 \times 5}$$

$$x = \frac{-4 \pm \sqrt{(16 - -4160)}}{10}$$

$$x = \frac{-4 \pm \sqrt{(16 + 4160)}}{10}$$

$$x = \frac{-4 \pm \sqrt{4176}}{10}$$

So $x = \dfrac{-4 + \sqrt{4176}}{10}$ and $x = \dfrac{-4 - \sqrt{4176}}{10}$

giving $x = 6.06$ (2 d.p.) and $x = -6.86$ (2 d.p.)

x cannot be -6.86 as this would give negative lengths.

So $x = 6.06$ and the lengths of the sides of the triangle are

10.06 cm, 11.12 cm and 15 cm.

> Using Pythagoras to set up the equation is straightforward, as long as you remember to expand the brackets correctly.

> Simplify.

> Since the question asked for answers correct to 2 d.p. you know the formula will be needed. Take care when substituting the values of a, b and c.

> Explain why the negative solution is discounted.

> Always answer the question you were asked with a final statement.

Example 15

Find the points of intersection of the quadratic graph $y = 3x^2 - x - 4$ and the straight line $y = 6 - 2x$.

Points of intersection are given by the equation

$$3x^2 - x - 4 = 6 - 2x$$

$$3x^2 + x - 10 = 0 \qquad \text{Simplify.}$$

$$(3x - 5)(x + 2) = 0$$

Either $\quad 3x - 5 = 0 \rightarrow 3x = 5 \rightarrow x = \frac{5}{3} = 1\frac{2}{3}$

or $\qquad x + 2 = 0 \rightarrow x = -2$

When $\qquad x = 1\frac{2}{3}$, $y = 6 - 2(1\frac{2}{3}) = 2\frac{2}{3}$

and when $\quad x = -2$, $y = 6 - 2(-2) = 10$

So the points of intersection are $(1\frac{2}{3}, 2\frac{2}{3})$ and $(-2, 10)$.

> You will see more questions like this in Chapter 28.

> The graphs will intersect when the expression $3x^2 - x - 4$ and the expression $6 - 2x$ both give the same value of y, so equate the two expressions.

> You only need to substitute the x values into one equation (the linear one is the easiest) to find the y-values.

> If you are asked for the points of intersection, you **must** work out the y-values as well as the x-values. Sometimes you are only asked for the x-values.

Exercise 24J

UAM **1** The sum of the square of a number and six times the number is 91. Find the two possible values of the number.

UAM **2** The length of a rectangular lawn is 5 m longer than the width. If the area of the lawn is 176 m² find the dimensions of the lawn.

Hint: let the width be x.

UAM **3** Two positive numbers have a difference of 3 and their squares have a sum of 317. Let one of the numbers be x.
Write down an equation in x based on this information and solve it to find the two numbers.

UAM **4** The sum of a positive number and three times its reciprocal is 5. Find two possible values of the number to 3 s.f.

Reminder:
the reciprocal of n is $\dfrac{1}{n}$

UAM **5** Find the lengths of the sides of the right-angled triangle shown in the diagram. Give your answer to 2 d.p.

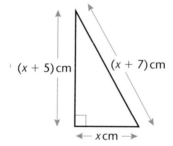

$(x + 5)$ cm $(x + 7)$ cm

x cm

UAM **6** The area of this triangle is 13 cm².

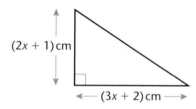

$(2x + 1)$ cm

$(3x + 2)$ cm

Calculate the value of x.

UAM **7** The rectangle in this diagram has a length of 22 cm and a width of 13 cm.
A triangle of base $(2x - 4)$ cm and height x cm is cut off and the area of the remaining shape is 248 cm².

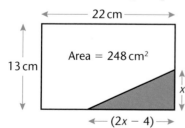

22 cm

Area = 248 cm²

13 cm

$(2x - 4)$

x

Calculate the value of x giving your answer to 3 s.f.

8 Find the points of intersection of the curve $y = 2x^2 + 3x - 5$ and the straight line $y = 2x + 10$.

9 Find the x-coordinates of the points of intersection of the curve $y = x^2 - 7x + 12$ and the straight line $y = 9 - 2x$.
Give your answers correct to 2 d.p.

UAM **10** Find the x-values of the points of intersection of the circle $x^2 + y^2 = 16$ and the straight line $y = 3 - x$.
Give your answers correct to 2 d.p.

Examination questions

1 **(a)** **(i)** Factorise $x^2 - 7x - 8$
    (ii) Hence solve the equation $x^2 - 7x - 8 = 0$
(3 marks)
AQA, Spec A, 1I, November 2003

2 Factorise $x^2 + 6x - 16$
(2 marks)
AQA, Spec A, 1I, June 2003

3 Factorise $y^2 - 9y + 14$
(2 marks)
AQA, Spec B, 2I, June 2003

4 Solve the equation
$x^2 - 10x - 5 = 0$
Give your answers to 2 decimal places.
(3 marks)
AQA, Spec A, 2H, November 2003

5 Find the quadratic equation whose solutions are the x-coordinates of the points of intersection of $y = x^2 - x - 6$ and $y = x + 2$.
(2 marks)
AQA, Spec A, 2H, November 2003

6 $x^2 - 6x + 13 = (x - a)^2 + b$
(a) Find the values of a and b.
(b) Hence find the minimum value of $x^2 - 6x + 13$.
(4 marks)
AQA, Spec B, 2H, June 2003

Summary of key points

Difference of two squares (grade B/A)

$$(x + a)(x - a) = x^2 - a^2$$

The difference of two squares means that any expression of the form $x^2 - a^2$ can be factorised into two factors that differ only in that the sign in one bracket is + and in the other is −
For example $9t^2 - y^2 = (3t)^2 - y^2 = (3t + y)(3t - y)$

Factorising quadratic expressions (grade B/A)

A quadratic expression of the form $x^2 + bx + c$ will often factorise into two brackets.
For example $x^2 + 10x + 9 = (x + 1)(x + 9)$

When factorising quadratics of the form $ax^2 + bx + c$ you first have to consider the factors of the coefficient of x^2.
Examples are $6x^2 + 11x + 4 = (3x + 4)(2x + 1)$ and
$3x^2 - 2x - 8 = (3x + 4)(x - 2)$

Another method of expanding a bracket is to use FOIL
For example, $3x^2 - 2x - 8 = (3x + 4)(x - 2)$
The   First terms in each bracket multiply to give $3x^2$,
the   Outside pair of terms   multiply to give $-6x$,
the   Inside pair of terms   multiply to give $+4x$,
the   Last terms in each bracket multiply to give -8
You can use FOIL to check that your factorising is correct. It acts in the same way as putting the terms into a grid because it generates all four terms of the expansion.

Solving quadratic equations (grade B/A)

Some quadratic equations can be solved by rearranging the terms.
For example $2t^2 - 72 = 0 \rightarrow 2t^2 = 72 \rightarrow t^2 = 36 \rightarrow t = \pm 6$

Some quadratic equations can be solved by factorising.
Rearrange the equation so that all the terms are on one side and are equated to zero.
Then, factorise the quadratic expression.
For example $x^2 = 4x + 21 \rightarrow x^2 - 4x - 21 = 0 \rightarrow (x + 3)(x - 7) = 0$
             Either $x + 3 = 0 \rightarrow x = -3$ or $x - 7 = 0 \rightarrow x = 7$
             So the solutions are $x = -3$ and $x = 7$

The quadratic formula (grade A)

The solutions of the quadratic equation $ax^2 + bx + c = 0$, where $a \neq 0$, are given by the formula: $x = \dfrac{-b \pm \sqrt{(b^2 - 4ac)}}{2a}$

The discriminant (grade A*)

The discriminant is the expression $b^2 - 4ac$ under the square root sign in the quadratic formula.

- If $b^2 - 4ac > 0$ there are two distinct solutions of the quadratic equation (if '$b^2 - 4ac$' is a perfect square then the quadratic will factorise).
- If $b^2 - 4ac = 0$ there is one solution (it is a repeated root).
- If $b^2 - 4ac < 0$ there are no real solutions because you cannot find the square root of a negative number.

Completing the square (grade A*)

Completing the square means rewriting an expression of the form $x^2 + bx + c$ as $(x + p)^2 + q$.
The value of p is always half of the value of b, the coefficient of x.
The value of q is found as in this example:

Put in the original number term.

$$x^2 - 6x + 2 = (x - 3)^2 - 3^2 + 2 = (x - 3)^2 - 9 + 2 = (x - 3)^2 - 7$$

p = half the coefficient of x

Subtract p^2 (3^2)

Solving problems using quadratic equations (grade A*)

The theory of quadratic equations can be applied to problems. Set up the quadratic equation, reading the question carefully, then solve the quadratic equation.
Remember to state the answer to the question you were asked, explaining, if necessary, why one solution (often a negative answer) must be discounted.

This chapter will show you how to:

- ✔ recognise linear, quadratic and cubic graphs by inspection
- ✔ recognise reciprocal and exponential graphs and use their associated properties
- ✔ solve simultaneous equations graphically
- ✔ solve quadratic equations graphically
- ✔ apply graphs to solving realistic problems

25.1 Non-linear equations

A straight line graph is called a **linear graph**. It represents a **linear equation** in which the highest power of x is x (x^1). Examples of linear equations include $y = 2x$, $y = -3x + 2$, $2x + 3y = 5$, $y = \frac{1}{3}x - 1$, $y = -2$ and $x = 4$.

> Linear functions have equations of the form $y = mx + c$ and give straight line graphs.

Graphs that have highest powers of x larger than 1 are called **non-linear graphs** – they do not form a straight line. Several non-linear graphs have easily recognisable features.

> **Key words:**
> linear graphs
> linear equation
> non-linear graphs

> Straight line or linear graphs and equations were dealt with in Chapter 16.

25.2 Graphs of quadratic functions

An algebraic expression in which the highest power of x is x^2 is called a **quadratic function**.

Examples of quadratic functions include $y = x^2 - 3$, $y = 3x^2 + 4x - 7$, $y = 4x^2 - x$, $y = -2x^2 + 3x - 4$ and $y = -7x^2$.

> **The graph of a quadratic function is a parabolic (U-shaped) curve that is symmetrical about a line parallel to the y-axis.**

Parabolic shapes occur naturally. The path of a ball after it has been thrown is an example of a parabola.

> **Key words:**
> quadratic function
> coefficients

> You met quadratic functions in Chapter 24.

> All quadratic graphs are parabolic, or U-shaped. The U shape can be the right way up or upside down.

This is the graph of the quadratic function $y = x^2$. It is symmetrical about the y-axis ($x = 0$) and touches the x-axis at only one point $(0, 0)$. In general quadratic functions may lie anywhere in the x–y plane.

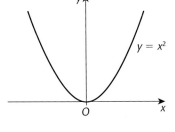

The general form of a quadratic function is given by

$$y = ax^2 + bx + c$$

The simplest quadratic equation is $y = x^2$.

$\neq$ means 'not equal to'.

where the **coefficients** a, b and c are constants (numbers) with $a \neq 0$.

You will find it useful to remember the following basic shapes:

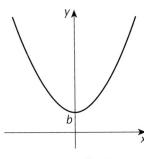

$y = ax^2 + b$
$a > 0, \quad b > 0$

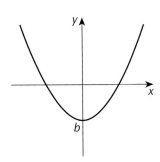

$y = ax^2 + b$
$a > 0, \quad b < 0$

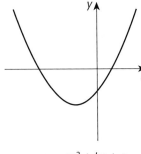

$y = ax^2 + bx + c$
$a > 0$

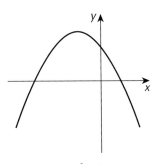

$y = ax^2 + bx + c$
$a < 0$

When the coefficient of x^2 is negative, the parabola is inverted.

When plotting the graph of a quadratic function, construct a table of values to calculate y for a given range of values of x.

Example 1

(a) Plot the graph of $y = 2x^2 - 5x - 3$ between x values of -2 and 4.

(b) Use your graph to find solutions to the equation
$2x^2 - 5x - 3 = 0$.

(a) When $x = -2$ $y = 8 + 10 - 3 = 15$

When $x = -1$ $y = 2 + 5 - 3 = 4$

When $x = 0$ $y = 0 - 0 - 3 = -3$

Substitute the given x-values into the equation to find the corresponding y-values...

... or complete the table of results.

x	−2	−1	0	1	2	3	4
$2x^2$	8	2	0	2	8	18	32
$-5x$	10	5	0	−5	−10	−15	−20
-3	−3	−3	−3	−3	−3	−3	−3
y	15	4	−3	−6	−5	0	9

Each row calculates one term in the equation of the graph you are plotting.

When working out the values of $2x^2$, do not forget that this means $2 \times x^2$; square first then multiply by 2. More care is needed with negative numbers.

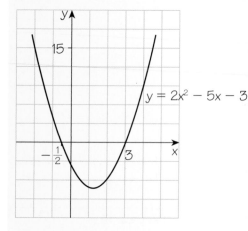

$y = 2x^2 - 5x - 3$

Plot the (x, y) points from your table of values and join them up with a *smooth* curve.

Do *not* join the points with straight lines. Marks will be lost in the exam if you do!

It takes practice to draw a smooth curve. Remember your curve should look symmetrical. It is important always to label your axes and label the curve with the equation.

(b) The curve crosses the x-axis at two positions $x = -\frac{1}{2}$ and $x = 3$. These are the solutions to the equation $2x^2 - 5x - 3 = 0$.

This quadratic equation can also be factorised to give $(2x + 1)(x - 3) = 0$ with solutions $x = -\frac{1}{2}$ and $x = 3$. Refer to Chapter 24 on factorising quadratics.

Some quadratic curves do not cross the x-axis at all. They still have a line of symmetry but do not have any solutions for $y = 0$.

In these cases, the quadratic expression cannot be factorised.

Guidelines to help you plot non-linear graphs

- Look at the range of values of x for which the graph is to be drawn.
- Put these integer values into the equation and work out the corresponding values for y.
- Place the results into a table.
- From the results identify the extreme values for $+x$ and $-x$ and $+y$ and $-y$ and draw a coordinate grid using these values.
- Plot the points from the table.
- Draw a smooth curve through all the points.
- Label the graph with the equation.

Exercise 25A

1 Draw the graph of $y = 3x^2$ by completing the table of values given.

x	-4	-3	-2	-1	0	1	2	3	4
y	48			3	0		12		

What is the equation of the line of symmetry?

2 Draw a graph of the function $y = -2x^2$ between the x values -4 and $+4$, by constructing a table of results.
What effect does the minus sign have on the graph?

3 Draw the graph of $y = x^2 - 12$ between $x = -4$ and $x = 4$.

4 For each of the following graphs make a table of values and draw the graph for the given range of values of x. State the line of symmetry of each graph.

  (a) $y = x^2 + x - 2$ for values of x from -3 to $+2$

  (b) $y = x^2 + 3x + 1$ for values of x from -5 to $+2$

  (c) $y = 2x^2 - 2x + 3$ for values of x from -3 to $+4$

  (d) $y = 2x^2 - 4x - 9$ for values of x from -2 to $+4$

5 Draw the graph of $y = x^2 - x - 12$. What are the solutions to the equation $x^2 - x - 12 = 0$?

> Look at Example 1(**b**).

6 For x values between -3 and $+4$ construct a table of results for the function $y = 3x^2 - 7x - 6$. What are the coordinates of the points of intersection between the curve and the x-axis?

> Remember: the x-axis is the line $y = 0$.

> Key words:
> cubic function

25.3 Graphs of cubic functions

A **cubic function** is one in which the highest power of x is x^3.

Typical examples of cubic functions include $x^3 - 4x + 2$, $-x^3 + 4$, $3x^3 + 5x^2 - 6x - 2$, $-8x^3$.

This is the graph of the cubic function $y = x^3$.

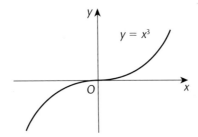

> Quadratic graphs, highest power x^2, bend once.
>
> Cubic graphs, highest power x^3, bend twice.
>
> This pattern continues. A curved graph always bends a number of times which is one less than its highest power.

This curve goes through the origin $(0, 0)$.
The graph of $y = x^3$ has rotational symmetry of order 2 about the origin.

The general form of a cubic function is given by

$$y = ax^3 + bx^2 + cx + d$$

where a, b, c and d are constants with $a \neq 0$.

Some shapes to remember include:

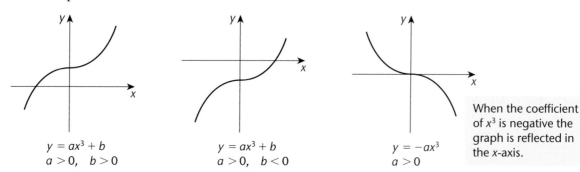

| $y = ax^3 + b$ | $y = ax^3 + b$ | $y = -ax^3$ |
| $a > 0, \quad b > 0$ | $a > 0, \quad b < 0$ | $a > 0$ |

When the coefficient of x^3 is negative the graph is reflected in the x-axis.

Example 2

Draw the graph of $y = x^3 - 2x^2 - 3x$ for $-2 \leqslant x \leqslant 4$.

Use the graph to find solutions to the equation $x^3 - 2x^2 - 3x = 0$.

x	-2	-1	0	1	2	3	4
x^3	-8	-1	0	1	8	27	64
$-2x^2$	-8	-2	0	-2	-8	-18	-32
$-3x$	6	3	0	-3	-6	-9	-12
y	-10	0	0	-4	-6	0	20

Complete a table of values to find values of y.

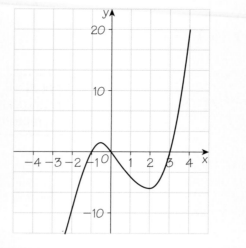

Draw an x–y grid with x values between -2 and 4 and y values between -10 and 20. Plot the points and join them up with a *smooth* curve.

At $x = -\frac{1}{2}$
$y = -\frac{1}{8} - 2 \times \frac{1}{4} - 3x - \frac{1}{2}$
$= \frac{7}{8}$

The curve crosses the x-axis at three points $(-1, 0)$, $(0, 0)$ and $(3, 0)$. The solutions of the equation $x^3 - 2x^2 - 3x = 0$ are $x = -1, 0$ or 3.

This cubic is easily factorised since x is a common factor and $x = 0$ must be one of the solutions. The remaining factor is quadratic and this gives the solutions $x = -1$ and $x = 3$.

Exercise 25B

1 Draw the graph of $y = -x^3$ taking values of x between -3 and $+3$.

2 Copy and complete the table for the cubic function $y = x^3 - 3x^2 - 5x$.

x	-2	-1	0	1	2	3
x^3						
$-3x^2$						
$-5x$						
y	-10		0			-15

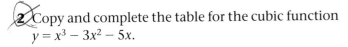

Draw the graph of this function for $-2 \leq x \leq 3$, and find the values for x when the curve crosses the x-axis.

3 Plot the graph of $y = x^3 - 3x - 2$ between $-3 \leq x \leq 3$ by constructing a table of values. Use your graph to estimate the value of x when $y = 0$.

4 Draw the graph of $y = x^3 - 2x^2 - 3x$ for x between -2 and $+4$. Use your graph to estimate the solutions to the equation $x^3 - 2x^2 - 3x = 0$.

5 Draw the graphs of the following cubic functions for $-3 \leq x \leq 3$ by constructing a table of x- and y-values.

(a) $y = \frac{1}{2}x^3 + 2$ (b) $y = (x - 1)^3$ (c) $y = x^3 + x^2 + x$

25.4 Graphs of reciprocal functions

Key words:
reciprocal function
asymptotes

A function in which the power of x is of the form x^{-1} or $\frac{1}{x}$ is called a **reciprocal function**.

Typical examples of reciprocal functions include $y = \frac{1}{x}, y = -\frac{3}{x}, y = \frac{2}{x} - 3$.

The key features of reciprocal graphs are that they tend towards certain values but never reach them.

You met reciprocals in Chapter 13.

These features are called **asymptotes**.

The graph of the simplest reciprocal function, $y = \dfrac{1}{x}$ (or $y = x^{-1}$) is:

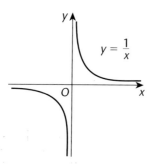

It is well worth remembering this shape.

When x tends to very large (positive or negative) values, y tends to $\dfrac{1}{\text{very large}}$ = very small values. In fact y tends to 0 but never reaches it. The x-axis is an asymptote.

Asymptotes are usually denoted by a dashed line, but a dashed line on the axis would not show up.

When x tends to very small (positive or negative) values, y tends to $\dfrac{1}{\text{very small}}$ = very large values. In fact y tends to infinity but never reaches it. The y-axis is another asymptote.

Graphs of reciprocal functions always have two asymptotes.

The general expression for a reciprocal function is $y = \dfrac{a}{x}$, where a is a positive or negative constant.

Examples of graphs of reciprocal functions are:

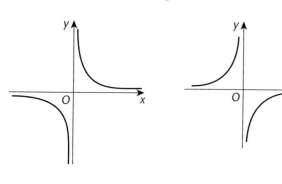

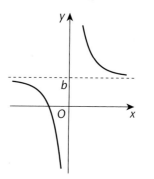

$$y = \frac{a}{x}$$

$$y = -\frac{a}{x}$$

$$y = \frac{a}{x} + b$$

$y = b$ is an asymptote on the graph of $y = \dfrac{a}{x} + b$.

$y = \dfrac{a}{x}$ is symmetrical about $y = -x$.

$y = -\dfrac{a}{x}$ is symmetrical about $y = x$.

$a > 0$ in these 3 sketch graphs.

Example 3

Draw the graph of the function $y = 2 - \dfrac{3}{x}$ for x values between -3 and $+3$.

x	-3	-2	-1	-0.5	-0.2	0.2	0.5	1	2	3
2	2	2	2	2	2	2	2	2	2	2
$-\dfrac{3}{x}$	1	1.5	3	6	15	-15	-6	-3	-1.5	-1
y	3	3.5	5	8	17	-13	-4	-1	0.5	1

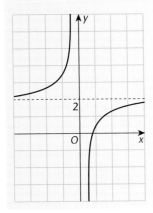

Begin by tabulating the results between these values. It is important to include smaller x values close to zero but not zero itself, which is undefined.

x values of -0.5, -0.2, 0.2 and 0.5 are typically used between -1 and $+1$.

The maximum and minimum values for y are 17 and -13.

The y-axis should be between -15 and $+20$.

The dashed line at $y = 2$ shows the horizontal asymptote. As x becomes large, $\dfrac{3}{x}$ tends to zero, so the value of y must approach 2.

The y-axis itself is the second (vertical) asymptote with equation $x = 0$.

Exercise 25C

1 For each of these graphs make a table of values and draw the graph for values of x from -3 to $+3$. Choose a sensible scale for each axis.

(a) $y = -\dfrac{2}{x}$　　　(b) $y = -\dfrac{5}{x}$　　　(c) $y = -\dfrac{3}{x} + 4$

(d) $y = -6 - \dfrac{4}{x}$　　(e) $y = \dfrac{6}{x}$　　　(f) $xy = 4$

2 Copy and complete the table of values for the function $y = -\dfrac{3}{x}$.

x	-3	-2	-1	-0.5	-0.2	0.2	0.5	1	2	3
y	1		3	6		-15		-3	-1.5	

Draw the graph over this x range. What are the equations of the two asymptotes?

3 Draw the graph of the function $y = \dfrac{3}{x} + 2$ over the range of x values between -3 and 3. How does the graph compare with that drawn in question 2?

4 Draw the graph $y = 4 - \dfrac{1}{x}$ between the x values 0 and 5.

What are the equations of the two asymptotes?

25.5 Graphs of exponential functions

Key words:
exponential function
power curve

An **exponential function** is a function of the form $y = a^x$ where a is a positive number.

Because the function power is x itself the graph is sometimes called the **power curve**. Exponential functions describe a very rapid and often dramatic increase in the values of y for small changes in the x value. Typical examples of exponential functions include $y = 2^x$, $y = 4^{3x}$, $y = (3.6)^{-x}$ and $y = 10^x$.

The basic shape of the curve is the same for all exponential functions.

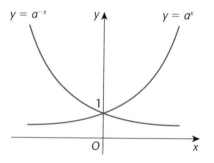

All exponential functions are positive, that is they lie above the x-axis. All exponential functions pass through $(0, 1)$, because $y = a^0 = 1$ for all a.

The inverse function $y = a^{-x}$, illustrating exponential decrease, is simply a reflection in the y-axis of the function $y = a^x$, as shown in the diagram.

Example 4

Draw the graph of the exponential function $y = (1.2)^x$ for x values between -2 and 4. Give the values to 2 d.p. where necessary.

x	−2	−1	0	1	2	3	4
y	0.69	0.83	1	1.20	1.44	1.73	2.07

Begin by constructing a table of values between $x = -2$ and $x = +4$.

You need a calculator for these calculations.

Decide on a sensible scale for each axis. Here the minimum and maximum values for y are 0.69 and 2.07. The y-axis should be between 0 and 2.5.

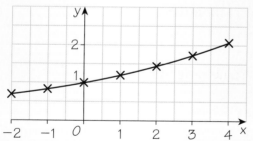

Plot these points and join them up with a *smooth* curve.

The smooth curve crosses the y-axis at $y = 1$ as expected. Do *not* join the points with straight lines.

Exercise 25D

1 Draw the graph of $y = (1.5)^x$ for x-values between -2 and $+4$ (give your values to 1 d.p.).

First construct a table.

2 Copy and complete the table below, and draw the graph of the exponential function $y = 10^{2x}$ for the x values shown in the table.

x	-0.5	-0.2	-0.1	0	0.1	0.2	0.5
y	0.1		0.6	1			10

3 Copy and complete the table of values for the function $y = (0.3)^x$ for x values between -3 and $+2$, giving your answers to 1 d.p.

x	-3	-2	-1	0	1	2
y		11.1	3.3			0.1

Draw the graph of this function for $-3 \leqslant x \leqslant 2$.

UAM 4 Certain bacteria are growing exponentially. The population N is given by the exponential function $N = (2a)^{\frac{t}{6}}$ where a is the population at the start and $t =$ time in hours. What is the population after 24 hours if the number of original bacteria was $a = 10$?

Exponential growth is growth that depends on the current size.

UAM 5 Radioactive radium has a half-life of 1600 years and it decays according to the exponential function $N = (2.71a)^{-\frac{0.693t}{1600}}$ where N is the amount of radium in grams, a is the original amount of radium in grams, and t is time in years. If the original amount of radium is 1 gram, how much radium is left after 100 years? Give your answer in grams to 3 s.f.

Half-life is used to describe the rate of decay of radioactive substances. It is the time taken for half the substance to decay.

25.6 Combined non-linear graphs

These are functions that are a combination of those functions described in the previous sections.

Combination functions involve linear, quadratic, cubic and/or reciprocal terms.

Example 5

Make a table of values for $y = \dfrac{1}{x} + x$ between $x = 0$ and $x = 5$.

Extend the table to negative values of x between 0 and -5.
Draw the graph of this function for $-5 \leqslant x \leqslant 5$.

x	0.1	0.2	0.5	1	2	3	4	5
$\frac{1}{x}$	10	5	2	1	0.5	0.3	0.25	0.2
y	10.1	5.2	2.5	2	2.5	3.3	4.3	5.2

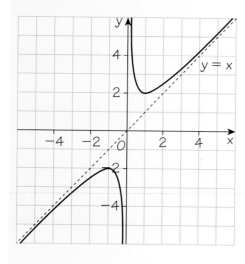

Do not include $x = 0$ as this is undefined (infinite).

As in the case of reciprocal graphs include x values close to zero.

Give the y values to 1 d.p.

For the corresponding negative x values the y values will also be negative but the same numerical value.

Note the rotational symmetry of the curve.

There are also two asymptotes at $x = 0$ (the y-axis) and along the line $y = x$. Remember: reciprocal graphs have two asymptotes.

As x gets larger (positive or negative) the $\frac{1}{x}$ term tends to zero so the graph is approximately $y = x$.

Exercise 25E

1 Copy and complete the table for the function $y = \dfrac{1}{x} - x$ between the x values $x = 0$ and $x = 5$.

x	0.1	0.2	0.5	1	2	3	4	5
$\frac{1}{x}$							)	
y	9.9		1.5	0		−3.8		

Draw the graph of this function.

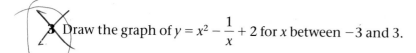

2 Draw the graph of the function $y = \dfrac{4}{x} - 3x$ between x values 0.1 and 5 (as in Example 5) by completing a table of results.

3 Draw the graph of $y = x^2 - \dfrac{1}{x} + 2$ for x between −3 and 3.

Remember to include x values close to zero but do not include zero itself.

25.7 Graphs that intersect

Linear graphs

Key words:
points of intersection

In Chapter 16, you used linear graphs to find solutions to simultaneous equations. You can use the same method to find solutions to linear and non-linear functions.

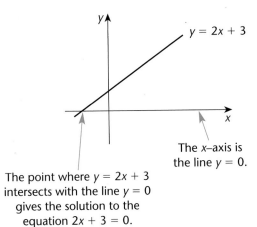

The point where $y = 2x + 3$ intersects with the line $y = 0$ gives the solution to the equation $2x + 3 = 0$.

The x–axis is the line $y = 0$.

Solutions to other equations, for example $y = 3$, can also be found by plotting the graph and finding the **points of intersection** with the particular line.

Example 6

Draw the graph of $y = x + 1$. From the graph find the x value when $x + 1 = 0$ and when $x + 1 = 3$.

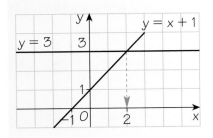

You can sketch this linear graph quickly by noting where the line crosses the x- and y-axes. When $x = 0$, $y = 1$. When $y = 0$, $x = -1$.

This is a quick and easy method for any straight line that is not parallel to the axes.

You need accurate drawings to obtain accurate solutions.

Solution of $x + 1 = 0$ is $x = -1$

The line crosses the x-axis at $x = -1$.

Solution of $x + 1 = 3$ is $x = 2$

You can solve $x + 1 = 3$ by finding where the graph crosses the line $y = 3$. This is called the point of intersection of the graph with the line $y = 3$. You can see that the solution is $x = 2$.

Example 7

Plot the graph of $y = 3x - 2$. Find the coordinates of the point of intersection where this line meets the line $y = 3$.

x	−1	0	1
3x	−3	0	3
−2	−2	−2	−2
y	−5	−2	1

Begin by constructing a table using x values of −1, 0 and 1.

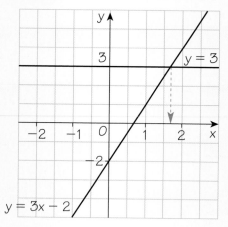

Plot the graph and the straight line $y = 3$.

Always draw your graphs as accurately as possible. In this example, the value of the point of intersection is not a whole number.

The exact coordinates are $(1\frac{2}{3}, 3)$ but examiners always allow for small errors.

The coordinates are (1.6, 3).

Quadratic graphs

When quadratic functions are plotted the curve may do one of the following:

it may cross the x-axis at two points

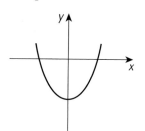

There are two solutions to $y = 0$.

it may just touch the x-axis

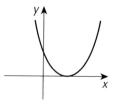

There is one solution to $y = 0$.

it may never cross or touch the x-axis.

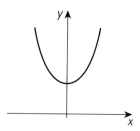

There are no solutions to $y = 0$.

Quadratic curves may also intersect other lines of the form $y = k$ (where k is any number). The method for finding solutions is to plot the line and look for points of intersection.

Example 8

On a coordinate grid with x values between -3 and $+3$ and y values between -5 and 10 draw the curve $y = x^2 - 2$.

(a) What are the solutions of the equation $x^2 - 2 = 0$?

(b) Draw the straight line $y = 3$. Write down the coordinates of both points where the curve and line intersect.

x	-3	-2	-1	0	1	2	3
x^2	9	4	1	0	1	4	9
-2	-2	-2	-2	-2	-2	-2	-2
y	7	2	-1	-2	-1	2	7

Begin by constructing a table of values between $x = 3$ and $x = -3$ and then plotting the curve.

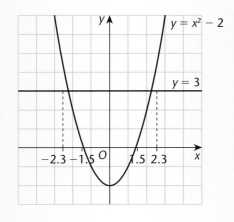

The curve crosses the x-axis at points $x = -1.5$ and $x = 1.5$.

Draw the line $y = 3$ on the graph. At the points of intersection draw two vertical dashed lines to the x-axis.

(a) Solutions to $x^2 - 2 = 0$ are $x = \pm 1.5$.

(b) The curve and the line $y = 3$ intersect at $(-2.3, 3)$ and $(2.3, 3)$.

Finding the solution of the point of intersection between $y = x^2 - 2$ and $y = 3$ is the same as solving the equation $x^2 - 2 = 3$. This rearranges to $x^2 = 5$.

The exact values are $(-\sqrt{5}, 3)$ and $(\sqrt{5}, 3)$ found by applying the quadratic formula from Chapter 24.

Example 9

Draw the graph of the quadratic function $y = x^2 - 2x - 4$ for values of x between $x = -2$ and $x = +4$.

(a) Use your graph to solve the equation $x^2 - 2x - 4 = 0$ from the graph.

(b) Use your graph to solve the equation $x^2 - 2x - 4 = 2x + 1$.

(c) Use your graph to find the positive solution of the equation $x^2 - x - 8 = 0$. Give your answer correct to 1 d.p.

x	−2	−1	0	1	2	3	4
x^2	4	1	0	1	4	9	16
−2x	4	2	0	−2	−4	−6	−8
−4	−4	−4	−4	−4	−4	−4	−4
y	4	−1	−4	−5	−4	−1	4

Complete a table of results for values of x between −2 and +4.

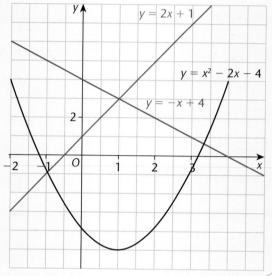

(a) From the graph the solutions of the equation

$$x^2 - 2x - 4 = 0 \text{ are } x = -1.2 \text{ and } x = 3.2$$

(b) From the graph, $x^2 - 2x - 4 = 2x + 1$ when $x = -1$

Points of intersection with the x-axis.

In part (b) draw the straight line $y = 2x + 1$. Where this line and the quadratic curve intersect gives the solution to the equation $x^2 - 2x - 4 = 2x + 1$.

You are solving

'curve' = 'line'.

To solve $x^2 - x - 8 = 0$, rearrange the equation so that it has the quadratic expression of the graph you have drawn on one side of the equation. The expression on the other side of the equation indicates the graph you need to draw.

(c)
$$x^2 - x - 8 = 0$$
$$x^2 - 2x - 4 + x - 4 = 0$$
$$x^2 - 2x - 4 = -x + 4$$

The point of intersection occurs when $x = 3.3$

Draw the line $y = -x + 4$ and find the point of intersection with the original quadratic curve. The line $y = -x + 4$ is shown on the graph.

A second solution exists at $x = -2.4$ but this is outside the range of the graph.

Exercise 25F

1 Draw the graph $y = 2x + 5$. It intersects the line $y = 3$ at a point P. What are the coordinates of P?

2 The line $y = 7$ intersects another line $y = 4x + 9$ at a point Q. What are the coordinates of Q? Where does the line $y = 4x + 9$ cross the x-axis?

3 Copy and complete the table for the equation $y = x^2 + 1$.

Draw this curve with x values from -2 to 2 and y values from 0 to 6. Draw the line $y = 3$ on the same graph. What are the coordinates of the points of intersection?

x	-2	-1	0	1	2
y	5				

4 Draw the graph of $y = x^2 + 4x + 2$ for values of x from -5 to $+1$. Use a scale of 1 cm for 1 unit on each axis.

(a) On the same axes draw the graph of the line $y = 5$.

(b) Write down the x values of the points of intersection of the curve and the line $y = 5$.

(c) What equation is solved by these x values?

(d) Use your graph to solve the equation $x^2 + 4x + 2 = 0$.

5 For each of the following, solve graphically the quadratic equations which are written alongside them. Give your answers correct to 1 d.p.

	Graph	Solve these equations	
(a)	$y = x^2 + x - 2$	$x^2 + x - 2 = 0$	$x^2 + x - 2 = 2$
(b)	$y = x^2 + 4x - 1$	$x^2 + 4x - 1 = 0$	$x^2 + 4x - 1 = -3$
(c)	$y = x^2 - 2x - 5$	$x^2 - 2x - 5 = 0$	$x^2 - 2x - 5 = -4$
(d)	$y = x^2 + 3x + 1$	$x^2 + 3x + 1 = 0$	$x^2 + 3x + 1 = 8$
(e)	$y = x^2 - 5x - 4$	$x^2 - 5x - 4 = 0$	$x^2 - 5x - 4 = -6$
(f)	$y = 2x^2 - 2x + 3$	$2x^2 - 2x + 3 = 0$	$2x^2 - 2x + 3 = 20$
(g)	$y = 2x^2 + 4x - 5$	$2x^2 + 4x - 5 = 0$	$2x^2 + 4x - 5 = 5$
(h)	$y = 2x^2 - 4x - 9$	$2x^2 - 4x - 9 = 0$	$2x^2 - 4x - 9 = -8$

Each solution will be given by the points of intersection of a quadratic graph and a straight line graph (which may be the x-axis).

6 Draw the graph of the function $y = x^2 - 2x - 4$ for x values between -3 and 4. Using the graph, find the solutions of $x^2 - 2x - 4 = 0$ giving your answers to 1 d.p. By drawing an appropriate linear graph on the same axes, solve the equation $x^2 - x - 6 = 0$.

Look at Example 9(c).

7 (a) Copy and complete the table of values for $y = x^2 - 4x + 3$.

x	-1	0	1	2	3	4	5
y	8	3		-1			

(b) Draw the graph of this function between -1 and 5 with y values between -3 and 10.

(c) Write down the solutions of $x^2 - 4x + 3 = 0$.

(d) By drawing an appropriate linear graph on the same axes, solve the equation $x^2 - 5x + 5 = 0$.

8 A stone is dropped down a well. The following table gives the distance from the top of the well after each second using the approximate formula $d = 5t^2$.

Time in seconds (t)	0	1	2	3
Distance in m (d)	0	5	20	45

Questions 8, 9 and 10 involve applications of quadratic graphs to solving problems. For each one, draw the graph, then use it to answer the question.

Draw a coordinate grid with x values between 0 and 4 and y values between 0 and 60. Plot the points and join them up with a smooth curve.

(a) How far did the stone drop in 1.5 seconds?

(b) How long did it take the stone to fall 35 m?

(c) If the well is 60 m deep, for how many seconds did the stone fall?

9 A car is accelerating from rest and the distance covered, s, is given by the expression $s = \frac{1}{2} at^2$ where a represents the acceleration in ms^{-2}, and t the time in seconds. If the car accelerates at a constant rate of $5\ ms^{-2}$, draw a graph of the distance covered for t values from 0 to 10 seconds.

ms^{-2} means metres per second per second, or m/s^2.

(a) What is the distance covered after 4.5 s?

(b) What is the time taken to cover 150 m?

10 Kinetic energy (in joules) is given by the equation $KE = \frac{1}{2} mv^2$, where m is the mass in kg and v the velocity in m/s. Work out the kinetic energy of a ball ($m = 1.6$ kg) travelling initially at 3 m/s.
Draw a graph of KE against v for v values between 0 and 4 m/s. What is the velocity of the ball if its KE is 10 joules?

25.8 Recognising non-linear graphs

You need to be able to recognise linear and
non-linear graphs and use them to describe real-life
situations.

Example 10

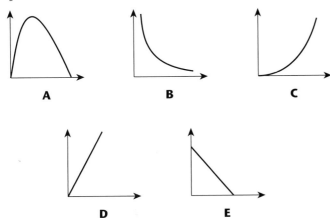

Which graph best matches the relationships below?
Describe the type of function used in each case.

(a) The circumference of a circle plotted against its diameter.

(b) The area of a circle as a function of its radius.

(c) The length of a rectangle of fixed area plotted against its width.

(a) $C = \pi d$ This is linear, so of the form $y = mx + c$.
As d increases so does C. The graph must
pass through $(0, 0)$.   Graph **D**

(b) $A = \pi r^2$ This is quadratic, so of the form $y = ax^2$.
As r increases so does A. The graph must
pass through $(0, 0)$.   Graph **C**

(c) $l = \dfrac{A}{w}$ This is reciprocal, so of the form $y = \dfrac{a}{x}$.

As l increases w decreases.   Graph **B**

Example 11

The following graphs describe the height (*y*-axis) of a liquid in a particular container as a function of its volume (*x*-axis).

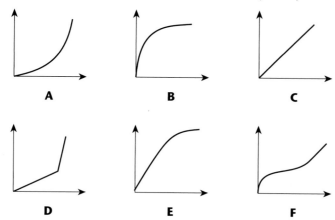

The four containers below are filled with liquid at a steady rate. Which graph best describes the relationship between the height of the liquid and the volume? Explain your answers.

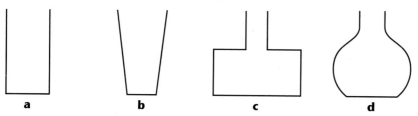

> The shape of the profile of the container helps you to find the correct graph. Sharp changes in the profile lead to sudden changes in the gradient of the graph. Smooth changes in profile lead to smooth changes in gradient.

(a) Straight sides so constant rate of increase in height   Graph C

(b) Rises more quickly at first then slows down gradually   Graph B

(c) Slow constant rate then sudden change to faster
constant rate   Graph D

(d) Faster at first, then slower (wider part of bowl) then
faster again   Graph F

Exercise 25G

1 Look at the following functions. Describe the graph of the function as either linear (L), quadratic (Q), cubic (Cu), reciprocal (R), exponential (E) or a combination (Co).

(a) $y = 2x^2 - 3$ (b) $y = 5 + x^3$ (c) $y = -\dfrac{3}{x}$

(d) $y = 5x^3 - 2x - 1$ (e) $y = 12a^x$ (f) $y = 4 - 3x$

(g) $y = \dfrac{1}{x} + x$ (h) $y = x + 2 - 2x^2$ (i) $y = x - 3 + \dfrac{2}{x}$

2 Match each equation to one of the graphs **A–E** below.

(a) $y = \dfrac{3}{x}$ (b) $y = 4x^3$ (c) $y = 4x + 3$

(d) $y = 3x^2 + 4$ (e) $y = 3^{4x}$

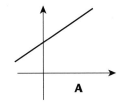
A

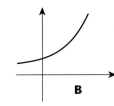
B

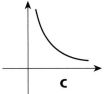
C

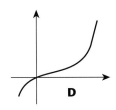
D

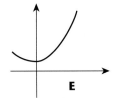
E

3 What type of graphs could be drawn from the following functions? Label them linear (L), quadratic (Q), cubic (Cu), reciprocal (R), exponential (E) or a combination (Co).

(a) $xy = -6$ (b) $3x + 2y = 4$ (c) $\dfrac{y}{3} = 2^x$

(d) $3x^2 + y = 2x - 4$ (e) $y = \dfrac{x^3 - 3x^2}{x}$ (f) $3x = xy - 2$

(g) $9x^3 = 4x^2 + y - 7$ (h) $xy = 2 + 4x^2$ (i) $x = \dfrac{y}{5}$

4 Sketch a graph to show how the depth of water changes as each of these tanks is filled at a steady rate.

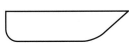
Tank A
cross–section

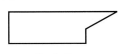
Tank B
cross–section

5 Water is poured into the following containers at a steady rate. Sketch a graph to show how the height of the liquid changes (*y*-axis) as a function of the volume (*x*-axis).

Examination questions

1 Copy and complete this table of values for $y = 2x - 1$.

x	-1	0	1	2	3
y	-3		1		

On a grid with x values between -2 and 4 and y values between -4 and 6 draw the graph of $y = 2x - 1$ for values of x between -1 and $+3$.
Find the coordinates of the point where the line $y = 2x - 1$ crosses the line $y = -2$.

(5 marks)
AQA, Spec A, 1I, June 2003

2 The line $y = -3$ crosses the line $y = x - 2$ at the point P. What are the coordinates of P?
The lines may be drawn on an x–y grid with x- and y-axes between -6 and 6.

(3 marks)
AQA, Spec A, 1I, November 2004

3 (a) Copy and complete the table of values for $y = 2x^2 - 4x - 1$.

x	-2	-1	0	1	2	3
y	15		-1		-1	5

(b) Draw the graph of $y = 2x^2 - 4x - 1$ for values of x from -2 to $+3$.

(c) An approximate solution of the equation is $x = 2.2$.

(i) Explain how you can find this from the graph.

(ii) Use your graph to write down another solution of this equation.

(6 marks)
AQA, Spec A, 1I, June 2004

4 (a) Copy and complete the table of values for $y = (0.8)^x$.

x	0	1	2	3	4
y	1	0.8	0.64		0.41

(b) Draw the graph of $y = (0.8)^x$ for values of x from 0 to 4.

(c) Use your graph to solve the equation $(0.8)^x = 0.76$.

(4 marks)
AQA, Spec A, 2H, June 2004

5 Liquid is poured at a steady rate into the bottle shown in the diagram.
As the bottle is filled, the height, h, of the liquid in the bottle changes.
Which of the five graphs below shows this change?
Give reasons for your choice.

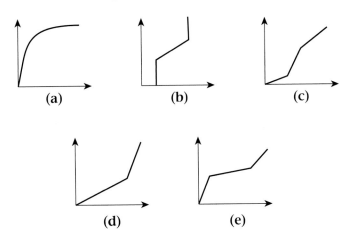

(a)   (b)   (c)

(d)   (e)

(2 marks)

AQA, Spec A, 2I, November 2003

6 (a) Copy and complete the table of values for $y = x^2 - 3x - 4$.

x	-3	-2	-1	0	1	2	3	4	5	6
y	14		0	-4	-6	-6		0	6	14

(b) On an x–y grid with x values between -4 and $+6$ and
y values between -8 and $+14$ draw the graph of $y = x^2 - 3x - 4$.

(c) Write down the solutions of $x^2 - 3x - 4 = 0$.

(d) By drawing an appropriate linear graph, write down the solutions
of $x^2 - 4x - 1 = 0$.   *(6 marks)*

AQA, Spec B, 3H, June 2002

Summary of key points

Linear functions (grade D)

Linear functions have equations of the form $y = mx + c$ and give straight line graphs.

Non-linear functions (grade C to A*)

A *quadratic function* is one in which the highest power of x is x^2. Quadratic graphs are parabolic (U-shaped) curves that are symmetrical about a line parallel to the y-axis.
The solutions of quadratic equations of the form $ax^2 + bx + c = 0$ are the values of x where the graph cuts the x-axis.

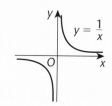

A *cubic function* is one in which the highest power of x is x^3. The solutions of cubic equations of the form $ax^3 + bx^2 + cx + d = 0$ are the values of x where the graph cuts the x-axis.

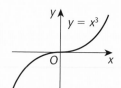

A *reciprocal function* is one in which the power of x is $\dfrac{1}{x}$ or x^{-1}.
$y = \dfrac{a}{x}$, where a is a positive or negative constant, is an example

of a reciprocal function. Reciprocal graphs have two asymptotes. An *asymptote* is a line that a graph approaches but which it never reaches.

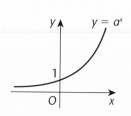

An *exponential function* is of the form $y = a^x$. Exponential functions are always positive and pass through the point (0, 1).

Solutions to quadratic equations can be found graphically by finding the points of intersection of the quadratic graph with the x-axis or with other linear graphs.

Graphs can be used to describe real-life situations.

This chapter will show you how to:

- ✔ recognise rational and irrational numbers
- ✔ handle terminating and recurring decimals
- ✔ write terminating and recurring decimals as fractions
- ✔ recognise and use surds
- ✔ manipulate and simplify surds

26.1 Definitions of rational and irrational numbers

Key words:
rational
irrational

All numbers in our number system are either rational or irrational.

A **rational** number is one that can be written in the form $\frac{a}{b}$ where a and b are integers and $b \neq 0$.

An **irrational** number is one that cannot be written in this way.

You will see more of these later in the chapter.

All integers are rational. For example, $7 = \frac{7}{1}$ and $-8 = \frac{-8}{1}$

All fractions are rational. For example, $\frac{1}{4}, \frac{5}{2}, \frac{8}{11}$

Some decimals are rational. For example, $2.79 = \frac{279}{100}, 0.3 = \frac{3}{10}$

Some, as you will see, are not.

26.2 Terminating and recurring decimals

Key words:
terminating
recurring

All rational numbers can be written as decimals, some of which are **terminating** decimals and some of which are not.

Here are two examples of terminating decimals:

$$\frac{7}{2} = 3.5 \qquad \frac{1}{8} = 0.125$$

Terminating means that the decimal ends – there is a finite number of decimal places.

To change a fraction into a decimal you can either do the division on your calculator or using short or long division methods.

In a non-calculator exam, you will have to use long or short division if you want to change a fraction into a decimal.

For example, $\frac{5}{8} = 0.625$

$$\begin{array}{r} 0.\ 6\ 2\ 5 \\ 8\overline{)5.\ ^5 0^2 0^4 0} \end{array}$$

Only those fractions that have a denominator with prime factors of *only* 2 and/or 5 can be converted to terminating decimals.

Fractions that do not satisfy this condition convert to decimals that repeat and do not terminate.
Look at these fractions written as decimals:

$\frac{1}{3} = 0.333333\ldots$ $\qquad \frac{5}{12} = 0.416666\ldots$

$\frac{8}{11} = 0.727272\ldots$ $\qquad \frac{3}{7} = 0.42857142857142\ldots$

When you work out the decimal answer on your calculator, the decimal places fill the whole of the calculator display.

If you do the division by short or long division methods you will see that the decimals never end.

This is what your calculator display will show when you work out 3 ÷ 7.

The decimals do not terminate – they are called **recurring** decimals.

Recurring decimals have either a repeating digit or a repeating pattern of digits.

Because they never end, you need to have a way of writing them which makes the repeating pattern clear.

When a single digit repeats, you put a dot over this digit.

$\frac{1}{3} = 0.333333\ldots$ is written as $0.\dot{3}$

$\frac{5}{12} = 0.416666\ldots$ is written as $0.41\dot{6}$

When more than one digit repeats, you put a dot over the *first* digit of the pattern and a dot over the *last* digit of the pattern.

$\frac{8}{11} = 0.727272\ldots$ is written as $0.\dot{7}\dot{2}$

$\frac{3}{7} = 0.42857142857142\ldots$ is written as $0.\dot{4}2857\dot{1}$

Some fractions give decimals where the pattern does not start repeating immediately.

$\frac{47}{110} = 0.4272727\ldots$ which is written as $0.4\dot{2}\dot{7}$

Exercise 26A

1 Use any non-calculator method to change these fractions into decimals. Then say if they are terminating or recurring decimals.

(a) $\frac{2}{3}$ (b) $\frac{4}{5}$ (c) $\frac{4}{9}$ (d) $\frac{5}{11}$ (e) $\frac{7}{8}$

(f) $\frac{7}{12}$ (g) $\frac{6}{25}$ (h) $\frac{3}{16}$ (i) $\frac{5}{6}$ (j) $\frac{5}{7}$

2 Without converting these fractions into decimals, decide which of the denominators indicate terminating decimals and which indicate recurring decimals.

(a) $\frac{3}{8}$ (b) $\frac{5}{12}$ (c) $\frac{7}{16}$ (d) $\frac{9}{11}$ (e) $\frac{11}{64}$

(f) $\frac{13}{20}$ (g) $\frac{15}{22}$ (h) $\frac{17}{24}$ (i) $\frac{19}{125}$ (j) $\frac{21}{128}$

> For a terminating decimal, the prime factors of the denominator can only be 2 or 5.

3 Convert the following fractions into decimals, and write down how many numbers there are in the recurring patterns.

(a) $\frac{7}{9}$ (b) $\frac{8}{11}$ (c) $\frac{5}{7}$ (d) $\frac{14}{27}$ (e) $\frac{142}{111}$

4 Investigate when happens when you multiply:

(a) a fraction that gives a terminating decimal by another fraction that gives a terminating decimal. Will your answer always be a terminating decimal?

(b) a fraction that gives a recurring decimal by another fraction that gives a recurring decimal. Will your answer always be a recurring decimal?

(c) a fraction that gives a terminating decimal by another fraction that gives a recurring decimal. Will the answer always be terminating, recurring or can't you tell?

5 Write out the full calculator display for the following recurring decimals, and the number of digits in the recurring pattern. Use them to answer the following questions.

$$\frac{1}{3} \quad \frac{1}{6} \quad \frac{1}{7} \quad \frac{1}{9} \quad \frac{1}{12} \quad \frac{1}{13}$$

(a) Is the number of digits in any of the recurring patterns larger than the denominator?

(b) How many recurring digits are there in the decimals of fractions with an even number denominator?

(c) Is there a connection between the denominators of the fractions with 1 recurring digit?

(d) Is there any connection between the denominators of the fractions with 6 recurring digits?

(e) Do your observations work with other fractions that give recurring decimals? Test them out on at least another four fractions each. What can you say about finding patterns from a small set of numbers?

26.3 Converting recurring decimals into fractions

To turn a decimal into a fraction is easy if the decimal terminates. Look at the place value of the last significant digit and write the fraction with this denominator. Simplify the fraction if possible.

2 decimal places means hundredths, 3 decimal places means thousandths, and so on.

For example, $0.48 = \frac{48}{100} = \frac{12}{25}$ $0.225 = \frac{225}{1000} = \frac{9}{40}$

When the decimal is a recurring decimal you need a special technique to find the equivalent fraction.

Example 1

Write each of these recurring decimals as fractions in their simplest form.

(a) 0.88888...

(b) 0.454545...

(c) 0.6108108108...

[handwritten annotations:] $0.45 \quad \frac{45}{99}$, $0.6 + 0.0108$, $= 0.6 + \frac{0.108}{10}$, $= \frac{3}{5} + \frac{\frac{108}{999}}{10}$

(a) Let

$$x = 0.8888...$$

[handwritten: $\frac{3}{5} + \frac{108}{999} \times \frac{1}{10}$]

$$10x = 8.8888...$$

Subtracting: $9x = 8$

Dividing both sides by 9: $x = \dfrac{8}{9}$

When 1 digit recurs multiply both sides by **10**.

When 2 digits recur multiply both sides by **100**.

When 3 digits recur multiply both sides by **1000**.

(b) Let

$$x = 0.454545...$$

$$100x = 45.454545...$$

Subtracting: $99x = 45$

Dividing both sides by 99: $x = \dfrac{45}{99}$

Dividing top and bottom by 9: $x = \dfrac{5}{11}$

continued ▶

(c) Let $x = 0.6108108108\ldots$

$$1000x = 610.810810810$$

Subtracting: $999x = 610.2$

Dividing both sides by 999: $x = \dfrac{610.2}{999}$

Multiplying top and bottom by 10: $x = \dfrac{6102}{9990}$

Dividing top and bottom by 9: $x = \dfrac{678}{1110}$

Dividing top and bottom by 6: $x = \dfrac{113}{185}$

> You must not leave your answer as $\dfrac{610.2}{999}$ because a fraction should consist of whole numbers.
>
> Notice that there is still quite a lot of work to do to get the answer into its simplest form.
>
> You will not score full marks if you do not simplify fully.

Example 1(c) can be done another way:

Let $x = 0.6108108108\ldots$

then $10x = 6.108108\ldots$

and $10\,000x = 6108.108108\ldots$

Subtracting: $9990x = 6102$

Dividing both sides by 9990: $x = \dfrac{6012}{9990}$

$x = \dfrac{113}{185}$

> This method requires *two* multiplications at the start but then has the advantage that *the recurring decimal pattern is the same.*
>
> When you subtract you automatically get whole numbers on the top and bottom of your fraction.
>
> Both methods give you the same answer.

You can choose whichever method you prefer, but if you choose the second one you must remember to multiply x by multiples of 10 which will give you the **same** pattern of recurring decimals.

Exercise 26B

1 Find the fractions that are equivalent to the following terminating decimals. Express the fractions in their simplest terms.

(a) 0.45   (b) 0.375   (c) 0.005

2 Find the fractions that are equivalent to the following recurring decimals.

(a) 0.5555...   (b) 0.8888...   (c) 0.212121...

(d) 0.454545...   (e) 2.636363...   (f) 0.320320320...

(g) 0.624624624...   (h) 0.15781578...   (i) 0.12351235...

(j) 0.16666...   (k) 0.0454545...   (l) 6.042042042...

(m) 2.161616...   (n) 15.088888...   (o) 4.0325325325...

3 Jane says '0.9999... is equivalent to 1'. Sally says she is wrong. How could Jane use fractions to help her explain her reasoning to Sally?

(handwritten annotation:)

(o) $4.0\dot{3}2\dot{5}$

$= 4 + 0.0\dot{3}2\dot{5}$

$= 4 + \dfrac{0.\dot{3}2\dot{5}}{10}$

$= 4 + \dfrac{325}{999} \times \dfrac{1}{10}$

$= \dfrac{8057}{1998}$

26.4 Surds

Key words:
surd

In Section 26.1 you met the definitions of rational and irrational numbers. Remember that a rational number is one that can be written in the form $\frac{a}{b}$, where a and b are integers and $b \neq 0$.

An irrational number does not terminate and has no recurring pattern.

Examples of irrational numbers are: $\sqrt{2}$, $\sqrt[3]{10}$, π and $4 + \sqrt{5}$.

You will see this if you enter them on a calculator:

$$\sqrt{2} = 1.414\ 213\ 562... \qquad \sqrt[3]{10} = 2.154\ 434\ 69...$$
$$\pi = 3.141\ 592\ 654... \qquad 4 + \sqrt{5} = 6.236\ 067\ 977...$$

When irrational numbers are written in a form using square roots and cube roots, they are called surds and they give the value *exactly*.

Surds are **exact** answers whereas their decimal equivalents are not.

Example 2

A cube has a volume of 10 cm³.

Calculate the length of one of its sides.

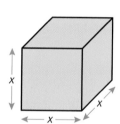

Let the length of a side be x cm

Volume = 10 cm³

$$x^3 = 10$$
$$x = \sqrt[3]{10}\ \text{cm}$$

Notice that to say that the length of the side is $x = 2.154\ 434\ 69...$ is both ridiculous and inexact.

$$\frac{2 \pm \sqrt{24}}{2} = \frac{2 \pm \sqrt{6} \times \sqrt{14}}{2}$$
$$= \frac{2 \pm 2\sqrt{16}}{2} = \frac{2(1 \pm \sqrt{6})}{2}$$

Example 3

Solve the quadratic equation $x^2 - 2x - 5 = 0$ giving your answers in surd form.

$$x^2 - 2x - 5 = 0$$
$$x = \frac{-b \pm \sqrt{b^2 - 4ac}}{2a}$$

Since 24 is not a perfect square, the roots of the equation will be irrational numbers.

Writing them in surd form will give their values **exactly**.

where $a = 1$, $b = -2$ and $c = -5$

$$x = \frac{2 \pm \sqrt{24}}{2}$$

You will see how to simplify this further in Example 7.

Exercise 26C

1 For each of the following, state whether they are rational or irrational numbers.

(a) $\sqrt{2}$ (b) $\frac{3}{5}$ (c) 0.85 (d) $\sqrt{25}$

(e) $\sqrt{20}$ (f) $0.6666...$ (g) $7 - \sqrt{16}$ (h) π

(i) $\sqrt[3]{27}$ (j) $\sqrt[3]{36}$ (k) $\sqrt{36} + 2$ (l) $7 - \sqrt{2}$

(m) $4 - \pi$ (n) π^2 (o) $\sqrt{3} - \sqrt{9}$ (p) $0.127\,845$ ——— **Not** recurring.

2 Give the length in surd form of the side of a square with area $2\ \text{cm}^2$.

3 Find the length of the hypotenuse in this right-angled triangle.
Give your answer in surd form.

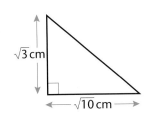

$\sqrt{3}\ \text{cm}$ $\sqrt{10}\ \text{cm}$

> Remember the length must be positive. In general when you see $\sqrt{10}$, for example, it means the positive square root.

4 Find the length of the side of a cube with a volume of $124\ \text{cm}^3$. Give your answer in surd form.

5 Solve the equation $x^2 - 4x - 11 = 0$, leaving your answer in surd form.

26.5 Rules for manipulating surds

There are rules you can use to manipulate and simplify surds. Here is a summary of those you need to know.

$$\sqrt{a} \times \sqrt{b} = \sqrt{ab} \qquad\qquad \frac{\sqrt{a}}{\sqrt{b}} = \sqrt{\frac{a}{b}}$$

$$m\sqrt{a} \times n\sqrt{b} = mn\sqrt{ab} \qquad m\sqrt{a} \div n\sqrt{b} = \frac{m}{n}\sqrt{\frac{a}{b}}$$

$$m\sqrt{a} + n\sqrt{a} = (m + n)\sqrt{a}$$

When you use these rules to simplify surds, look for factors that are square numbers (see Example 5).

Example 4

Work out each of the following, leaving your answers in surd form.

(a) $\sqrt{3} \times \sqrt{5}$ (b) $\sqrt{63} \div \sqrt{21}$ (c) $2\sqrt{11} \times 5\sqrt{2}$ (d) $12\sqrt{30} \div 3\sqrt{6}$

(a) $\sqrt{3} \times \sqrt{5} = \sqrt{3 \times 5} = \sqrt{15}$

(b) $\sqrt{63} \div \sqrt{21} = \sqrt{63 \div 21} = \sqrt{3}$

(c) $2\sqrt{11} \times 5\sqrt{2} = 2 \times 5 \times \sqrt{11 \times 2} = 10\sqrt{22}$

(d) $12\sqrt{30} \div 3\sqrt{6} = 12 \div 3 \times \sqrt{30 \div 6} = 4\sqrt{5}$

Example 5

Simplify each of the following, leaving your answers in surd form.

(a) $\sqrt{72}$ (b) $\sqrt{\dfrac{96}{50}}$ (c) $\sqrt{147} - \sqrt{48}$ (d) $\sqrt{252} - \sqrt{28}$

(a) $\sqrt{72} = \sqrt{36 \times 2} = \sqrt{36} \times \sqrt{2} = 6\sqrt{2}$

(b) $\sqrt{\dfrac{96}{50}} = \dfrac{\sqrt{96}}{\sqrt{50}} = \dfrac{\sqrt{16 \times 6}}{\sqrt{25 \times 2}} = \dfrac{\sqrt{16} \times \sqrt{6}}{\sqrt{25} \times \sqrt{2}} = \dfrac{4\sqrt{6}}{5\sqrt{2}} = \dfrac{4}{5}\sqrt{\dfrac{6}{2}} = \dfrac{4}{5}\sqrt{3}$

> Finding factors that are square numbers is the key to all these answers.

(c) $\sqrt{147} - \sqrt{48} = \sqrt{49 \times 3} - \sqrt{16 \times 3}$
$= \sqrt{49} \times \sqrt{3} - \sqrt{16} \times \sqrt{3}$
$= 7\sqrt{3} - 4\sqrt{3} = 3\sqrt{3}$

(d) $\sqrt{252} - \sqrt{28} = \sqrt{36 \times 7} - \sqrt{4 \times 7}$
$= \sqrt{36} \times \sqrt{7} - \sqrt{4} \times \sqrt{7}$
$= 6\sqrt{7} - 2\sqrt{7}$
$= 4\sqrt{7}$

Exercise 26D

1 Simplify these expressions:

(a) $\sqrt{3} \times \sqrt{3}$ (b) $\sqrt{7} \div \sqrt{7}$ (c) $\sqrt{5} \times \sqrt{5} \div \sqrt{5}$

2 Simplify the following, leaving your answers in surd form.

(a) $\sqrt{2} \times \sqrt{5}$ (b) $\sqrt{6} \div \sqrt{2}$ (c) $\sqrt{5} \times \sqrt{8}$

(d) $\sqrt{24} \div \sqrt{6}$ (e) $\sqrt{7} \times \sqrt{3}$ (f) $\sqrt{30} \div \sqrt{5}$

3 Simplify each of these, giving your answer in surd form.

(a) $3\sqrt{5} \times 2\sqrt{3}$ (b) $6\sqrt{10} \div 3\sqrt{2}$ (c) $14\sqrt{6} \div 7\sqrt{2}$

(d) $20\sqrt{18} \div 5\sqrt{6}$ (e) $4\sqrt{5} \times 2\sqrt{17}$ (f) $4\sqrt{6} \times 3\sqrt{11}$

4 Write each of the following surds in the form $\sqrt{a \times b}$, where a is a square number. There may be more than one possible answer.

(a) $\sqrt{50}$ (b) $\sqrt{32}$ (c) $\sqrt{24}$ (d) $\sqrt{27}$

(e) $\sqrt{45}$ (f) $\sqrt{72}$ (g) $\sqrt{200}$ (h) $\sqrt{1000}$

5 Simplify each of the following surds in the form $a\sqrt{b}$.

(a) $\sqrt{50}$ (b) $\sqrt{32}$ (c) $\sqrt{75}$ (d) $\sqrt{48}$

(e) $\sqrt{54}$ (f) $\sqrt{300}$ (g) $\sqrt{164}$ (h) $\sqrt{294}$

6 Simplify each of the following, leaving your answer in surd form.

(a) $\sqrt{\frac{32}{50}}$ (b) $\sqrt{\frac{24}{64}}$ (c) $\sqrt{\frac{27}{63}}$ (d) $\sqrt{\frac{48}{98}}$

7 Simplify

(a) $\sqrt{18} + \sqrt{8}$ (b) $\sqrt{20} + \sqrt{45}$ (c) $\sqrt{75} - \sqrt{48}$

(d) $\sqrt{150} - \sqrt{96}$ (e) $\sqrt{63} + \sqrt{175}$ (f) $\sqrt{162} - \sqrt{72}$

26.6 Rationalising the denominator

Key words:
rationalise

When the denominator of a fraction is a surd, remove the surd from the denominator by multiplying the top and bottom of the fraction by the appropriate square root.

This process is called **rationalising** the denominator.

For fractions of the form $\dfrac{a}{\sqrt{b}}$ you multiply numerator and denominator by $\sqrt{b}$ and simplify.

Example 6

Simplify:

(a) $\dfrac{1}{\sqrt{5}}$ (b) $\dfrac{12}{\sqrt{6}}$

(a) $\dfrac{1}{\sqrt{5}} = \dfrac{1 \times \sqrt{5}}{\sqrt{5} \times \sqrt{5}} = \dfrac{\sqrt{5}}{5}$

Notice that the aim is to get the denominator to be an integer (i.e. rational).

(b) $\dfrac{12}{\sqrt{6}} = \dfrac{12 \times \sqrt{6}}{\sqrt{6} \times \sqrt{6}} = \dfrac{12\sqrt{6}}{6} = 2\sqrt{6}$

Exercise 26E

1 Simplify

(a) $\dfrac{1}{\sqrt{2}}$ (b) $\dfrac{1}{\sqrt{11}}$ (c) $\dfrac{6}{\sqrt{3}}$ (d) $\dfrac{8}{\sqrt{2}}$

2 Simplify

(a) $\dfrac{1}{2\sqrt{3}}$ (b) $\dfrac{1}{5\sqrt{7}}$ (c) $\dfrac{10}{4\sqrt{5}}$ (d) $\dfrac{6}{8\sqrt{3}}$

3 Simplify each of the following

(a) $\dfrac{\sqrt{3}}{\sqrt{5}}$ (b) $\dfrac{\sqrt{10}}{2\sqrt{6}}$ (c) $\dfrac{\sqrt{3}}{\sqrt{12}}$ (d) $\dfrac{\sqrt{8}}{\sqrt{18}}$

4 Simplify

(a) $\dfrac{\sqrt{72}}{3\sqrt{2}}$ (b) $\dfrac{2\sqrt{27}}{3\sqrt{3}}$ (c) $\dfrac{4\sqrt{2}\times\sqrt{3}}{2\sqrt{2}}$ (d) $\dfrac{8\sqrt{2}\times 2\sqrt{15}}{4\sqrt{8}}$

5 Multiply these fractions, simplifying your answer.

(a) $\dfrac{1}{\sqrt{2}}\times\dfrac{1}{\sqrt{3}}$ (b) $\dfrac{4}{\sqrt{3}}\times\dfrac{3}{\sqrt{5}}$ (c) $\dfrac{\sqrt{3}}{\sqrt{5}}\times\dfrac{\sqrt{3}}{\sqrt{6}}$ (d) $\dfrac{\sqrt{2}}{\sqrt{3}}\times\dfrac{\sqrt{2}}{\sqrt{12}}$

26.7 Applications of surds

You will often be asked to give answers to problems in surd form. This saves you some final calculation, and also provides the answer in a more accurate form. You will also find it useful to use surds throughout the process of solving a problem. You can then cancel and simplify at the final stage.

Example 7

Solve the quadratic equation $x^2 - 2x - 5 = 0$ giving your answers in surd form.

> This is Example 3, but with the final simplified answer.

The solution for Example 3 was:

$$x = \dfrac{2\pm\sqrt{24}}{2} = \dfrac{2\pm\sqrt{4\times 6}}{2} = \dfrac{2\pm\sqrt{4}\times\sqrt{6}}{2}$$

> $\sqrt{24}\equiv\sqrt{4}\times\sqrt{6}$

$$= \dfrac{2\pm 2\sqrt{6}}{2} = 1\pm\sqrt{6}$$

> Divide throughout by 2.

Example 8

In the right-angled triangle ABC, $AB = \sqrt{15}$ cm and $BC = \sqrt{12}$ cm.
Calculate the length of AC, giving your answer in its simplest form.

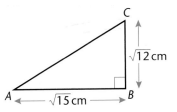

$AC^2 = AB^2 + BC^2$
$AC^2 = (\sqrt{15})^2 + (\sqrt{12})^2$
$AC^2 = 15 + 12$
$AC^2 = 27$
$AC = \sqrt{27}$
$AC = \sqrt{9 \times 3}$
$AC = \sqrt{9} \times \sqrt{3}$
$AC = 3\sqrt{3}$ cm

> Use Pythagoras' theorem.

Example 9

The sides of a square are of length $(7 - \sqrt{2})$ cm.
Calculate the area of the square.

$\text{Area of square} = (7 - \sqrt{2}) \times (7 - \sqrt{2})$
$= 49 - 7\sqrt{2} - 7\sqrt{2} + 2$
$= 51 - 14\sqrt{2}$ cm^2

> Notice that when you square the bracket you get 4 terms.
>
> Notice also that $(-\sqrt{2}) \times (-\sqrt{2}) = +2$

Example 10

The area of a triangle is 20 cm^2. The base is of length $5\sqrt{2}$ cm.
Calculate the height of the triangle.

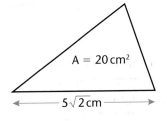

$\text{Area of triangle} = \frac{1}{2} \times \text{base} \times \text{height}$
$20 = \frac{1}{2} \times 5\sqrt{2} \times h$
$40 = 5\sqrt{2} \times h$
$\dfrac{40}{5\sqrt{2}} = h$

$h = \dfrac{40 \times \sqrt{2}}{5\sqrt{2} \times \sqrt{2}}$

$h = \dfrac{40 \times \sqrt{2}}{5 \times 2}$

$h = \dfrac{40\sqrt{2}}{10}$

$h = 4\sqrt{2}$ cm

> Multiplying by 2.

> Divide by $5\sqrt{2}$.

> Rationalise the denominator.

Exercise 26F

1 Find the area of this rectangle, giving your answer as a surd in its simplest form.

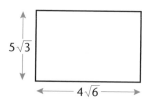

$5\sqrt{3}$

$4\sqrt{6}$

2 In this right-angled triangle, find the length of the hypotenuse. Give your answer as a surd in its simplest form.

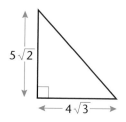

$5\sqrt{2}$

$4\sqrt{3}$

3 Solve the equation $x^2 - 6x - 9 = 0$, leaving your answer in surd form.

4 The area of a parallelogram is 27 cm². If the height is $\sqrt{3}$ what will be the length of the base?

5 **(a)** Which of the following are irrational numbers?

$$1.666... \qquad 2\sqrt{3} \qquad \tfrac{3}{4} \qquad 9 - \sqrt{4}$$

(b) Find a pair of irrational numbers, a, b, such that $a \times b$ is rational.

6 Find the area of a circle with a radius of $\dfrac{3}{\sqrt{2}}$ units.

7 Simplify the following:

(a) $\dfrac{2}{\sqrt{7}}$ **(b)** $\sqrt{90} + \sqrt{98}$ **(c)** $\dfrac{2\sqrt{10} \times 3\sqrt{10}}{3\sqrt{5} \times \sqrt{2}}$

8 A square has a side length of $5 + \sqrt{3}$ units.

(a) Find its perimeter.   **(b)** Find its area.

9 Find fractions which are equivalent to the following decimals.

(a) 0.458   **(b)** 0.7777...   **(c)** 0.450 450...

10 A rectangle has a length of $3\sqrt{2}$ and a height of $2\sqrt{3}$.

(a) Find its perimeter.

(b) Find its area.

(c) Find the length of the diagonal.

11 By first changing these recurring decimals into fractions, show that

$$0.\dot{4} + 0.\dot{5} = 1$$

12 Find the area of this isosceles triangle.

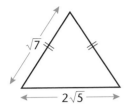

13 Calculate:

$$(\sqrt{11} + 3)(\sqrt{11} - 3)$$

 14 Find the length AB, given that ADC and ACB are right angles.

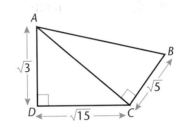

15 Find the value of y that makes these true.

(a) $\sqrt{3} \times \sqrt{y} = \sqrt{15}$ (b) $\sqrt{3} \times \sqrt{y} = 9$ (c) $2\sqrt{6} \times 3\sqrt{y} = 72$

(d) $\dfrac{1}{3\sqrt{y}} = \dfrac{\sqrt{11}}{33}$ (e) $\left(\dfrac{5}{\sqrt{y}}\right)^2 = \dfrac{1}{4}$ (f) $\dfrac{\sqrt{y} \times \sqrt{18}}{\sqrt{3}} = 2\sqrt{3}$

Examination questions

1 (a) You are given that $\sqrt{12} + \sqrt{27} = a\sqrt{3}$ where a is an integer.
Find the value of a.
(b) Find the value of $(m + p)^2$ when $m = \sqrt{2}$ and $p = \sqrt{8}$

(4 marks)
AQA, Spec A, 1H, June 2003

2 Express $0.\dot{4}\dot{8}$ as a fraction in its simplest form.

(2 marks)
AQA, Spec A, 2H, June 2003

3 (a) Express $0.\dot{4}\dot{2}$ as a fraction in its simplest form.
(b) Hence, or otherwise, express $0.7\dot{4}\dot{2}$ as a fraction.
Write this fraction in its simplest form.

(5 marks)
AQA, Spec B, 3H, March 2004

4 (a) By rationalising the denominator, simplify $\dfrac{15}{\sqrt{5}}$

(b) Show that $(\sqrt{3} + \sqrt{12})^2 = 27$

(4 marks)
AQA, Spec B, 3H, June 2003

Summary of key points

A *rational* number is one that can be written in the form $\dfrac{a}{b}$ where a and b are integers and $b \neq 0$. An *irrational* number is one that cannot be written in this way.

Converting fractions to decimals (grade C/B)

Only those fractions that have a denominator with prime factors of *only* 2 and/or 5 can be converted to terminating decimals.

Decimals which do not terminate are called *recurring* decimals.

When a single digit repeats, you put a dot over this digit.
When more than one digit repeats, you put a dot over the *first* digit of the pattern and a dot over the *last* digit of the pattern.

For example, $\frac{8}{11} = 0.727272\ldots$ is written as $0.\dot{7}\dot{2}$,
$\frac{3}{7} = 0.42857142857142\ldots$ is written as $0.\dot{4}2857\dot{1}$.

Converting terminating decimals to fractions (grade C/B)

To turn a terminating decimal into a fraction, look at the place value of the last significant digit and write the fraction with this denominator. Simplify if possible.

For example, $0.48 = \frac{48}{100} = \frac{12}{25}$, $\quad 0.225 = \frac{225}{1000} = \frac{9}{40}$

Converting recurring decimals to fractions (grade A/A*)

When the decimal is a recurring decimal you need a special technique to find the equivalent fraction.

When **1** digit recurs multiply both sides by **10**
When **2** digits recur multiply both sides by **100**
When **3** digits recur multiply both sides by **1000**

Manipulating surds (grade A/A*)

An irrational number does not terminate and there is no recurring pattern.

For example, $\sqrt{2} = 1.414\,213\,562\ldots$ $\quad \sqrt[3]{10} = 2.154\,434\,69\ldots$

Surds are irrational numbers written in a form using square roots and cube roots.

$$\sqrt{a} \times \sqrt{b} = \sqrt{ab} \qquad \frac{\sqrt{a}}{\sqrt{b}} = \sqrt{\frac{a}{b}} \qquad m\sqrt{a} \times n\sqrt{b} = mn\sqrt{ab}$$

$$m\sqrt{a} \div n\sqrt{b} = \frac{m}{n}\sqrt{\frac{a}{b}} \qquad m\sqrt{a} + n\sqrt{a} = (m+n)\sqrt{a}$$

When the denominator of a fraction is a surd, remove the surd from the denominator by multiplying the top and bottom of the fraction by the appropriate square root.

For fractions of the form $\dfrac{a}{\sqrt{b}}$ you multiply numerator and denominator by $\sqrt{b}$ and simplify.

This chapter will show you how to:

✔ draw graphs of sine, cosine and tangent
✔ solve trigonometric equations using the symmetry properties of the trigonometric graphs
✔ find sine, cosine and tangent of 30°, 45° and 60° using right-angled triangles
✔ find areas of triangles using the trigonometrical formula for area
✔ find lengths and angles in triangles using the sine and cosine rules
✔ work out areas of segments of circles

27.1 Graphs of cos x, sin x and tan x

Key words:
quadrant
period
discontinuity
asymptote

Look at this circle, centre O, of radius 1.

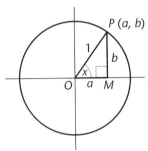

$P(a, b)$ is any point in the first quadrant of the circle.
Triangle OPM is right-angled and angle $POM = x°$

Using the definitions of sine and cosine that you learnt in Chapter 22, Section 22.3

$$\cos x = \frac{\text{adjacent}}{\text{hypotenuse}} = \frac{OM}{OP} = \frac{a}{1} = a$$

and $\sin x = \frac{\text{opposite}}{\text{hypotenuse}} = \frac{PM}{OP} = \frac{b}{1} = b$

These definitions can be used to map out the graphs of sine and cosine for angles of any size.

In other words $\cos x = a$ and $\sin x = b$

Consider what happens to the values of a and b as P moves around the circle from a starting point, N, 1 unit to the right of the origin, the position where $x = 0°$.
x will increase from 0°, to 90°, to 180°, to 270° and will complete one revolution when $x = 360°$.

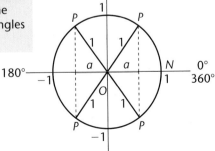

Drawing the graph of cos *x*

In the first **quadrant** , $a = 1$ when $x = 0°$
then *a* gradually decreases from 1 to 0
as *x* increases from 0° to 90°.
$a = 0$ when $x = 90°$.

In the second quadrant, $a = 0$ when $x = 90°$
then *a* gradually decreases from 0 to -1
as *x* increases from 90° to 180°.
$a = -1$ when $x = 180°$.

In the third quadrant, $a = -1$ when $x = 180°$
then *a* gradually increases from -1 to 0
as *x* increases from 180° to 270°.
$a = 0$ when $x = 270°$.

In the fourth quadrant, $a = 0$ when $x = 270°$
then *a* gradually increases from 0 to 1
as *x* increases from 270° to 360°.
$a = 1$ when $x = 360°$.

Then *P* is back at the starting point.

If you carry on, with *x* increasing yet further, to 450°, 540°, 630°
and 720°, the above pattern is repeated.
You can go on like this 'ad infinitum'.

> 'ad infinitum' means
> 'forever' or 'to infinity'.

The graph you will get for values of *x* plotted against the values of *a*
(where $a = \cos x$) is the one you can obtain from a graphical
calculator if you use it to show the graph of $y = \cos x$.

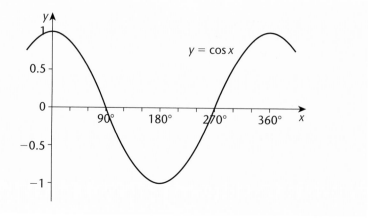

> Notice that the graph has a
> maximum value of $+1$ and
> a minimum value of -1.

You can draw a cosine graph by using your calculator to compile a
table of values similar to the one shown here. Values of *x* from 0° to
360° in intervals of 30° will give a good graph.

$x°$	0	30	60	90	120	150	180	210	240	270	300	330	360
$\cos x°$	1	0.87	0.5	0	−0.5	−0.87	−1	−0.87	−0.5	0	0.5	0.87	1

This table shows how the value of a (where $a = \cos x$) gradually decreases from 1 to 0 then to −1 then increases again to 0 and back to 1 as P moves round the circle.

You can see that the graph has symmetrical properties. A vertical line through $x = 180°$ is an axis of symmetry. This means that $\cos x = \cos(360° - x)$

If you drew the graph for negative values of x (by letting P travel round the circle from N in a clockwise direction) you would find that the y-axis is a line of symmetry. This means that $\cos x = \cos(-x)$

Other symmetrical relationships can easily be seen, for example,

$$\cos(180° - x) = -\cos x$$
$$\cos(180° - x) = \cos(180° + x)$$

The graph has a **period** of 360° (i.e. it repeats itself every 360°).

Drawing the graph of sin x

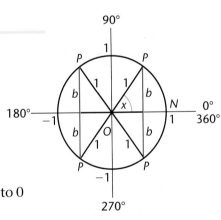

In the first quadrant, $b = 0$ when $x = 0°$
then b gradually increases from 0 to 1 as x increases from 0° to 90°.
$b = 1$ when $x = 90°$.

In the second quadrant, $b = 1$ when $x = 90°$
then b gradually decreases from 1 to 0 as x increases from 90° to 180°.
$b = 0$ when $x = 180°$.

In the third quadrant, $b = 0$ when $x = 180°$
then b gradually decreases from 0 to −1 as x increases from 180° to 270°.
$b = -1$ when $x = 270°$.

In the fourth quadrant, $b = -1$ when $x = 270°$
then b gradually increases from −1 to 0 as x increases from 270° to 360°.
$b = 0$ when $x = 360°$.

Then P is back at the starting point.

As before, you can carry on and the cycle will repeat itself.

The graph you will get for values of x plotted against the values of b (where $b = \sin x$) is the one you can obtain from a graphical calculator if you use it to show the graph of $y = \sin x$.

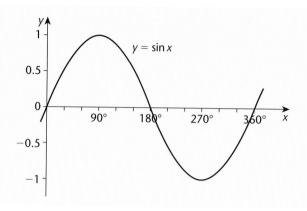

Notice that the graph has a maximum value of $+1$ and a minimum value of -1.

This graph also has symmetrical properties.
Some of the more obvious ones are:

$$\sin x = \sin(180° - x)$$
$$\sin(360° - x) = -\sin x$$
$$\sin(90° - x) = \sin(90° + x)$$

You can draw a graph from a table of values in the same way as for $y = \cos x$

If you drew the graph for negative values of x, then you would see that
$$\sin(-x) = -\sin x$$

The graph has a period of $360°$ (i.e. it repeats itself every $360°$).

Drawing the graph of tan x

The graph of $y = \tan x$ is a little more difficult to deduce.

$$\tan x = \frac{\text{opposite}}{\text{adjacent}} = \frac{b}{a}$$

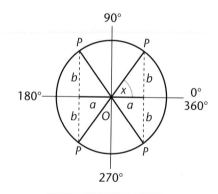

$a = 1$ and $b = 0$ when $x = 0°$

so when $x = 0°$ $\quad \tan x = \dfrac{0}{1} = 0$

$a = 0$ and $b = 1$ when $x = 90°$

so when $x = 90°$ $\quad \tan x = \dfrac{1}{0}$ which is undefined

$a = -1$ and $b = 0$ when $x = 180°$

so when $x = 180°$ $\quad \tan x = \dfrac{0}{-1} = 0$

$a = 0$ and $b = -1$ when $x = 270°$

so when $x = 270°$ $\quad \tan x = \dfrac{-1}{0}$ which is undefined

$a = 1$ and $b = 0$ when $x = 360°$

so when $x = 360°$ $\quad \tan x = \dfrac{0}{1} = 0$

Dividing by zero gives an answer which is undefined. The value is infinite.

If you look at the value of tan x as x increases from 0° to 90° you will see that the value of tan x increases very rapidly close to $x = 90°$. For example, tan 70° ≈ 2.75, tan 80° ≈ 5.67, tan 85° ≈ 11.4, tan 88° ≈ 28.6, tan 89° ≈ 57.3. You can see the effect of this when you look at the graph and see how it tends to infinity as x nears 90°.

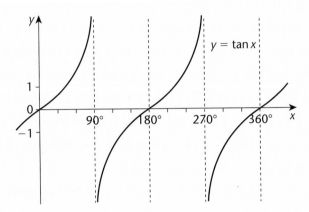

You can use a graphical calculator to plot a graph of $y = \tan x$.

The points where the graph tends to infinity are called **discontinuities** .

The vertical lines such as $x = 90°$ and $x = 270°$ are called **asymptotes** .

Notice that this graph also has symmetrical properties, for example, $\tan(180° - x) = -\tan x$ and $\tan(-x) = -\tan x$.

The period of the tangent graph is 180° (not 360°). (This means it repeats itself every 180°.)

The symmetry properties of the sine, cosine and tangent curves are important because you can use them to solve trigonometric equations, as you will see in the next section.

27.2 Solving trigonometric equations

Key words:
trigonometric equation

A **trigonometric equation** is an equation involving a trigonometric function.

Because sine and cosine have a period of 360° and tangent has a period of 180° and all the graphs continue *ad infinitum*, there will be an infinite number of solutions to any trigonometric equation.
You need to look at the range of values required for the solution and make sure you state answers in the correct range.

Example 1

You are given that sin 34° = 0.5592.
For values of x in the range $0° \leqslant x \leqslant 360°$,
use the symmetry of the sine curve to answer the following:

(a) Find another solution of the equation sin x = 0.5592.

(b) Solve the equation sin x = −0.5592.

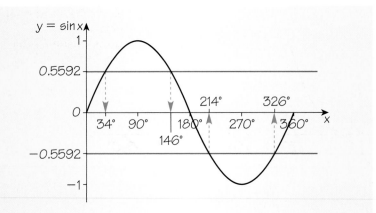

Notice how the given value of 34° is really important in finding other solutions. The symmetry properties of the curve are the key to finding the correct solutions.

(a) Another solution of sin x = 0.5592 is x = 180° − 34° = 146°

(b) The solutions to sin x = −0.5592 are x = 180° + 34° = 214°

and   x = 360° − 34° = 326°

The question only asks for solutions in the range $0° \leqslant x \leqslant 360°$.

Example 2

Solve the equation cos x = 0.3090 for $0° \leqslant x \leqslant 360°$.
Give your answers correct to the nearest degree.

One solution of cos x = 0.3090 is x = cos⁻¹(0.3090) which

gives x = 72°

$y = \cos x$

The other solution is x = 360° − 72° = 288°

Use the inverse cos function on your calculator to find one solution.
cos⁻¹(0.3090) = 72.001 02
... rounds to 72°

Remember to check that your calculator is set to **degree** mode.

Sketch the graph of $y = \cos x$.

Example 3

Find all the values of x that satisfy the equation $\tan x = -1.327$ for $0° \leqslant x \leqslant 540°$.
Give your answers to the nearest degree.

One solution of $\tan x = -1.327$ is $x = \tan^{-1}(-1.327)$ which gives

$x = -52.999... = -53°$

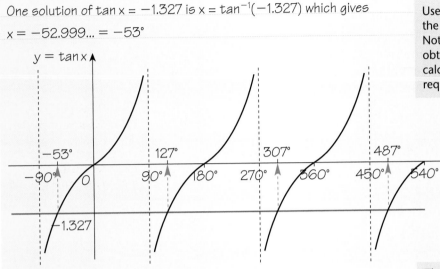

There are three solutions in the range $0° \leqslant x \leqslant 540°$

they are $x = -53° + 180° = 127°$

$x = 127° + 180° = 307°$

$x = 307° + 180° = 487°$

Use the tan⁻¹ function on the calculator.
Notice that the solution obtained from the calculator is not in the required region.

The period of 180° enables you to find the other solutions quite easily.

Example 4

Find the solution to the equation $3 \sin x = 1$ in the range $0° \leqslant x \leqslant 540°$.

$3 \sin x = 1$

$\sin x = \frac{1}{3}$

$x = \sin^{-1}\left(\frac{1}{3}\right)$

$x = 19.4712°... = 19.5°$

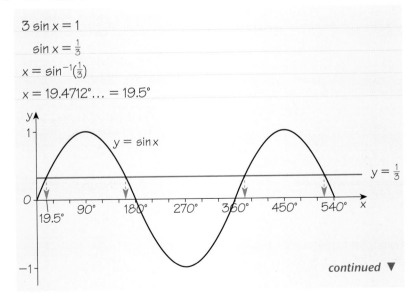

continued ▼

Begin by rearranging the equation to make $\sin x$ the subject.
Use your calculator to work out the inverse function to find one solution.
Other solutions can be found by using the symmetry of the sine graph.

The next solution is $(720 + 19.5) = 739.5$ which is outside the range of angle required.

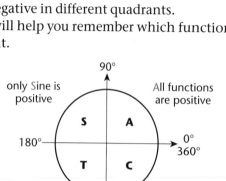

From the symmetry of the graph,

$$x = 19.5°, (180° − 19.5°), (360° + 19.5°), (540° − 19.5°)$$

so the full set of solutions is

$$x = 19.5°, 160.5°, 379.5° \text{ and } 520.5°$$

You can see from the graphs of sin, cos and tan that they are positive and negative in different quadrants.

This diagram will help you remember which function is positive in which quadrant.

90°

only Sine is positive | All functions are positive

S **A**

180° — 0° 360°

T **C**

only Tangent is positive | only Cosine is positive

270°

A method of remembering this is to start from the 4th quadrant going anti-clockwise and spell out the word **CAST**.

Useful results

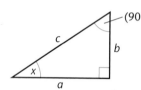

$$\sin x = \frac{b}{c} \quad \cos x = \frac{a}{c}$$

$$\cos (90° − x) = \frac{b}{c} \quad \sin (90° − x) = \frac{a}{c}$$

So, $\sin x = \cos(90° − x)$ and $\cos x = \sin(90° − x)$

These results mean that you can interchange sin and cos for acute angles as long as you remember to change the angle by subtracting it from 90°.

Exercise 27A

1 Find two angles in the range $0° \leqslant x \leqslant 360°$ that satisfy each of the following equations. Give your answers to the nearest degree.

(a) $\sin x = 0.8$ (b) $\tan x = 3.5$ (c) $\cos x = 0.45$

(d) $\tan x = -1.2$ (e) $\sin x = -0.24$ (f) $\cos x = -0.866$

Your calculator may give a negative value for x. Use your knowledge of the period of trigonometric graphs to extend them into the required range.

2 Find all solutions of the equation $\sin x = \frac{2}{3}$ for $0° \leqslant x \leqslant 360°$. Give your answers correct to the nearest degree.

3 Find all the solutions to the equation $\cos x = 0.5$ in the range $0° \leqslant x \leqslant 360°$.

4 Find all the solutions to the equation $\tan x = 2.05$ in the range $-180° \leqslant x \leqslant 180°$. Give your answers correct to the nearest degree.

5 You are given that cos 38° = 0.7880.

(a) Find another value of x in the range $0° \leqslant x \leqslant 360°$ for which cos x = 0.7880.

(b) Find all values of x in this range that satisfy the equation cos x = −0.7880.

(c) Find all solutions of the equation sin x = 0.7880 in this range.

6 Given that sin 67° = 0.9205

(a) find all values of x in the range $-360° \leqslant x \leqslant 360°$ that satisfy the equation sin x = −0.9205

(b) find all solutions of the equation cos x = 0.9205 in this range.

7 Solve the equation 4 cos x = 1 for values of x in the range $0° \leqslant x \leqslant 360°$. Give your solutions correct to 1 d.p.

8 Solve the equation 7 sin x = −4 for values of x in the range $-180° \leqslant x \leqslant 180°$. Give your solutions correct to 1 d.p.

27.3 Sine, cosine and tangent of 30°, 45° and 60°

Key words:
surd form

The graphs of sin x, cos x and tan x allow you to find the values of these functions easily when x = 0°, 90°, 180°, ... For other values you need to use a calculator. However, the sine, cosine and tangent of 30°, 45° and 60° can also be found without a calculator. Consider the two right-angled triangles:

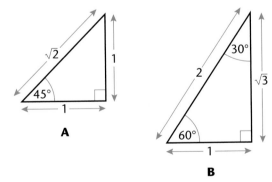

A is an isosceles right-angled triangle.
Hypotenuse of
$\mathbf{A} = \sqrt{1^2 + 1^2} = \sqrt{2}$

B is half an equilateral triangle of side 2.
Height of $\mathbf{B} = \sqrt{2^2 - 1^2} = \sqrt{3}$

The values of the side lengths are written either as integers or in **surd form** . Leaving the lengths in this way provides a method of evaluating sine, cosine and tangent of 30°, 45° and 60° **exactly** and is often useful in answering questions that involve these angles.

Remember: a surd is an irrational number, such as $\sqrt{5}$.
$\sqrt{5}$ = 2.236... on your calculator. It is not finite and cannot be written as an exact decimal.

(Handwritten notes:)

1) $-\frac{1}{2}$

2) $-\frac{1}{2}$.

3) $\tan 135 = \tan(180 - 45) = -\tan 45 = -1.$
$\frac{\sin 45}{-\cos 45} = -1.$

4) $\sin 300 = \sin(360 - 60) = -\sin 60 = -\frac{\sqrt{3}}{2}.$

5) $\cos 225 = \cos(180 + 45) = -\cos 45 = -\frac{1}{\sqrt{2}}$

6) $\tan 330 = \tan(360 - 30) = -\tan 30 = -\frac{1}{\sqrt{3}}.$

Example 5

Using appropriate right-angled triangles work out the exact value of
(a) sin 30° (b) cos 45° (c) tan 60°.

(a) Using triangle **B**

$$\sin 30° = \frac{opposite}{hypotenuse} = \frac{1}{2}$$

(b) Using triangle **A**

$$\cos 45° = \frac{adjacent}{hypotenuse} = \frac{1}{\sqrt{2}} = \frac{\sqrt{2}}{2}$$

Rationalise the denominator (see Section 26.6).

(c) Using triangle **B**

$$\tan 60° = \frac{opposite}{adjacent} = \frac{\sqrt{3}}{1} = \sqrt{3}$$

The following table gives you a complete set of results written in surd form.

x	0°	30°	45°	60°	90°
$\sin x$	0	$\frac{1}{2}$	$\frac{1}{\sqrt{2}}$	$\frac{\sqrt{3}}{2}$	1
$\cos x$	1	$\frac{\sqrt{3}}{2}$	$\frac{1}{\sqrt{2}}$	$\frac{1}{2}$	0
$\tan x$	0	$\frac{1}{\sqrt{3}}$	1	$\sqrt{3}$	∞

You do not need to remember these – only how to work them out using the two triangles.

∞ means 'infinity'.

Exercise 27B

1 In the diagram below work out the lengths of a and b. Leave your answer in surd form.

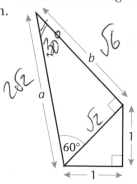

2 A child's slide is 3 m long and is elevated at an angle of 30° to the horizontal. How high is the slide above the ground?

3 A 5 m long flag pole is being erected and is kept in place using guy ropes at a 45° angle. What is the length of the guy rope? Give your answer as a surd in its simplest form.

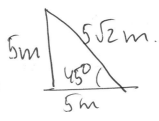

4 The cross section of a tent forms an equilateral triangle *ABC*, as shown below.

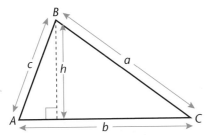

If the width of the tent (*AC*) is 1.8 m calculate
(a) the vertical height of the tent
(b) the slant height.

Leave your answer to part (a) in surd form.

5 In this triangle the cosine of angle *x* is $\frac{3}{4}$.
Work out the values of sin *x* and tan *x*, leaving your answers in surd form.

6 In this triangle the tangent of angle *x* is $\frac{7}{11}$.
Work out the values of sin *x* and cos *x*, leaving your answers in surd form.

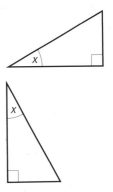

27.4 The area of a triangle using trigonometry

The area of a triangle is given by the expression $\frac{1}{2}bh$ where *b* is the base length and *h* is the perpendicular height. However, you can also work out the area of a triangle if you know the lengths of any two sides *and* the size of the angle between them.

The diagram below shows a general triangle *ABC* with side lengths given by *a*, *b* and *c*.

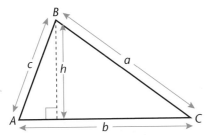

Notice how to label the sides of the triangle.
a is the side opposite angle *A*, *b* is opposite angle *B* and *c* is opposite angle *C*.

The area of the triangle is $\frac{1}{2} \times base \times height = \frac{1}{2}bh$

The height h is given by the trigonometric function $\sin C = \dfrac{h}{a}$ giving $h = a \sin C$.

Substituting this value of h into the original expression gives Area $= \frac{1}{2}b \times a \sin C$, which we can write as

Area $ABC = \frac{1}{2}ab \sin C$.

Similarly it can be shown that

$$\text{Area } ABC = \tfrac{1}{2}ac \sin B.$$
and
$$\text{Area } ABC = \tfrac{1}{2}bc \sin A.$$

> Notice that the two sides involved are a and b and the angle in between is C.

> This formula is given to you at the beginning of the exam paper.

> When the triangle is right-angled use the simple area formula
> Area $= \frac{1}{2} \times b \times h$

Example 6

Find the area of the triangle *PQR*.

From the diagram $p = 14$ cm and $q = 8$ cm.

Area $PQR = \frac{1}{2}pq \sin R$

$\qquad = \frac{1}{2} \times 14 \times 8 \times \sin 60°$

$\qquad = 56 \times 0.8660$

$\qquad = 48.4974$

$\qquad = 48.5$

Area $PQR = 48.5$ cm^2

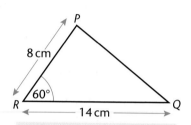

> $\sin 60°$ also equals $\dfrac{\sqrt{3}}{2}$ and using this would give the area $= \frac{1}{2} \times 14 \times 8 \times \dfrac{\sqrt{3}}{2}$
> $= 28\sqrt{3}$ cm^2 (this is exact).

Example 7

The area of a triangle *XYZ* is 90 cm^2. If $XY = 15$ cm and $XZ = 21.5$ cm work out angle *X* to the nearest degree.

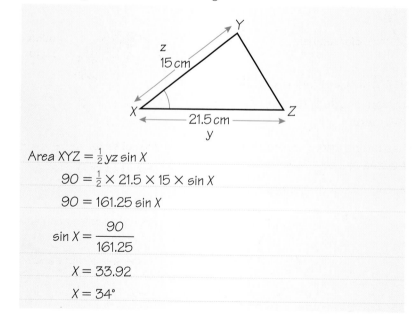

> Begin by sketching the triangle.

Area $XYZ = \frac{1}{2}yz \sin X$

$\quad 90 = \frac{1}{2} \times 21.5 \times 15 \times \sin X$

$\quad 90 = 161.25 \sin X$

$\sin X = \dfrac{90}{161.25}$

$\quad X = 33.92$

$\quad X = 34°$

Exercise 27C

1 Find the area of each triangle, giving your answers to 1 d.p.

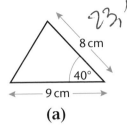

23,1

8 cm

40°

9 cm

(a)

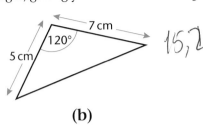

7 cm

120°

5 cm

15,2

(b)

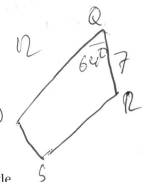

12

Q

64°

7

P

12

S

2 The area of a triangle *RST* is 64 cm². If all three angles of the triangle are acute and *RS* = 13 cm, *ST* = 15 cm, calculate the angle *S* to 2 d.p.

3 *PQRS* is a parallelogram with *PQ* = 12 cm, *QR* = 7 cm and angle *Q* = 64°. Work out the area of *PQRS*. Give your answer to 3 s.f.

> Draw a sketch first.

4 A triangular field *ABC* has an area of 8320 m². If two sides of the field are 120 m and 410 m long what is the angle between them?

5 A tetrahedron is constructed from four equilateral triangles of side length 12 cm. What is the total surface area of the tetrahedron?

27.5 Segment of a circle

> **Key words:**
> segment

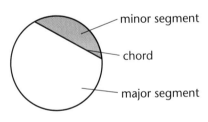

minor segment

chord

major segment

Inadient = $\frac{160}{51}$.

This diagram shows a circle with a chord, cutting the circle into two **segments** . The smaller portion is called the minor segment and the larger one is the major segment.

The chord divides a sector into a triangle and a segment. This leads to a method for finding the area of a segment.

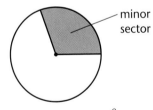

minor sector

Area of sector $= \dfrac{\theta}{360} \times \pi r^2$

(See Chapter 7)

Area of segment = area of sector − area of triangle

$$= \frac{\theta}{360} \times \pi r^2 - \frac{1}{2}r^2 \sin \theta$$

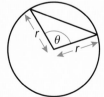

r *θ* *r*

> Area of triangle
> $= \frac{1}{2}ab \sin C$
>
> So the area of a triangle formed by the radii of a circle $= \frac{1}{2}r^2 \sin \theta$.

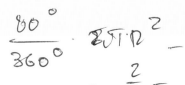

$$\frac{80°}{360°} \cdot \pi r^2 - \frac{1}{2} \sin 80° \cdot 100 =$$

$$= \frac{2}{9} \pi \cdot 100 - 50 \sin 80 °$$

Example 8

(a) Find the arc length and area of the minor sector of this circle.

(b) Find the area of the minor segment cut off by the chord of this circle.

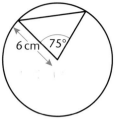

(a) Arc length $= \frac{75}{360} \times 2 \times \pi \times 6$

$\qquad = 7.85 \text{ cm (3 s.f)} \quad \text{or} \quad 2.5\pi \text{ cm}$

Sector area $= \frac{75}{360} \times \pi \times 6^2$

$\qquad = 23.5619\ldots$

$\qquad = 23.6 \text{ cm}^2 \text{ (3 s.f)} \quad \text{or} \quad 7.5\pi \text{ cm}^2$

> Part **(a)** of this question could be on the non-calculator paper or the calculator paper.

(b) Area of segment $=$ area of sector $-$ area of triangle

$\qquad = 23.5619\ldots - \frac{1}{2} \times 6^2 \times \sin 75°$

$\qquad = 23.5619\ldots - 17.3866\ldots$

$\qquad = 6.1752\ldots$

$\qquad = 6.18 \text{ cm}^2 \text{ (3 s.f.)}$

> Use the non-rounded version of the sector area and only round to 3 s.f. at the end.

Exercise 27D

1 Find the area of the shaded segments.
Give your answers to 1 d.p.

(a)

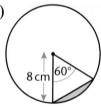

(b)

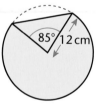

(c)

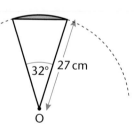

2 An Australian sound board is made out of a rectangular piece of wood 30 cm by 4 cm. It consists of a rectangle with a segment of a circle at each end.

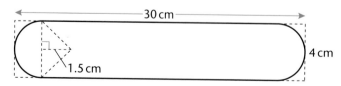

> Hint: Use Pythagoras to find the radius first.

(a) Find the area of the board.

(b) Find the perimeter.

27.6 The sine rule

This is a very useful trigonometric rule that allows you to work out side lengths or angles in any triangle.

Consider the triangle ABC with h, the perpendicular from B to the line AC, meeting the line at the point M.

From triangle ABM, $\sin A = \dfrac{h}{c}$ or $h = c \sin A$.

From triangle BCM, $\sin C = \dfrac{h}{a}$ or $h = a \sin C$.

So $c \sin A = a \sin C$

or $\dfrac{c}{\sin C} = \dfrac{a}{\sin A}$

This result is called the   **sine rule** .

It can easily be shown that both of these are equal to $\dfrac{b}{\sin B}$.

The full version of the sine rule is then written as:

$$\frac{a}{\sin A} = \frac{b}{\sin B} = \frac{c}{\sin C}$$

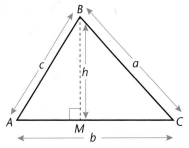

This is another formula that is given to you at the start of the exam paper.

Use the sine rule when you are given an opposite side and angle (e.g. a and A)

Example 9

In triangle PQR calculate the angle R to 3 significant figures.

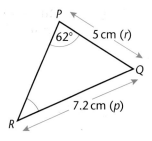

Using the sine rule:

$$\frac{p}{\sin P} = \frac{r}{\sin R}$$

$$\frac{\sin P}{p} = \frac{\sin R}{r}$$

$$\sin R = r\frac{\sin P}{p}$$

$$\sin R = 5 \times \frac{\sin 62°}{7.2}$$

$$= \frac{5 \times 0.8829}{7.2}$$

$$\sin R = 0.6131\ldots$$

$$R = 37.8182\ldots$$

$$R = 37.8° \text{ (correct to 3 s.f.)}$$

The sine rule can be inverted and this is useful when you want to work out an angle.

It makes it easier to rearrange the formula so that the unknown quantity is the numerator.

Exercise 27E

1 In triangle *ABC*, find the length *c* to 3 significant figures.

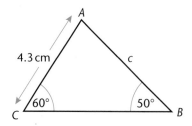

2 In the triangle *PQR*, angle *R* = 120°.
Find angle *Q* to 3 s.f.

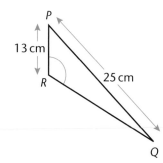

3 In triangles *ABC* find the length of each side marked with a letter.

(a)

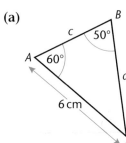

(b)

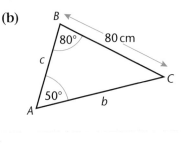

4 In a triangle *XYZ*, *XY* = 11 cm, *YZ* = 7.5 cm and the angle at
X = 28°. Work out the two possible values for the angle at *Y*.

5 Nab End is due north of Midwich. Jim sets off from Midwich (*M*)
and walks 8 km on a bearing 047° to reach Park Hill (*P*). He then
walks to Nab End (*N*) on a bearing of 310°.

  (a) Sketch the triangle *MNP*.

  (b) Find angle *MNP*.

  (c) Use the sine rule to find the distance from Park Hill to
  Nab End. Give your answer correct to 3 s.f.

6 Triangle *RST* has angle *TRS* = 31°, side *RS* = 8 cm and side
ST = 5 cm.

  (a) Sketch two possible triangles that fit the description above.

  (b) Find the two possible angles for ∠*STR*.

Remember:
$\sin(180 - x) = \sin x$

Key words:
cosine rule

27.7 The cosine rule

This is another useful trigonometric formula that can be used on any triangle.

Consider the triangle ABC with h the perpendicular from B to the line AC, meeting the line at the point M.

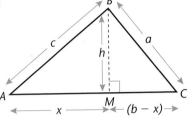

Using Pythagoras' theorem in triangle BCM gives

$$a^2 = h^2 + (b - x)^2$$
$$a^2 = h^2 + b^2 - 2bx + x^2$$
$$a^2 = h^2 + x^2 + b^2 - 2bx$$

If you now apply Pythagoras' theorem in triangle ABM you obtain

$$c^2 = h^2 + x^2$$

Substituting $h^2 + x^2 = c^2$ in the expression for a^2 gives

$$a^2 = c^2 + b^2 - 2bx$$

Now for triangle ABM we know that $\cos A = \dfrac{x}{c}$ and $x = c \cos A$

Substituting for x we obtain the expression

$$a^2 = b^2 + c^2 - 2bc \cos A$$

This is the **cosine rule** .

The cosine rule is given to you at the start of the exam paper.

Use the cosine rule when you are given two sides and the angle between them.

When the angle A is 90°, the cosine rule simply reduces to Pythagoras' theorem for a right-angled triangle.

The equation can also be rearranged to give the angle:

$$\cos A = \frac{b^2 + c^2 - a^2}{2bc}$$

Note that this version will calculate the size of an angle if you know all three sides.
The angle you are finding is opposite the side whose square you subtract in the numerator.

Example 10

In triangle PQR, $PQ = 15$ cm, $QR = 5$ cm and $PR = 12$ cm. Calculate the angle at Q.

Begin with a quick sketch.

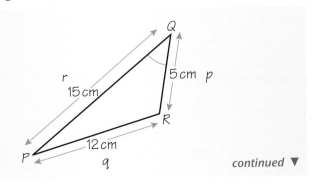

continued ▼

Using the cosine rule:

$$\cos Q = \frac{r^2 + p^2 - q^2}{2rp}$$

$$\cos Q = \frac{15^2 + 5^2 - 12^2}{2(15)(5)}$$

$$= \frac{106}{150}$$

$$\cos Q = 0.7066$$

$$Q = 45.04°$$

2(15)(5) means $2 \times 15 \times 5$
Brackets used in this way always mean 'multiply'.

Example 11

A yacht sets sail from port P and travels due East for 24 km towards a lighthouse L before turning on a bearing of 155°. The yacht travels a further 40 km towards a marker buoy B, before returning directly to port. How long is the journey from the buoy to port?

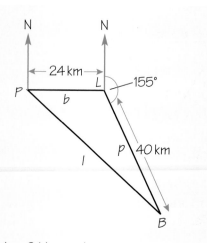

Begin with a sketch, marking on the diagram all the lengths and angles given.

Let $p = 40$ km, $b = 24$ km and

angle $PLB = (360° - 90° - 155°) = 115°$.

Let the distance required be l.

Using the cosine rule:

$$l^2 = b^2 + p^2 - 2bp \cos PLB$$

$$l^2 = 24^2 + 40^2 - (2 \times 24 \times 40) \cos 115°$$

$$l^2 = 2176 - (-811.42)$$

$$l^2 = 2987.42$$

$$l = 54.7 \text{ km (correct to 3 s.f.)}$$

Remember: for obtuse angles the cosine of an angle is negative, e.g. $\cos 120° = -\cos 60° = -0.5°$.

Your calculator automatically works out the value of l^2, you need not worry about the negative cosine.
Remember to take the square root to find l.

Exercise 27F

1 In the triangle ABC shown below $a = BC = 9$ cm, $c = AB = 7$ cm and angle $B = 65°$. Work out the length of b (AC) correct to 3 s.f.

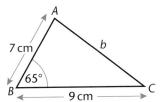

2 A triangular piece of land with corners at FGH is being fenced off as shown opposite. Only two sides have been completely fenced with $FG = 50$ m (h), $GH = 80$m (f) and the angle at G is $115°$. Work out how much more fencing is needed to complete the job.

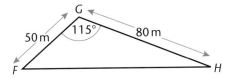

3 In the triangle shown, find angle C to 3 s.f.

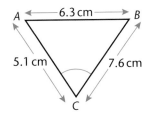

4 In triangle ABC, $AB = 14$ cm, $BC = 25$ cm and $AC = 15$ cm. Find the size of the angle at A (to 1 d.p.).

5 In a triangle, the side lengths are 3 cm, $\sqrt{8}$ cm and $\sqrt{3}$ cm. Find the sizes of all the angles in the triangle.

6 On a snooker table the cue and black balls are situated as shown in the diagram.

Find the distance the cue ball has to travel before hitting the black ball. Give your answer to the nearest cm.

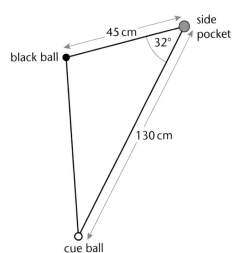

Examination questions

1 (a) Sketch the graph of $y = \sin x$ for values of x from $0°$ to $360°$.

(b) One solution of the equation $\sin x = 0.92$ is $x = 67°$.
Use your sketch graph to find another solution of this equation.

(c) Use your sketch graph to work out the value of $\sin 293°$.

(4 marks)
AQA June 2003

2 The diagram shows a triangle ABC. $AB = 6$ cm,
$BC = 5$ cm and angle $B = 75°$.
You are given that $\sin 75° = 0.966$ to
3 significant figures.
Calculate the area of the triangle.
Give your answer to a suitable degree of accuracy.

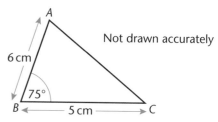

Not drawn accurately

(3 marks)
AQA June 2004

3 In triangle ABC, $AB = 11$ cm, $BC = 9$ cm
and $CA = 10$ cm.
Find the area of triangle ABC.

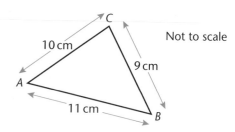

Not to scale

(5 marks)
AQA November 2003

4 Two ships, A and B, leave port at 1300 hours.
Ship A travels at a constant speed of
18 km per hour on a bearing of $070°$.
Ship B travels at a constant speed of
25 km per hour on a bearing of $152°$.

Calculate the distance between A and B at
1400 hours.

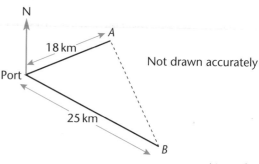

Not drawn accurately

(4 marks)
AQA June 2003

Summary of key points

Graphs of trigonometric functions (grade A/A*)

The three basic *trigonometric functions* are sine ($\sin x$), cosine ($\cos x$) and tangent ($\tan x$), where x is the angle in degrees.

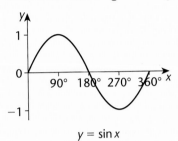

$$y = \sin x$$

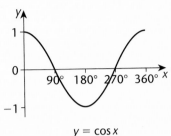

$$y = \cos x$$

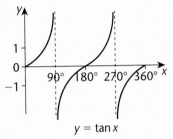

$$y = \tan x$$

These are some of the results you can deduce from symmetry:

$$\sin x = \sin(360 + x),$$
$$\cos x = \cos(360 + x),$$
$$\tan x = \tan(180 + x),$$

$$\sin(180 - x) = \sin x$$
$$\cos(180 - x) = -\cos x$$
$$\tan(180 - x) = -\tan x$$

$$\sin(-x) = -\sin x$$
$$\cos(-x) = \cos x$$
$$\tan(-x) = -\tan x$$

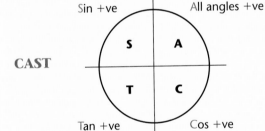

CAST

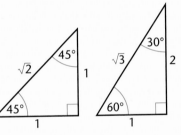

You can use these right-angled triangles to work out sin, cos and tan of 30°, 45° and 60° in surd form.

Trigonometric equations (grade A/A*)

Trigonometric equations are equations involving trigonometric functions of the form

$$\sin x = p \qquad \cos x = q \qquad \tan x = r$$

Segment of a circle (grade A)

Area of segment = area of sector − area of triangle

$$= \frac{\theta}{360} \times \pi r^2 - \tfrac{1}{2}r^2 \sin \theta$$

Triangle formulae (grade A/A*)

Area of triangle $= \tfrac{1}{2}ab \sin C$

Sine rule $\qquad \dfrac{a}{\sin A} = \dfrac{b}{\sin B} = \dfrac{c}{\sin C}$

Cosine rule $\quad a^2 = b^2 + c^2 - 2bc \cos A$

$$\cos A = \frac{b^2 + c^2 - a^2}{2bc}$$

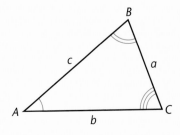

This chapter will show you how to:

✔ solve simultaneous equations where one is linear and one is quadratic
✔ find algebraically the intersection of a circle and a straight line
✔ simplify, add, subtract, multiply and divide algebraic fractions and use them to solve equations
✔ change of subject of a formula where the variable appears on both sides of the equation

28.1 Simultaneous equations ... one linear, one quadratic

Key words:
substitution

When two straight lines intersect they do so in only one point. In Chapter 5, Section 5.6, you learned how to solve two linear equations simultaneously to find the point of intersection.

When a quadratic graph and a linear graph are plotted on the same axes there are three possible situations:
- the linear graph **intersects** the quadratic graph at two points,
- the linear graph **touches** the quadratic graph at **one point**,
- the linear graph and the quadratic graph **do not intersect**.

This diagram shows the quadratic graph of $y = x^2 - 2x - 7$ and the graph of the straight line $y = x + 3$.

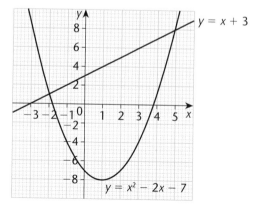

There are two points of intersection, $(-2, 1)$ and $(5, 8)$.

To find whether a quadratic graph and a linear graph intersect at 2, 1 or 0 points by a graphical method is time consuming. Unless any points of intersection are integer values, the answers will only be approximate.

> **To find solutions by an algebraic method means you have to find values of x and y that satisfy a quadratic equation and a linear equation simultaneously.**

> **To solve one linear and one quadratic simultaneous equation you need to use the method of substitution .**

You cannot use the method of elimination as for two linear equations because you cannot add or subtract the equations to eliminate a variable.

Example 1

Find any points of intersection of the quadratic graph $y = x^2 - 2x - 7$ and the graph of the straight line $y = x + 3$.

$$y = x + 3 \qquad (1)$$
$$y = x^2 - 2x - 7 \qquad (2)$$

Substitute the expression for y from equation (1) into equation (2):

$$x + 3 = x^2 - 2x - 7$$
$$x^2 - 3x - 10 = 0$$
$$(x - 5)(x + 2) = 0$$
$$x = 5 \quad \text{or} \quad x = -2$$

Substitute these x values into the linear equation (1):

When $x = 5$, $\quad y = 5 + 3 = 8$

When $x = -2$, $\quad y = -2 + 3 = 1$

There are two points of intersection $(5, 8)$ and $(-2, 1)$.

These are the graphs on page 508.

Always start with the linear equation and rearrange it (if necessary) to make x or y the subject.

This will nearly always be y because the linear equation is usually given in the form $y = \ldots$

Since both equations are given in the form $y = \ldots$ this method is equivalent to equating the two expressions.

Solve the resulting quadratic in x then substitute in the linear equation to find the y values.

This method is much quicker (and more accurate) than drawing graphs to find solutions. You can see that the coordinates of the two points of intersection are the same as those obtained by the graphical method.

The next example shows you how to handle substitution in a slightly different way.

Example 2

Find any points of intersection of the graph of $y^2 = 4x + 13$ and the graph of the straight line $y = 2x - 1$.

$$y = 2x - 1 \qquad (1)$$
$$y^2 = 4x + 13 \qquad (2)$$

From (1) $\qquad y + 1 = 2x$

so $\qquad 2y + 2 = 4x$

Substitute this expression for $4x$ in (2):

$$y^2 = 2y + 2 + 13$$
$$y^2 - 2y - 15 = 0$$
$$(y + 3)(y - 5) = 0$$
$$y = -3 \quad \text{or} \quad y = 5$$

Substituting these y values into the linear equation (1):

When $y = -3$, $\quad -3 = 2x - 1$, giving $x = -1$

When $y = 5$, $\qquad 5 = 2x - 1$, giving $x = 3$

There are two points of intersection $(-1, -3)$ and $(3, 5)$.

> Notice that the quadratic is of a different type.
>
> One way to solve this is to spot that one equation contains $2x$ and the other one contains $4x$.
>
> Since these are the only x terms in each equation you can easily eliminate x and solve the resulting quadratic in y.

Here is another solution ... equally good!

$$y = 2x - 1 \qquad (1)$$
$$y^2 = 4x + 13 \qquad (2)$$

Square equation (1) and substitute for y^2 in equation (2):

From (1) $\qquad y^2 = (2x - 1)^2 = 4x^2 - 2x - 2x + 1$
$$= 4x^2 - 4x + 1$$

Substitute in (2):

$$4x^2 - 4x + 1 = 4x + 13$$
$$4x^2 - 8x - 12 = 0$$
$$x^2 - 2x - 3 = 0$$
$$(x + 1)(x - 3) = 0$$
$$x = -1 \quad \text{or} \quad x = 3$$

From equation (1) $\quad y = -3 \quad$ when $x = -1$

and $\qquad\qquad\qquad y = 5 \quad$ when $x = 3$

The two solutions are $(-1, -3)$ and $(3, 5)$.

> Remember that squaring a linear expression will give a quadratic with 3 terms.
>
> Always put a bracket round the linear term before you square it. This will help you to remember the correct method for expanding the bracket.

> Divide each term by 4.

Exercise 28A

Use the method of substitution to solve each of these pairs of simultaneous equations.

State clearly the points of intersection of the straight line graph and the quadratic graph in each case.

1 $y = x + 1$
$y = x^2 - 5$

2 $y = x + 4$
$y = x^2 - 2x$

3 $y = 4x$
$y = x^2 - 2x + 5$

4 $y = 1 - x$
$y = 2x^2 - 5$

5 $y = 2x - 1$
$y = x^2 + 4x$

6 $y = 2x - 3$
$y = 2x^2 - x - 23$

7 $y = 5x + 2$
$y = x^2 + x - 10$

8 $3y = x + 6$
$y^2 = 2x + 7$

28.2 The equation of a circle, centre (0, 0)

Look at this diagram of a circle with its centre at the origin.

If the coordinates of point P are (x, y), then, using Pythagoras' theorem

$$x^2 + y^2 = r^2$$

This is true for any point on the circumference and it is the general equation of any circle with its centre at the origin.

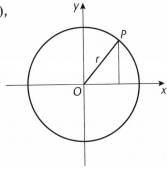

To find any points of intersection between a straight line and a circle you use the methods of the last section.

Example 3

(a) Show that the points of intersection of the straight line
$y = x + 4$ and the circle $x^2 + y^2 = 9$ are given by the equation
$$2x^2 + 8x + 7 = 0$$

(b) Solve this equation to find the x values of the points of intersection, leaving your answers in surd form.

(a)
$$y = x + 4 \qquad (1)$$
$$x^2 + y^2 = 9 \qquad (2)$$

Square equation (1):
$$y^2 = (x + 4)^2 = x^2 + 4x + 4x + 16$$
$$= x^2 + 8x + 16$$

Substitute for y^2 in equation (2):
$$x^2 + x^2 + 8x + 16 = 9$$
$$2x^2 + 8x + 7 = 0$$

(b) $a = 2 \quad b = 8 \quad c = 7$

$$x = \frac{-8 \pm \sqrt{8^2 - 4(2)(7)}}{2(2)}$$

$$= \frac{-8 \pm \sqrt{8}}{4}$$

$$= \frac{-8 \pm 2\sqrt{2}}{4}$$

$$x = \frac{-4 \pm \sqrt{2}}{2}$$

Suppose the straight line had been $y = x + 5$.

Verify for yourself that the quadratic that will need to be solved is
$$2x^2 + 10x + 16 = 0$$
What happens when you try to solve this equation? Why does this happen?

(**Hint:** try drawing a sketch of the two graphs.)

Solving $2x^2 + 8x + 7 = 0$ by the formula.

$$x = \frac{-b \pm \sqrt{b^2 - 4ac}}{2a}$$

Notice that $4(2)(7)$ means $4 \times 2 \times 7$.

Exercise 28B

Find the points of intersection of the circle and the straight line in each of the following.

1 $x^2 + y^2 = 29$
$y = x - 7$

2 $x^2 + y^2 = 2$
$y = 2x + 1$

3 $x^2 + y^2 = 10$
$x + 2y = 5$

4 $x^2 + y^2 = 25$
$y = x + 3$

5 $x^2 + y^2 = 9$
$x + y = 2$

6 $x^2 + y^2 = 16$
$2x - y = 1$

In questions **7** and **8** say whether the circle and the straight line have

(**a**) 2 points of intersection,

(**b**) 1 point of intersection (the line is a tangent to the circle),

(**c**) 0 points of intersection.

7 $x^2 + y^2 = 8$
$y = x - 4$

8 $x^2 + y^2 = 12$
$x + y = 6$

You do not need to solve completely. Use the discriminant.
See Chapter 24.

28.3 Simplifying algebraic fractions

You will need to know:
- how to factorise ... common factor, quadratic, difference of two perfect squares

Algebraic fractions have letters, or combinations of letters and numbers, in their numerator and denominator.

$$\frac{4}{a} \quad \frac{b}{2} \quad \frac{2pq}{r} \quad \frac{x+6}{x} \quad \frac{4}{x-5} \quad \frac{18x^3}{3x} \quad \frac{2x-6}{10x}$$ are all algebraic fractions.

Algebraic fractions can sometimes be simplified. Look for a common factor of the numerator and the denominator.

$\dfrac{x+6}{x}$ is an algebraic fraction that cannot be simplified.

> You cannot cancel the x terms ... **never** cancel across a $+$ or a $-$ sign.

$\dfrac{18x^3}{3x}$ can be simplified. $\dfrac{18x^3}{3x} = \dfrac{18 \times x \times x \times x}{3 \times x} = 6x^2$

> You can see that there are common factors of 3 and x.

$\dfrac{2x-6}{10x}$ can be simplified. $\dfrac{2x-6}{10x} = \dfrac{2(x-3)}{10x} = \dfrac{x-3}{5x}$

> Factorise then divide top and bottom by 2.

To simplify algebraic fractions:

(1) factorise
(2) divide by the HCF (or cancel common factors).

> Remember to factorise first.

Example 4

Simplify the following, where possible.

(a) $\dfrac{2x}{x+3}$ **(b)** $\dfrac{3ab}{6a^2}$ **(c)** $\dfrac{4x-16}{8}$

(d) $\dfrac{x^2+3x-4}{x+4}$ **(e)** $\dfrac{m^2-25}{m^2-3m-10}$

(a) $\dfrac{2x}{x+3}$ *cannot be simplified*

> Always look for a common factor first.

(b) $\dfrac{3ab}{6a^2} = \dfrac{3 \times a \times b}{6 \times a \times a} = \dfrac{b}{2a}$

> Then look for other forms of factorising such as quadratics and the difference of two perfect squares.

(c) $\dfrac{4x-16}{8} = \dfrac{4(x-4)}{8} = \dfrac{x-4}{2}$

(d) $\dfrac{x^2+3x-4}{x+4} = \dfrac{(x+4)(x-1)}{(x+4)} = x-1$

> All of these are present in these examples.

(e) $\dfrac{m^2-25}{m^2-3m-10} = \dfrac{(m+5)(m-5)}{(m-5)(m+2)} = \dfrac{m+5}{m+2}$

> Remember that you can only cancel whole brackets.

Exercise 28C

Simplify the following expressions, where possible.

1 $\dfrac{3x + 6}{9}$ **2** $\dfrac{4x + 5}{2}$ **3** $\dfrac{4x}{2x - 10}$

4 $\dfrac{4a^2}{6ab}$ **5** $\dfrac{3a + 2b}{a}$ **6** $\dfrac{10 - 5f}{2f}$

7 $\dfrac{ab + 2a}{a^2}$ **8** $\dfrac{2p + 4}{3p + 6}$ **9** $\dfrac{5 - p}{5 + p}$

10 $\dfrac{x + 2}{x^2 - x - 6}$ **11** $\dfrac{x^2 + 7x + 6}{x + 1}$ **12** $\dfrac{y^2 - y - 20}{y^2 + y - 12}$

13 $\dfrac{x^2 - 36}{x^2 - 8x + 12}$ **14** $\dfrac{m^2 - 1}{5m^2 - 2m - 3}$ **15** $\dfrac{2x^2 - 5x - 12}{3x^2 - 7x - 20}$

28.4 Multiplying and dividing algebraic fractions

To multiply numerical fractions you multiply the numerators and multiply the denominators, simplifying if possible.

To divide numerical fractions you invert the second fraction and multiply, simplifying if possible.

The rules are exactly the same for algebraic fractions, but always remember to look for any factors before you simplify.

Example 5

Work out the following, simplifying your answers.

(a) $\dfrac{2pq}{r} \times \dfrac{pr^2}{3q^2}$ **(b)** $\dfrac{4}{a} \div \dfrac{b}{2}$ **(c)** $\dfrac{2x + 6}{x^2 + 4x} \div \dfrac{x^2 + 2x - 3}{x^2 - 16}$

(a) $\dfrac{2pq}{r} \times \dfrac{pr^2}{3q^2} = \dfrac{2 \times p \times \cancel{q}}{\cancel{r}} \times \dfrac{p \times \cancel{r} \times r}{3 \times \cancel{q} \times q} = \dfrac{2p^2r}{3q}$

(b) $\dfrac{4}{a} \div \dfrac{b}{2} = \dfrac{4}{a} \times \dfrac{2}{b} = \dfrac{8}{ab}$

(c) $\dfrac{2x + 6}{x^2 + 4x} \div \dfrac{x^2 + 2x - 3}{x^2 - 16} = \dfrac{2x + 6}{x^2 + 4x} \times \dfrac{x^2 - 16}{x^2 + 2x - 3}$

$= \dfrac{2\cancel{(x + 3)}}{x\cancel{(x + 4)}} \times \dfrac{\cancel{(x + 4)}(x - 4)}{\cancel{(x + 3)}(x - 1)}$

$= \dfrac{2(x - 4)}{x(x - 1)}$

In **(c)**, invert the second fraction and change ÷ to ×.

Then factorise and cancel any common factors.

This example is grade A*.

Exercise 28D

Work out the following, simplifying your answers.

1 $\dfrac{a}{4} \times \dfrac{a}{3}$ **2** $\dfrac{b}{2} \div \dfrac{a}{5}$ **3** $\dfrac{pq}{r} \times \dfrac{r^2}{p}$ **4** $\dfrac{x}{y^2} \div \dfrac{x}{yz}$

5 $\dfrac{y}{3} \times \dfrac{y+2}{2}$ **6** $\dfrac{x+1}{4} \times \dfrac{1}{3x+3}$

7 $\dfrac{10a^2}{3b} \div \dfrac{5a}{6b^2}$ **8** $\dfrac{x^2+5x}{x^2+2x} \times \dfrac{x+2}{4}$

9 $\dfrac{2x-3}{14} \div \dfrac{4x-6}{7}$ **10** $\dfrac{m^2-9}{4} \times \dfrac{8}{m+3}$

11 $\dfrac{a^2-2a}{a^2-1} \div \dfrac{3a-6}{a^2+4a+3}$ **12** $\dfrac{y^2-16}{y^2-y} \div \dfrac{y^2-9y+20}{2y-2}$

28.5 Addition and subtraction of algebraic fractions

When adding or subtracting algebraic fractions, the rules are the same as for numerical fractions. But *before you find the LCM* you must *factorise* the numerators and/or denominators if possible.

Look back at Chapter 2 if you need a reminder.

Look at a numerical example and an algebraic example side by side ... you will see the similar approach.

Example 6

Work out   **(a)** $\dfrac{2}{5} + \dfrac{3}{7}$ **(b)** $\dfrac{2}{a} + \dfrac{3}{b}$ **(c)** $\dfrac{x+4}{5} + \dfrac{x-6}{4}$

(a) $\dfrac{2}{5} + \dfrac{3}{7}$

$= \dfrac{14}{35} + \dfrac{15}{35}$

$= \dfrac{29}{35}$

(b) $\dfrac{2}{a} + \dfrac{3}{b}$

$= \dfrac{2b}{ab} + \dfrac{3a}{ab}$

$= \dfrac{2b+3a}{ab}$

(c) $\dfrac{x+4}{5} + \dfrac{x-6}{4}$

$= \dfrac{4(x+4)}{20} + \dfrac{5(x-6)}{20}$

$= \dfrac{4(x+4) + 5(x-6)}{20}$

$= \dfrac{4x+16+5x-30}{20} = \dfrac{9x-14}{20}$

Just as $5 \times 7 = 35$ is the LCM of 5 and 7 then $a \times b = ab$ is the LCM of a and b.

Notice the equivalent fractions $\dfrac{2}{a} = \dfrac{2b}{ab}$ and $\dfrac{3}{b} = \dfrac{3a}{ab}$.

The LCM of 4 and 5 is 20. Notice the equivalent fractions on the first line of the solution.

Example 7

Simplify **(a)** $\dfrac{5}{2x} + \dfrac{3}{8y}$ **(b)** $\dfrac{4}{x+1} - \dfrac{3}{x-5}$

(a) $\dfrac{5}{2x} + \dfrac{3}{8y}$

$= \dfrac{20y}{8xy} + \dfrac{3x}{8xy}$

$= \dfrac{20y + 3x}{8xy}$

Notice that in (a) the LCM of $2x$ and $8y$ is $8xy$.

You do not need to use $2x \times 8y = 16xy$ although it is not wrong to do so … you will need to simplify later if you do.

(b) $\dfrac{4}{(x+1)} - \dfrac{3}{(x-5)}$

$= \dfrac{4(x-5)}{(x+1)(x-5)} - \dfrac{3(x+1)}{(x+1)(x-5)}$

$= \dfrac{4(x-5) - 3(x+1)}{(x+1)(x-5)}$

$= \dfrac{4x - 20 - 3x - 3}{(x+1)(x-5)}$

$= \dfrac{x - 23}{(x+1)(x-5)}$

> When the denominator has more than one term, put brackets round the expression. It helps you to spot the LCM and get the correct equivalent fractions.

> Take care with the signs when you expand the brackets in the numerator.

Exercise 28E

Simplify each of the following.

1 $\dfrac{2x}{3} + \dfrac{x}{5}$ **2** $\dfrac{3}{a} + \dfrac{4}{b}$ **3** $\dfrac{5}{x} - \dfrac{3}{2y}$ $\quad \dfrac{10y - 3x}{2yx}$ **4** $\dfrac{3}{4x} - \dfrac{2}{3x^2}$ $\quad 12x^2$

5 $\dfrac{6}{5x} + \dfrac{3}{2xy}$ **6** $\dfrac{x-1}{3} + \dfrac{x+5}{4}$

7 $\dfrac{2x-1}{5} + \dfrac{x+2}{4}$ **8** $\dfrac{x+6}{2} - \dfrac{x+1}{3}$

9 $\dfrac{y-3}{5} - \dfrac{y-2}{3}$ **10** $\dfrac{2m+3}{5} - \dfrac{3m-4}{2}$

11 $\dfrac{2}{x+3} + \dfrac{5}{x-1}$ **12** $\dfrac{3}{x+4} - \dfrac{1}{x+3}$

28.6 Equations involving algebraic fractions

You can use algebraic fractions to solve quite complex equations.

The equation in the next example can be solved in two ways.

The methods of solution are quite similar to each other and you can use either … just choose the one you prefer.

Example 8

Solve the equation $\dfrac{3x+1}{2} - \dfrac{x+4}{3} = 5$

Method A

$$\frac{3x+1}{2} - \frac{x+4}{3} = 5$$

$$\frac{3(3x+1) - 2(x+4)}{6} = 5$$

$$\frac{9x+3-2x-8}{6} = 5$$

$$\frac{7x-5}{6} = 5$$

$$7x - 5 = 30$$

$$7x = 35$$

$$x = 5$$

Method B

$$\frac{3x+1}{2} - \frac{x+4}{3} = 5$$

The LCM of the terms in the denominator is 6

Multiply all terms by 6

$$\frac{6(3x+1)}{2} - \frac{6(x+4)}{3} = 5 \times 6$$

$$3(3x+1) - 2(x+4) = 30$$

$$9x + 3 - 2x - 8 = 30$$

$$7x - 5 = 30$$

$$7x = 35$$

$$x = 5$$

Mutiplying all terms by the LCM of the denominators (in this case 6) will give you an equation with no fractions in it.

Example 9

Solve the equation $\dfrac{4}{2x-1} - \dfrac{1}{x+1} = 1$

Method A

$$\frac{4}{2x-1} - \frac{1}{x+1} = 1$$

$$\frac{4(x+1) - 1(2x-1)}{(2x-1)(x+1)} = 1$$

$$\frac{4x+4-2x+1}{(2x-1)(x+1)} = 1$$

$$\frac{2x+5}{(2x-1)(x+1)} = 1$$

Notice the equivalent fractions. Take care with the signs when expanding the numerator.

$$2x+5 = (2x-1)(x+1)$$

$$2x+5 = 2x^2 + x - 1$$

$$0 = 2x^2 - x - 6$$

$$0 = (2x+3)(x-2)$$

Either $\quad 2x+3 = 0 \quad$ giving $x = -1.5$

or $\quad\quad x-2 = 0 \quad$ giving $x = 2$

Multiply both sides by $(2x-1)(x+1)$ then expand and rearrange to get a quadratic equation.

These questions often have two solutions.

Method B

$$\frac{4}{2x-1} - \frac{1}{x+1} = 1$$

The LCM of the terms in the denominator is $(2x-1)(x+1)$

Multiply all terms by $(2x-1)(x+1)$

$$\frac{4(2x-1)(x+1)}{(2x-1)} - \frac{(2x-1)(x+1)}{(x+1)} = (2x-1)(x+1)$$

$$4(x+1) - (2x-1) = (2x-1)(x+1)$$

$$4x+4-2x+1 = 2x^2 + x - 1$$

$$0 = 2x^2 - x - 6$$

$$0 = (2x+3)(x-2)$$

Either $\quad 2x+3 = 0 \quad$ giving $x = -1.5$

or $\quad\quad x-2 = 0 \quad$ giving $x = 2$

Multiply throughout by the LCM of the terms in the denominator i.e. by $(2x-1)(x+1)$.

Then simplify each term, expand and rearrange to get a quadratic equation.

This is a standard method with which you need to be familiar.

Exercise 28F

Solve each of these equations. *even*

1 $\dfrac{x+1}{2}+\dfrac{x-4}{3}=5$

2 $\dfrac{x+5}{4}+\dfrac{x-1}{2}=3$

3 $\dfrac{x-2}{3}-\dfrac{x+1}{4}=1$

4 $\dfrac{2x-3}{3}+\dfrac{3x+1}{4}=12$

5 $\dfrac{5x+1}{4}-\dfrac{x-4}{6}=2$

6 $\dfrac{2x-5}{2}-\dfrac{4x-1}{3}=0.5$

7 $\dfrac{3}{x+1}+\dfrac{2}{2x-3}=1$

8 $\dfrac{9}{x-2}+\dfrac{5}{x+2}=2$

9 $\dfrac{7}{x-3}-\dfrac{6}{x-1}=2$

10 $\dfrac{9}{2x-7}+\dfrac{6}{x-1}=3$

11 $\dfrac{11}{4x-5}-\dfrac{8}{x+1}=1$

12 $\dfrac{7}{2x-1}+\dfrac{11}{x+1}=6$

28.7 Changing the subject of a formula

In this section you will see how to change the subject of a formula when the subject appears more than once and on both sides of the formula.

You use the same methods as for solving equations. You will see that a pattern emerges for the order in which you do the operations.

In this section each example also includes a traditional equation so that you can see the similarity in the method.

> You learned how to change the subject of a formula in Chapter 12.
> In all of the questions the subject only appeared once.

Example 10

(a) Solve the equation $2(2m+1)=7(m-1)$.
(b) Make m the subject of the formula $a(bm+c)=d(m-e)$.

(a) $2(2m+1)=7(m-1)$	**(b)** $a(bm+c)=d(m-e)$	
$4m+2=7m-7$	$abm+ac=dm-de$	Expand.
$2+7=7m-4m$	$ac+de=dm-abm$	Rearrange.
$9=3m$	$ac+de=m(d-ab)$	Factorise.
$\dfrac{9}{3}=m$	$\dfrac{ac+de}{d-ab}=m$	Divide through by $(d-ab)$.
$3=m$		

Example 11

(a) Solve the equation $5 = \dfrac{y+3}{y-4}$.

(b) Make y the subject of the formula $t = \dfrac{hy+3}{ky-4}$.

(a)

$$5 = \frac{y+3}{y-4}$$

$$5(y-4) = y+3$$

$$5y - 20 = y + 3$$

$$5y - y = 3 + 20$$

$$4y = 23$$

$$y = \frac{23}{4}$$

$$y = 5.75$$

(b)

$$t = \frac{hy+3}{ky-4}$$

$$t(ky-4) = hy+3$$

$$tky - 4t = hy + 3$$

$$tky - hy = 3 + 4t$$

$$y(tk - h) = 3 + 4t$$

$$y = \frac{3+4t}{tk-h}$$

Multiply through by $(ky - 4)$.

Expand.

Rearrange.

Factorise.

Divide through by $(tk - h)$.

Example 12

(a) Solve the equation $\sqrt{\dfrac{2a}{a-1}} = 3$.

(b) Make a the subject of the formula $\sqrt{\dfrac{2a}{a-x}} = 3x$.

(a)

$$\sqrt{\frac{2a}{a-1}} = 3$$

$$\frac{2a}{a-1} = 9$$

$$2a = 9(a-1)$$

$$2a = 9a - 9$$

$$9 = 9a - 2a$$

$$9 = 7a$$

$$\frac{9}{7} = a$$

(b)

$$\sqrt{\frac{2a}{a-x}} = 3x$$

$$\frac{2a}{a-x} = 9x^2$$

$$2a = 9x^2(a-x)$$

$$2a = 9x^2 a - 9x^3$$

$$9x^3 = 9x^2 a - 2a$$

$$9x^3 = a(9x^2 - 2)$$

$$\frac{9x^3}{9x^2 - 2} = a$$

Square both sides.

Multiply through by $(a - x)$.

Expand.

Rearrange.

Factorise.

Divide through by $(9x^2 - 2)$.

You can see the pattern of the order in which you do the operations. The order is always the same, the only extra step you might need is to take the square root at the end ... and only when the letter you want as the subject is squared.

To change the subject of a formula, follow this order of
operations.
(1) Square both sides
(2) Multiply through by ...
(3) Expand
(4) Rearrange
(5) Factorise
(6) Divide through by ...
(7) Square root both sides

Exercise 28G

Make y the subject of each of these formulae.

1 $my + b = d - my$ **2** $py + 3 = 7 + qy$ **3** $ay - x = w + by$

4 $w(y + a) = k(y + b)$ **5** $h(ay - c) = m(y + d)$ **6** $\dfrac{h - y}{h + y} = k$

7 $\dfrac{y + 3}{y - 2} = \dfrac{d}{e}$ **8** $\sqrt{\dfrac{2y + a}{y}} = m$ **9** $\sqrt{\dfrac{3y}{m - y}} = 2w$

10 $T = 2k\sqrt{\dfrac{y}{g}}$ **11** $ky^2 + 2e = f - hy^2$ **12** $m(2y^2 + a) = h(w - 3y^2)$

Examination questions

1 (a) Copy and complete the table of
values for $y = x^2 - 4x - 2$

x	-2	-1	0	1	2	3	4	5	6
y	10	3	-2	-5		-5	-2	3	10

(b) Draw the graph $y = x^2 - 4x - 2$
for values of x between -2 and 6.

(c) Use your graph to write down the solutions of the equation $x^2 - 4x - 2 = 0$

(d) By drawing an appropriate linear graph, write down the solutions of $x^2 - 5x - 3 = 0$

(7 marks)
AQA, Spec B, 3H, June 2003

2 (a) Make c the subject of the formula $E = mc^2$

(b) Make m the subject of the formula $E = mgh + \frac{1}{2}mv^2$ *(4 marks)*
AQA, Spec A, 2H, November 2004

3 Simplify fully $\dfrac{x^2 - 16}{3x^2 + 10x - 8}$ *(4 marks)*
AQA, Spec A, 1H, November 2003

4 Solve the equation $\dfrac{x}{x + 1} - \dfrac{2}{x - 1} = 1$ *(5 marks)*
AQA, Spec A, 2H, June 2003

Summary of key points

Simultaneous equations (grade A/A*)

When a quadratic graph and a linear graph are plotted on the same axes there are three possible situations:

- the linear graph *intersects* the quadratic graph in *two points*,
- the linear graph *touches* the quadratic graph at *one point*,
- the linear graph and the quadratic graph *do not intersect*.

To find solutions by an algebraic method, you have to find values of x and y that satisfy a quadratic equation and a linear equation simultaneously.

To solve one linear and one quadratic simultaneous equation you need to use the *method of substitution*.

Equation of a circle (grade A/A*)

The equation of a circle with centre $(0, 0)$ and radius r is $x^2 + y^2 = r^2$

Algebraic fractions (grade A/A*)

To simplify algebraic fractions:

(1) factorise

(2) divide by the HCF (or cancel common factors).

Remember to factorise *first*.

The rules for *adding and subtracting* algebraic fractions are the same as for numerical fractions, but *before you find the LCM* you must *factorise the numerators and/or denominators* if it is possible to do so.

Changing the subject of a formula

To change the subject of a formula when the subject appears more than once and on both sides of the formula follow this order of operations:

(1) Square both sides

(2) Multiply through by ...

(3) Expand

(4) Rearrange

(5) Factorise

(6) Divide through by ...

(7) Square root both sides

This chapter will show you how to:

✔ understand and use function notation
✔ use transformations to change the position and/or shape of a graph
✔ recognise how graphs are related by looking at their equations
✔ apply this knowledge to graphs of algebraic and trigonometric functions

29.1 Function notation and mappings

Key words:
mapping
function

You are familiar with writing equations to represent x–y relationships and the graphs of these relationships.

For example, $y = 3x^2 - 4$

There are other notations that can be used to relate x and y.

Two alternatives for the expression $y = 3x^2 - 4$ are:

$$f : x \rightarrow 3x^2 - 4 \quad \text{and} \quad f(x) = 3x^2 - 4$$

$f : x \rightarrow 3x^2 - 4$ is known as a **mapping** because it maps numbers on the x-axis onto corresponding numbers on the y-axis.

$f(x) = 3x^2 - 4$ is called **function** notation. $f(x)$ is read as 'f of x' and means a function of x. You can think of $f(x)$ in terms of a series of function boxes.

For the function $f(x) = 3x^2 - 4$, they will look like this:

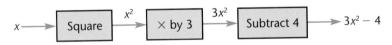

$$x \rightarrow \boxed{\text{Square}} \xrightarrow{x^2} \boxed{\times \text{ by } 3} \xrightarrow{3x^2} \boxed{\text{Subtract 4}} \rightarrow 3x^2 - 4$$

As you follow the instruction in each of the boxes in turn, in the correct order, you turn a value of x into a value of $f(x)$, or y.

> A mapping simply turns an x value into a y value. It is just like calculating y from an equation or from a table of values before you plot a graph.

> $y = 3x^2 - 4$, $f : x \rightarrow 3x^2 - 4$ and $f(x) = 3x^2 - 4$ are equivalent expressions. You can use whichever one you prefer but you need to be prepared to meet any of them in an exam.

Example 1

$f(x) = 8 - x^2$. Find the value of $f(0)$, $f(2)$ and $f(-3)$ and the values of x for which $f(x) = -17$.

$f(0) = 8 - 0^2 = 8 - 0 = 8$

$f(2) = 8 - 2^2 = 8 - 4 = 4$

$f(-3) = 8 - (-3)^2 = 8 - 9 = -1$

> Simply replace x by the number given and calculate the value of the expression.

continued ▼

If $f(x) = -17$ then $8 - x^2 = -17$

$$8 + 17 = x^2$$
$$x^2 = 25$$
$$x = \pm 5$$

> Set up an equation to answer the last part. Notice that there will be **two** solutions in this case.

Exercise 29A

1 $f(x) = x^2 + 3$
   (a) Find the values of (i) $f(4)$ (ii) $f(-3)$ (iii) $f(0)$ (iv) $f(\frac{1}{2})$
   (b) Find the values of x for which (i) $f(x) = 28$ (ii) $f(x) = 4x$

2 $f(x) = 2x^2 - 3x + 1$
   (a) Find the values of (i) $f(0)$ (ii) $f(2)$ (iii) $f(-2)$ (iv) $f(9)$
   (b) Find the values of x for which (i) $f(x) = 0$ (ii) $f(x) = 10$

29.2 Transformations of graphs

> **Key words:**
> one-way stretch

You will need to know:
- the transformations of translation and reflection
- the shapes of graphs such as $y = x^2$, and $y = x^3$

A translation is described by a column vector and a reflection is described by a mirror line.

There is another transformation that you have not previously met that is important for the work in this section. It is called a **one-way stretch** . These diagrams explain what you need to know.

> Look back at Chapter 18 if you need to refresh your memory.

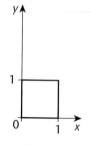

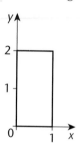

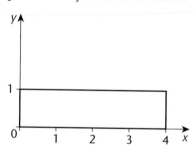

> You can see that a one-way stretch is exactly that ... a stretch in one direction only.

This is a square of side 1 unit placed at the origin.

This is a one-way stretch of the unit square, from the x-axis, parallel to the y-axis, of scale factor 2.

This is a one-way stretch of the unit square, from the y-axis, parallel to the x-axis, of scale factor 4.

This section looks at transformations that change the position and/or shape of a graph and at how you can see that graphs are related by looking at their equations.

The results that follow apply to *all* graphs, but the graph of $f(x) = x^2$ or $y = x^2$ is used to illustrate the first four important results.

Translations of graphs

$y = f(x) \rightarrow y = f(x) + a$

Look at the graphs of $y = x^2$, $y = x^2 + 3$ and $y = x^2 - 2$.

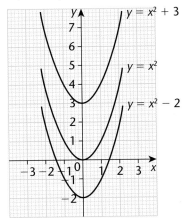

You can see that the graph of $y = x^2$ has been translated up 3 units to become the graph of $y = x^2 + 3$, then down 2 units to become the graph of $y = x^2 - 2$.

The same type of transformation will be needed for $y = x^2$ to be transformed into other graphs of the form $y = x^2 + a$, where a is any positive or negative number.

So the general result is:

$$y = f(x) \rightarrow y = f(x) + a \text{ represents a translation of } \begin{pmatrix} 0 \\ a \end{pmatrix}$$

If $a > 0$ the translation is vertically upwards and if $a < 0$ it is vertically downwards.

$y = f(x) \rightarrow y = f(x + a)$

Look at the graphs of $y = x^2$, $y = (x + 3)^2$ and $y = (x - 2)^2$.

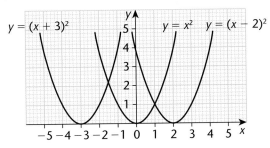

You can see that the graph of $y = x^2$ is translated 3 units to the left to become the graph of $y = (x + 3)^2$ and 2 units to the right to become the graph of $y = (x - 2)^2$.

Notice that the $+3$ in the equation means a shift to the **left**.
The -2 in the equation means a shift to the **right**.

This may surprise you because you might have expected it to be the other way round.

If you think about the value of x that gives $y = 0$ in each equation, you will see that the lowest points of the graphs are $(-3, 0)$ and $(2, 0)$ respectively.
This explains the directions of the translations.

The same type of transformation will be needed for $y = x^2$ to be transformed into other graphs of the form $y = (x + a)^2$, where a is any positive or negative number.
So the general result is:

$$y = f(x) \rightarrow y = f(x + a) \text{ represents a translation of } \begin{pmatrix} -a \\ 0 \end{pmatrix}$$

If $a > 0$ the translation is to the left and if $a < 0$ it is to the right.

Exercise 29B

1 Using axes from -4 to 4 for x and -10 to 20 for y, draw the graph of $y = x^2$.
On the same axes draw the graphs of
(a) $y = x^2 + 5$ **(b)** $y = x^2 - 4$ **(c)** $y = (x + 1)^2$ **(d)** $y = (x - 3)^2$.
State clearly in each case how the graph is obtained from the graph of $y = x^2$.

2 Write down the equation of the graph that can be obtained from the graph of $y = x^2$ by a translation of
(a) $\begin{pmatrix} 0 \\ 2 \end{pmatrix}$ **(b)** $\begin{pmatrix} -6 \\ 0 \end{pmatrix}$ **(c)** $\begin{pmatrix} 0 \\ -5 \end{pmatrix}$ **(d)** $\begin{pmatrix} -4 \\ 0 \end{pmatrix}$.

Stretches of graphs

$y = f(x) \rightarrow y = af(x)$
Look at the graphs of $y = x^2$, $y = 2x^2$ and $y = \frac{1}{2}x^2$.

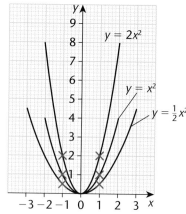

You can see that the graph of $y = x^2$ has been stretched parallel to the y-axis.
Points on the x-axis do not move.
For the graph of $y = 2x^2$, the y-coordinates of the graph of $y = x^2$ have been **doubled**.

For the graph of $y = \frac{1}{2}x^2$, the y-coordinates of the graph of $y = x^2$ have been **halved**.

The same type of transformation is needed for $y = x^2$ to be transformed into other graphs of the form $y = ax^2$, where a is any positive whole number or a fraction.

So the general result is:

> $y = f(x) \rightarrow y = af(x)$ **represents a one-way stretch from the x-axis, parallel to the y-axis, of scale factor a.**

The y-coordinates of the original graph are multiplied by a to get the transformed graph.

$y = f(x) \rightarrow y = f(ax)$

Look at the graphs of $y = x^2$, $y = (2x)^2$ and $y = (\frac{1}{2}x)^2$.

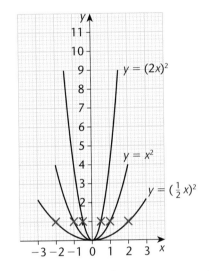

You can see that the graph of $y = x^2$ has been stretched parallel to the x-axis.

Points on the y-axis do not move.

For the graph of $y = (2x)^2$, the x-coordinates of the graph of $y = x^2$ have been **halved**.

For the graph of $y = (\frac{1}{2}x)^2$, the x-coordinates of the graph of $y = x^2$ have been **doubled**.

The same type of transformation is needed for $y = x^2$ to be transformed into other graphs of the form $y = (ax)^2$, where a is any positive whole number or a fraction.

So the general result is:

> $y = f(x) \rightarrow y = f(ax)$ **represents a one-way stretch from the y-axis, parallel to the x-axis, of scale factor $\frac{1}{a}$.**

The x-coordinates of the original graph are multiplied by $\frac{1}{a}$ to get the transformed graph.

An interesting fact ...

There is an obvious connection between the two results involving one-way stretches.

Look at $y = (2x)^2$, which can be written as $y = 4x^2$.
Is this a one-way stretch parallel to the x-axis of scale factor $\frac{1}{2}$?
Or is it a one-way stretch parallel to the y-axis of scale factor 4?
The answer is that it is both of these!
If you look closely at the graphs, together with the definitions of
the transformations, you will see that either description fits.

Similarly $y = (\frac{1}{2}x)^2$ can be written as $y = \frac{1}{4}x^2$.
So this can be either ...

      a one-way stretch parallel to the x-axis of scale factor 2

or   a one-way stretch parallel to the y-axis of scale factor $\frac{1}{4}$.

> Either description will gain full marks ... as long as it is correct!

Exercise 29C

1 Describe the transformation that maps the graph of $y = x^2$ onto each of these graphs.

    (a) $y = 3x^2$ **(b)** $y = (3x)^2$ **(c)** $y = \frac{1}{3}x^2$ **(d)** $y = (\frac{1}{3}x)^2$.

2 Write down the equation of the graph that can be obtained from the graph of $y = x^2$ by a transformation of

    (a) a one-way stretch from the x-axis parallel to the y-axis of scale factor

        (i) 4   **(ii)** 2.5   **(iii)** $\frac{1}{5}$

    (b) a one-way stretch from the y-axis parallel to the x-axis of scale factor

        (i) $\frac{1}{4}$ **(ii)** 4   **(iii)** 1.5

Reflections of graphs

$y = f(x) \rightarrow y = -f(x)$ and $y = f(x) \rightarrow y = f(-x)$
The next diagram shows the graph of $y = x^3 - 2$ or $f(x) = x^3 - 2$.
It also shows the graph of two related functions.

The function $-f(x)$
This is simply the negative of the original function. So, in this
example, $-f(x) = -(x^3 - 2)$
      or $-f(x) = -x^3 + 2$

The function $f(-x)$
This is obtained from the original function by replacing x with $-x$.
So, in this example, $f(-x) = (-x)^3 - 2$
      or $f(-x) = -x^3 - 2$

> Remember that cubing a negative gives a negative.

So, the three graphs are $y = f(x)$ or $y = x^3 - 2$
                           $y = -f(x)$ or $y = -x^3 + 2$
                           $y = f(-x)$ or $y = -x^3 - 2$

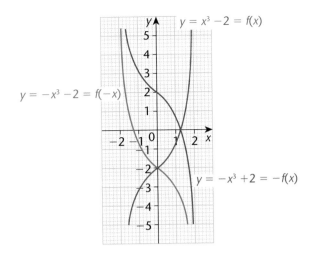

$y = -f(x)$ is a reflection of $y = f(x)$ in the x-axis.

$y = f(-x)$ is a reflection of $y = f(x)$ in the y-axis.

Do not confuse the two. Putting a $-$ in front of the function makes a positive y value negative, hence reflection in the x-axis.

Replacing x with $-x$, makes a positive x value negative, hence reflection in the y-axis.

Exercise 29D

1 Draw the graph of the straight line $y = 2x + 1$ for values of x from -3 to 3.

   (**a**) (**i**) Reflect the graph of $y = 2x + 1$ in the x-axis.

      (**ii**) Write down and simplify the equation of the reflected graph.

   (**b**) (**i**) Reflect the graph of $y = 2x + 1$ in the y-axis.

      (**ii**) Write down and simplify the equation of the reflected graph.

2 Draw the graph of $y = x^2 - 4x$ for values of x from 0 to 3.

   (**a**) (**i**) Reflect the graph of $y = x^2 - 4x$ in the x-axis.

      (**ii**) Write down and simplify the equation of the reflected graph.

   (**b**) (**i**) Reflect the graph of $y = x^2 - 4x$ in the y-axis.

      (**ii**) Write down and simplify the equation of the reflected graph.

3 Write down and simplify the equations of each of these graphs when they are reflected in (**i**) the x-axis (**ii**) the y-axis.

You do not need to draw the graphs.

   (**a**) $y = 3x - 2$ (**b**) $y = 1 - 4x$ (**c**) $y = x^2 + 6x$

   (**d**) $y = x^3 + 4$ (**e**) $y = 2x^2 - 5$ (**f**) $y = x^2 + 3x - 1$

The next two questions test your knowledge of all the transformations you have learnt.

4 Copy this graph of $y = f(x)$ y

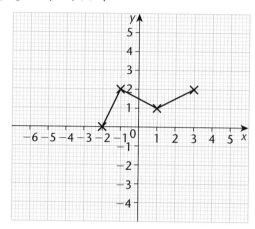

On the same axes draw the graphs of

(a) $y = f(x) + 2$ **(b)** $y = 2f(x)$

(c) $y = -f(x)$ **(d)** $y = f(x + 3)$

(e) $y = 2f(x) - 3$ **(f)** $y = -f(x) + 4$

> Work out what transformation (or combination of transformations) the equation represents before you draw the graph.

5 Copy this graph of $y = g(x)$

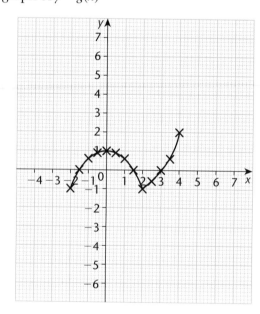

On the same axes draw the graphs of

(a) $y = g(x) + 3$ **(b)** $y = 2g(x)$

(c) $y = -g(x)$ **(d)** $y = g(x + 2)$

(e) $y = 2g(x) + 2$ **(f)** $y = -g(x) - 3$

> Work out what transformation (or combination of transformations) the equation represents before you draw the graph.

29.3 Combinations of transformations

You will need to know:
- how to complete the square for a function.

If you know the equation of the function you are starting with and the equation of the function you want at the end of the transformation, you can work out the steps of the transformation.

In this section you will see examples of questions involving combinations of transformations applied to algebraic functions.

Example 2

(a) Complete the square for the function $f(x) = x^2 - 8x + 13$.

(b) Hence or otherwise, describe how the graph of $y = x^2 - 8x + 13$ can be obtained from the graph of $y = x^2$.

(c) Draw a sketch graph to illustrate your answer.

(a) $f(x) = x^2 - 8x + 13$

$f(x) = (x - 4)^2 - 16 + 13$

$f(x) = (x - 4)^2 - 3$

Completing the square was covered in Chapter 24. Look back if you need a reminder.

(b) Starting with $y = x^2$ as $y = f(x)$

$y = (x - 4)^2$ is $y = f(x - 4)$

and represents a translation of $\begin{pmatrix} 4 \\ 0 \end{pmatrix}$

$y = (x - 4)^2 - 3$ is $y = f(x - 4) - 3$

so represents a further translation of $\begin{pmatrix} 0 \\ -3 \end{pmatrix}$

So, $y = x^2 - 8x + 13$ can be obtained from $y = x^2$ by a combination of two translations $\begin{pmatrix} 4 \\ 0 \end{pmatrix}$ and $\begin{pmatrix} 0 \\ -3 \end{pmatrix}$ which can

be written as one translation of $\begin{pmatrix} 4 \\ -3 \end{pmatrix}$

This means you start with a graph of $y = x^2$, and then find out how to transform it.

Two translations can always be combined into one in this way.

continued ▼

(c)

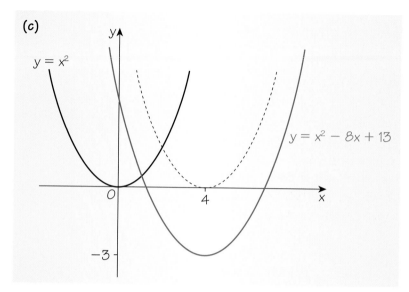

You don't need a table of values to plot a sketch graph.
It is exactly what it says it is … a **neat sketch** showing the main features, such as any points where the graph crosses the axes and, in this case, the lowest point of the graph.

Example 3

Describe the combined transformation that takes the graph of $y = x^2$ onto the graph of $y = 2(x + 1)^2$.
Draw a sketch graph to illustrate your answer.

Starting with $y = x^2$ as $y = f(x)$

$y = (x + 1)^2$ is $f(x + 1)$

and represents a translation of $\begin{pmatrix} -1 \\ 0 \end{pmatrix}$

$y = (x + 1)^2 \rightarrow y = 2(x + 1)^2$ is $f(x + 1) \rightarrow 2f(x + 1)$

and represents a one-way stretch parallel to the y-axis of scale factor 2.

So the combined transformation is

a translation of $\begin{pmatrix} -1 \\ 0 \end{pmatrix}$ followed by

a one-way stretch parallel to the y-axis of scale factor 2.

Remember:
$f(x) \rightarrow f(x + a)$ is a translation of $\begin{pmatrix} -a \\ 0 \end{pmatrix}$.

$f(x) \rightarrow af(x)$ is a one-way stretch parallel to the y-axis, of scale factor a.

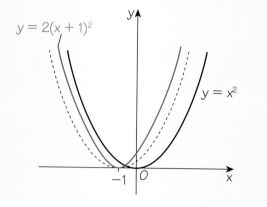

This sketch graph shows these transformations.

Example 4

Show how the graph of $y = x^2$ can be transformed into the graph of $y = 3(x - 2)^2 + 5$, illustrating your answer with a sketch graph.

Starting with $y = x^2$ as $y = f(x)$

$y = x^2 \rightarrow y = (x - 2)^2$ is $f(x) \rightarrow f(x - 2)$

which is a translation of $\begin{pmatrix} 2 \\ 0 \end{pmatrix}$

$y = (x - 2)^2 \rightarrow y = 3(x - 2)^2$ is $f(x - 2) \rightarrow 3f(x - 2)$

which represents a one-way stretch parallel to the y-axis of

scale factor 3.

$y = 3(x - 2)^2 \rightarrow y = 3(x - 2)^2 + 5$

is $3f(x - 2) \rightarrow 3f(x - 2) + 5$

and represents a translation of $\begin{pmatrix} 0 \\ 5 \end{pmatrix}$

So the combined transformation is

a translation of $\begin{pmatrix} 2 \\ 0 \end{pmatrix}$ followed by

a one-way stretch parallel to the y-axis of scale factor 3

followed by a translation of $\begin{pmatrix} 0 \\ 5 \end{pmatrix}$

Note that you can't combine the translations $\begin{pmatrix} 2 \\ 0 \end{pmatrix}$ and $\begin{pmatrix} 0 \\ 5 \end{pmatrix}$ as they are separated by a stretch.

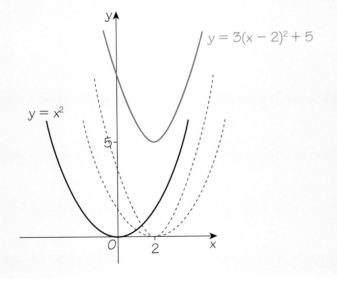

This sketch graph shows these transformations.

If you know the sequence of transformations, you can work out the equation of the final graph.

Example 5

$y = x^2$ is transformed into another graph by the following transformations carried out in this order:

1 A translation of $\begin{pmatrix} 4 \\ 0 \end{pmatrix}$

2 A one-way stretch of scale factor 2 from the x-axis, parallel to the y-axis

3 A reflection in the x-axis

4 A translation of $\begin{pmatrix} 0 \\ -1 \end{pmatrix}$

Draw sketch graphs to show each stage of the transformation and deduce the equation of the curve after all four transformations.

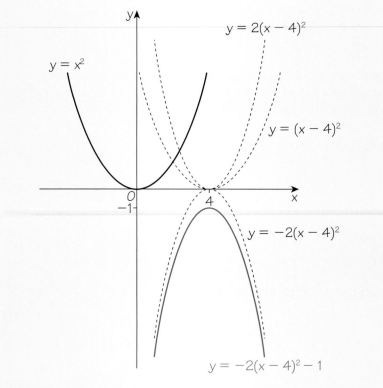

Start by drawing the graph of $y = x^2$.

Step 1: $y = x^2 \rightarrow y = (x - 4)^2$

Step 2: $y = (x - 4)^2 \rightarrow y = 2(x - 4)^2$

Step 3: $y = 2(x - 4)^2 \rightarrow y = -2(x - 4)^2$

Step 4: $y = -2(x - 4)^2 \rightarrow y = -2(x - 4)^2 - 1$

So $y = -2(x - 4)^2 - 1$ is the equation of the curve after all four transformations.

$f(x) \rightarrow -f(x)$ is a reflection in the x-axis.

$f(x) \rightarrow f(x) + a$ is a translation of $\begin{pmatrix} 0 \\ a \end{pmatrix}$. In this case, $a = -1$.

Exercise 29E

1 For each of these graphs
    (i)   Complete the square.
    (ii) Describe how the graph is obtained from the graph of $y = x^2$.
    (iii) Draw a sketch graph to illustrate your answer.
    (a) $y = x^2 - 4x - 3$ **(b)** $y = x^2 - 6x + 11$
    (c) $y = x^2 + 2x + 6$ **(d)** $y = x^2 + 10x + 19$

2 Describe the combination of transformations that takes the
graph of $y = x^2$ onto each of these graphs.
For each one, draw a sketch graph to illustrate your answer.
    (a) $y = (x + 2)^2 - 3$ **(b)** $y = (x - 3)^2 + 2$ **(c)** $y = 2x^2 + 3$
    (d) $y = 2(x - 3)^2$ **(e)** $y = 3(x + 2)^2$ **(f)** $y = -x^2 + 3$
    (g) $y = -3x^2$ **(h)** $y = -2x^2 - 3$

3 Show how the graph of $y = x^2$ can be transformed into each of
these graphs, illustrating your answer with a sketch graph.
    (a) $y = 2(x + 1)^2 + 4$ **(b)** $y = 3(x + 4)^2 - 2$
    (c) $y = 4(x - 3)^2 - 7$ **(d)** $y = -2(x - 4)^2 + 5$

4 In each of these, the graph of $y = x^2$ is transformed into another
graph by the given transformations carried out in the order stated.
For each one, draw sketch graphs to illustrate **each stage** of the
transformation.
Deduce the equation of the graph **after each stage**.

    (a) **(i)**   A translation of $\begin{pmatrix} -2 \\ 0 \end{pmatrix}$

         (ii) A reflection in the x-axis

         (iii) A translation of $\begin{pmatrix} 0 \\ 5 \end{pmatrix}$

    (b) **(i)**   A translation of $\begin{pmatrix} -3 \\ 0 \end{pmatrix}$

         (ii) A one-way stretch from the x-axis parallel to the y-axis
of scale factor 2

         (iii) A translation of $\begin{pmatrix} 0 \\ 6 \end{pmatrix}$

    (c) **(i)**   A translation of $\begin{pmatrix} 4 \\ 0 \end{pmatrix}$

         (ii) A one-way stretch from the x-axis parallel to the y-axis
of scale factor 2

         (iii) A reflection in the x-axis

    (d) **(i)**   A reflection in the x-axis

         (ii) A one-way stretch from the y-axis parallel to the x-axis
of scale factor 2

         (iii) A translation of $\begin{pmatrix} 0 \\ -3 \end{pmatrix}$

(e) (i)   A translation of $\begin{pmatrix} -5 \\ 0 \end{pmatrix}$

   (ii)   A one-way stretch from the *x*-axis parallel to the *y*-axis of scale factor 3

   (iii)   A reflection in the *x*-axis

   (iv)   A translation of $\begin{pmatrix} 0 \\ 2 \end{pmatrix}$

29.4 Transformations of trigonometric graphs

Key words:
amplitude
period

You will need to know:
- the shapes of the graphs $y = \sin x$ and $y = \cos x$.

When transformations are applied to the trigonometric graphs, particularly to $y = \sin x$ and $y = \cos x$, the results enable you to see the connection between the two graphs, and to use their symmetry properties to solve simple trigonometric equations.

The **period** of a trigonometric graph is the number of degrees after which it repeats itself.

The **amplitude** is the maximum height of the waveform above the *x*-axis.

These examples show you how the standard transformations can alter the amplitue, period and position of a trigonometric graph.

For example, for $y = \sin x$

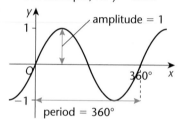

You need to be familiar with these. They are frequently tested on the higher level papers.

Example 6

Show graphically how $y = \sin x$ can be transformed into $y = 2 \sin x$.

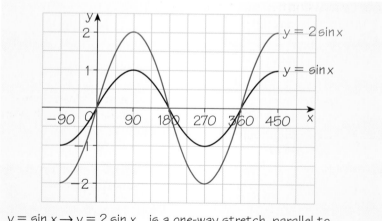

$y = \sin x \rightarrow y = 2 \sin x$ is a one-way stretch, parallel to the y-axis, of scale factor 2

The period of the transformed graph is still 360° but the amplitude has been multiplied by 2.

$f(x) \rightarrow af(x)$ gives a one-way stretch, parallel to the *y*-axis, of scale factor *a*.

Example 7

Show graphically how $y = \sin x$ can be transformed into $y = \sin\left(\frac{1}{2}x\right)$.

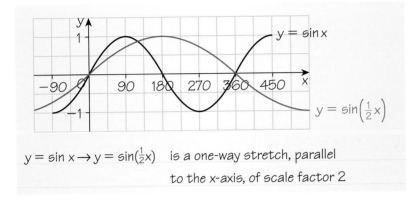

The x-coordinates are multiplied by 2. This makes the period of the transformed graph 720°.

$y = \sin x \rightarrow y = \sin\left(\frac{1}{2}x\right)$ is a one-way stretch, parallel to the x-axis, of scale factor 2

$f(x) \rightarrow f(ax)$ gives a stretch, parallel to the x-axis, of $\frac{1}{a}$.

Example 8

Show graphically how $y = \cos x$ can be transformed into $y = \cos(2x)$.

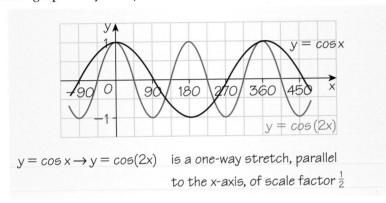

The x-coordinates are multiplied by $\frac{1}{2}$. This makes the period of the transformed graph 180°.

$y = \cos x \rightarrow y = \cos(2x)$ is a one-way stretch, parallel to the x-axis, of scale factor $\frac{1}{2}$

Example 9

Show graphically how $y = \cos x$ can be transformed into $y = \cos x + 1$.

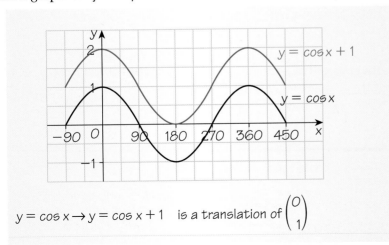

The y-coordinates of the transformed graph increase by 1.

Amplitude and period remain the same.

$y = \cos x \rightarrow y = \cos x + 1$ is a translation of $\begin{pmatrix} 0 \\ 1 \end{pmatrix}$

Example 10

Show graphically how $y = \cos x$ can be transformed into $y = \cos(x + 90°)$.

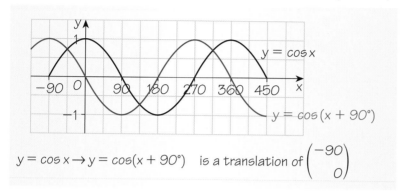

$y = \cos x \rightarrow y = \cos(x + 90°)$ is a translation of $\begin{pmatrix} -90 \\ 0 \end{pmatrix}$

The translation turns the $y = \cos x$ graph into a **reflection** in the x-axis of the $y = \sin x$ graph. So, $y = \cos(x + 90°)$ is the same graph as $y = -\sin x$.

Amplitude and period remain the same.

Exercise 29F

1 Draw sketch graphs to show each of these transformations of the graphs of $y = \sin x$ or $y = \cos x$.
  (i) For each one, start by drawing the graph of $y = \sin x$ or $y = \cos x$ then draw the transformed graph **on the same axes** using values of x in the range $-90°$ to $450°$.
  (ii) Describe each transformation in words.
  (iii) State the period and amplitude of each transformed graph.

  (a) $y = \cos x \rightarrow y = -\cos x$ **(b)** $y = \sin x \rightarrow y = \sin 2x$

  (c) $y = \sin x \rightarrow y = \sin x - 2$ **(d)** $y = \cos x \rightarrow y = 2\cos x$

  (e) $y = \cos x \rightarrow y = \cos(\frac{1}{2}x)$ **(f)** $y = \sin x \rightarrow y = \sin(x - 90°)$

  (g) $y = \sin x \rightarrow y = \frac{1}{2}\sin x + 1$ There are **two** transformations – what are they?

  (h) $y = \cos x \rightarrow y = 1 - \cos(x - 90°)$ There are **three** transformations – what are they?

29.5 Applications of trigonometric graphs

You can use the symmetry properties of the graphs of $y = \sin x$ and $y = \cos x$ to solve simple trigonometric equations.

Example 11

You are given that $\sin 65° = 0.9063$.
Use the graph of $y = \sin x$ to find values of x in the range $0° \leqslant x \leqslant 360°$ for which
(a) $\sin x = 0.9063$ **(b)** $\sin x = -0.9063$.

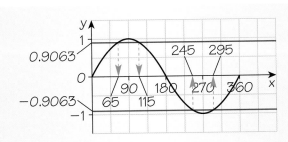

(a) A horizontal line drawn through $y = 0.9063$ intersects the graph of $y = \sin x$ at the points $x = 65°$ (the given value)

and $x = 115°$ (by symmetry).

(b) A horizontal line drawn through $y = -0.9063$ intersects the graph of $y = \sin x$ at the points $x = 245°$ (by symmetry)

and $x = 295°$ (by symmetry).

Notice that the solutions occur 25° either side of the maximum and minimum points of the graph.

Example 12

Consider a graph of $y = \cos x$ for values of x in the range $0° \leqslant x \leqslant 360°$.

(a) If $\cos 54° = 0.5878$ state another value of x in this range for which $\cos x = 0.5878$.

(b) On a separate diagram, sketch the graph of $y = \cos 2x$.

(c) Hence solve the equation $\cos 2x = 0.5878$ for values of x in the range $0° \leqslant x \leqslant 360°$.

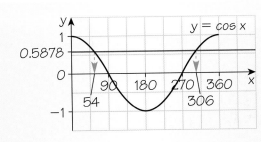

(a) A horizontal line drawn through $y = 0.5878$ intersects the graph of $y = \cos x$ again at $x = 306°$ (by symmetry).

(b)

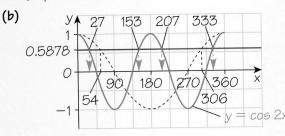

The dashed curve shows $y = \cos x$.

continued ▼

(c) Since cos 54° = 0.5878,

one solution of cos 2x = 0.5878 is x = 27°

A horizontal line drawn at y = 0.5878 will intersect the

graph of y = cos 2x **four** times altogether.

One solution is $\qquad$ x = 27°

so the others are $\quad$ x = 180° − 27° = 153°

$\qquad\qquad\qquad$ x = 180° + 27° = 207°

$\qquad\qquad\qquad$ x = 360° − 27° = 333°

cos(2 × 27°) = cos 54°

Notice that the four solutions illustrate the period of 180° of the y = cos 2x graph.

$\qquad$ 27° + 180° = 207°

and $\quad$ 153° + 180° = 333°

Exercise 29G

1 You are given that cos 34° = 0.8290
Draw the graph of $y = \cos x$ for $0° \leq x \leq 360°$ and use it to find values of x in this range for which
(a) $\cos x = 0.8290$ $\qquad\qquad$ **(b)** $\cos x = -0.8290$

2 You are given that sin 71° = 0.9455
Draw the graph of $y = \sin x$ for $0° \leq x \leq 360°$ and use it to find values of x in this range for which
(a) $\sin x = 0.9455$ $\qquad\qquad$ **(b)** $\sin x = -0.9455$

3 Draw the graph of $y = \sin x$ for $0° \leq x \leq 360°$.
(a) Given that sin 42° = 0.6691 state another value of x in this range for which $\sin x = 0.6691$
(b) On the same axes, draw the graph of $y = \sin 2x$.
(c) Hence solve the equations **(i)** $\sin 2x = 0.6691$ and
(ii) $\sin 2x = -0.6691$ for values of x in the range $0° \leq x \leq 360°$.

Examination questions

1 The diagram shows the graph of $y = \sin x°$ for $0 \leq x \leq 360$.

Copy the graph.
On the same axes sketch the following graphs:
(a) $y = 2 \sin x°$ for $0 \leq x \leq 360$
(b) $y = \sin 2x°$ for $0 \leq x \leq 360$
(c) $y = 2 + \sin x°$ for $0 \leq x \leq 360$

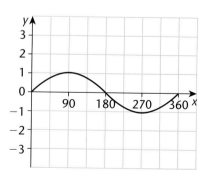

(3 marks)
AQA June 2004

2 The diagrams, **which are not drawn to scale**, show the graph
of $y = x^2$ and four other graphs A, B, C and D.
A, B, C and D represent four different transformations of $y = x^2$.
Find the equation of each of the graphs A, B, C and D.

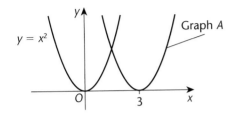

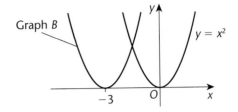

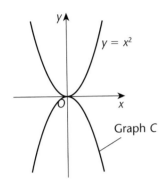

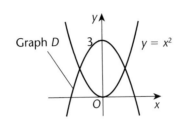

(4 marks)
AQA, Spec A, 1H, November 2003

3 The sketch below is of the graph of $y = x^2$

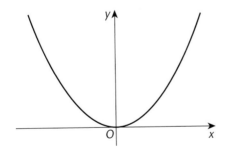

Copy or trace the graph.
Use your copy as a guide to sketch

(a) $y = x^2 + 2$

Repeat for

(b) $y = (x - 3)^2$

(c) $y = \frac{1}{2}x^2$

(3 marks)

Summary of key points

Transformations of graphs (grade A*)

$y = f(x) \rightarrow y = f(x) + a$ represents a translation of $\begin{pmatrix} 0 \\ a \end{pmatrix}$

If $a > 0$ the translation is vertically upwards and if $a < 0$ it will be vertically downwards.

$y = f(x) \rightarrow y = f(x + a)$ represents a translation of $\begin{pmatrix} -a \\ 0 \end{pmatrix}$

If $a > 0$ the translation is to the left and if $a < 0$ it will be to the right.

$y = f(x) \rightarrow y = af(x)$ represents a one-way stretch from the x-axis, parallel to the y-axis, of scale factor a.

The y-coordinates of the original graph are multiplied by a.

$y = f(x) \rightarrow y = f(ax)$ represents a one-way stretch from the y-axis, parallel to the x-axis of scale factor $\dfrac{1}{a}$.

The x-coordinates of the original graph are multiplied by $\dfrac{1}{a}$.

$y = -f(x)$ is a reflection of $y = f(x))$ in the x-axis.

$y = f(-x)$ is a reflection of $y = f(x)$ in the y-axis.

Combinations of transformations (grade A*)

If you know the equation of the function you are starting with and the equation of the function you want at the end of the transformation, you can work out the steps of the transformation.

For example, to transform the graph of $y = x^2$ onto the graph of $y = 2(x + 1)^2$

$y = x^2 \rightarrow y = (x + 1)^2$ is a translation of $\begin{pmatrix} -1 \\ 0 \end{pmatrix}$

$y = (x + 1)^2 \rightarrow y = 2(x + 1)^2$ is a one-way stretch parallel to the y-axis of scale factor 2.

So the combined transformation is:

a translation of $\begin{pmatrix} -1 \\ 0 \end{pmatrix}$ followed by a one-way stretch parallel to the y-axis of scale factor 2.

In the same way, if you know the sequence of the transformations, you can work out the equation of the final graph.

Transformations of trigonometric graphs (grade A*)

Transformations can be applied to the trigonometric graphs, particularly to $y = \sin x$ and $y = \cos x$.

They can alter the *amplitude* of the graph, the *period* of the graph or its *position* relative to the origin.

You can use the symmetry properties of the graphs of $y = \sin x$ and $y = \cos x$ to solve trigonometric equations.

30 Vectors

This chapter will show you how to:

- ✔ define a vector quantity
- ✔ use the correct notation for vectors
- ✔ add and subtract vectors
- ✔ recognise displacement vectors and position vectors
- ✔ solve problems using vector geometry

30.1 Definition and representation of a vector

Key words:
vector
scalar
directed line segment

A **vector** quantity is one which has magnitude (size) and direction.

Examples are: Displacement (16 km on a bearing of 065°)
Velocity (110 km/h due east)
Acceleration (5 m/s² along the *x*-axis)
Force (150 N acting vertically upwards)

Displacement is distance in a particular direction.

Quantities that have magnitude but no particular direction are called **scalar** quantities.

Velocity and speed are measured in the same units. Velocity is speed in a particular direction.

Examples are: Mass of car (1470 kg)
Length of room (3.6 m)
Speed of train (125 mph)

A vector can be represented by a **directed line segment** which is simply a line with an arrow on it.

Length of line = magnitude of vector.
Direction of arrow = direction of vector.

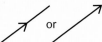

The arrow can be at the end of the line or in the middle of it.

30.2 Vector notation

You will recall using vectors in Chapter 18 (Transformations) and also in Chapter 29 (Transformations of graphs). They were column vectors of the form $\begin{pmatrix} 3 \\ -2 \end{pmatrix}$ where the numbers were the components in the *x* and *y* directions.

There are other ways of writing vectors and you need to be familiar with all of them.

The vector shown is written as $\overrightarrow{AB}$ which shows you that it is a displacement from A to B.

If lower case letters are used, it will be in **bold** print in a textbook, but since you cannot write in bold print you should <u>underline</u> the letter when you write it.

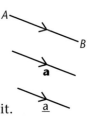

> Both the capital letter notation and the bold lower case notation will be used in the rest of this chapter.

30.3 Multiplication by a scalar

Vectors are equal if they have the same magnitude and direction.

So, for example,

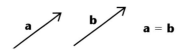

$\mathbf{a} = \mathbf{b}$

Vectors can be multiplied by scalars (numbers).

These examples show you some of the outcomes.

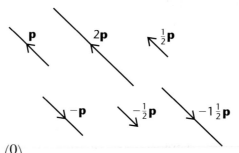

> All of these vectors are parallel since they are written **in terms of p only**, but they have different lengths and the negative ones point in the opposite direction.

Example 1

$$\mathbf{a} = \begin{pmatrix} 3 \\ -1 \end{pmatrix} \quad \mathbf{b} = \begin{pmatrix} -4 \\ 2 \end{pmatrix} \quad \mathbf{c} = \begin{pmatrix} 0 \\ 4 \end{pmatrix}$$

(a) Draw these vectors on a square grid: $\mathbf{a}$ $\mathbf{b}$ $\mathbf{c}$ $-\mathbf{a}$ $2\mathbf{b}$ $\frac{1}{2}\mathbf{c}$ $-1\frac{1}{2}\mathbf{b}$ $-2\mathbf{c}$

(b) What is the column vector notation for **(i)** $2\mathbf{a}$ **(ii)** $\frac{1}{2}\mathbf{b}$ **(iii)** $-\mathbf{c}$?

(a)

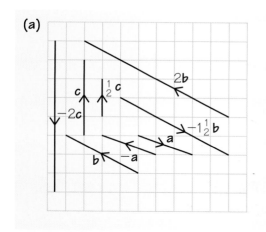

$a = \begin{pmatrix} 3 \\ -1 \end{pmatrix}$ which is 3 squares right and 1 square down.

$b = \begin{pmatrix} -4 \\ 2 \end{pmatrix}$ which is 4 squares left and 2 squares up.

$c = \begin{pmatrix} 0 \\ 4 \end{pmatrix}$ which is 4 squares up.

continued ▶

(b) (i) $2\mathbf{a} = 2\begin{pmatrix} 3 \\ -1 \end{pmatrix} = \begin{pmatrix} 6 \\ -2 \end{pmatrix}$

(ii) $\frac{1}{2}\mathbf{a} = \frac{1}{2}\begin{pmatrix} -4 \\ 2 \end{pmatrix} = \begin{pmatrix} -2 \\ 1 \end{pmatrix}$

(iii) $-\mathbf{c} = -\begin{pmatrix} 0 \\ 4 \end{pmatrix} = \begin{pmatrix} 0 \\ -4 \end{pmatrix}$

Exercise 30A

1 $\mathbf{a} = \begin{pmatrix} -2 \\ 5 \end{pmatrix}$ $\qquad \mathbf{b} = \begin{pmatrix} 0 \\ -3 \end{pmatrix}$ $\qquad \mathbf{c} = \begin{pmatrix} 4 \\ 2 \end{pmatrix}$

(a) Draw these vectors on a square grid: **a** **b** **c** $-\mathbf{a}$ $\frac{1}{2}\mathbf{c}$ $-2\mathbf{b}$

(b) Write these as column vectors: $3\mathbf{a}$ $-4\mathbf{b}$ $-1\frac{1}{2}\mathbf{c}$

2 $\mathbf{p} = \begin{pmatrix} 2 \\ -4 \end{pmatrix}$ $\qquad \mathbf{q} = \begin{pmatrix} 1 \\ 3 \end{pmatrix}$ $\qquad \mathbf{r} = \begin{pmatrix} -3 \\ -2 \end{pmatrix}$

(a) Draw these vectors on a square grid: **p** **q** **r** $\frac{1}{2}\mathbf{p}$ $2\mathbf{q}$ $-\mathbf{r}$

(b) Write these as column vectors: $5\mathbf{r}$ $-2\mathbf{q}$ $1\frac{1}{2}\mathbf{p}$

30.4 Adding and subtracting vectors

Key words:
displacement vector
resultant

Look at this diagram showing three **displacement vectors**.

Displacement vectors are 'free' vectors as opposed to 'position' vectors.

You will see the difference between the two in the next section.

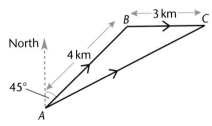

They represent a journey from A to B of 4 km North-east, followed by a journey from B to C of 3 km due East.

You can see that you can go directly from A to C, the final destination is the same.

This can be expressed in vector notation as $\overrightarrow{AB} + \overrightarrow{BC} = \overrightarrow{AC}$ and forms the basis of vector addition.

The vector $\overrightarrow{AC}$ is called the **resultant** of vectors $\overrightarrow{AB}$ and $\overrightarrow{BC}$ because going from A to C is the same result as going from A to B then from B to C.

Look at this parallelogram *PQRS*.

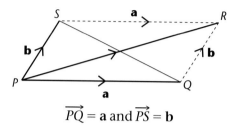

$$\overrightarrow{PQ} = \mathbf{a} \text{ and } \overrightarrow{PS} = \mathbf{b}$$

A parallelogram has opposite sides that are equal and parallel, so this means that $\overrightarrow{SR} = \mathbf{a}$ and $\overrightarrow{QR} = \mathbf{b}$.

Using vector addition,

$$\overrightarrow{PR} = \overrightarrow{PQ} + \overrightarrow{QR}$$
$$= \mathbf{a} + \mathbf{b}$$

and
$$\overrightarrow{SQ} = \overrightarrow{SR} + \overrightarrow{RQ}$$
$$= \mathbf{a} + -\mathbf{b}$$
$$= \mathbf{a} - \mathbf{b}$$

and
$$\overrightarrow{QS} = \overrightarrow{QR} + \overrightarrow{RS}$$
$$= \mathbf{b} + -\mathbf{a}$$
$$= \mathbf{b} - \mathbf{a}$$

> Notice that the order of the letters is important.
> $\overrightarrow{SR} = \mathbf{a}$ so $\overrightarrow{RS} = -\mathbf{a}$
> $\overrightarrow{QR} = \mathbf{b}$ so $\overrightarrow{RQ} = -\mathbf{b}$

> Notice the 'nose to tail' order of the capital letters: *PQ* followed by *QR* is the same as *PR*.
> This is always true for vector addition.

> Notice that the **diagonals** of the parallelogram are represented by the vectors $\mathbf{a} + \mathbf{b}$ and either $\mathbf{a} - \mathbf{b}$ or $\mathbf{b} - \mathbf{a}$ depending on the direction.

These results illustrate the *triangle law* or *parallelogram law* of vector addition.

Example 2

$$\mathbf{a} = \begin{pmatrix} 3 \\ -2 \end{pmatrix} \quad \mathbf{b} = \begin{pmatrix} -4 \\ 5 \end{pmatrix} \quad \mathbf{c} = \begin{pmatrix} 0 \\ 6 \end{pmatrix}$$

Write these as column vectors:

(a) $\mathbf{a} + \mathbf{b}$ (b) $\mathbf{b} - \mathbf{a}$ (c) $2\mathbf{b} + 3\mathbf{c}$ (d) $\mathbf{c} - 2\mathbf{a}$

(a) $\mathbf{a} + \mathbf{b} = \begin{pmatrix} 3 \\ -2 \end{pmatrix} + \begin{pmatrix} -4 \\ 5 \end{pmatrix} = \begin{pmatrix} -1 \\ 3 \end{pmatrix}$

(b) $\mathbf{b} - \mathbf{a} = \begin{pmatrix} -4 \\ 5 \end{pmatrix} - \begin{pmatrix} 3 \\ -2 \end{pmatrix} = \begin{pmatrix} -7 \\ 7 \end{pmatrix}$

(c) $2\mathbf{b} + 3\mathbf{c} = \begin{pmatrix} -8 \\ 10 \end{pmatrix} + \begin{pmatrix} 0 \\ 18 \end{pmatrix} = \begin{pmatrix} -8 \\ 28 \end{pmatrix}$

(d) $\mathbf{c} - 2\mathbf{a} = \begin{pmatrix} 0 \\ 6 \end{pmatrix} - \begin{pmatrix} 6 \\ -4 \end{pmatrix} = \begin{pmatrix} -6 \\ 10 \end{pmatrix}$

Example 3

$$\mathbf{a} = \begin{pmatrix} 2 \\ -1 \end{pmatrix} \quad \mathbf{b} = \begin{pmatrix} 4 \\ 3 \end{pmatrix}$$

On a square grid draw diagrams to illustrate the vectors

$$\mathbf{a} + \mathbf{b} \quad \mathbf{b} - \mathbf{a} \quad 2\mathbf{a} - \mathbf{b}$$

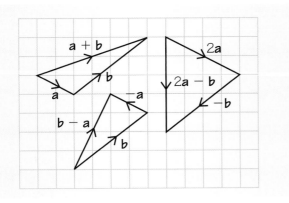

Exercise 30B

1 $\mathbf{d} = \begin{pmatrix} -4 \\ 1 \end{pmatrix} \quad \mathbf{e} = \begin{pmatrix} 2 \\ -2 \end{pmatrix} \quad \mathbf{f} = \begin{pmatrix} -1 \\ 5 \end{pmatrix}$

Write these as column vectors $\mathbf{d} + \mathbf{e} \quad \mathbf{f} - \mathbf{d} \quad 2\mathbf{f} + \mathbf{e} \quad \mathbf{d} - 3\mathbf{e}$
$\mathbf{d} + \mathbf{e} + \mathbf{f} \quad 3\mathbf{f} + 2\mathbf{d} \quad 4\mathbf{e} - 3\mathbf{d} \quad 2\mathbf{d} - \mathbf{e} - 2\mathbf{f}$

2 $\mathbf{u} = \begin{pmatrix} -2 \\ 1 \end{pmatrix} \quad \mathbf{v} = \begin{pmatrix} 3 \\ 4 \end{pmatrix} \quad \mathbf{w} = \begin{pmatrix} -4 \\ 0 \end{pmatrix}$

On a square grid draw diagrams to illustrate these vectors
$\mathbf{u} + \mathbf{v} \quad \mathbf{v} - \mathbf{u} \quad \mathbf{w} - 2\mathbf{u} \quad 2\mathbf{v} + \mathbf{w} \quad \mathbf{u} - \mathbf{v} - \mathbf{w} \quad \mathbf{v} + \mathbf{w} - 3\mathbf{u}$

30.5 Position vectors

You will need to know:
- Pythagoras' theorem
- how to use the tangent ratio to calculate an angle

> **Key words:**
> position vector
> free vector
> modulus

Look at these vectors drawn on the *x–y* plane.

They are all equal but only one of them *starts at the origin*.

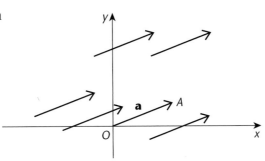

A vector that starts at the origin is known as a **position vector** .

So the vector labelled $\overrightarrow{OA}$, or **a**, is the position vector of A.

The other vectors are displacement vectors or **free vectors** because they 'float' in space and are not connected to the origin.

You can see that if $A = (4, 2)$ then $\overrightarrow{OA} = \begin{pmatrix} 4 \\ 2 \end{pmatrix}$ so the coordinates of the point are the same as the components of the position vector.

This enables you to calculate the magnitude of the vector and its direction.

Using Pythagoras' theorem,

$$OA^2 = 4^2 + 2^2$$
$$= 20$$

so $|OA| = \sqrt{20}$
$$= \sqrt{4 \times 5}$$
$$= 2\sqrt{5}$$

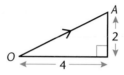

> Notice the answer has been left in surd form ... this gives the exact value.

$|OA|$ is called the **modulus** of $\overrightarrow{OA}$ and is the notation used to represent the magnitude of a vector.

Alternatively, you can write $|\mathbf{a}| = 2\sqrt{5}$

To find the direction of a vector, use trigonometry.

If θ is the angle between OA and the x-axis,

$$\tan \theta = \tfrac{2}{4}$$
$$\theta = \tan^{-1} (0.5)$$
$$\theta = 26.6° \text{ (3 s.f.)}$$

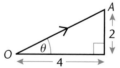

> You can use the same methods to calculate the magnitude (length) and direction of any free vector.

Example 4

The diagram shows a grid of congruent parallelograms.

The origin is labelled O and the position vectors of points A and B are given by $\overrightarrow{OA} = \mathbf{a}$ and $\overrightarrow{OB} = \mathbf{b}$.

Write, in terms of **a** and **b**,

(a) $\overrightarrow{OP}$ (b) $\overrightarrow{OV}$ (c) $\overrightarrow{OR}$ (d) $\overrightarrow{OG}$

(e) $\overrightarrow{AB}$ (f) $\overrightarrow{NE}$ (g) $\overrightarrow{FB}$ (h) $\overrightarrow{UG}$

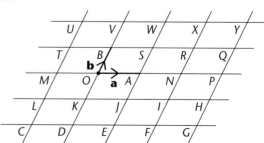

(a) $\overrightarrow{OP} = 3\mathbf{a}$

(b) $\overrightarrow{OV} = 2\mathbf{b}$

$\overrightarrow{OP}, \overrightarrow{OV}, \overrightarrow{OR}, \overrightarrow{OG}$, are all position vectors because they start at the origin O.

The others are free vectors.

(c) $\overrightarrow{OR} = \overrightarrow{ON} + \overrightarrow{NR}$
    $= 2\mathbf{a} + \mathbf{b}$

(d) $\overrightarrow{OG} = \overrightarrow{OP} + \overrightarrow{PG}$
    $= 3\mathbf{a} - 2\mathbf{b}$

(e) $\overrightarrow{AB} = \overrightarrow{AO} + \overrightarrow{OB}$
    $= -\mathbf{a} + \mathbf{b}$

(f) $\overrightarrow{NE} = \overrightarrow{NA} + \overrightarrow{AE}$
    $= -\mathbf{a} - 2\mathbf{b}$

There are alternative solutions. For example,

(g) $\overrightarrow{FB} = \overrightarrow{FD} + \overrightarrow{DB}$
    $= -2\mathbf{a} + 3\mathbf{b}$

(h) $\overrightarrow{UG} = \overrightarrow{UY} + \overrightarrow{YG}$
    $= 4\mathbf{a} - 4\mathbf{b}$

(g) $\overrightarrow{FB} = \overrightarrow{FR} + \overrightarrow{RB}$
    $= 3\mathbf{b} - 2\mathbf{a}$

Example 5

Using the parallelogram grid for Example 4, answer the following.

(a) Write down the position vector equal to $3\mathbf{a} - \mathbf{b}$.

(b) Write down three other vectors equal to $3\mathbf{a} - \mathbf{b}$.

(c) Write down a vector parallel to $\overrightarrow{CP}$.

(d) Write down three vectors that are half the size of CP and in the opposite direction.

(e) Write down all the vectors that are equal to $3(\mathbf{a} + \mathbf{b})$.

(f) Write down, in terms of $\mathbf{a}$ and $\mathbf{b}$, the vectors
$$\overrightarrow{CK} \quad \overrightarrow{CA} \quad \overrightarrow{CR} \quad \overrightarrow{CY}.$$

What is the connection between these answers and the positions of C, K, A, R and Y on the grid?

(a) $\overrightarrow{OH}$

(b) $\overrightarrow{VQ} \quad \overrightarrow{TN} \quad \overrightarrow{MI}$

(c) $\overrightarrow{CP} = 2\mathbf{b} + 4\mathbf{a}$

A parallel vector is $\overrightarrow{LQ}$

d) New vector $= -\mathbf{b} - 2\mathbf{a}$
   $\overrightarrow{QA} \quad \overrightarrow{PJ} \quad \overrightarrow{HE}$

e) $\overrightarrow{CR} \quad \overrightarrow{LX} \quad \overrightarrow{DQ} \quad \overrightarrow{KY}$

f) $\overrightarrow{CK} = \mathbf{a} + \mathbf{b}$
   $\overrightarrow{CA} = 2\mathbf{a} + 2\mathbf{b}$
   $\overrightarrow{CR} = 3\mathbf{a} + 3\mathbf{b}$
   $\overrightarrow{CY} = 4\mathbf{a} + 4\mathbf{b}$

C, K, A, R and Y are all points on a straight line.

You can choose *any* vectors from the grid that involve 3 steps in the same direction as $\mathbf{a}$ and one in the opposite direction to $\mathbf{b}$.

Again, there is more than one choice. An alternative is $\overrightarrow{MY}$.

Multiply by $-\frac{1}{2}$.
There are many examples of $-\mathbf{b} - 2\mathbf{a}$ that you could choose here.

Exercise 30C

1 Draw a separate diagram for each of these position vectors. Find the magnitude and direction (the angle made with the positive *x*-axis) of each position vector.

(a) $\begin{pmatrix} 0 \\ 4 \end{pmatrix}$ (b) $\begin{pmatrix} 3 \\ 5 \end{pmatrix}$ (c) $\begin{pmatrix} 6 \\ -2 \end{pmatrix}$ (d) $\begin{pmatrix} -4 \\ 6 \end{pmatrix}$ (e) $\begin{pmatrix} -3 \\ -7 \end{pmatrix}$

> When measuring angles from the positive *x*-axis, those angles measured anticlockwise from 0° to 180° are positive and those measured clockwise from 0° to 180° are negative.

2 The diagram shows a grid of congruent parallelograms. The origin is labelled *O* and the position vectors of points *A* and *B* are given by $\overrightarrow{OA} = \mathbf{a}$ and $\overrightarrow{OB} = \mathbf{b}$.

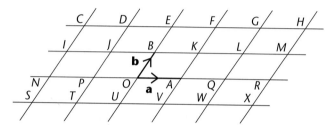

Write, in terms of **a** and **b**,

(a) $\overrightarrow{OR}$ (b) $\overrightarrow{OU}$ (c) $\overrightarrow{OC}$ (d) $\overrightarrow{OX}$ (e) $\overrightarrow{GP}$

(f) $\overrightarrow{IW}$ (g) $\overrightarrow{HR}$ (h) $\overrightarrow{MJ}$ (i) $\overrightarrow{UF}$ (j) $\overrightarrow{ET}$

3 Use the parallelogram grid for question 2 to answer the following.

(a) Write down the position vector equal to 3**a** + **b**.

(b) Write down three other vectors equal to 3**a** + **b**.

(c) Write down three vectors parallel to $\overrightarrow{ND}$.

(d) Write down three vectors that are half the size of $\overrightarrow{TM}$ and in the opposite direction.

(e) Write down the vectors $\overrightarrow{DI}$, $\overrightarrow{FP}$ and $\overrightarrow{HU}$. Explain the connection between these vectors.

4 Here is another grid of congruent parallelograms. The origin is labelled *O* and the position vectors of points *A* and *B* are given by $\overrightarrow{OA} = \mathbf{a}$ and $\overrightarrow{OB} = \mathbf{b}$. On a copy of the grid, mark the points *C* to *L* where

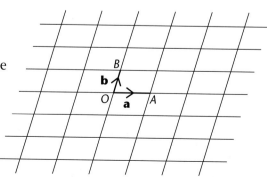

(a) $\overrightarrow{OC} = -\mathbf{a} + 2\mathbf{b}$ (b) $\overrightarrow{OD} = 3\mathbf{a} + \mathbf{b}$

(c) $\overrightarrow{OE} = -\mathbf{a} - \mathbf{b}$ (d) $\overrightarrow{OF} = \mathbf{a} - 3\mathbf{b}$

(e) $\overrightarrow{OG} = 2\mathbf{a} - 2\mathbf{b}$ (f) $\overrightarrow{OH} = 3\mathbf{b}$

(g) $\overrightarrow{OI} = \mathbf{a} + \frac{3}{2}\mathbf{b}$ (h) $\overrightarrow{OJ} = -\frac{3}{2}\mathbf{a} - 2\mathbf{b}$

(i) $\overrightarrow{OK} = 2\mathbf{a} + \frac{1}{2}\mathbf{b}$ (j) $\overrightarrow{OL} = \frac{5}{2}\mathbf{a} - \frac{3}{2}\mathbf{b}$

30.6 Vector geometry

Key words:
collinear

The rules for the addition and subtraction of vectors and the multiplication of a vector by a scalar are the key elements of questions on vector geometry.

Some questions require you to *prove* geometric results.

Showing that lines are parallel or that points are **collinear** (lie on a straight line) are typical examples.

Examples 6, 7 and 8 are typical of the questions you can expect on the higher level GCSE papers.

Example 6

$OABC$ is a quadrilateral with $\overrightarrow{OA} = \mathbf{a}$, $\overrightarrow{OB} = \mathbf{b}$ and $\overrightarrow{OC} = \mathbf{c}$. M is the mid-point of AB and P is the point on BC such that $BP : PC = 3 : 1$.

Find expressions for these vectors, giving your answers in their simplest form.

(a) $\overrightarrow{AB}$ (b) $\overrightarrow{AM}$ (c) $\overrightarrow{OM}$ (d) $\overrightarrow{BC}$ (e) $\overrightarrow{BP}$ (f) $\overrightarrow{MP}$

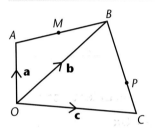

(a) $\overrightarrow{AB} = \overrightarrow{AO} + \overrightarrow{OB} = -\mathbf{a} + \mathbf{b}$ or $\mathbf{b} - \mathbf{a}$

(b) $\overrightarrow{AM} = \frac{1}{2}\overrightarrow{AB} = \frac{1}{2}(\mathbf{b} - \mathbf{a}) = \frac{1}{2}\mathbf{b} - \frac{1}{2}\mathbf{a}$

(c) $\overrightarrow{OM} = \overrightarrow{OA} + \overrightarrow{AM} = \mathbf{a} + \frac{1}{2}\mathbf{b} - \frac{1}{2}\mathbf{a} = \frac{1}{2}\mathbf{a} + \frac{1}{2}\mathbf{b}$

(d) $\overrightarrow{BC} = \overrightarrow{BO} + \overrightarrow{OC} = -\mathbf{b} + \mathbf{c}$ or $\mathbf{c} - \mathbf{b}$

(e) $\overrightarrow{BP} = \frac{3}{4}\overrightarrow{BC} = \frac{3}{4}(\mathbf{c} - \mathbf{b}) = \frac{3}{4}\mathbf{c} - \frac{3}{4}\mathbf{b}$

As $BP : PC = 3 : 1$ then $BP = \frac{3}{4}BC$.

You will often need to interpret ratios in this way.

(f) $\overrightarrow{MP} = \overrightarrow{MB} + \overrightarrow{BP} = (\frac{1}{2}\mathbf{b} - \frac{1}{2}\mathbf{a}) + (\frac{3}{4}\mathbf{c} - \frac{3}{4}\mathbf{b})$

$\quad = \frac{1}{2}\mathbf{b} - \frac{3}{4}\mathbf{b} - \frac{1}{2}\mathbf{a} + \frac{3}{4}\mathbf{c}$

$\quad = \frac{3}{4}\mathbf{c} - \frac{1}{4}\mathbf{b} - \frac{1}{2}\mathbf{a}$

Since M is the mid-point of AB then $\overrightarrow{MB} = \overrightarrow{AM}$.

Example 7

In triangle ABC points P and Q are the mid-points of sides AB and AC respectively. $\overrightarrow{AP} = \mathbf{x}$ and $\overrightarrow{AQ} = \mathbf{y}$.

Prove that PQ is parallel to BC and half its length.

This is a proof of what is known as the mid-point theorem.

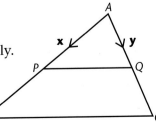

$$\vec{PQ} = \vec{PA} + \vec{AQ} = -\mathbf{x} + \mathbf{y} = \mathbf{y} - \mathbf{x}$$

$$\vec{AB} = 2\mathbf{x} \quad \text{and} \quad \vec{AC} = 2\mathbf{y}$$

So $\vec{BC} = \vec{BA} + \vec{AC} = -2\mathbf{x} + 2\mathbf{y} = 2\mathbf{y} - 2\mathbf{x}$

So $\vec{BC} = 2(\mathbf{y} - \mathbf{x}) = 2\vec{PQ}$

... which means that BC is twice the length of PQ and that BC and PQ have the same direction or ... PQ is parallel to BC and half its length.

> This is the most important point in the final statement of the proved result.

Remember that when vectors are equal they are equal in two ways ... their lengths are equal and they have the same direction.

Example 8

The diagram shows a trapezium $OABC$ in which OC is parallel to AB and $OC = \frac{2}{3}AB$. $\vec{OA} = \mathbf{a}$ and $\vec{OB} = \mathbf{b}$.

P is the point on AB such that $AP : PB = 1 : 2$ and M is the mid-point of OB.

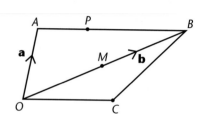

(a) Find these vectors in terms of $\mathbf{a}$ and $\mathbf{b}$, giving your answers in their simplest form.

    (i) $\vec{AB}$ **(ii)** $\vec{AP}$ **(iii)** $\vec{PM}$ **(iv)** $\vec{MC}$

(b) What do your answers to $\vec{PM}$ and $\vec{MC}$ tell you about the relationship between points P, M and C?

(a) (i) $\vec{AB} = \vec{AO} + \vec{OB} = -\mathbf{a} + \mathbf{b} = \mathbf{b} - \mathbf{a}$

(ii) $\vec{AP} = \frac{1}{3}\vec{AB} = \frac{1}{3}(\mathbf{b} - \mathbf{a})$ or $\frac{1}{3}\mathbf{b} - \frac{1}{3}\mathbf{a}$

> As $AP : PB = 1 : 2$ then $AP = \frac{1}{3}AB$.

(iii) $\vec{PM} = \vec{PA} + \vec{AO} + \vec{OM}$

$= (-\frac{1}{3}\mathbf{b} + \frac{1}{3}\mathbf{a}) + (-\mathbf{a}) + (\frac{1}{2}\mathbf{b})$

$= -\frac{1}{3}\mathbf{b} + \frac{1}{2}\mathbf{b} + \frac{1}{3}\mathbf{a} - \mathbf{a}$

$= \frac{1}{6}\mathbf{b} - \frac{2}{3}\mathbf{a}$

(iv) $\vec{MC} = \vec{MO} + \vec{OC} = \vec{MO} + \frac{2}{3}\vec{AB}$

$= -\frac{1}{2}\mathbf{b} + \frac{2}{3}(\mathbf{b} - \mathbf{a})$

$= -\frac{1}{2}\mathbf{b} + \frac{2}{3}\mathbf{b} - \frac{2}{3}\mathbf{a}$

$= \frac{1}{6}\mathbf{b} - \frac{2}{3}\mathbf{a}$

continued ▶

(b) You can see from parts **(a)(iii)** and **(iv)** that

$$\overrightarrow{PM} = \overrightarrow{MC}$$

So PM and MC are equal in length and have the same

direction

... but M **is a point in common** to both of these line

segments so if the directions are the same points P, M and

C must all be part of the same straight line.

So the conclusion is that:

P, M and C are collinear and M is the mid-point of PC.

> Remember that when vectors are equal they are equal in two ways ... their **lengths** are equal and they have the same **direction**.
>
> When you have a **common point** and vectors are **equal** (or where one is a multiple of the other) then you can deduce that the points lie in a straight line.

Exercise 30D

Diagrams in this exercise are not drawn accurately.

1 *OPQ* is a triangle.
A, B and C are the mid-points of *OP*, *OQ* and *PQ* respectively.
$\overrightarrow{OA} = \mathbf{a}$ and $\overrightarrow{OB} = \mathbf{b}$
Find these vectors in terms of **a** and **b**, simplifying your answers

(a) $\overrightarrow{OP}$ $2a$ (b) $\overrightarrow{OQ}$ $2b$ (c) $\overrightarrow{PQ}$ $-2a+2b$ (d) $\overrightarrow{AB}$ $-a+b$

(e) $\overrightarrow{PC}$ $\frac{1}{2}(-2a+2b)$ (f) $\overrightarrow{OC}$ (g) $\overrightarrow{BP}$ (h) $\overrightarrow{QA}$

$2a+\frac{1}{2}(-2a+2b)$ $b+2a$ $-2b+a$

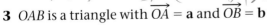

2 *OFG* is a triangle.
A and C are the mid-points of *OF* and *FG* respectively.
B lies on *OG* such that $OB:BG = 1:2$
$\overrightarrow{OA} = \mathbf{a}$ and $\overrightarrow{OB} = \mathbf{b}$
Find these vectors in terms of **a** and **b**, simplifying
your answers

(a) $\overrightarrow{AF}$ a (b) $\overrightarrow{AB}$ $-a+b$ (c) $\overrightarrow{OG}$ $3b$ (d) $\overrightarrow{FO}$ $-2a$

(e) $\overrightarrow{FG}$ (f) $\overrightarrow{GA}$ (g) $\overrightarrow{BF}$ (h) $\overrightarrow{OC}$

$-2a+3b$ $-3b+a$ $-b+2a$ $2a+\frac{1}{2}(3b-2a)$

3 *OAB* is a triangle with $\overrightarrow{OA} = \mathbf{a}$ and $\overrightarrow{OB} = \mathbf{b}$

M is the mid-point of *OB* and P is the point on *AB* such
that $AP:PB = 2:1$

Find expressions for these vectors in terms of **a** and **b**,
simplifying your answers

(a) $\overrightarrow{OM}$ b (b) $\overrightarrow{AB}$ $-a+2b$ (c) $\overrightarrow{AP}$ $\frac{2}{3}(-a+b)$ (d) $\overrightarrow{OP}$ $a+\frac{2}{3}(-b-a)$

(e) $\overrightarrow{BA}$ (f) $\overrightarrow{MA}$ (g) $\overrightarrow{MP}$

$-2b+a$ $-b+a$ $-b+a+\frac{2}{3}(2b-a)$

$=\dfrac{-3b+3a+2b-2a}{3} = \dfrac{a-b}{3}$

4 *OAB* is a triangle.

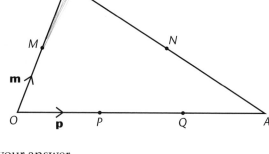

M is the mid-point of *OB*, N is the mid-point of *AB*.
P and Q are points of trisection of *OA*.
$\overrightarrow{OP} = \mathbf{p}$ and $\overrightarrow{OM} = \mathbf{m}$

(a) Find expressions for these vectors in terms
of **p** and **m**, simplifying your answers.

  (i) $\overrightarrow{OA}$ *3p* (ii) $\overrightarrow{OB}$ *2m* (iii) $\overrightarrow{BA}$ *-2m+3p*
  (iv) $\overrightarrow{MN}$ (v) $\overrightarrow{PQ}$ (vi) $\overrightarrow{MP}$
  (vii) $\overrightarrow{MQ}$ (viii) $\overrightarrow{NQ}$ (ix) $\overrightarrow{PN}$

(b) What kind of quadrilateral is *PQNM*? Explain your answer.

5 *ABCDEF* is a regular hexagon with centre *O*.
$\overrightarrow{AB} = \mathbf{b}$ and $\overrightarrow{AC} = \mathbf{c}$

Find these vectors in terms of **b** and **c**,
simplifying your answers

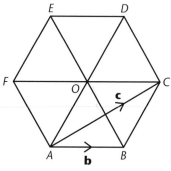

(a) $\overrightarrow{BC}$ *-b+c* (b) $\overrightarrow{AO}$ *-b+c* (c) $\overrightarrow{AD}$ *-2b+2c*
(d) $\overrightarrow{EC}$ (e) $\overrightarrow{AF}$ (f) $\overrightarrow{AE}$

-(-b+c)+b+b
= 3b - c

-b-b+c
= -2b+c

6 *OPQR* is a trapezium with *PQ* parallel to *OR*.
$\overrightarrow{OP} = -2\mathbf{a} + 3\mathbf{b}$ and $\overrightarrow{OQ} = 4\mathbf{a} + 5\mathbf{b}$

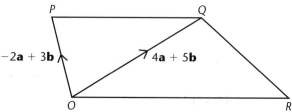

(a) Find $\overrightarrow{PQ}$ in terms of **a** and **b**,
simplifying your answer.

(b) $\overrightarrow{QR} = 5\mathbf{a} + k\mathbf{b}$ (where *k* is a number
to be determined).

Find $\overrightarrow{OR}$ in terms of **a** and **b** and *k* and hence work out the
value of *k*.

7 In the diagram $\overrightarrow{OP} = 2\mathbf{a}$, $\overrightarrow{PA} = \mathbf{a}$, $\overrightarrow{OB} = 3\mathbf{b}$ and $\overrightarrow{BR} = \mathbf{b}$
Q lies on *AB* such that *AQ : QB* = 2 : 1

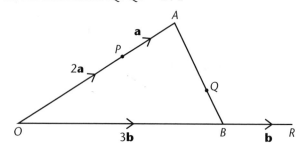

(a) Find expressions for these vectors in terms of **a** and **b**,
simplifying your answers.

  (i) $\overrightarrow{AB}$ (ii) $\overrightarrow{AQ}$ (iii) $\overrightarrow{QB}$ (iv) $\overrightarrow{PQ}$ (v) $\overrightarrow{QR}$ (vi) $\overrightarrow{PR}$

(b) Explain clearly the relationship between points *P*, *Q* and *R*.

UAM **8** *OATB* is a quadrilateral.

P, Q, R and *S* are the mid-points of *OA, AT, TB* and *OB* respectively.

$\overrightarrow{OA} = \mathbf{a}$, $\overrightarrow{OB} = \mathbf{b}$ and $\overrightarrow{OT} = \mathbf{t}$

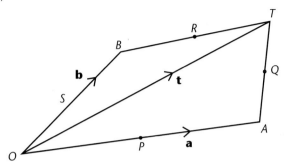

(a) Find, in terms of **a** and **b**, simplifying your answers.

(i) $\overrightarrow{AT}$ (ii) $\overrightarrow{QT}$ (iii) $\overrightarrow{TB}$ (iv) $\overrightarrow{TR}$ (v) $\overrightarrow{PS}$ (vi) $\overrightarrow{QR}$

(b) Explain why your answers for vectors *PS* and *QR* show that *PQRS* is a parallelogram.

UAM **9** The diagram shows a quadrilateral *OAGB*.

$\overrightarrow{OA} = \mathbf{a}$, $\overrightarrow{OB} = 2\mathbf{b}$ and $\overrightarrow{OG} = 3\mathbf{a} + 2\mathbf{b}$

(a) Find expressions for these vectors in terms of **a** and **b**, simplifying your answers.

(i) $\overrightarrow{AG}$ (ii) $\overrightarrow{BG}$

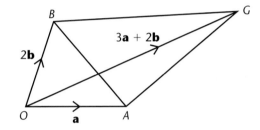

(b) What kind of quadrilateral is *OAGB*? Give a reason for your answer.

(c) Point *P* lies on *AB* such that $AP:PB = 1:3$ Find expressions for these vectors in terms of **a** and **b**, simplifying your answers.

(i) $\overrightarrow{AB}$ (ii) $\overrightarrow{AP}$ (iii) $\overrightarrow{OP}$

(d) Describe, as fully as possible, the position of *P*.

UAM **10** The diagram shows a triangle *OAB*

M is the mid-point of *OA*.

P lies on *BM* and $BP = \frac{3}{4}BM$

$\overrightarrow{OA} = 2\mathbf{a}$ and $\overrightarrow{OB} = 2\mathbf{b}$

(a) Find expressions for these vectors in terms of **a** and **b**, simplifying your answers.

(i) $\overrightarrow{BM}$ (ii) $\overrightarrow{BP}$ (iii) $\overrightarrow{OP}$

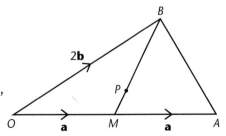

(b) *N* lies on *OB* such that $ON = \frac{2}{5}OB$

Q lies on *AN* such that $AQ = \frac{5}{8}AN$

Find expressions for these vectors in terms of **a** and **b**, simplifying your answers.

(i) $\overrightarrow{ON}$ (ii) $\overrightarrow{AN}$ (iii) $\overrightarrow{AQ}$ (iv) $\overrightarrow{OQ}$

(c) What can you deduce about points *P* and *Q*?

Examination questions

1

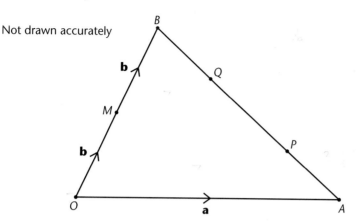

Not drawn accurately

OAB is a triangle where *M* is the mid-point of *OB*.
P and *Q* are points on *AB* such that $AP = PQ = QB$.
$\overrightarrow{OA} = \mathbf{a}$ and $\overrightarrow{OB} = 2\mathbf{b}$.

(a) Find, in terms of **a** and **b**, expressions for
  (i) $\overrightarrow{BA}$ **(ii)** $\overrightarrow{MQ}$ **(iii)** $\overrightarrow{OP}$

(b) What can you deduce about quadrilateral *OMQP*?
  Give a reason for your answer.

(7 marks)
AQA, Spec A, Paper 1H, June 2003

2 *ABCDEF* is a regular hexagon with centre *O*.
$\overrightarrow{OA} = \mathbf{a}$ and $\overrightarrow{AB} = \mathbf{b}$

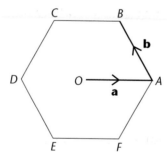

Diagram drawn accurately

(a) Find expressions in terms of **a** and **b**, for
  (i) $\overrightarrow{OB}$ **(ii)** $\overrightarrow{AC}$ **(iii)** $\overrightarrow{EC}$

(b) The positions of points *P* and *Q* are given by the vectors
  $\overrightarrow{OP} = \mathbf{a} - \mathbf{b}$ $\overrightarrow{OQ} = \mathbf{a} + 2\mathbf{b}$

  (i) Draw and label the positions of points *P* and *Q* on the diagram.
  (ii) Hence, or otherwise, deduce an expression for $\overrightarrow{PQ}$.

(6 marks)
AQA, Spec A, Paper 1H, June 2004

Summary of key points

Vectors (grade A/A*)

A *vector* quantity is one which has *magnitude* (size) and *direction*.

Scalar quantities are quantities that have magnitude but no particular direction.

A vector can be represented by a *directed line segment* which is simply a line with an arrow on it.

> Length of line = magnitude of vector.
>
> Direction of arrow = direction of vector.

Vectors are equal if they have the same magnitude and direction.

Vectors can be written as

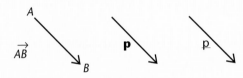

Vectors can be multiplied by scalars (numbers).

Vectors can be added and subtracted using the *parallelogram law*.

$$\overrightarrow{OP} = \overrightarrow{OA} + \overrightarrow{AP} = a + b$$
$$\overrightarrow{BA} = \overrightarrow{BO} + \overrightarrow{OA} = -b + a = a - b$$
$$\overrightarrow{AB} = \overrightarrow{AO} + \overrightarrow{OB} - a + b = b - a$$

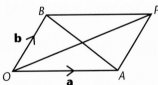

A vector that starts at the origin is known as a *position vector*.

The length of a vector is sometimes written using *modulus* notation, such as: $|OA| = 2\sqrt{5}$ or $|a| = 2\sqrt{5}$

Vector geometry involves the rules for the addition and subtraction of vectors and the multiplication of a vector by a scalar.

Remember that when vectors are equal they are equal in *two* ways ... their *lengths* are equal and they have the same *direction*.

31 Proof

This chapter will show you how to:

✔ tell the difference between 'verify' and 'prove'
✔ prove results using simple, step by step chains of reasoning
✔ use a counter example to disprove a statement
✔ handle proof of simple geometrical statements
✔ handle more difficult proofs, geometrical and algebraic

31.1 Prove v verify

Key words:
verify
proof

Proof questions test the 'Using and Applying' part of what you learn in mathematics.

You can expect to have to:

- find examples that match a general statement or give counter examples which disprove a statement
- show how one statement follows from another
- explain your reasoning or criticise a piece of faulty reasoning
- know the difference between a practical demonstration of something and a proof.

Much of the work is based on the properties of numbers. You need to know about odd and even numbers, prime numbers, factors and multiples and square and cube numbers.

You also need to know the meanings of certain words that often occur in questions of this type – words such as *consecutive* (one after the other), *integer* (whole number) and *product* (multiply).

You will need to remember these important results for addition and multiplication of whole numbers:

Odd + Odd = Even	Odd × Odd = Odd
Odd + Even = Odd	Odd × Even = Even
Even + Odd = Odd	Even × Odd = Even
Even + Even = Even	Even × Even = Even

Here are some examples which will help you to understand what you need to do when asked to prove a result or explain something.

Example 1a

n is any integer.

Explain why $2n + 1$ must be an odd number.

An integer is a whole number.

> Try $n = 1$ $2n + 1 = 2 \times 1 + 1 = 2 + 1 = 3$
>
> Try $n = 2$ $2n + 1 = 2 \times 2 + 1 = 4 + 1 = 5$
>
> Try $n = 3$ $2n + 1 = 2 \times 3 + 1 = 6 + 1 = 7$
>
> The answer is always an odd number.

This is not a proof because the explanation is incomplete.

All you have done is show that the result works when $n = 1, 2, 3$.

This is to **verify** which means to check that something is true by substituting numbers into an expression or a formula.

Will it work for $n = 4, 5, 6, 7, ...$?

You cannot possibly test it for all numbers, it would take forever!

Look at the question again.

Example 1b

n is any integer.

Explain why $2n + 1$ must be an odd number.

> $2n = 2 \times n$
>
> Multiplying any whole number by 2 gives an even number, so $2n$ is an even number.
>
> So $2n + 1 =$ even number $+ 1 =$ odd number
>
> $2n + 1$ must always be an odd number.

This is a **proof**. You are showing that it doesn't matter what number you start with (*n* could stand for any whole number), the result will always be an odd number.

You should be able to see from this example that there is a huge difference between verifying a result (simply checking with numbers) and proving a result.

Verifying is a practical demonstration. Proof shows how one statement follows from another using simple chains of reasoning.

Here is another example showing **(a)** verifying and **(b)** proof.

Example 2a

Explain why the sum of any three consecutive integers is always a multiple of 3.

Consecutive means one after the other.

Try 1, 2 and 3 $1 + 2 + 3 = 6 = 2 \times 3$

Try 5, 6 and 7 $5 + 6 + 7 = 18 = 6 \times 3$

Try 10, 11 and 12 $10 + 11 + 12 = 33 = 11 \times 3$

You always get a multiple of 3.

Multiple of 3 ... a number in the 3 times table.

Example 2b

Explain why the sum of any three consecutive integers is always a multiple of 3.

If x is one of the whole numbers the others are $(x + 1)$ and $(x + 2)$

So $x + (x + 1) + (x + 2) = 3x + 3 = 3(x + 1)$

$3(x + 1)$ means $3 \times (x + 1)$

So the answer must be a multiple of 3.

This is even easier if you choose $(x - 1)$, x and $(x + 1)$.

You can see from the two examples that proof usually involves some algebra skills.

In Example 1b, you simply needed to know that $2n$ means $2 \times n$, a very basic algebra fact.

In Example 2b, you had to collect together like terms and take out a common factor, skills you learnt in Chapter 11.

Notice that proof questions often use the word 'explain' in the question.

You only get full marks if you prove the result as in Examples 1b and 2b.

You may get some marks for verifying (giving numerical examples) but answers which just check the result using numbers often score no marks at all!

Exercise 31A

 1 p is an odd number and q is an even number.

  (a) Explain why $p + q - 1$ is always an even number.

  (b) Explain why $pq + 1$ is always an odd number.

UAM **2** n is a positive integer.
Explain why $n(n + 1)$ must be an even number.

UAM **3** x is an odd number.
Explain why $x^2 + 1$ is always an even number.

UAM **4** If b is an even number, prove that $(b - 1)(b + 1)$ is an odd number.

UAM **5** x is an odd number and y is an even number.
Explain why $(x - y)(x + y)$ is an odd number.

UAM **6** Explain why the sum of 4 consecutive numbers is always an even number.

31.2 Proof by counter example

Key words:
counter example

Proof by **counter example** asks you to give an example to show that a statement is incorrect. You do this by finding one example where the stated result does not work.

You can substitute numbers into an expression or formula until you find a case where the result is not true.

Example 3

Tony says that when n is an even number, $\frac{1}{2}n + 3$ is always even.
Give an example to show that he is wrong.

> Even numbers are 2, 4, 6, …
>
> Try $n = 2$ $\frac{1}{2}n + 3 = \frac{1}{2}(2) + 3 = 1 + 3 = 4$ … even
> Try $n = 4$ $\frac{1}{2}n + 3 = \frac{1}{2}(4) + 3 = 2 + 3 = 5$ … odd
>
> When $n = 4$ the result is not true, so Tony is wrong.

Example 4

Simon says that when you square a number you never get an answer which is smaller than the original number.

Give a counter example to show that Simon is wrong.

Try 1	$1^2 = 1 \times 1 = 1 \dots$	the same
Try 2	$2^2 = 2 \times 2 = 4 \dots$	bigger
Try 3	$3^2 = 3 \times 3 = 9 \dots$	bigger

Trying 4, 5, ... will be even bigger answers than these.

Squaring numbers bigger than 3 will never give a smaller number.

Try -2 $(-2)^2 = (-2) \times (-2) = 4 \dots$ bigger

negative $\times$ negative = positive.

Try a decimal number smaller than 1, such as 0.5

$(0.5)^2 = 0.5 \times 0.5 = 0.25 \dots$ which is **smaller**

The square of 0.5 is smaller than 0.5, so Simon is wrong.

Example 3 is very easy, you can spot a counter example on only your second attempt.

Example 4 is much harder.
You might think that Simon is correct when you try 1, 2, 3, ... because the answers are all bigger than the original number (and getting much bigger all the time).
A negative number (e.g. -2) does not work either.
You need to spot that the answer must come from a decimal number smaller than 1.

You can see that it can take quite a bit of work before you find a counter example. Do not give up too soon!

Exercise 31B

1 Sam says that when k is an even number, $k^2 + \frac{1}{2}k$ is always odd. Give an example to show that he is wrong. $4^2 + \frac{1}{2} \times 4 = 18$

2 Heather says that $m^3 + 2$ is never a multiple of 3. Give an example to show that she is wrong. $1^3 + 2 = 3$

3 p is an odd number and q is an even number. Andrew says that $p + q - 1$ cannot be a prime number. Explain why he is wrong. $1 + 2 - 1 = 2$

4 a and b are both prime numbers. Give an example to show that $a + b$ is not always an even number. $3 \times 7 = 21$

5 Ian says that $n^2 + 3n + 1$ is a prime number for all values of n. Give a counter example to show that he is wrong. $0^3 + 3 \times 0 + 1 = 1$

6 Give a counter example to each of these statements.

   (a) The square root of any number is always smaller than the original number.

   (b) The cube of any number is always bigger than the square of the same number.

31.3 Proof in geometry

If you have to prove a result in geometry you must never just find particular examples that work or check the result by using a protractor to measure angles.

You have to use step by step reasoning, showing clearly how one statement follows from another.

Diagrams are not usually drawn accurately.

Example 5

Prove that the sum of the angles of a triangle is 180°.

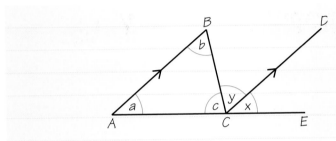

This result was proved in Chapter 4.

Extend side AC to E.
Draw line CD parallel to AB.
Label the angles as shown on the diagram.

$x = a$ (corresponding angles)

$y = b$ (alternate angles)

$x + y + c = 180°$ (sum of angles on a straight line at point C)

So $a + b + c = 180°$

The sum of the angles of $\triangle ABC$ is 180°.

Look back at angle properties and parallel line properties in the early part of Chapter 4 if you need to remind yourself of these facts.

Notice how each step is clearly stated with a valid reason and how the statements follow on from each other.

Geometry proofs must be set out in this way.

Exercise 31C

UAM **1** Prove that the exterior angle of a triangle is equal to the sum of the two opposite interior angles.

UAM **2** Prove that the sum of the interior angles of a quadrilateral is 360°.

UAM **3** Prove that each interior angle of a regular pentagon is 108°.

31.4 Proof using congruent triangles

Chapter 23 began with the statement of three important circle properties, none of which were proved at the time.

We begin this section with a proof of two of these results.
Both proofs involve congruent triangles.

The four conditions for triangles being congruent are:

SSS	SAS
ASA/SAA	RHS

Look back at Chapter 19 if you need to refresh your memory.

Example 6

Prove that the perpendicular from the centre of a circle to a chord bisects the chord.

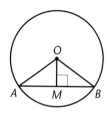

Given	O is the centre of the circle and the line OM is
	perpendicular to the chord AB
To prove	AM = MB

Proof

In triangles OAM and OBM,

1   OA = OB   **(both radii)**

2   OM = OM   **(common side)**

3   angle OMA = angle OMB   **(both 90°, given)**

So   triangles $\dfrac{OAM}{OBM}$ are congruent **(RHS)**

Therefore   AM = BM

Therefore   line OM bisects line AB

Begin a proof by breaking it down into individual steps. Use the word **'Proof'** at the start. When giving a proof it is important to give your reasons for the steps that you take. Use words like **'therefore'**, **'so'**, **'then'** to show what follows on from your last statement. Do not be afraid to write a few words. Use the correct mathematical language wherever possible.

Lettering the triangles in the correct order, one above the other, is very important because it allows you to 'pick off' the equal sides and angles.

Example 7

Prove that tangents drawn to a circle from an external point are equal in length.

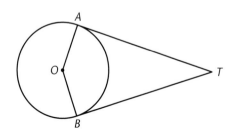

Given   A circle, centre O, and tangents from T
drawn to touch the circle at points A and B.

To prove   TA = TB

Proof

Draw line TO, in triangles TAO and TBO.

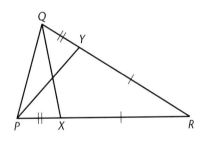

1 TO = TO   (common side)
2 OA = OB   (both radii)
3 angle TAO = angle TBO   (both 90°, tangent &
                                radius meet at 90°)

So   triangles $\dfrac{TAO}{}$ are congruent (RHS)

Therefore   TA = TB

Here is another example using congruent triangles.

Example 8

Prove that PY = QX.

Given   Triangle PQR is isosceles with RQ = RP.
Also, QY = PX.

Proof

In triangles PQY and QPX,

1 angle PQY = angle QPX   (base angles of
                                isosceles △RPQ)

2 QY = PX   (given)

3 QP = QP   (common side)

So triangles $\dfrac{PQY}{QPX}$ are congruent (SAS)

Therefore PY = QX

Always try to identify the
appropriate triangles and
attempt to satisfy one of
the conditions for
congruency: SSS, SAS, ASA
and RHS.

Always number your facts 1, 2 and 3 so that it is clear what you are stating.

Always give a reason for each fact.

Write the two congruent triangles above one another in your final statement so that you can see which points correspond.

Don't forget the proofs of the circle theorems that were given in detail in Chapter 23, Section 23.2. You could be asked to prove any of them in your GCSE exam.

Exercise 31D

In all of these questions try to set out your work as in Examples 6, 7 and 8.

Proof needs to be thorough.

All steps should follow logically.

UAM **1** *ABCD* is a parallelogram.

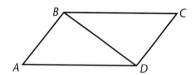

Prove that triangle *DAB* is congruent to triangle *BCD*.

UAM **2** *PQRS* is a quadrilateral with *PQ* = *PS* and *QR* = *SR*.

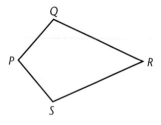

Prove that angle *PQR* = angle *PSR*.

Draw the line *PR*.

UAM **3** In the diagram *ABCD* is a parallelogram. *AX* bisects angle *A* and *CY* bisects angle *C*.

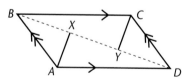

Prove that the triangles *ABX* and *CDY* are congruent.

UAM **4** *PQRS* is a rectangle. A point *X* lies outside of the rectangle such that *QX* = *RX*. Prove that *PX* = *SX*.

Draw a diagram showing this information.

31.5 Proof using algebraic manipulation

In Section 31.1 you saw the difference between verify and proof.

Examples 1a and 1b, 2a and 2b illustrated the difference for two fairly straightforward questions.

The examples in this section use the same methods but the questions become gradually more difficult.

Example 9

Prove that the product of two odd numbers is always odd.

> Let the two odd numbers be $2n + 1$ and $2m + 1$
>
> The product $= (2n + 1)(2m + 1)$
>
> $\qquad = 4nm + 2n + 2m + 1$
>
> $\qquad = 2(2nm + n + m) + 1$
>
> which is of the form $2k + 1$ (where $k = 2nm + n + m$)
>
> so it must be an odd number.

Remember ... do *not* substitute numbers to check the result ... this would be 'verify' not 'prove'.

Example 1 showed you that a number of the form $2n + 1$ is always odd.

Example 10

The sequence of numbers 1, 3, 6, 10, 15, ... is known as the triangle number sequence.

A formula for the nth triangular number is $\frac{1}{2}n(n + 1)$.

Prove that the sum of any two consecutive triangular numbers is always a perfect square.

> The nth triangular number is $\frac{1}{2}n(n + 1)$
>
> The $(n + 1)$th triangular number is $\frac{1}{2}(n + 1)(n + 2)$
>
> Their sum $= \frac{1}{2}n(n + 1) + \frac{1}{2}(n + 1)(n + 2)$
>
> $\qquad = \frac{1}{2}(n + 1)\,[n + n + 2]$
>
> $\qquad = \frac{1}{2}(n + 1)[2n + 2]$
>
> $\qquad = \frac{1}{2}(n + 1)2(n + 1)$
>
> $\qquad = (n + 1)(n + 1)$
>
> $\qquad = (n + 1)^2$ which is a perfect square.

The pattern of dots explains the name of the sequence. Look back at Chapter 14, Exercise 14C, question 2.

Remember that consecutive means one after the other.

Take out a common factor of $\frac{1}{2}(n + 1)$ then factorise the term in the square bracket. This brings in a '2' which cancels out the $\frac{1}{2}$ leading to the final expression.

Example 11

Prove that the difference between the squares of two consecutive odd numbers is always a multiple of 8.

> Choose $(2n - 1)$ and $(2n + 1)$ as the two consecutive odd numbers.
>
> Difference between the squares $= (2n + 1)^2 - (2n - 1)^2$
>
> $\qquad = (4n^2 + 4n + 1) - (4n^2 - 4n + 1)$
>
> $\qquad = 4n^2 + 4n + 1 - 4n^2 + 4n - 1$
>
> $\qquad = 8n \qquad$ which is a multiple of 8.

> You can have $(2n + 1)$ and $(2n + 3)$ if you prefer – the algebra skills required are the same.

> Squaring the two brackets (3 terms, remember!) then changing the signs when the 2nd bracket is removed.

Notice that in Examples 9, 10 and 11, the algebra skills of expanding brackets, factorising and simplifying are crucial.

> These are A/A* questions.

You will need to make sure that you master these skills if you want to succeed at algebraic proofs at this level.

Exercise 31E

UAM **1** Prove that
  (a) the product of two even numbers is even
  (b) the sum of an odd number and an even number is odd
  (c) the difference between the squares of two consecutive numbers is odd
  (d) the sum of three consecutive odd numbers is odd.

UAM **2** Here is a 6×6 square grid. One 2×2 square is highlighted.

1	2	3	4	5	6
7	8	**9**	**10**	11	12
13	14	**15**	**16**	17	18
19	20	21	22	23	24
25	26	27	28	29	30
31	32	33	34	35	36

  (a) In this 2×2 square, work out
    (i) 'bottom left $\times$ top right'
    (ii) 'top left $\times$ bottom right'
    (iii) the difference between these two answers.
  (b) Try this for two more 2×2 squares taken from the 6×6 square grid.
      What do you notice?
  (c) **Prove** that your result will be the same for any 2×2 square taken from the 6×6 square grid.

> Let n be the number in the top left corner of any 2×2 square and write down expressions for the other three numbers in the 2×2 square in terms of n.

UAM **3** **(a)** Repeat question 2 for an 8×8 square grid, testing any two 2×2 squares to obtain a numerical answer.
Prove your result as in part **(c)** of question 2.

(b) What will be the difference if you repeat the process in an $n \times n$ square grid?

UAM **4** A, B, C and D are four consecutive integers.

(a) Verify that $B^2 + C^2 - D^2 = A^2 - 4$ for any four consecutive integers of your choice.

(b) **Prove** that $B^2 + C^2 - D^2 = A^2 - 4$ for any four consecutive integers.

> Let $A = n$ and start by writing B, C and D in terms of n.

UAM **5** A, B and C are three consecutive odd numbers.

(a) Verify that $A^2 + B^2 - C^2 = B \times (C - 10)$ for any three consecutive odd numbers of your choice.

(b) If $A = 2n - 1$, write down expressions for B and C in terms of n.

(c) Hence **prove** that $A^2 + B^2 - C^2 = B \times (C - 10)$ for any three consecutive odd numbers.

UAM **6** A Fibonacci sequence is formed by adding together the previous two terms to get the next term.
So, starting with 1 and 4, the sequence will be

$$1, \quad 4, \quad 5, \quad 9, \quad 14, \quad 23, \ldots$$

In the questions that follow,

u_n means the nth term, so, for example, $u_3 = 5$

and $\quad$ S_n means the sum of the first n terms, so, for example, $S_3 = 1 + 4 + 5 = 10$.

(a) Write down the first 10 terms of the sequence.

(b) Verify that $S_4 = u_6 - u_2$.

(c) Verify that $S_5 = u_7 - u_2$.

(d) Verify that $S_6 = u_8 - u_2$.

(e) Let the first term of a Fibonacci sequence $= a$ and the second term $= b$.

Use algebra to write down the first eight terms of the sequence and prove that the sum of the first six terms is equal to the difference between the 8th term and the 2nd term.

Examination questions

1 In the diagram, the lines AC and BD intersect at E.
AB and DC are parallel and $AB = DC$.

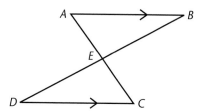

Prove that triangles ABE and CDE are congruent.

(4 marks)
AQA June 2003

2 $ABCD$ and $PQRS$ are squares. Angle $DAP = y$.

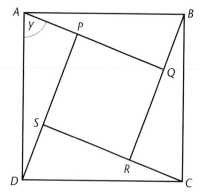

Prove that triangles ABQ and DAP are congruent.

(4 marks)
AQA June 2004

3 Annie, Bert and Charu are investigating the number sequence
$$21, 40, 65, 96, 133, \ldots$$

(a) Annie has found the following pattern.

1st term	$1 \times 2 + 3^2 + 2 \times 5 = 21$
2nd term	$2 \times 3 + 4^2 + 3 \times 6 = 40$
3rd term	$3 \times 4 + 5^2 + 4 \times 7 = 65$
4th term	$4 \times 5 + 6^2 + 5 \times 8 = 96$
5th term	$5 \times 6 + 7^2 + 6 \times 9 = 133$

Complete the nth term for Annie's pattern.

nth term $n \times (n + 1) +$ _____ $+$ _____ $\times$ _____

(b) Bert has found this formula for the nth term
$$(3n + 1)(n + 3) + 5$$

Charu has found this formula for the nth term
$$(2n + 3)^2 - (n + 1)^2$$

Prove that these two formulae are equivalent.

(5 marks)
AQA, Spec A, 2H, November 2004

Summary of key points

Questions on proof can be set at various grade levels depending on the content of the material. The proofs involving odd and even numbers at the start of this chapter will typically be at grades D or C, but proofs involving congruent triangles or algebraic manipulation or the circle theorem proofs in Chapter 23 will all be at grades A or A*.

Odd + Odd = Even	Odd × Odd = Odd
Odd + Even = Odd	Odd × Even = Even
Even + Odd = Odd	Even × Odd = Even
Even + Even = Even	Even × Even = Even

You need to know the difference between 'prove' and 'verify'.

Verify means to check that something is true by substituting numbers into an expression or a formula. It is a practical demonstration of a result.

In a *proof* you have to show how one statement follows from another using simple chains of reasoning.

Proof questions often have the word 'explain' in the question and usually involve some algebra skills.

In proof by *counter example* you find *one* example where the stated result does not work. You can substitute numbers into an expression or formula until you find a case where the result is not true.

To prove a result in geometry, you have to use step by step reasoning, showing clearly how one statement follows from another.

When you prove a result using congruent triangles remember that the four conditions for triangles being congruent are:

   SSS, SAS, ASA/SAA, RHS

Use the word *'Proof'* at the start and break it down into individual steps.

Always *number your facts 1, 2 and 3* so that it is clear what you are stating.

Always *give a reason* for each fact using the correct mathematical language wherever possible.

Write the two congruent *triangles above one another* in your final statement *so that you can see which points correspond.*

Don't forget the proofs of the circle theorems that were given in detail in Chapter 23, Section 23.2.

Proving a result by algebraic manipulation depends on the skills of expanding brackets, factorising and simplifying. You will also need a knowledge of basic mathematical facts, for example that a number of the form $2n$ is always even because it is a multiple of 2 and so $(2n + 1)$ is always odd.

Examination practice

Higher Paper 1 (non calculator)

1 (a) Jack and Jill share £160 in the ratio 3 : 1
How much does each receive? *(2 marks)*

(b) What percentage of the £160 does Jack receive? *(2 marks)*

2 Solve **(a)** $2(x + 3) = 32$ *(3 marks)*

(b) $6x - 3 = 9 - 2x$ *(3 marks)*

3 The scatter graph shows marks scored by 20 pupils on English and History exam papers.

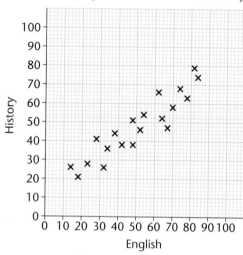

(a) Write down the lowest mark on the History paper. *(1 mark)*

(b) Describe the relationship between the English and the History marks. *(1 mark)*

(c) On a copy of the scatter graph, draw a line of best fit. *(1 mark)*

(d) Jenny was absent for the English exam but scored 62 on the History paper. Use your line of best fit to estimate her English exam mark. *(1 mark)*

4 Draw triangle A at (1, 1), (1, 3) and (2, 1).
Enlarge triangle A by scale factor 2 with (0, 0) as the centre of enlargement. *(3 marks)*

5 The mountain K2 is 8611 metres high, to the nearest metre.
What is its smallest possible height in kilometres? *(2 marks)*

6 In the diagram, AB is parallel to CD.

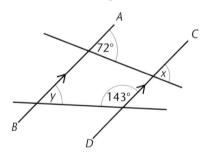

(a) State the value of x.
Give a reason for your answer. *(2 marks)*

(b) Find the value of y. *(2 marks)*

7 Simplify

  (a) $w^5 \times w^4$ **(b)** $x^{10} \div x^2$ **(c)** $(y^3)^4$ *(3 marks)*

8 Use approximations to estimate the value of $\dfrac{407 \times 6.03}{0.298}$ *(3 marks)*

9 Make y the subject of the formula $h = 7y + 30$ *(2 marks)*

10 The diagram shows a right-angled triangle ABC.
   $AB = 12$ cm and $AC = 14$ cm.

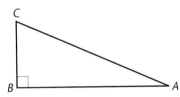

   Calculate the length of BC.
   Leave your answer as a square root.   *(3 marks)*

11 **(a)** Complete the table of values for $y = 2x^2 - 6x + 1$ *(2 marks)*

x	-1	0	1	2	3	4
y	9	-3	1	9		

  (b) Draw the graph of $y = 2x^2 - 6x + 1$ for values of x from -1 to $+4$ *(2 marks)*
  (c) Use your graph to solve the equation $2x^2 - 6x + 1 = 0$ *(2 marks)*

12 The table shows the distances travelled to school by 50 pupils living in a town.

Distance travelled, d (km)	Frequency
$0 < d \leqslant 2$	14
$2 < d \leqslant 4$	19
$4 < d \leqslant 6$	7
$6 < d \leqslant 8$	8
$8 < d \leqslant 10$	2

Use the mid-points of the class intervals to calculate an estimate of the mean distance
travelled to school by these pupils.   *(3 marks)*

13 Here are six numbers written in standard form.

   1.7×10^4 1.42×10^6 6.71×10^0 5.2×10^{-4} 4.8×10^{-1} 3.4×10^{-3}

  (a) Write down the largest number.   *(1 mark)*
  (b) Write down the smallest number.   *(1 mark)*
  (c) Write 3.4×10^{-3} as an ordinary number.   *(1 mark)*
  (d) Work out $1.7 \times 10^4 \div 0.1$
     Give your answer in standard form.   *(1 mark)*

14 Triangles *ABC* and *ACD* are similar.
AB = 4 cm and *AC* = 6 cm.

Calculate the length of *AD*.

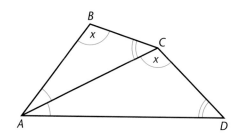

(3 marks)

15 The box plot shows the heights, in centimetres, of plants in a nursery.

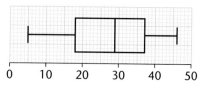

(a) Write down the median height. *(1 mark)*

(b) Work out the interquartile range of the heights. *(2 marks)*

16 Solve the simultaneous equations

$$2x - 3y = 4$$
$$3x + y = 17$$

You **must** show your working.
Do not use trial and improvement. *(3 marks)*

17 (a) Points *P*, *Q*, *R* and *S* lie on a circle.
PQ = *QR*
Angle *PQR* = 112°

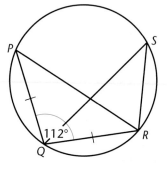

Explain why angle *QSR* = 34° *(2 marks)*

(b) The diagram shows a circle, centre *O*.
TA is a tangent to the circle at *A*.
Angle *BAC* = 56° and angle *BAT* = 75°

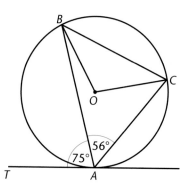

(i) Calculate angle *BOC*. *(1 mark)*
(ii) Calculate angle *OCA*. *(3 marks)*

18 Find the equation of the straight line passing through the point (0,4)
which is perpendicular to the line $y = \frac{2}{5}x + 1$ *(2 marks)*

19 The diagram shows a sector of a circle of radius 18 centimetres.

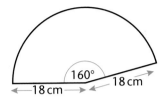

Find the perimeter of the sector.
Give your answer in terms of π. *(3 marks)*

20 The diagram shows the graph of an equation of the form $y = x^2 + bx + c$

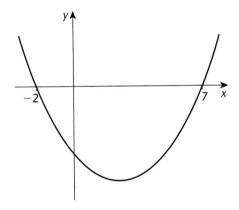

Find the values of b and c.
You must show your method. (3 marks)

21 A village has a population of 1200.
The population is classified by age as shown in the table below.

Age (years)	0–10	11–25	26–40	41–60	61–80	81+
Number of people	133	194	327	241	210	95

A stratified sample of 150 is planned.
Calculate the number of people that should be sampled from each age group. *(3 marks)*

22 (a) Prove that $0.\overline{67} = \frac{67}{99}$ *(2 marks)*

(b) Hence, or otherwise, express $0.2\overline{67}$ as a fraction.
Give your answer in its simplest form. *(2 marks)*

23 The area of this rectangle is 45 cm².

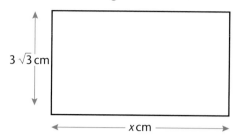

$3\sqrt{3}$ cm

x cm

Find the value of x, writing your answer in the form $a\sqrt{b}$ where a and b are integers. *(3 marks)*

24 In triangle OAB, M lies on AB such that $AM = \frac{2}{3}AB$.
$\overrightarrow{OA} = \mathbf{a}$ and $\overrightarrow{OB} = \mathbf{b}$

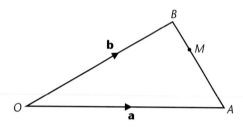

Find $\overrightarrow{OM}$ in terms of $\mathbf{a}$ and $\mathbf{b}$
Give your answer in its simplest form. *(3 marks)*

25 (a) n is a positive integer.
    (i) Explain why $n(n-1)$ must be an even number. *(1 mark)*
    (ii) Explain why $2n-1$ must be an odd number. *(1 mark)*
(b) Expand and simplify $(2n-1)^2$ *(2 marks)*
(c) Prove that the square of any odd number is always 1 more than a multiple of 8. *(3marks)*

26 Trainees in a company have to sit and pass an exam before they are offered a job with the company.
Results from several years show that the probability of passing the exam is 0.7
Three people are chosen at random from the current group of trainees.
What is the probability that exactly two of them pass the exam? *(3 marks)*

27 (a) (i) Factorise $x^2 - 6x + 9$ *(2 marks)*
    (ii) Hence, or otherwise, solve the equation
$$(y-4)^2 - 6(y-4) + 9 = 0$$
(2 marks)
(b) Simplify $\dfrac{x^2 - 49}{x^2 + 7x}$ *(3 marks)*

Higher Paper 2 (calculator)

1 The table shows the amounts needed to make jam tarts.

Amount for 24 jam tarts	Amount for 30 jam tarts
Flour	220 g
Lard	50 g
Butter	50 g
Jam	440 g

Calculate the amounts needed to make 30 jam tarts. *(3 marks)*

2 The diagram shows a trapezium.

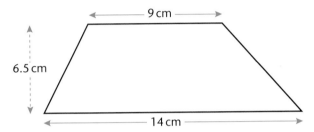

Calculate the area of the trapezium. *(2 marks)*

3 These are the times recorded, to the nearest minute, by 20 runners in a 10 km race.

45, 52, 50, 37, 43, 48, 51, 38, 40, 39

49, 46, 50, 42, 48, 41, 50, 45, 39, 43

(a) Construct an ordered stem-and-leaf diagram to represent the data. *(3 marks)*

(b) What is the median time? *(1 mark)*

4 Draw axes, 2 cm for 1 unit of x for $0 \leqslant x \leqslant 4$

and 1 cm for 1 unit of y for $0 \leqslant y \leqslant 12$

Draw the graph of $y = 2x + 3$ for values of x from 0 to 4 *(3 marks)*

5 Peter has a part-time job.

He is paid £40.50 for working from 8.30 a.m. to 4 p.m.

(a) How much is this per hour? *(2 marks)*

Peter is saving for a remote controlled car that costs £60.

(b) What percentage is £40.50 of £60? *(2 marks)*

6 Use your calculator to work out $\sqrt{\dfrac{22514}{(18.3)^2}}$

(a) Write down the full calculator display. *(1 mark)*

(b) Give your answer to 3 significant figures. *(1 mark)*

7 (a) Factorise **(i)** $10x - 4$ *(1 mark)*
  (ii) $y^2 + 6y$ *(1 mark)*

  (b) Expand and simplify $2(x + 3) + 3(2x - 5)$ *(2 marks)*

  (c) Solve the inequality $7w < 5w + 13$ *(2 marks)*

8 Draw axes, 1 cm for 1 unit for both x and y, for $-7 \leqslant x \leqslant 7$ and $-7 \leqslant y \leqslant 7$
  Draw triangle A at $(-2, 3)$, $(-1, 3)$ and $(-1, 6)$

  (a) Reflect triangle A in the line $y = 1$, and label your image B. *(2 marks)*

  (b) Rotate triangle A 90°, clockwise, about the origin. Label your image C. *(3 marks)*

9 Lisa, Meg and Nia share £3680 in the ratio $7 : 5 : 4$
  How much does each of them receive? *(3 marks)*

10 A spinner has five possible outcomes, scores of 1, 2, 3, 4 and 5.
  The spinner is biased and the probabilities of it landing on each of the numbers are:

Number	1	2	3	4	5
Probability	$\frac{1}{5}$	p	$\frac{1}{5}$	$\frac{1}{5}$	$\frac{2}{11}$

  (a) Work out the value of p. *(2 marks)*

  (b) If the spinner is spun twice, what is the probability that it lands on 3 both times? *(2 marks)*

11 A cylinder has a diameter of 22 cm and a height of 35 cm.
  Calculate the volume of the cylinder.
  Give your answer to a sensible degree of accuracy. *(5 marks)*

12 Use trial and improvement to solve the equation
  $$x - \frac{1}{x} = 4$$

  Copy and complete this table. The first trial has been done for you.

x	$x - \dfrac{1}{x}$	Comment
4	3.75	Too low

  Give your answer to 1 decimal place. *(4 marks)*

13 Jim invests £3500 at 2.8% compound interest.
  How many years will it take for his investment to exceed £4000? *(3 marks)*

14 Solve the equation $\dfrac{x - 4}{3} - \dfrac{x + 2}{5} = 2$ *(4 marks)*

15

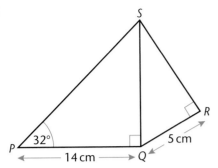

PQS and QRS are right-angled triangles.
PQ = 14 cm, QR = 5 cm and angle SPQ = 32°

Calculate the size of angle SQR.   *(4 marks)*

16 Distances in the solar system can be measured in kilometres (km) or in astronomical units (AU).
Jupiter is 778 million km from the Sun.
In astronomical units, Jupiter is 5.2 AU from the Sun.
How many km are there in 1 AU?
Give your answer to a sensible degree of accuracy.   *(3 marks)*

17 The diagram shows a straight line graph.

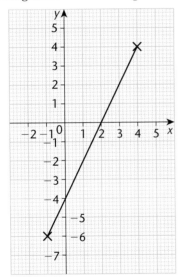

(a) What is the equation of this straight line?   *(3 marks)*

(b) This line is translated by the vector $\begin{pmatrix} 0 \\ 7 \end{pmatrix}$
Work out the equation of the new line.   *(2 marks)*

18 The table shows the cost of electricity at the end of every three month period and some 4-point moving averages.

Year	2003				2004				2005			
Quarter	1	2	3	4	1	2	3	4	1	2	3	4
Cost (£)	91	87	80	90	99	91	88	98	103	95	96	
4-point moving average		87	89	90	92	94	95					

(a) Work out the next two 4-point moving averages.   *(2 marks)*

(b) Use the trend of the moving averages to predict the cost of electricity at the end of the 4th quarter of 2005.   *(3 marks)*

19 Paint is sold in cylindrical tins.
Two of the sizes you can buy are shown in the diagram.

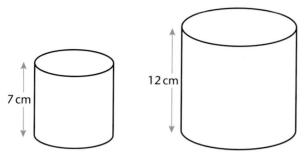

The smaller tin is similar in shape to the larger tin.
The smaller tin is 7 cm tall and holds 250 m*l* of paint.
The larger tin is 12 cm tall.
How much paint does the larger tin hold? *(2 marks)*

20 *ABCD* is a quadrilateral.
$BC = 7$ cm, $CD = 10$ cm and $AD = 5$ cm.
Angle $BCD = 68°$ and angle $BAD = 90°$.

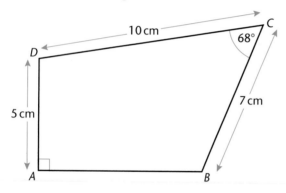

Calculate the perimeter of the quadrilateral. *(5 marks)*

21 (a) Solve the equation $x^2 - 5x + 1 = 0$
Give your answers to 2 decimal places. *(3 marks)*

(b) Make *y* the subject of this formula
$$k(y - a) = my + b$$ *(4 marks)*

22 Prove that the line drawn from the centre of a circle to the mid-point of a chord is perpendicular to the chord.

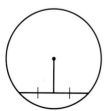

(4 marks)

23 A games machine is on sale in a shop for a certain price (the original price).
This original price is then increased by 20%
Later, when the shop has a sale, the games machine is advertised at '20% OFF'
Is the sale price more than, the same as or less than the original price?   *(3 marks)*

24 The histogram shows the lengths (in minutes) of 80 telephone calls.

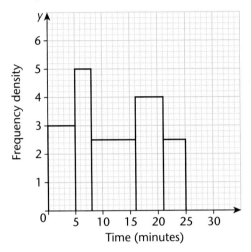

(a) Work out the median length of telephone call.   *(2 marks)*

(b) Work out the interquartile range of the lengths of these telephone calls.   *(2 marks)*

25 The diagram shows the graph of $y = f(x)$

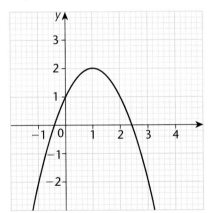

On three separate sets of axes, sketch the graphs of these transformations of $y = f(x)$

(a) $y = f(x) - 2$ (b) $y = f(x - 2)$ (c) $y = f(2x)$ *(3 marks)*

26 A cone has a curved surface area of 120 cm, measured to the nearest square centimetre.
The base radius of the cone is 5 cm, measured to the nearest centimetre.
What is the upper bound of its slant height?   *(3 marks)*

Answers

Chapter 1

Exercise 1A

1. (a) 73   (b) 89   (c) 512
   (d) 15   (e) 4   (f) 300
2. (a) 16, 16, 25, 25, 100, 100
   (b) $8, -8, 27, -27, 125, -125$
   (c) 4, impossible, 5, impossible
   (d) $2, -2, 4, -4$
3. (a) true   (b) false   (c) false
   (d) true   (e) true
4. (a) 69.302   (b) 2847.2896   (c) 2.3
   (d) 93.464   (e) 14   (f) 16
5. (a) 64, 4096 and others
   (b) They have the form n^6 where n is an integer.

Exercise 1B

1. (a) $2^2 \times 5$ (b) $2^3 \times 3^2$
   (c) $3^2 \times 5$ (d) $2^4 \times 3^2$
   (e) $2^2 \times 3^3 \times 5$ (f) $2^3 \times 5^3$
   (g) $2 \times 3^2 \times 5^2 \times 7$ (h) $3^3 \times 7^2$
2. 3 and 11
3. 2 or 3
4. (a) 2 and 3   (b) 2, 3, 5   (c) 2 and 3

Exercise 1C

1. (a) 16   (b) 180   (c) 24
   (d) 15   (e) 21   (f) 8
2. 40 is the HCF of 40 and 160;
   8 is the HCF of 40, 72 and 160
   66 is the HCF of 462 and 132;
   12 is the HCF of 132 and 72.
3. (a) No common prime factors
   (b) No common prime factors
   (c) 17 is a common prime factor.
4. 48 cm

Exercise 1D

1. (a) 48   (b) 160   (c) 54
   (d) 2310   (e) 840   (f) 540
2. 3 min 2 sec
3. 90 minutes
4. (a) $\frac{41}{60}$ (b) $\frac{37}{144}$ (c) $5\frac{41}{72}$
5. 180 seconds or 3 minutes

Examination style questions

1. (a) $2^2 \times 7$ (b) 84
2. (a) $a = 2, b = 3$ (b) 27
3. (a) 5   (b) 3×5^2
4. 30

Chapter 2

Exercise 2A

1. (a) $\frac{3}{4}$ (b) $\frac{3}{5}$ (c) $\frac{2}{3}$ (d) $\frac{3}{10}$
   (e) $\frac{5}{6}$ (f) $\frac{2}{3}$ (g) $\frac{5}{6}$ (h) $\frac{3}{5}$

2. $\frac{1}{2}, \frac{1}{4}$
3. (a) $\frac{1}{6}, \frac{1}{3}, \frac{1}{2}$

Exercise 2B

1. (a) $\frac{1}{4}, \frac{3}{8}, \frac{1}{2}$ (b) $\frac{4}{10}, \frac{1}{2}, \frac{3}{5}$ (c) $\frac{1}{3}, \frac{3}{4}, \frac{5}{6}$
2. (a) $\frac{3}{5} < \frac{5}{8}$ (b) $\frac{1}{3} > \frac{3}{10}$ (c) $\frac{5}{6} > \frac{4}{9}$
   (d) $\frac{3}{4} < \frac{5}{6}$ (e) $\frac{3}{5} < \frac{5}{7}$ (f) $\frac{1}{4} < \frac{2}{5}$
3. (a) $\frac{1}{4}, \frac{3}{5}, \frac{7}{10}$ (b) $\frac{7}{12}, \frac{2}{3}, \frac{5}{6}$
   (c) $\frac{17}{40}, \frac{5}{8}, \frac{7}{10}$ (d) $\frac{13}{25}, \frac{57}{100}, \frac{14}{20}$

Exercise 2C

1. (a) $1\frac{1}{2}$ (b) $\frac{1}{20}$ (c) $4\frac{19}{24}$
   (d) $2\frac{23}{30}$ (e) $\frac{17}{35}$ (f) $2\frac{1}{4}$
   (g) $5\frac{5}{12}$ (h) $\frac{34}{45}$ (i) $2\frac{23}{40}$

Exercise 2D

1. $\frac{19}{24}$
2. (a) $15\frac{2}{5}$ cm²   (b) $26\frac{23}{45}$ cm²
3. £8.00

Exercise 2E

1. (a) $\frac{3}{7}$ (b) 6   (c) 4
   (d) 11   (e) $\frac{15}{4}$ or $3\frac{3}{4}$ (f) $\frac{37}{33}$ or $1\frac{4}{33}$

Exercise 2F

1. (a) $\frac{1}{4}$ (b) 6   (c) $\frac{1}{15}$ (d) 20
   (e) $1\frac{1}{3}$ (f) $4\frac{1}{2}$ (g) $\frac{5}{7}$ (h) $\frac{3}{13}$
   (i) $\frac{7}{17}$ (j) 0.4 or $\frac{2}{5}$ (k) 2.5 or $2\frac{1}{2}$ (l) $\frac{8}{9}$
   (m) $\frac{13}{83}$ (n) $\frac{11}{136}$
2. $\frac{10}{21}, \frac{1}{2}, \frac{6}{11}, \frac{8}{13}, \frac{11}{15}, \frac{3}{4}, \frac{7}{9}, \frac{4}{5}, \frac{5}{6}$

Exercise 2G

1. (a) $\frac{5}{4}$ or $1\frac{1}{4}$ (b) $2\frac{2}{7}$ (c) $\frac{6}{7}$ (d) $\frac{8}{21}$
   (e) $\frac{9}{16}$ (f) 8   (g) $1\frac{5}{12}$

Exercise 2H

1. 2.5 m²
2. $\frac{1}{2}$
3. £17
4. (a) 3   (b) 0.45 m or $\frac{9}{20}$ m
5. $\frac{2}{5}$
6. $2\frac{68}{75}$ litres
7. $2\frac{1}{2}$ m

Examination style questions

1. (a) $\frac{13}{40}$ (b) $\frac{1}{40}$
2. 5

Chapter 3

Exercise 3A

1 (a) $\frac{3}{20}$ (b) $\frac{9}{25}$ (c) $\frac{2}{5}$ (d) $\frac{39}{50}$ (e) $\frac{1}{8}$
2 207%, 210%, 232%, 234%, 240%
3

Percentage	Fraction	Decimal
60%	$\frac{3}{5}$	0.6
72%	$\frac{18}{25}$	0.72
62.5%	$\frac{5}{8}$	0.625
102.5%	$1\frac{1}{40}$	1.025
$12\frac{1}{2}\%$	$\frac{1}{8}$	0.125

4 0.4, 42%, $43\frac{3}{4}\%$, 0.438, $\frac{9}{20}$

Exercise 3B

1 $\frac{1}{3}$ off
2 £125 with 60% off is cheaper.
3 '$\frac{1}{4}$ off the price' is better value.

Exercise 3C

1 English: 55%, Maths: 53%, French: 57%, History: 52%, Geography: 54%
2 14%
3 26%
4 31.25%
5 (a) 47% (b) 29% (c) 52%
6 12.5%
7 3%

Exercise 3D

1 (a) 31.5 (b) 64.96 (c) 2788 ml
   (d) 326.4 (e) £145.125 (f) 815.24 ml
2 £436.80
3 329
4 £30.10, £18.92
5 £4816.50
6 £922.35
7 Jane has the greater salary.
8 (a) 80.22 kg (b) 81.8244 kg
9 128 960
10 Tom: £268.88, Tina: £243.75, Tracy: £260.63

Exercise 3E

1 (a) (i) £22.75 (ii) £7.00
   (b) (i) £152.75 (ii) £46.99
2 £148.05
3 The red car (£4465)
4 £21.61
5 £45.46
6 The first method is cheaper by £54.61

Exercise 3F

1 18%
2 3.3%
3 4%
4 (a) 1% (b) 3953
5 (a) 9% (b) 25%
6 (a) 43% (b) 46%

Exercise 3G

1 14%
2 265%
3 30% profit, 29% loss, 25% profit, 5% loss, 49% profit, 4% loss
4 15%
5 38%
6 67%
7 (a) 6.2% (b) £179,000

Exercise 3H

1 £380
2 £14 000
3 £114.89
4 16.7 cm
5 6 kg
6 £74.47
7 (a) 980 (b) 1065
8 (a) £14 500 (b) £12 340.43
9 £1946.25
10 (a) 40 cm (b) 40.2 cm

Exercise 3I

1 £146.02
2 £9192.73
3 2740
4 The first method is best.
5 7 years
6 (a) £6.50 (b) £2301.65 (c) £5271.33
7 3 years
8 (a) £15 799.23 (b) £1410.38 (c) £123.33
9 (a) 12 years (b) 11 years
10 8%

Examination style questions

1 (a) £3025 (b) £1200
2 £76.50
3 (a) 6.3% (b) 2800

Chapter 4

Exercise 4A

1 38° **2** 59° **3** 56° **4** 39° **5** 24°
6 128°, 15° **7** 45° **8** 60° **9** 18°

Exercise 4B

1 62°
2 109°
3 65°
4 $d = e = 110°$
5 105°
6 $g = 63°, h = 117°$
7 $i = 74°, j = 106 = 106°$
8 $l = n = p = 53°, m = 127°$
9 $q = 82°, r = 75°$

Exercise 4C

1 (a) (i) 065° (ii) 245°
   (b) (i) 115° (ii) 295°
   (c) (i) 225° (ii) 045°
   (d) (i) 325° (ii) 145°

2 Answers will need to be measured individually.

(a)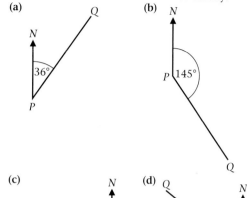

(b)

(c)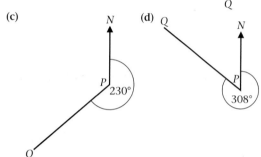

(d)

(e)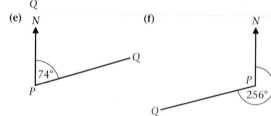

(f)

(g)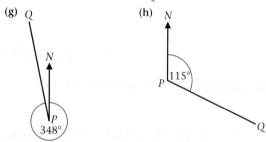

(h)

3 (a) 055° (b) 110° (c) 015°
    (d) 235° (e) 290° (f) 195°

(d) parallelograms, rhombuses
(e) squares, rectangles
(f) parallelograms, rectangles, rhombuses, squares
(g) kites
(h) kites, parallelograms, rectangles
(i) parallelograms, rhombuses, rectangles, squares
(j) trapeziums, rhombuses, parallelograms, rectangles, squares
(k) rectangles, squares
(l) kites, rhombuses, squares

Exercise 4G

1 $a = 92°$ **2** $b = 55°, c = 125°$
3 $d = 69°$ **4** $e = 135°$
5 $f = 64°$ **6** $g = 137°$
7 $h = 100°, i = 105°$ **8** $j = 153°, k = 45°$
9 $l = 116°, m = 64°$
10 $n = 120°, p = 40°$
11 $q = 77°, r = 75°$ **12** $t = 72°$
13 $u = 105°, w = 75°$ **14** $x = 64°, y = 116°$
15 160°
16 No. If such a polygon existed and had n sides, then
$$130 = \frac{80(n-2)}{n}$$ so that $130n = 180n - 360$ so that
$50n = 360$ which is impossible since n must be an integer.

Examination style questions

1 122°
2 12 sides
3 $a = 135°, b = 67.5°, c = 45°$

Chapter 5

Exercise 5A

1 (a) $h = 2$ (b) $b = 11$ (c) $x = 1$
   (d) $k = 7$ (e) $y = 20$ (f) $a = 50$
   (g) $f = 27$ (h) $w = 80$ (i) $n = 6$
2 (a) $b = 3$ (b) $t = 5$ (c) $p = 3$
   (d) $g = 7$ (e) $x = 4$ (f) $n = 3$
   (g) $m = 13$ (h) $y = 60$ (i) $k = 7$
   (j) $h = 4$ (k) $r = 8$ (l) $a = 8$
3 (a) $t = 24$ (b) $y = 18$ (c) $f = 36$
   (d) $r = 33$ (e) $q = 36$ (f) $g = 72$
   (g) $s = 112$ (h) $b = 150$ (i) $n = 8$
   (j) $m = 20$ (k) $v = 48$ (l) $u = 900$
4 (a) $t = 6$ (b) $p = 3$ (c) $r = 392$
   (d) $m = 60$ (e) $x = 80$ (f) $g = 650$
   (g) $s = 60$ (h) $d = 11$ (i) $j = 6$

Exercise 5B

1 (a) $w = 4$ (b) $u = 5$ (c) $t = 8$
   (d) $m = 3$ (e) $d = 2$ (f) $g = 9$
   (g) $s = 3$ (h) $b = 0$ (i) $a = 5$
2 (a) $r = 9$ (b) $h = 28$ (c) $f = 30$
   (d) $x = 2$ (e) $d = 110$ (f) $a = 9$
3 (a) $x = 8$ (b) $t = 5$ (c) $w = 10$
   (d) $a = 2$ (e) $n = 4$ (f) $q = 4$
   (g) $y = 10$ (h) $e = -6$ (i) $p = -3$

Exercise 4D

1 $d = 62°, e = 138°$ **2** $f = 53°, g = 127°$
3 $h = 60°$ **4** $i = 36°, j = 108°, k = 72°$
5 $l = 75°, m = 30°$ **6** $n = 134°, p = 118°, q = 108°$

Exercise 4E

1 $a = 96°, b = 40°$ **2** $c = d = 78°$
3 $e = 47°, f = 80°$ **4** $g = 38°, h = 66°$
5 $i = 270°$ **6** $j = 78°, k = 51°$
7 $l = m = 32°, n = 62°$ **8** $p = 27°$

Exercise 4F

1 (a) kites, rhombuses, squares
   (b) rhombuses, squares
   (c) trapeziums

Exercise 5C

1 (a) $b = 4\frac{1}{2}$ (b) $c = 1\frac{2}{7}$ (c) $y = 3\frac{1}{3}$
   (d) $j = 2\frac{5}{7}$ (e) $g = 1\frac{2}{3}$ (f) $p = 1\frac{1}{2}$

2 (a) $k = 1.25$ (b) $e = 1.5$ (c) $i = 2.2$
    (d) $s = 0.2$ (e) $m = 1.75$ (f) $a = 7.5$
3 (a) $k = -2$ (b) $f = -1$ (c) $m = -3$
    (d) $g = -5$ (e) $w = -2$ (f) $s = -2$
    (g) $u = -6$ (h) $f = -2$ (i) $v = -5$
    (j) $x = -6$ (k) $a = -1$ (l) $n = -5$
4 (a) $m = -4$ (b) $q = -3$ (c) $x = -10$
    (d) $b = -24$ (e) $e = -12$ (f) $t = -36$

Exercise 5D

1 (a) $x = 2$ (b) $u = 4$ (c) $r = 3$
    (d) $p = 5$ (e) $w = 7$ (f) $b = 4$
2 (a) $a = -2$ (b) $m = -3$ (c) $d = -2$
    (d) $y = -4$ (e) $t = -1.5$ (f) $r = -2$
3 (a) $x = 2.5$ (b) $z = 7.2$ (c) $m = \frac{1}{3}$
    (d) $h = -3.5$ (e) $k = 4.5$ (f) $c = -3.5$

Exercise 5E

1 45 m
2 91, 92, 93
3 Stephen 15 yrs, Susan 22 yrs, Sarah 24 yrs
4 $6x + 4 = 22$, $x = 3$, side lengths are 10 m, 5 m and 7 m
5 6
6 $x = 22°$ and angles are 44°, 110° and 26°
7 100°, 80°
8 £48

Exercise 5F

1 (a) $a = 4, b = 1$ (b) $x = 3, y = -2$ (c) $b = -2, c = -1$
2 (a) $s = 5, t = 3$ (b) $u = 2, v = -2$ (c) $p = -1, q = -3$
3 (a) $x = 1, y = 3$
    (b) $a = 1.5, b = 0.5$
    (c) $p = -1, q = 2$
4 (a) $x = 2, y = 1$ (b) $a = 3, b = -4$
    (c) $x = 5.6, y = 3.2$ (d) $p = -1, q = -2$
    (e) $e = 3, f = 2$ (f) $x = -2, y = -3$
5 (a) $x = 3, y = 1$ (b) $x = 2, y = 4$
    (c) $a = 1, b = -2$ (d) $g = 6, h = -3$
    (e) $w = 4, z = 7$ (f) $x = -2, y = 3$
6 75, 41
7 apple 15 p, banana 18 p
8 (a) $x = 4, y = 6$ (b) $a = -2, b = 3$
    (c) $g = 4, h = 1$ (d) $w = 1, z = 2$
    (e) $x = 4, y = -1$ (f) $x = 3, y = -1$

Exercise 5G

1 (a) $5.5 < 6.4$
    (b) $12.7 > 12.6$
    (c) $0 < 0.01$
    (d) $1112 < 1121$
2 (a) F   (b) T   (c) F
    (d) F   (e) T   (f) T
3 (a) 5, 6, 7, 8   (b) 8, 9, 10, 11, 12
    (c) 1, 2, 3, 4, 5   (d) 95, 96, 97, 98, 99, 100
    (e) -1, 0, 1, 2, 3   (f) $-9, -8, -7, -6, -5$
4 (a) 3.3, 3.4, 3.5, 3.6, 3.7, 3.8, 3.9, 4.0
    (b) 20.8, 20.9, 21.0, 21.1, 21.2
    (c) 0.3, 0.4, 0.5, 0.6, 0.7, 0.8, 0.9
    (d) 78.9, 79.0, 79.1, 79.2, 79.3, 79.4
5 1(a) $4 < b < 9$ 1(b) $7 < r \leq 12$
    1(c) $0 < p \leq 5$ 1(d) $94 < k \leq 100$
    1(e) $-2 < x \leq 3$ 1(f) $-9 \leq q < -4$
    2(a) $3.2 < h < 4.1$ 2(b) $20.7 < t \leq 21.2$
    2(c) $0.3 \leq y \leq 0.9$ 2(d) $78.9 \leq m < 79.5$

6

Exercise 5H

1 (a) $m \leq 3$ (b) $v > \frac{3}{2}$ (c) $x \leq 2.4$
    (d) $w \leq -4$ (e) $a > -9$ (f) $t < -3$
2 (a) $x > 2$ (b) $s < 2$ (c) $u \leq 5$
    (d) $a \geq 10$ (e) $q > 4$ (f) $r \leq 2$
    (g) $n \geq -2$ (h) $p \leq -3$ (i) $y > -4$
3 (a) $q > 4$ (b) $c \leq 10$ (c) $v \leq 20$
    (d) $n \leq 2$ (e) $x \geq 7$ (f) $g \leq 1$
    (g) $m \leq -5$ (h) $w > -4$ (i) $k \leq 3$
    (j) $t \leq 8$ (k) $f > 1$ (l) $y \leq 3$
4 (a) $3 < z < 10$ (b) $3 \leq c < 4$
    (c) $2 < x \leq 6$ (d) $20 < b \leq 15$
    (e) $0 < h < 9$ (f) $\frac{3}{4} < t \leq \frac{5}{2}$
    (g) $-2 < p \leq 1$ (h) $-2.5 \leq q \leq -0.5$
    (i) $-2 \leq t \leq 4$ (j) $1 < y < 12$
    (k) $-4.5 < r < 2.5$ (l) $-2.4 \leq d \leq 2.4$

Exercise 5I

1 $x < 3$ **2** $y < -27$ **3** $z \geq 4$
4 $h \leq -5$ **5** $k \geq 4.5$ **6** $t < 2.25$
7 $d > 7.5$ **8** $e \leq -2.5$ **9** $f \leq -2.4$

Examination style questions

1 (a) $x = 36$ (b) $y = 7$ (c) $c = 2$
2 (a) $r = 1$ (b) $s = \frac{1}{2}$ (c) $x = 32$ (d) $y = -3$
3 (a) $x = 4$ (b) $x < 7$
4 $x = -3, y = 5$
5 (a) $x \geq -1$ (b) $x < 2$ (c) $-1, 0, 1$

Chapter 6

Exercise 6A

1 (a) 3900 mm   (b) 540 cm   (c) 123 mm
    (d) 4.05 km   (e) 6.125 m   (f) 30.3 cm
2 (a) 5050 kg   (b) 8.2 kg
    (c) 560 g   (d) 16.12 tonnes
3 (a) 7.45 l (b) 0.8 l (c) 304 ml   (d) 53 cl
4 (a) 7.008 m   (b) 9.5 cl   (c) 40 g
    (d) 5 m   (e) 5.05 tonnes   (f) 0.2 l
5 0.06772 tonnes
6 2.83 m, 2.798 m, 2 m

Exercise 6B

1 (a) 5 cm   (b) 3 inches
  (c) 60 cm   (d) 14 inches
  (e) 24 km   (f) 20 miles
  (g) 4 km   (h) 1.25 miles
2 (a) 2 *l*   (b) 8.75 pints
  (c) 31.5 *l*   (d) 3 gallons
  (e) 0.5 gallons   (f) 18.9 *l*
3 (a) 6.6 pounds   (b) 14.3 pounds
  (c) 5 kg   (d) 24 kg
  (e) 212.5 g   (f) 13.6 ounces
4 (a) 13.125 cm   (b) 1.52 inches
  (c) 6.4 kg   (d) 1.87 gallons
  (e) 1.32 pounds   (f) 23.68 km
  (g) 1.43 *l*   (h) 14.6 miles
5 The $1\frac{1}{2}$ *l* carton holds most.
6 No

Exercise 6C

1 (a) 30 000 cm^2 (b) 5 200 000 cm^3
  (c) 7.05 m^2 (d) 500 000 cm^3
  (e) 4 000 cm^2 (f) 0.3 m^3
2 (a) 100
  (b) 10 000 000 000 or 10^{10}
  (c) 9
3 (a) 1 000 000 000 or 10^9
  (b) 1 000 000 000 or 10^9
  (c) 1728
4 3.25 km^2
5 (a) 159 600 cm^2
  (b) 15.96 m^2
  (c) 16 m^2
  (d) £168
6 (a) 248 625 cm^3 (b) no

Exercise 6D

1 (a) 5.8   (b) 7.4   (c) 2.2   (d) 5.7
  (e) 4.3   (f) 15.8   (g) 11.3   (h) 17.2
  (i) 145.1   (j) 522.0
2 (a) 3.26   (b) 6.54   (c) 0.88   (d) 0.03
  (e) 11.06   (f) 4.01   (g) 3.90   (h) 2.31
  (i) 0.00   (j) 5.13
3 (a) 1.255   (b) 5.293   (c) 4.127
  (d) 0.001   (e) 0.000
4 (a) 15.2   (b) 15.15   (c) 15.153   (d) 15
5 (a) 16
  (b) 15.6
  (c) 15.63
  (d) 16, 15.6, 15.63, the answers are the same – rounding
     does not always differentiate between results.

Exercise 6E

1 (a) 3000   (b) 20 000   (c) 300 000
  (d) 80 000   (e) 5
2 (a) 6800   (b) 33 000   (c) 150 000
  (d) 270 000   (e) 5.3
3 (a) 0.04   (b) 0.04   (c) 0.006
  (d) 0.004   (e) 0.5
4 (a) 0.57   (b) 0.000 38   (c) 0.0019
  (d) 0.20   (e) 0.0020
5 (a) 300 000
  (b) 290 000
6 (a) 44 m
  (b) 43.6 m

Exercise 6F

1 (a) 2 hr 6 min   (b) 3 hr 48 min
  (c) 5 hr 39 min   (d) 4 hr 42 min
  (e) 3.25 hr   (f) 6.42 hr
  (g) 4.13 hr   (h) 0.2 hr
2 58.8 mph
3 275 km
4 10.74 hr or 10 hr 44 min
5 (a) 9.6 mph   (b) 7 miles
  (c) 4.2 mph (to 1 dp)
6 25.2 kph
7 (a) 4 h 29 m   (b) 1289 miles
  (c) 287.7 mph (to 1 dp)
8 The dog is faster.
9 (a) 780 s   (b) 6.41 m/s (to 2 dp)
  (c) 23.07 km/h (to 2 dp)
10 (a) 3.6 km/h or 1 m/s   (b) 90.2 km
  (c) 29.25 km/h

Exercise 6G

1 (a) 15 cm^3
  (b) 82 g
  (c) 2.11 g/cm^3 to 3 sf
  (d) 4800 g or 4.8 kg
  (e) 4600 cm^3 to 3 sf
2 938 g to 3 sf
3 0.0114 kg/cm^3 to 3 sf
4 (a) 113 cm^3 to 3 sf   (b) 5.10 cm to 3 sf
5 (a) 0.667 m^3 to 3 sf   (b) 2230 kg to 3 sf
6 31 250 cm^3
7 34 200 kg
8 321 g to 3 sf
9 £626.32
10 7.94 cm

Exercise 6H

1 (a) 64   (b) 3   (c) 99
  (d) 10 000   (e) 5   (f) 2.5
2 (a) 63.7868   (b) 3.15 to 3 sf   (c) 92.4703
  (d) 10 600 to 3 sf   (e) 5.86 to 3 sf   (f) 1.67 to 3 sf
3 £60 000   4 8p   5 £1600
6 (a) 33 *l*   (b) £27
7 (a) 2000   (b) 6000 kg
8 1.1354
9 0.021 524
10 16 598

Exercise 6I

1 (a) 3.5 cm to 4.5 cm
  (b) 44.5 kg to 45.5 kg
  (c) 172.5 g to 173.5 g
  (d) 339.5 mm to 340.1 mm
  (e) 3.55 m to 3.65 m
  (f) 12.25 *l* to 12.35 *l*
  (g) 122.45 g to 122.55 g
  (h) 0.55 cm to 0.65 cm
  (i) 7.235 km to 7.245 km
  (j) 9.665 min to 9.675 min
  (k) 4.915 g to 4.925 g
  (l) 7.015 ml to 7.025 ml
2 (a) 6.5 m to 7.5 m
  (b) 6.95 m to 7.05 m
  (c) 6.995 to 7.005
  (d) 6.9995 to 7.0005
  (e) Each is to a different degree of accuracy.

Exercise 6J

1. (a) 14.05 cm   (b) 20.9475 cm
  (c) 2.3 cm   (d) 1.3 cm
2. (a) 3.55 cm to 3.65 cm, 7.15 cm to 7.25 cm
  (b) 25.3825 cm^2
3. 4 minutes
4. 21.8 m^2
5. (a) 15.15 to 15.25 seconds
  (b) 99.5 m to 100.5 m
  (c) 6.63 to 3 sf
6. 4.15 cm to 4.17 cm (to 3 sf)
7. (a) 68.934375 cm^3, 63.738125 cm^3
  (b) 7.835 kg to 7.845 kg
  (c) 0.123 kg/cm^3 to 3 sf
8. 55.7 min to 3 sf

Exercise 6K

1. 1.8 m, 1.5%
2. 1.67% to 3 sf
3. (a) 24.75 cm^2 (b) 23.75%
4. 3.31% to 3 sf
5. (a) 2.45 m, 2.35 m   (b) 4.9 m, 4.7 m
  (c) 0.1 m   (d) 2.1% to 3 sf
6. 1.02% to 3 sf
7. (a) 17.4% to 3 sf   (b) 27.1% to 3 sf
8. (a) 1.25 g/cm^3 to 1.28 g/cm^3 (to 3 sf)
  (b) 3.85% (to 3 sf)

Examination style questions

1. Gemma is right. $30/(4 \times 0.5) = 15$
2. 8.18 s to 3 sf
3. $300 \times 4/0.2 = 6000$

Chapter 7

Exercise 7A

1. 11.4 cm^2
2. 69 cm^2
3. 114 cm^2
4. (a) $\frac{5}{2}x^2 + x$ (b) $10x + 2$ (c) 1.4 cm
5. (a) 12 cm   (b) 5.2 cm

Exercise 7B

1. (a) 46.34 cm^2 (b) 36.4 cm
2. 11.1 cm
3. (a) 3.0 m   (b) 27.4 m

Exercise 7C

1. (a) 8.38 cm, 33.51 cm^2 (b) 24.04 cm, 126.71 cm^2
  (c) 17.45 cm, 43.63 cm^2
2. 30°
3. 15π cm^2

Exercise 7D

1

plan   front   side

2
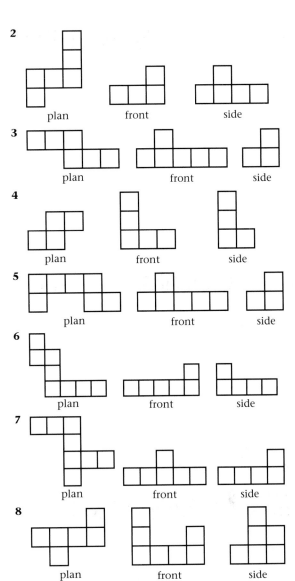

2
plan   front   side

3
plan   front   side

4
plan   front   side

5
plan   front   side

6
plan   front   side

7
plan   front   side

8
plan   front   side

Exercise 7E

1
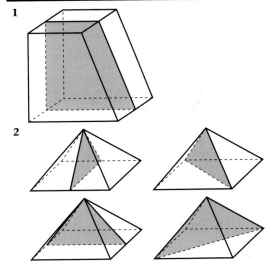

2

3

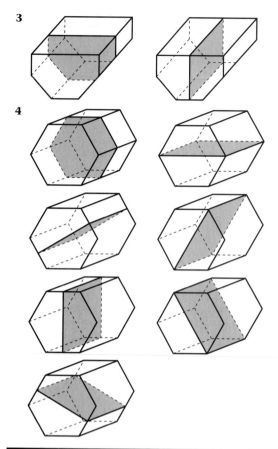

4

Exercise 7F

1 (a) 324   (b) 5700   (c) 4115 to 4 sf
2 (a) 354 cm²   (b) 1879 cm²   (c) 1691 cm²
3 6

Exercise 7G

1 (a) 101 cm³ to 3 sf   (b) 1030 cm³ to 3 sf
    (c) 5.09 cm³ to 3 sf
2 (a) 151 cm²   (b) 599 cm²
    (c) 19.2 cm² all to 3 sf
3 (a) 13 345 cm³   (b) 112 kg to 3 sf

Exercise 7H

1 (a) 56 cm³   (b) 12 cm³   (c) 254 cm³ to 3 sf
2 245 cm²   **3** 6 cm

Exercise 7I

1 (a) 7240 cm³   (b) 48 400 cm³ both to 3 sf
2 (a) 1810 cm²   (b) 6420 cm² both to 3 sf
3 87.4 cm to 3 sf

Exercise 7J

1 (a) area   (b) volume   (c) area
    (d) none   (e) none   (f) length
2 316.54 cm³   **3** p is a length, t is a number

Examination style questions

1 (a) 24π cm²   (b) 40π cm²   (c) 240π cm³

Chapter 8

Exercises 8A–8F

Students' own constructions

Exercise 8G

1

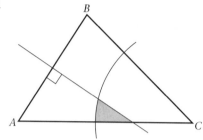

2

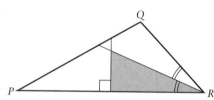

Exercise 8H

1

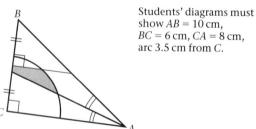

$BX = BY = 3$ cm

2

Students' diagrams must show $AB = 10$ cm, $BC = 6$ cm, $CA = 8$ cm, arc 3.5 cm from C.

Examination style questions

1

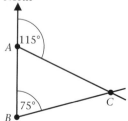

2

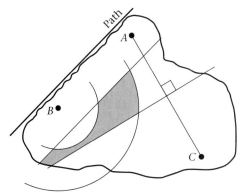

Chapter 9

Exercise 9A

1 (a) quantitative   (b) quantitative
   (c) qualitative   (d) qualitative
   (e) quantitative   (f) quantitative
2 (a) discrete   (b) discrete
   (c) continuous   (d) continuous
   (e) continuous   (f) discrete
3 (a) primary   (b) secondary
   (c) secondary   (d) secondary
   (e) primary

Exercise 9B

1

Mark	Tally	Frequency
1–10		0
11–20	||||| ||||| |||||		16
21–30	||||| ||||| |||	13
	Total	29

2

Mass m (kg)	Tally	Frequency
$40 \leqslant m < 50$	||	2
$50 \leqslant m < 60$	||||	4
$60 \leqslant m < 70$	||||| ||	7
$70 \leqslant m < 80$	|||||	5
$80 \leqslant m < 90$	||	2
	Total	20

3

Time (min)	Tally	Frequency
10–14		0
15–19	||||| ||||| |||||	15
20–24	||||| ||||| ||||	14
25–29	||||| ||||| ||||| ||	17
30–34	|||	3
35–40			1
	Total	50

4

Amount, m (£)	Tally	Frequency
$0 \leqslant m < 20$	|||||	5
$20 \leqslant m < 40$	|||||	5
$40 \leqslant m < 60$	|||||	5
$60 \leqslant m < 80$	|||||	5
$80 \leqslant m < 100$	||||| |||	8
	Total	28

5

Number on dice	Tally	Frequency
1		
2		
3		
4		
5		
6		
	Total	

6

Colour of car	Tally	Frequency
Black		
Blue		
Red		
Grey		
Green		
Other		
	Total	

Exercise 9C

1 (a) random   (b) biased
   (c) random   (d) random
   (e) random   (f) biased
2 12 (Year 7), 13 (Year 8), 12 (Year 9), 13 (Year 10) and
   13 (Year 11)
4 (b) (i) 10% (to the nearest whole percentage)
      (ii) 9, 11, 13, 9, 6

Exercise 9D

1

	Left-handed	Right-handed	Total
Girls	6	44	50
Boys	9	41	50
Total	15	85	100

Left-handed 15%

2

	Poynton (Central)	Poynton (West & East)	Total
Males	3522	2743	6265
Females	3270	3898	7168
Total	6792	6641	13 433

Females 53%

3 (a)

		Level 3	4	5	6	7	8	Total
English	Boys	11	28	34	31	15	1	120
	Girls	4	20	36	43	22	5	130
Total		15	48	70	74	37	6	250

(b) 250　　(c) 67.5%　　(d) 16.9%

4 (a) 649 km　　(b) 536 km　　(c) 1529 km
5 (a) £1231　　　　　(b) £1326
6 (a) 121　　　　　　(b) 122
　　(c) 305　　　　　　(d) 3%

Examination style questions

1

Age	Frequency
1–10	4
11–20	8
21–30	5
31–40	2
41–50	1

8
2 (a) 7　　　　　　　　(b) 9
5 (b) 320
6 (b) 4 (0–12), 7 (13–24), 16 (25–40), 13 (41–60), 10 (61+)

Chapter 10

Exercise 10A

1 (a)
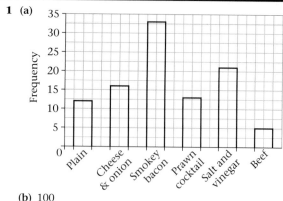

(b) 100

2 (a)
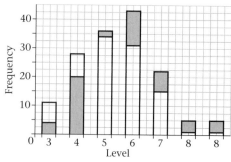

(b) Girls achieved better at the higher levels.
(c) Yes, compound bar charts allow direct comparisons.

3
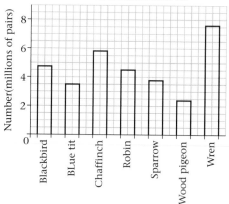
Common birds in the UK

4
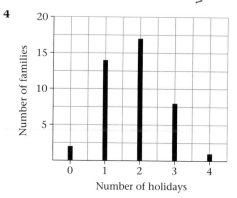

5
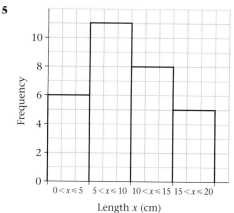

6

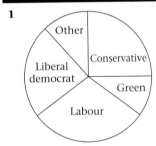

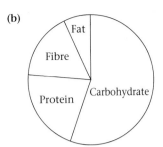

(b)

(c) **(i)** brand B
   (ii) brand B

6

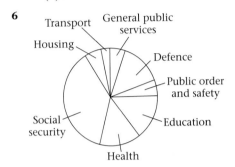

Exercise 10B

1

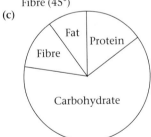

2 Art (96), English (45), French (42), Geography (63), History (57), Maths (108), PE (78) and Science (51)
3 Stayed in 6th form (41), Went to college (20), Got a full-time job (31), On a training scheme (18), Out of work (14) and Other (11)
4 **(a)** 1 g = 3.6°
   (b) Protein (54°), Carbohydrate (225°), Fat (36°) and Fibre (45°)
   (c)

5 **(a)** 1 g of ingredient = $\dfrac{360°}{\text{Total weight of cereal}}$

$= \frac{360}{720}$ g = 0.5°;

Ingredient	Amount in g	Seated angle calculation	Angle
Carbohydrate	400	400 × 0.5 =	200
Protein	150	150 × 0.5 =	75
Fibre	120	120 × 0.5 =	60
Fat	50	50 × 0.5 =	25
Total	720	Total angle	360

Exercise 10C

1

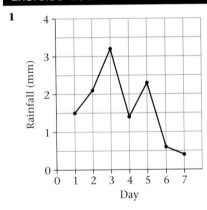

2

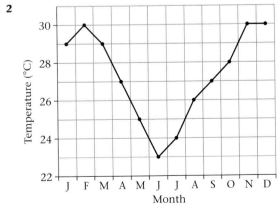

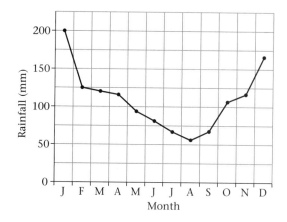

Exercise 10D

1
```
0 | 9
1 | 1 2 4 5 6 7 7 8 8 9 9 9
2 | 0 1 1 2 3 3 4
```

2 (a)
```
0 | 9
1 | 2 4 7 8 9
2 | 5 6 9 9
3 | 1 2 3 3 7 7
4 | 0 3 3
```
(b) 2

3 (a) 18
(b)
```
0 | 8
1 | 2 3 4 4 7 8 9
2 | 0 1 1 5 9
3 | 2 4 5 8
4 | 1
```

4 (a)
```
1 | 2 5 8
2 | 0 0 3 4 5 6 7 8 9
3 | 0 0 0 0 1 2 3 4 5 5 5 6 6 7 8 8 9
4 | 0 0 0 2 2 3 4 4 5 6
5 | 0 1
```
(b) 9

Exercise 10E

1 (a) 3   **(b)** 1 and 4   **(c)** 2
2 (a) negative   **(b)** zero   **(c)** positive
    (d) positive   **(e)** positive
3 (a)(b)

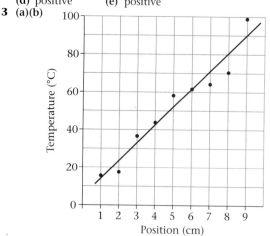

(c) Positive correlation. The closer to the heat the hotter the bar.

4 (a)

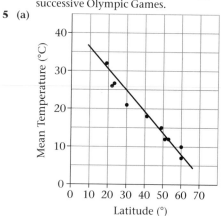

(c) positive correlation
(d) More female competitors are taking part in successive Olympic Games.

5 (a)

(b) negative correlation
(c) temperature becomes colder

6 (a)(b)

(c) positive correlation
(d) between 27 and 28 marks

Exercise 10F

1 (a) 165 cm   **(b)** 17 kg   **(c)** 5 s
    (d) 17.5 m^2 **(e)** 150 ml

2 (a)

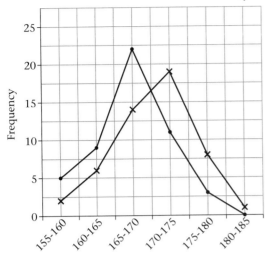

(b) Boys tend to be slightly taller.

3 (a)

Catch (kg)	Frequency	Mid-point values
0.1–0.5	6	0.3
0.6–1.0	6	0.8
1.1–1.5	9	1.3
1.6–2.0	4	1.8
2.1–2.5	3	2.3
2.6–3.0	4	2.8

(b)

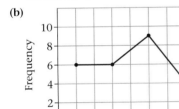

4 (a)

Temperature (°C)	Frequency	Mid-point values
0–3	6	1.5
4–7	14	5.5
8–11	11	9.5
12–15	15	13.5
16–19	8	17.5
20–23	3	21.5
24–27	3	25.5

(b)

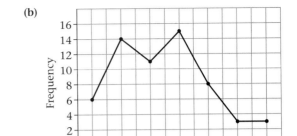

5

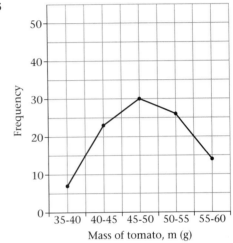

Exercise 10G

1

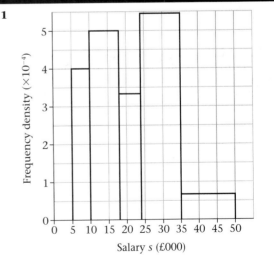

2

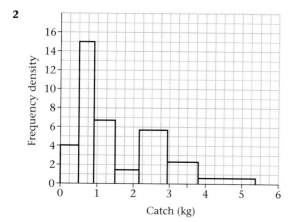

Catch (kg)

3

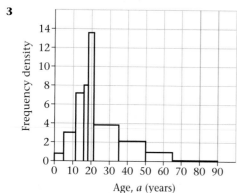

Age, *a* (years)

4

Sponsor money *s* (£)	Frequency
$0 \leq s < 1$	5
$1 \leq s < 3$	14
$3 \leq s < 5$	8
$5 \leq s < 10$	10
$10 \leq s < 15$	5

Examination style questions

1
```
0 | 4 8 9
1 | 0 1 2 2 4 5 6 8 9
2 | 0 2 3 3 6 7
```

2 (a)(b)

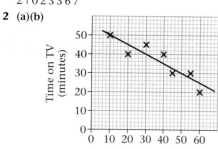

Time on computers (minutes)

(c) Negative correlation

3

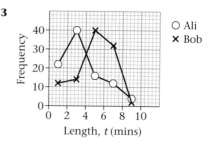

Length, *t* (mins)

4

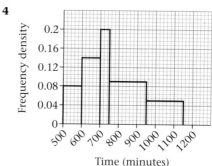

Time (minutes)

Chapter 11

Exercise 11A

1 (a) $10k$ (b) $18b$ (c) $20a$
   (d) $6ab$ (e) $12cd$ (f) $5xy$
2 (a) $100 + 14 = 114$ (b) $200 + 30 = 230$
   (c) $420 + 18 = 438$ (d) $120 - 6 = 114$
   (e) $350 - 28 = 322$ (f) $320 - 24 = 296$
3 (a) $5p + 30$ (b) $3x + 3y$
   (c) $4u + 4v + 4w$ (d) $2y - 16$
   (e) $63 - 7z$ (f) $8a - 8b + 48$
4 (a) $6c + 18$ (b) $20t + 15$ (c) $10p + 2q$
   (d) $6a - 3b$ (e) $18c - 12d$ (f) $14x + 7y - 21$
   (g) $18a - 24b + 6c$ (h) $2x^2 + 6x + 4$ (i) $4y^2 - 12y - 40$
5 A & F   B & K   C & I   D & L   E & G   H & J

Exercise 11B

1 $b^2 + 4b$ **2** $5a + a^2$ **3** $k^2 - 6k$
4 $9m - m^2$ **5** $2a^2 + 3a$ **6** $4g^2 + g$
7 $2p^2 + pq$ **8** $t^2 + 5tw$ **9** $m^2 + 3mn$
10 $2x^2 - xy$ **11** $4r^2 - rt$ **12** $a^2 - 4ab$
13 $2t^2 + 10t$ **14** $3x^2 - 24$ **15** $5k^2 + 5kl$
16 $6a^2 + 12a$ **17** $8g^2 + 2gh$ **18** $15p^2 - 10pq$
19 $6xy + 15xz$ **20** $r^3 + r$ **21** $a^3 + 3a$
22 $t^3 - 7t$ **23** $2p^3 + 6pq$ **24** $4x^3 + 4x^2$

Exercise 11C

1 $-4k - 8$ **2** $-6x - 18$ **3** $-15n + 5 = 5 - 15n$
4 $-12t - 20$ **5** $3 - 12p$ **6** $14 - 6x$
7 $5y + 22$ **8** $5k + 21$ **9** $2a + 18$
10 $7t - 16$ **11** $8y + 19$ **12** $7x + 40$
13 $8x + 7$ **14** $13n - 5$ **15** $5x - 21$
16 $14x - 8$ **17** $2b - 5$ **18** $4m + 2$
19 $4t + 4$ **20** $2k - 14$ **21** $2a + 23$
22 $4p + 14$ **23** $2g - 8$ **24** $-4w - 5$
25 $x^2 + 7x + 8$ **26** $2x^2 - 2x + 12$

Exercise 11D

1 (a) $g = 2$ (b) $k = 2$ (c) $s = 3$ (d) $n = 7$
(e) $f = 10$ (f) $v = 10$ (g) $m = 6\frac{1}{2}$ (h) $w = 2\frac{1}{2}$
2 (a) $t = 2$ (b) $r = 3$ (c) $b = 4\frac{1}{2}$ (d) $w = 6\frac{1}{2}$
(e) $x = 2\frac{1}{2}$ (f) $y = -1\frac{1}{2}$ (g) $k = 3\frac{1}{2}$ (h) $a = 9\frac{1}{2}$

Exercise 11E

1 (a) $a = 3$ (b) $d = 5$ (c) $x = 2$ (d) $p = 2$
(e) $t = 6$ (f) $b = 10$ (g) $g = 2$ (h) $k = 1$
(i) $y = -2$ (j) $r = -3$
2 (a) $b = 3$ (b) $a = 2$ (c) $x = 4$ (d) $p = 5$
(e) $s = 3$ (f) $t = 1$ (g) $w = -3$ (h) $y = -4$
3 (a) $d = 2$ (b) $y = 20$ (c) $x = 3$
(d) $a = -2$ (e) $c = -4$ (f) $b = -6\frac{1}{2}$
4 (a) $x = -7$ (b) $x = 2$ (c) $x = 3$
(d) $x = -7$ (e) $x = -2$ (f) $x = 4$
5 (a) $x = 3$ (b) $x = 6$ (c) $x = 1$
(d) $x = 4$ (e) $x = -13$ (f) $x = 12\frac{1}{2}$

Exercise 11F

1 (a) $x \leq 11$ (b) $m > -1.5$ (c) $w > 2$ (d) $y \leq -3$
(e) $p < -10$ (f) $k > -1$ (g) $x \geq 5.5$ (h) $n \geq 4$
2 (a) $-2 \leq x \leq 4$ $-2, -1, 0, 1, 2, 3, 4$
(b) $-2 < y < 5$ $-1, 0, 1, 2, 3, 4$
(c) $-2 \leq n < 4\frac{1}{4}$ $-2, -1, 0, 1, 2, 3, 4$
(d) $-2 < m \leq 5$ $-1, 0, 1, 2, 3, 4, 5$
(e) $-3\frac{1}{2} < t < 2\frac{1}{2}$ $-3, -2, -1, 0, 1, 2$
(f) $3 < x < 6$ $4, 5$
(g) $-2\frac{1}{5} \leq y \leq 1\frac{1}{5}$ $-2, -1, 0, 1$
(h) $-2\frac{3}{4} < x \leq 4$ $-2, -1, 0, 1, 2, 3, 4$

Exercise 11G

1 (a) $\square = x$ (b) $\square = a$ (c) $\square = 6$ (d) $\square = 4$
(e) $\square = 4$ (f) $\square = 3$ (g) $\square = 2$ (h) $\square = 4$
2 (a) $5(p + 4)$ (b) $2(a + 6)$ (c) $3(y + 5)$
(d) $7(b + 3)$ (e) $4(q + 3p)$ (f) $6(k + 4l)$
3 (a) $4(t - 3)$ (b) $3(x - 3)$ (c) $5(n - 4)$
(d) $2(b - 4)$ (e) $6(a - 3b)$ (f) $7(k - 1)$
4 (a) $y(y + 7)$ (b) $x(x + 5)$ (c) $n(n + 1)$
(d) $x(x - 7)$ (e) $p(p - 8)$ (f) $a(a - b)$
5 (a) $2(3p + 2)$ (b) $2(2a + 5)$ (c) $2(3 - 2t)$
(d) $4(3m - 2n)$ (e) $5(5x + 3y)$ (f) $3(4y - 3z)$
6 (a) $3x(x - 2)$ (b) $x(8x - y)$ (c) $4a(2 + b)$
(d) $p^2(p - 5)$ (e) $3t^2(t + 2)$ (f) $5y(2z - 3y)$
(g) $6a(3a + 2b)$ (h) $4p(4p - 3q)$
7 (a) $2ab^2(2 + 3b)$ (b) $5x(2y - x)$
(c) $3p^2q(1 - 2pq)$ (d) $2n(4mn^2 + 2n - 3m^2)$
(e) $6hk(h - 2k^2 - 3hk)$

Exercise 11H

1 $a^2 + 9a + 14$ 2 $x^2 + 4x + 3$
3 $x^2 + 10x + 25$ 4 $t^2 + 3t - 10$
5 $x^2 + 3x - 28$ 6 $n^2 + 3n - 40$
7 $x^2 + x - 20$ 8 $p^2 - 16$
9 $x^2 - 13x + 36$ 10 $h^2 - 11h + 24$
11 $y^2 - 6y + 9$ 12 $a^2 + 11a + 28$
13 $m^2 + m - 56$ 14 $q^2 + 13q + 42$
15 $20 - d - d^2$ 16 $x^2 - 11x + 24$
17 $x^2 - 19x + 84$ 18 $y^2 - 10y - 96$

Exercise 11I

1 (a) $x^2 + 10x + 25$ (b) $x^2 + 12x + 36$ (c) $x^2 - 6x + 9$
(d) $x^2 + 2x + 1$ (e) $x^2 - 8x + 16$ (f) $x^2 - 10x + 25$
(g) $x^2 + 14x + 49$ (h) $x^2 - 16x + 64$ (i) $x^2 + 6x + 9$
(j) $x^2 + 4x + 4$ (k) $x^2 - 10x + 25$ (l) $x^2 + 2ax + a^2$
2 (a) $(x + \boxed{6})^2 = x^2 + \boxed{12}x + 36$
(b) $(x - \boxed{7})^2 = x^2 - \boxed{14}x + 49$
(c) $(x + \boxed{9})^2 = x^2 + 18x + \boxed{81}$
(d) $(x - \boxed{10})^2 = x^2 - 20x + \boxed{100}$
3 (a) $x^2 - 16$ (b) $x^2 - 25$ (c) $x^2 - 4$
(d) $x^2 - 121$ (e) $x^2 - 9$ (f) $x^2 - 1$
(g) $x^2 - 81$ (h) $x^2 - a^2$ (i) $t^2 - x^2$

Exercise 11J

1 $3a^2 + 14a + 8$ 2 $5x^2 + 13x + 6$
3 $6t^2 + 19t + 15$ 4 $8y^2 + 30y + 7$
5 $12x^2 + 28x + 15$ 6 $4x^2 - x - 3$
7 $6z^2 + 11z - 10$ 8 $7y^2 - y - 8$
9 $3n^2 + 19n - 40$ 10 $6b^2 - 7b - 5$
11 $7p^2 - 25p - 12$ 12 $6z^2 - 17z + 12$
13 $10x^2 - 23x + 9$ 14 $4y^2 - 12y + 9$
15 $12a^2 + 23a + 10$ 16 $9x^2 + 24x + 16$
17 $4x^2 - 28x + 49$ 18 $25 - 40x + 16x^2$
19 $4x^2 - 1$ 20 $9y^2 - 4$
21 $25n^2 - 16$ 22 $9x^2 - 25$
23 $1 - 4x^2$ 24 $9t^2 - 4x^2$

Examination style questions

1 (a) $7(x + 2)$ (b) $10m - 3$
2 $22a + 3c$ 3 $w = 5$
4 (a) $4(c + 3)$ (b) $x(x + 5)$
5 $a(2a - 1)$ 6 $4y(3y - 2)$
7 $3x(x - 2y)$ 8 $x^2 - 2x - 3$
9 $x^2 + 8x + 16$ 10 $4x^2 + 17x - 15$
11 (a) $7x + 17$ (b) $y^2 - 6x + 8$ (c) $4t^2 - 25$

Chapter 12

Exercise 12A

1 $C = 500 - 48y$ 2 $t = 7r + 5s$
3 $t = 45w + 20$ 4 $t = 65w + 30$
5 $p = 8x + 6$ or $p = 2(4x + 3)$

Exercise 12B

1 (a) 4 (b) 3 (c) 5 (d) 76
(e) 30 (f) 37 (g) 84 (h) 27
(i) -5 (j) 4 (k) 6 (l) 3

2
x	1	2	3	4	5
$x^2 + 2x$	3	8	15	24	35

3
x	3	4	3.5	3.7	3.8
$x^3 - x$	24	60	39.375	46.953	51.072

4 (a) 90 (b) -18 (c) 31
(d) 5 (e) 7 (f) 8

Exercise 12C

1 (a) $a = 6$ (b) $a = 4$ (c) $a = 6$ (d) $a = 12.5$
2 (a) $b = 22$ (b) $b = 19$ (c) $b = 24$ (d) $b = 20$
3 (a) $A = 32$ (b) $A = 88$ (c) $A = 30$ (d) $A = 75$
4 (a) $s = 26$ (b) $s = 110$ (c) $s = 50$ (d) $s = 24$
5 (a) $v = 7$ (b) $v = 10$ (c) $v = 11$ (d) $v = 13$
6 (a) $94.2\,\text{cm}^2$ (b) $176\,\text{cm}^2$
7 (a) $337\,\text{cm}^2$ (b) $219\,\text{cm}^2$

Exercise 12D

1 (a) $w = 3$ (b) $w = 4$ (c) $w = 6$ (d) $w = 4$
2 (a) $R = 9$ (b) $R = 7$ (c) $I = 8$ (d) $I = 6$
3 (a) $l = 5$ (b) $l = 9$ (c) $w = 13$ (d) $w = 11.5$
4 (a) $u = 6$ (b) $a = 5$ (c) $t = 9$ (d) $u = -8$

Exercise 12E

1 (a) $a = d - 8$ (b) $a = t - 12$
   (c) $a = k + 6$ (d) $a = w + 7$

2 (a) $w = \dfrac{P}{4}$ (b) $w = \dfrac{a}{3}$ (c) $w = \dfrac{A}{l}$ (d) $w = \dfrac{h}{k}$

3 (a) $x = \dfrac{y + 6}{5}$ (b) $x = \dfrac{y + 7}{4}$

   (c) $x = \dfrac{y - 1}{2}$ (d) $x = \dfrac{y - 5}{6}$

4 (a) $59\,°\text{F}$ (b) $C = \dfrac{F - 32}{1.8}$

   (c) $20\,°\text{C}$ (d) $27.77 \ldots \approx 28\tilde{\ }\,°\text{C}$

5 (a) $r = \dfrac{p - 2t}{4}$ (b) $r = \dfrac{v - 4h}{7}$

   (c) $r = \dfrac{w + 2s}{3}$ (d) $r = \dfrac{y + 5p}{6}$

6 (a) $a = \dfrac{v - u}{t}$ (b) $t = \dfrac{v - u}{a}$

7 (a) $a = 2(b - 6)$ (b) $a = 2(b - 7)$ (c) $a = 3(b + 1)$

   (d) $a = 4(b + 3)$ (e) $a = \dfrac{b}{2} - 1$ (f) $a = \dfrac{b}{3} + 5$

8 (a) $x = \dfrac{2y}{3}$ (b) $x = \dfrac{3y + 5}{2}$

   (c) $x = y(z + 5)$ (d) $x = \dfrac{q(3p - s)}{2}$

9 (a) $w = K^2 - t$ (b) $w = A^2 + a$

   (c) $w = \left(\dfrac{h - l}{2}\right)^2$ (d) $w = \dfrac{(T - 5)^2}{r}$

10 (a) $r = \pm\sqrt{g(t + m)}$ (b) $r = \pm\sqrt{4(h - 3a)}$

    (c) $r = \pm\sqrt{\dfrac{3V}{\pi h}}$ (d) $r = \pm\sqrt{\dfrac{A}{\pi} + s^2}$

11 $h = \dfrac{A}{2\pi r} - r$

12 $l = g\left(\dfrac{T}{2\pi}\right)^2$ or $l = \dfrac{gT^2}{4\pi^2}$

Exercise 12F

1 $x = 4.6$ **2** $x = 2.8$ **3** $x = 3.2$
4 $x = 6.1$ **5** $x = 10.3$

Examination style questions

1 $500 - 22x$ **2** $x = 2.4$
3 $x = \pm\sqrt{(w - y)}$ **4** $t = 3(u - 5)$
5 $y = 5n + 2$

Chapter 13

Exercise 13A

1 (a) 4100 (to 3 sf)   (b) 2690
   (c) $-2\,100\,000$ (d) 595
   (e) 3 (f) 2.59
   (g) 3.16 (h) -2.05
   (i) 0.0197 (j) 0.0278
2 (a) 4^5 (b) 3^9 (c) $\sqrt[5]{9.4}$
3 (a) 144 (b) 796 (c) $83\,100$
4 $221\,000\,000$ (to 3 s.f.)
5 These answers are to 3 sf.
   (a) $253\,000$ (b) $0.000\,047\,5$
   (c) 9040 (d) $127\,000\,000$
   (e) 171 (f) -71.90
6 (a) 1024 (b) $15\,625$ (c) 256
   (d) $16\,807$ (e) 1024 (f) 7168
7 She should take the first option.

Exercise 13B

1 (a) 12^9 (b) 45^7 (c) 6^{12} (d) 5^{11} (e) 7^{16}
2 (a) a^5 (b) b^7 (c) c^6 (d) d^{10}
   (e) x^4 (f) y^3 (g) z^5 (h) w^4
3 (a) $6a^7$ (b) $15b^9$ (c) $4c^8$ (d) $9d^8$
5 (a) p^3 (b) q^4 (c) $r^1 = r$ (d) s^3

   (e) $\dfrac{1}{e^3}$ (f) $\dfrac{1}{f^4}$ (g) $\dfrac{1}{g}$ (h) $\dfrac{1}{h^5}$

6 (a) $5x^3$ (b) $2y^4$ (c) $4r$ (d) $\dfrac{1}{5s^3}$

Exercise 13C

1 (a) 6^7 (b) 13^{12} (c) 25^{17}
   (d) 5^{17} (e) 33^{15} (f) 12^{18}
2 (a) m^5 (b) a^7 (c) n^6 (d) u^{10} (e) t^9
3 (a) d^4 (b) a^3 (c) r^5 (d) e^4 (e) t^7
4 (a) $6h^7$ (b) $15e^9$ (c) $4g^8$ (d) $9r^5$ (e) $24e^4$
5 (a) a^3 (b) t^4 (c) $e^1 = e$ (d) s^3 (e) $t^0 = 1$
6 (a) e^{-3} (b) f^{-4} (c) g^{-1} (d) h^{-3} (e) w^{-5}
7 (a) $5x^3$ (b) $3y^4$ (c) $3r$

   (d) $\dfrac{1}{4s^3} = \dfrac{1}{4}s^{-3}$ (e) $2t^3$

8 (a) h^9 (b) e^8 (c) $24c^9$
9 (a) d^2 (b) $3a^3$ (c) $2r^2$

Exercise 13D

1 (a) 12 (b) 1 (c) 1 (d) 9.7

2 (a) $\dfrac{1}{4^2}$ (b) $\dfrac{1}{3^5}$ (c) $\dfrac{1}{5^3}$ (d) $\dfrac{1}{2^{10}}$

3 (a) $\frac{1}{16}$ (b) $\frac{1}{243}$ (c) $\frac{1}{125}$ (d) $\frac{1}{1024}$

4 (a) 3^{-2} (b) 6^{-3} (c) 10^{-9} (d) 8^{-4}
5 (a) 4 (b) 3.43 (c) $\frac{1}{81}$ (d) 2560
6 $2, 4, 8, 16$, in general $\dfrac{1}{2^{-n}} = 2^n$

Exercise 13E

1 (a) $\dfrac{1}{x^3}$ (b) $\dfrac{1}{x^4}$ (c) $\dfrac{1}{x}$ (d) $\dfrac{1}{x^5}$

   (e) $\dfrac{5}{x^3}$ (f) $\dfrac{3}{x^2}$ (g) $\dfrac{6}{x^4}$ (h) $\dfrac{4}{x}$

2 (a) $\frac{1}{8}$ (b) $\frac{1}{16}$ (c) $\frac{1}{2}$ (d) $\frac{1}{32}$
   (e) $\frac{5}{8}$ (f) $\frac{3}{4}$ (g) $\frac{3}{8}$ (h) 2

3 (a) k^{-2} (b) e^{-5} (c) g^{-1}
   (d) x^{-n} (e) $6p^{-1}$ (f) $5a^{-3}$
   (g) $4d^{-2}$ (h) $8h^{-m}$ (i) $3x^{-2}$
   (j) $4a^{-4}$ (k) $3b^{-1}$ (l) $\frac{2}{3}y^{-3}$

4 (a) $e^{-3}; \dfrac{1}{e^3}$ (b) $8s^{-1} = \dfrac{8}{s}$

   (c) $4f^{-2} = \dfrac{4}{f^2}$ (d) $\dfrac{4}{5}t^{-4} = \dfrac{4}{5t^4}$

   (e) $n^{-2} = \dfrac{1}{n^2}$ (f) $3m^{-3} = \dfrac{3}{m^3}$

   (g) $2a^{-2} = \dfrac{2}{a^2}$ (h) $\dfrac{3}{2}q^{-1} = \dfrac{3}{2q}$

Exercise 13F

1 (a) $8a^2b^3$ (b) $21f^5g^3$
   (c) $20x^4y^3$ (d) $36p^6q^5$
   (e) $56r^3s^6$ (f) $74c^4d^5$
2 (a) b^2 (b) $3g$
   (c) $2y^2$ (d) $3p^2q^3$
   (e) $3r^2s$ (f) $75c^4d^5$
3 (a) $2x$ (b) $4jk^2$
   (c) b^2 (d) $3m^3$
   (e) $\dfrac{2}{3e}$ (f) $4q^2$

4 A–H; B–J; C–L; D–I; E–G; F–K
5 (a) $2xy$ (b) $3b^2$

   (c) $2c^{-1}d^{-1} = \dfrac{2}{cd}$ (d) $\dfrac{3m}{2n} = \dfrac{3mn^{-1}}{2}$

6 (a) $2rst$ (b) $4x^3y^2z^{-1} = \dfrac{4x^3y^2}{z}$

   (c) $\dfrac{a^2c^3}{3b} = \dfrac{1}{3}a^2b^{-1}c^3$

7 (a) $\frac{1}{9}$ (b) 16 (c) 270 (d) $\frac{2}{27}$

Exercise 13G

1 (a) p^6 (b) q^8
   (c) r^{12} (d) f^{15}
   (e) d^{12}
2 (a) $8j^6$ (b) $16m^{12}$
   (c) $27w^{15}$ (d) $125x^{12}$
   (e) $49d^{10}$
3 (a) 8^8 (b) 8^{-2}
   (c) not possible (d) 8^{14}
   (e) 8^{15} (f) 8^3
   (g) 8^{-2} (h) 8^{-1}
4 (a) 7^{-5} (b) 7^{-6}
   (c) 7^{-6} (d) not possible
   (e) 7^7 (f) 7^3
   (g) 7^{-10} (h) 7^{-3}

Exercise 13H

1 (a) 5 (b) 12
   (c) 3 (d) 2
   (e) 6 (f) 5
   (g) 10 (h) $\frac{4}{5}$ or 0.8
2 (a) $875^{1/2}$ (b) $184^{1/4}$
   (c) $89^{1/3}$ (d) $864^{1/8}$
3 (a) $\frac{1}{5}$ or 0.2 (b) $\frac{1}{2}$ or 0.5
   (c) $\frac{1}{3}$ or 0.33 (d) $\frac{1}{2}$ or 0.5

Exercise 13I

1 (a) 25 (b) 16
   (c) 27 (d) 216
   (e) 243 (f) $\frac{1}{49}$ or 0.02
   (g) 128 (h) $\frac{1}{3125}$ or 0.00032
2 (a) $52^{3/2}$ (b) $79^{2/3}$ (c) $143^{3/5}$ (d) $728^{3/4}$
3 (a) $2401^{3/4}$ (b) $256^{3/4}$ (c) $3125^{4/5}$
4 (a) 108 (b) 8 (c) 2512.9
5 (a) 7 (b) -5 (c) $\frac{1}{4}$ (d) 4
   (e) $\frac{1}{3}$ (f) 2 (g) 2 (h) -3
7 (a) a = 2 b = 3
   (b) a = 3 b = 10
   (c) a = 3 b = 4

Exercise 13J

1 (a) 3×10^6 (b) 7.4×10^3
   (c) 3.2×10^4 (d) 6.035×10^5
   (e) 1.08×10^2 (f) 6.8×10^1
   (g) 6.505×10^2 (h) 9.99×10^1
2 (a) 5×10^{-4} (b) 6×10^{-3}
   (c) 4×10^{-1} (d) 1.2×10^{-4}
   (e) 7.17×10^{-2} (f) 1.975×10^{-4}
   (g) 9.009×10^{-1} (h) 1.0003×10^{-3}
3 $0.9 \to 9 \times 10^{-1}$, $900\,000 \to 9 \times 10^5$, $900 \to 9 \times 10^2$, $0.09 \to 9 \times 10^{-2}$, $90 \to 9 \times 10^1$, $9000 \to 9 \times 10^3$
4 (a) 1.2735×10^4 km (b) 6.50×10^8 tonnes
5 2×10^{-4} km

Exercise 13K

1 (a) 50 000 (b) 3800 (c) 0.006
   (d) 7 260 000 000 (e) 0.008 492 (f) 4 370 000
   (g) 0.0001006 (h) 6 238.7
2 (a) 148 800 000 (b) 0.01 (c) 0.000 000 000 03
3 (a) 1.23×10^4 (b) 8.0×10^6 (c) 1.7×10^{-1}
   (d) 2.5×10^{-5} (e) 1.8×10^7 (f) 1.25×10^{-1}
   (g) 2.16×10^3 (h) 2×10
4 (a) 5.566×10^5 (b) 6.02×10^3
   (c) 4.32×10^{-2} (d) 7.59×10^{-3}
   (e) 3.9928×10^2 (f) 6.67952×10^3
5 7832450 is bigger by 152 450 bytes

Exercise 13L

1 (a) 8.8×10^8 (b) 9.9×10^2 (c) 4.2×10
   (d) 1.4×10^5 (e) 2.226×10^8 (f) 4×10
2 2.592×10^{10}
3 1.098×10^{12} km^3
4 (a) 1.284×10^7 (b) 5.6496×10^{13}
   (c) 1.7424×10^{14} (d) 1.748×10^7
   (e) 3.24×10^{-1} (to 3 s.f.)
5 1.17×10^{-2} tonnes or 11.7 kg
6 2.43×10^{13} km^2
7 (a) 3.78×10^5 cm^2 (b) 5680 cm
8 (a) 1.08×10^{-1} g or 0.108 g
   (b) 1080 g/cm^3
9 2.5×10^{13} miles
10 1.25×10^{-7} cm

Examination style questions

1 (a) m^7 (b) p^3 (c) q^8
2 (a) 144 (b) $6x^4y^2$
3 (a) x^3 (b) y^7
4 $8x^3y^6$
5 (a) 1 (b) 4 (c) $\frac{1}{3}$

Chapter 14

Exercise 14A

1 (a) $5, 7, 9, 11$ **(b)** 2 **(c)** 43

2 (a) $8, 11, 14, 17; 3; 95$

    (b) $1\frac{1}{2}, 2, 2\frac{1}{2}, 3; \frac{1}{2}; 16$

    (c) $1, -2, -5, -8; -3; -86$

    (d) $1, \frac{1}{2}, 0, -\frac{1}{2}; -\frac{1}{2}; -13\frac{1}{2}$

3 (a) 20

    (b) 21st

    (c) If $2n + 4 = 35$ then $n = 15\frac{1}{2}$ but $15\frac{1}{2}$ is not a whole number

4 (a) 17

    (b) 20th

    (c) If $3n - 1 = 90$ then $3n = 91$ but 91 is not a multiple of 3

5 (a) $2, 5, 10, 17$ **(b)** 145

6 (a) $-2, 1, 6, 13$ **(b)** 166

7 (a) $\frac{2}{1}, \frac{3}{2}, \frac{4}{3}, \frac{5}{4}$ or $2, 1\frac{1}{2}, 1\frac{1}{3}, 1\frac{1}{4}$

    (b) $\frac{10}{9} = 1\frac{1}{9}$

8 (a) 5th

    (b) If $n^2 + 10 = 55$ then $n^2 = 45$ but 45 is not a square number

Exercise 14B

1 $3n + 1; 151$ **2** $2n + 3; 103$

3 $4n + 3; 203$ **4** $5n - 2; 248$

5 $3n + 3; 153$ **6** $7n - 1; 349$

7 $10n - 8; 492$ **8** $9n + 7; 457$

9 $n + 4; 54$ **10** $3n - 7; 143$

11 $29 - 4n; -171$ **12** $25 - 5n; -225$

Exercise 14C

1 (a)

    (b) (i) 27 **(ii)** $n^2 + 2$

    (c) The differences increase by 2 each time.

2 (a) 55

    (b) Differences are consecutive whole numbers. Differences increase by 1 each time.

    (c) 20

3 (a) **(i)** $7 \times 10 = 70$ **(ii)** $20 \times 23 = 460$ **(iii)** $n(n + 3)$

    (b) 10×13 perimeter $= 46$

    (c) $114\,\text{cm}$

Exercise 14D

1 $n^2 + 3; 228$ **2** $n^2 - 2; 223$

3 $n^2 + 6; 231$ **4** $n^2 - 5; 220$

5 $2n^2; 450$ **6** $3n^2; 675$

7 $1 - n^2; -224$ **8** $\dfrac{n^2 + 1}{2}; 113$

Examination style questions

1 (a) **(i)** $5, 9, 13$

    (ii) no; if $4n + 1 = 122$, then $4n = 121$, but 121 is not divisible by 4

    (b) $3n + 1$

2 (a) 21 **(b)** $4n + 1$ **(c)** 50

Chapter 15

Exercise 15A

1 (a) P(given homework) $= 0.4$

    (b) P(it will snow tomorrow) $= 1\%$

    (c) P(throwing two consecutive 6's) $= 0.02\dot{7}\,(0.028)$

2 (a) P(2) $= \frac{1}{6}$ **(b)** P(even) $= \frac{1}{2}$ **(c)** P(<4) $= \frac{1}{2}$

    (d) P(3 or 5) $= \frac{1}{3}$ **(e)** P(not 6) $= \frac{5}{6}$

3 (a) $\frac{3}{100}$ **(b)** $\frac{11}{100}$

4 (a) $\frac{11}{16}$ **(b)** $\frac{5}{16}$ **(c)** 0 **(d)** 1

5 Student results for tossing a coin 100 times

    P(H) $=$ P(T) $= \frac{1}{2}$

    Toss the coin more than 100 times

6 (a)

	Tally	Frequency	Relative frequency
Red			
Blue			
	Total	50	

    Student results

    (b) Student results

    (c) P(Red) $= \frac{7}{10}$ P(Blue) $= \frac{3}{10}$

Exercise 15B

1 (a) P(vowel) $= \frac{5}{26}$ **(b)** P(consonant) $= \frac{21}{26}$

2 (a) P(I) $= \frac{4}{11}$ **(b)** P(S) $= \frac{4}{11}$ **(c)** P(M or P) $= 1$

3 (a) P(3) $= \frac{1}{6}$ **(b)** $\frac{5}{6}$

    (c) P(3 or 4) $= \frac{1}{3}$ **(d)** P(not 2 or 3) $= \frac{2}{3}$

4 (a) P(square number) $= \frac{5}{30} = \frac{1}{6}$

    (b) P(prime number) $= \frac{10}{30} = \frac{1}{3}$

    (c) P(>10) $= \frac{20}{30} = \frac{2}{3}$

5 (a) P(R) $= \frac{2}{14} = \frac{1}{7}$ **(b)** P(B or G) $= \frac{9}{14}$

    (c) P(not Y) $= \frac{11}{14}$ **(d)** P(not G or R) $= \frac{7}{14} = \frac{1}{2}$

6 (a) **(i)** P(Y) $= \frac{3}{8}$

    (ii) P(not R or Y) $= \frac{2}{8} = \frac{1}{4}$

    (b) 4 Green, 6 Red and 6 Yellow

Exercise 15C

1 (a) HHH, HHT, HTH, HTT, THH, THT, TTH, TTT

    (b)

	HH	HT	TH	TT
H	HHH	HHT	HTH	HTT
T	THH	THT	TTH	TTT

    (c) A sample space diagram avoids mistakes.

2

		Ordinary dice					
		1	**2**	**3**	**4**	**5**	**6**
Tetrahedral dice	**1**	1, 1	1, 2	1, 3	1, 4	1, 5	1, 6
	2	2, 1	2, 2	2, 3	2, 4	2, 5	2, 6
	3	3, 1	3, 2	3, 3	3, 4	3, 5	3, 6
	4	4, 1	4, 2	4, 3	4, 4	4, 5	4, 6

$0 + 1 = 1$ $0 + -1 = -1$ $0 + -2 = -2$
$1 + 1 = 2$ $1 + -1 = 0$ $1 + -2 = -1$
$2 + 1 = 3$ $2 + -1 = 1$ $2 + -2 = 0$
P(negative total) $= \frac{3}{9} = \frac{1}{3}$

3

		Colour		
		Blue (B)	**Green (G)**	**Red (R)**
Number	**4**	B, 4	G, 4	R, 4
	7	B, 7	G, 7	R, 7
	9	B, 9	G, 9	R, 9
	11	B, 11	G, 11	R, 11

P(winning) $= \frac{4}{12} = \frac{1}{3}$

4

		Throw 1					
		1	**2**	**3**	**4**	**5**	**6**
Throw 2	**1**	2	3	4	5	6	7
	2	3	4	5	6	7	8
	3	4	5	6	7	8	9
	4	5	6	7	8	9	10
	5	6	7	8	9	10	11
	6	7	8	9	10	11	12

(a) P(3,3) $= \frac{1}{36}$
(b) P(sum < 5) $= \frac{6}{36} = \frac{1}{6}$
(c) $\frac{1}{12}$

5

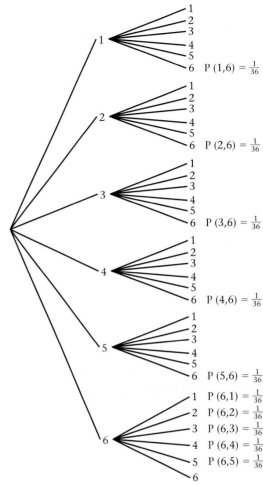

P(one six) $= \frac{10}{36} = \frac{5}{18}$

Exercise 15D

1 (Red, Head)   (Blue, Head)   (Green, Head)
(Red, Tail)   (Blue, Tail)   (Green, Tail)
P(Red, Tail) $= \frac{1}{6}$

2 (a)

		Dice			
		1	**2**	**3**	**4**
Spinner	Red (R)	R, 1	R, 2	R, 3	R, 4
	White (W)	W, 1	W, 2	W, 3	W, 4
	Blue (B)	B, 1	B, 2	B, 3	B, 4

P(Red or White, even) $= \frac{4}{12} = \frac{1}{3}$

3 (b) (i) 0.32   (ii) 0.08
4 (b) (i) $\frac{25}{64}$ (ii) $\frac{17}{32}$ (iii) $\frac{39}{64}$

Exercise 15E

1 (a)

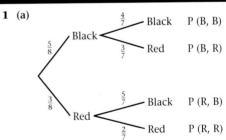

(b) P(Red,Red) $= \frac{3}{28}$
2 (a) P(B, B) $= \frac{1}{6}$ (b) P(B, G or G, B) $= \frac{5}{9}$
3 (a) P(different) $= \frac{4}{8} = \frac{1}{2}$ (b) P(both roses) $= \frac{1}{8}$
4 (b) (i) $\frac{3}{7}$ (ii) $\frac{10}{21}$ (iii) $\frac{11}{21}$
5 (b) (i) $\frac{2}{9}$ (ii) $\frac{14}{45}$ (iii) $\frac{31}{45}$ (iv) $\frac{17}{45}$

Examination style questions

1 (a) 0.3
  (b) 400 red, 200 blue

2 (a) $\frac{3}{10}$
  (b) 68

3 (a)

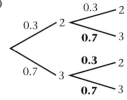

  (b) 0.09

4 (a) (i) $\frac{51}{100}$
    (ii) Yes; the relative frequencies are close to $\frac{1}{2}$ red, $\frac{1}{3}$ blue and $\frac{1}{6}$ green, which are the probabilities that you would get from a fair dice.
  (b) 10 is too few to be significant.

5 (a)

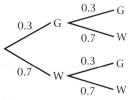

  (b) $\frac{29}{50}$ or 0.58

6 $\frac{13}{28}$

Chapter 16

Exercise 16A

1

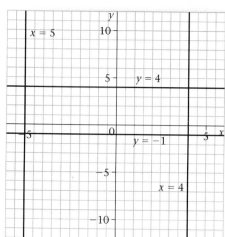

2 (a) (i)

x	-2	0	2
y	-5	-3	-1

(ii)

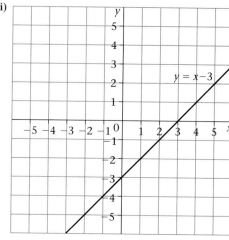

(b) (i)

x	-3	0	1
y	-3	3	5

(ii)

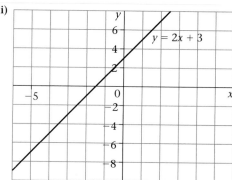

(c) (i)

x	-2	0	5
y	6	4	-1

(ii)

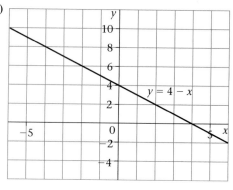

(d) (i)

x	−1	0	2
y	−4	−1	5

(ii)

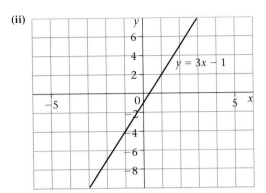

$y = 3x - 1$

(e) (i)

x	−2	0	3
y	6	2	−4

(ii)

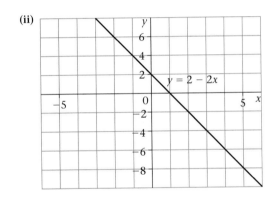

$y = 2 - 2x$

(f) (i)

x	−4	0	4
y	3	1	−1

(ii)

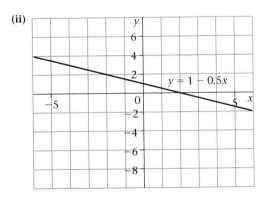

$y = 1 - 0.5x$

3

x	−3	0	3
y	−5	−2	1

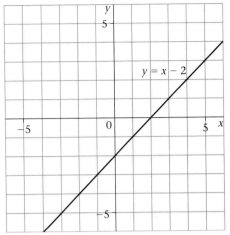

$y = x - 2$

4

x	−5	0	3
y	−7	3	9

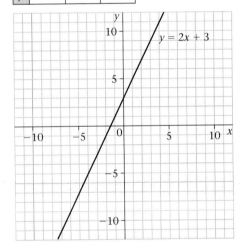

$y = 2x + 3$

5

x	−4	0	2
y	8	4	2

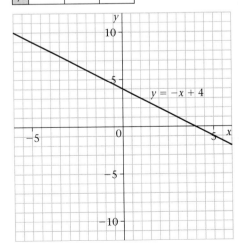

$y = -x + 4$

6

x	-3	0	3
y	-10	-1	8

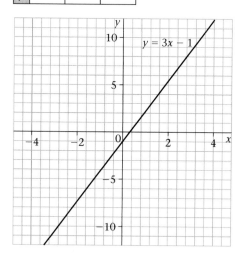

7

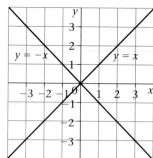

Lines cross over at 90° (right angles).

8

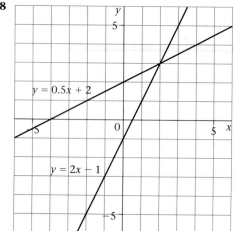

$(2, 3)$

Exercise 16B

1 AB$(2\frac{1}{2}, 7\frac{1}{2})$ CD$(5, 1\frac{1}{2})$ EF$(-3, 6\frac{1}{2})$ GH$(\frac{1}{2}, 2\frac{1}{2})$ IJ$(8, 10)$
2 (a) AB$(4\frac{1}{2}, 1)$ (b) CD$(7, 5\frac{1}{2})$
  (c) EF$(3\frac{1}{2}, 6)$ (d) GH$(-1, 5)$
  (e) IJ$(\frac{1}{2}, -\frac{1}{2})$

Exercise 16C

1 (a) $\frac{4}{8} = \frac{1}{2}$
   (b) $\frac{-3}{9} = -\frac{1}{3}$
   (c) $\frac{-4}{1} = -4$
2 $y = -2x + 14$
3 $y = \frac{1}{3}x + 6$
4 (a) (i) 1 (ii) $y = x - 3$
   (b) (i) -1 (ii) $y = -x - 10$
   (c) (i) $-\frac{1}{5}$ (ii) $y = -\frac{1}{5}x + \frac{21}{5}$
   (d) (i) -13 (ii) $y = -13x - 20$
5 (a) $-\frac{1}{6}$ (b) $\frac{1}{4}$
   (c) $-\frac{3}{2}$ (d) 4
   (e) $-\frac{2}{3}$ (f) $\frac{5}{16}$
   (g) $-\frac{5}{3} = -1\frac{2}{3}$ (h) $\frac{4}{5} = 0.8$
6 $y = \frac{1}{2}x$
7 $y = -\frac{1}{4}x + 7$
8 (a), (b), (e)
9 (b), (d)

Exercise 16D

1 (a) $m = 3, c = 2$
   (b) $m = -1, c = 3$
   (c) $m = 4, c = -7$
   (d) $m = -3, c = -4$
   (e) $m = \frac{2}{3}, c = 8$
   (f) $m = 0.8, c = -0.3$
2 (a) $y = 3x + 6, m = 3, c = 6$
   (b) $y = -4x - 2, m = -4, c = -2$
   (c) $y = -3x + 1, m = -3, c = 1$
   (d) $y = -2x + 7, m = -2, c = 7$
   (e) $y = x + 3, m = 1, c = 3$
   (f) $y = -3x - 4, m = -3, c = -4$
   (g) $y = -\frac{3}{2}x + 1, m = -\frac{3}{2}, c = 1$
   (h) $y = -\frac{1}{3}x - \frac{2}{3}, m = -\frac{1}{3}, c = -\frac{2}{3}$
   (i) $y = -\frac{3}{4}x - \frac{1}{2}, m = -\frac{3}{4}, c = -\frac{1}{2}$
   (j) $y = \frac{1}{2}x - 4, m = \frac{1}{2}, c = -4$
   (k) $y = \frac{3}{2}x - \frac{3}{4}, m = \frac{3}{2}, c = -\frac{3}{4}$
   (l) $y = -2x + 3, m = -2, c = 3$

Exercise 16E

1

x	0	1	2	3	4	5
$x - y = 1$	-1	0	1	2	3	4
$x + 2y = 4$	2	1.5	1	0.5	0	-0.5

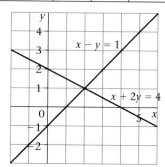

$x = 2$ and $y = 1$

2

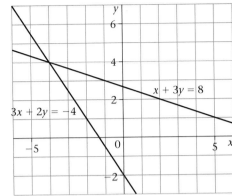

$x = -4$ and $y = 4$

3 $x = -1$ and $y = 0$

4 $x = 2$ and $y = 1$; the two lines are perpendicular

Exercise 16F

1 (a) $y \leqslant 4$

　(b) $x > -2$

　(c) $y \leqslant 2x + 6$

　(d) $x + y - 4$

2 (a)

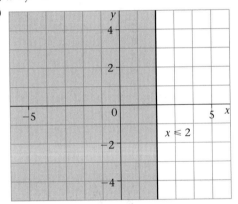

　(b)

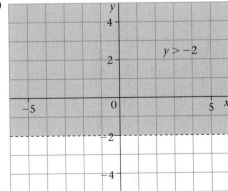

(c)

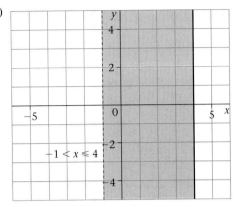

(d)

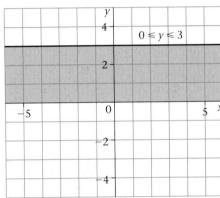

(e)

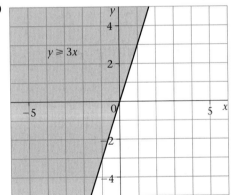

(f)

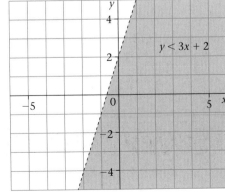

(g)

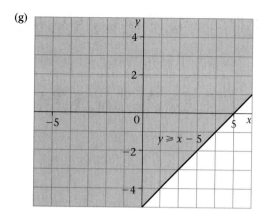

$y \geq x - 5$

(h)

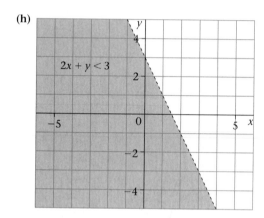

$2x + y < 3$

3 (a)

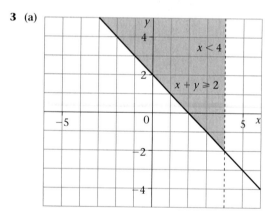

$x < 4$

$x + y \geq 2$

(b)

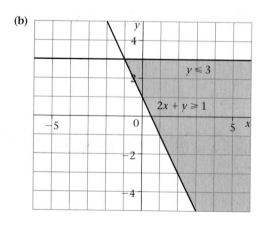

$y \leq 3$

$2x + y \geq 1$

4

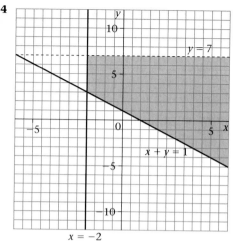

$y = 7$

$x + y = 1$

$x = -2$

5 $x \leq 0, y > 2, y < x + 5$

Exercise 16G

1

miles(x)	0	5	10	20	50	100
km(y)	0	8	16	32	80	160

Conversion graph
km–miles

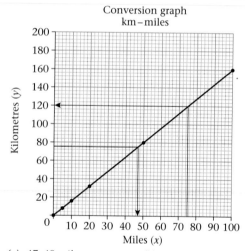

(a) 47–48 miles   **(b)** 117–118 km

2

cm	0	2.5	5	7.5	12.5	15	25	30
inches	0	1	2	3	5	6	10	12

Conversion graph
cm–inches

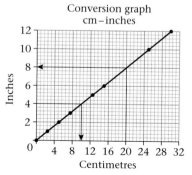

(a) 10 cm   **(b)** 8 inches

Exercise 16H

1 (a) 08:30   (b) 15 min
  (c) 1 hr 15 min   (d) 1.6 kph
2 (a) 5 hrs   (b) return journey
  (c) line is steeper   (d) 20 kph
3 (a)

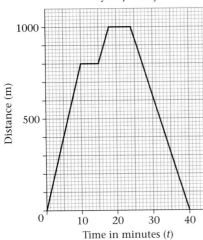

Glyn's journey

(b) 1.3 m/s   (c) 0.83 m/s

4 (a)

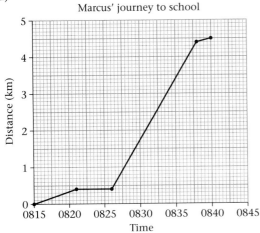

Marcus' journey to school

(b) 222.2 m/min   (c) 13.3 km/h

5 (a)

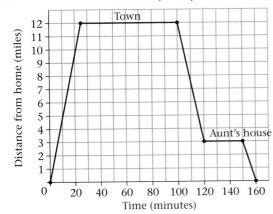

Alison's journey

(b) 28.8 mph

Exercise 16I

1 (a) 2.5 m/s^2 (b) 625 m
2 (a) 500 m   (b) $16\frac{2}{3}$ km/h (16.7 km/h)
  (c) 187.5 m   (d) $1\frac{2}{3}$ (1.7) m/s^2

Examination style questions

1

x	-1	0	1	2	3
y	-3	-1	1	3	5

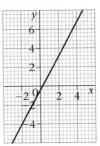

2 $(-1, -3)$
3 (a) 140 km   (b) 100 km/h
4 $x = 3, y = 2$
5

$y = x + 2$
$y = 4$
R
$x = -3$

6 (a) $y = -3x + 9$ (b) -9 (c) $\frac{1}{3}$

Chapter 17

Exercise 17A

1 (a) 125 ml cranberry and 75 ml orange juice
  (b) 225 ml
2 (a) 144   (b) 70
3 (a) 7.5 kg   (b) 0.8 kg
4 (a) 7.5 kg   (b) 24 kg peat, 9 kg sand
5 (a) 120 g meat, 60 g mushrooms, 20 g onions
  (b) 400 ml

Exercise 17B

1 (a) 3 : 5   (b) 8 : 5   (c) 4 : 3
  (d) 12 : 7   (e) 4 : 7   (f) 4 : 1
2 (a) 5 : 1   (b) 1 : 500   (c) 20 : 1
  (d) 3 : 20   (e) 3 : 20   (f) 7 : 40
  (g) 14 : 5   (h) 9 : 40   (i) 1 : 20

3 (a) 2:1   (b) 8:9
   (c) 6:7   (d) 2:1
   (e) 3:1   (f) 4:1
4 40:8:1
5 6:1
6 16:6:3

Exercise 17C

1 1680
2 (a) 245 cm   (b) 385 cm
3 (a) 5:4   (b) 240 g
   (c) 540 g
4 (a) 96   (b) 114
5 50

Exercise 17D

1 (a) 5:1   (b) 3.5:1   (c) 2.25:1
   (d) 2.4:1   (e) 15:1   (f) 1.5:1
   (g) 12.5:1   (h) 2.4:1   (i) 5:1
2 (a) 1:1.5   (b) 1:1.2   (c) 1:2.67
   (d) 1:1.6   (e) 1:5.6   (f) 1:4.5
   (g) 1:0.25   (h) 1:0.3

Exercise 17E

1 (a) 30:20   (b) 9:21
   (c) 25:20   (d) 24:40
2 (a) £180:£20   (b) 50 g:75 g
   (c) 350 m:250 m   (d) 160 l:90 l
3 (a) £84:£36   (b) £30:£90
   (c) £72:£48   (d) £50:£70
4 18
5 24
6 400 ml orange juice, 480 ml wine, 80 ml lime juice.
7 cement 250 kg, sand 750 kg, gravel 1000 kg
8 Paul £102, John £85, Sarah £68

Exercise 17F

1 £1.11   **2** 75 miles
3 1.62 kg   **4** £5.25
5 £12.30   **6** £95.20
7 225 g
8 (a) 3.75 km   (b) 36 min

Exercise 17G

1 12 days   **2** 4 hrs
3 14 horses   **4** 10 days
5 5 pumps   **6** 500 people
7 3 days   **8** 40 books
9 (a) 6 hrs   (b) 200 mph
10 (a) 12 days   (b) 16 men

Exercise 17H

1 £16.45
2 108 min or 1 hr 48 min
3 288
4 £75
5 2 hrs
6 £1.02
7 (a) 72   (b) 6 days
8 (a) 128 min or 2 hours 8 min
   (b) 30

Exercise 17I

1 (a) 1 km   (b) 1.75 km
   (c) 200 m   (d) 3.125 km
2 (a) 2 cm   (b) 10 cm
   (c) 7.5 cm   (d) 6.25 cm
3 (a) 75 km   (b) 33 km
4 (a) 80 cm   (b) 64 cm   (c) 32 cm
5 (a) 325 km   (b) 6 cm
6 (a) 300 m   (b) 1.2 km   (c) 650 m
   (d) 2.25 km   (e) 1.6 km

Exercise 17J

1 (a) $y = 16.5$ (b) $x = 4$
2 (a) $w = 28$ (b) $t = 4.5$
3 $a = 11.3, b = 4.9$
4 (a)

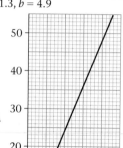

   (b) $A = 2.4 l$ (c) 45.6 cm²   (d) 17.5 cm
5 $d = \dfrac{50m}{29}$
6 no; $\dfrac{4.8}{3} = \dfrac{11.52}{7.2} = \dfrac{32}{20} = 1.6$, but $\dfrac{18.25}{12.5} = 1.46$
7 (a) $e = 32w$ (b) 112 mm   (c) 5.7 kg

Exercise 17K

1 (a) $y = 0.5$ (b) $x = 36$
2 $a = 12.5, b = 30$
3

l	0.2	0.5	0.6	1.5	3.6	7.5
h	120	48	40	16	$6\frac{2}{3}$	3.2

4 (a) $wz = 36$ (b) $z = 24$ (c) 6 and −6
5 yes; $0.6 \times 18 = 2.4 \times 4.5 = 9 \times 1.2 = 30 \times 0.36 = 10.8$
6 (a) $PV = 288$ (b) 115.2 cm³   (c) 1.92 bar
   (d)

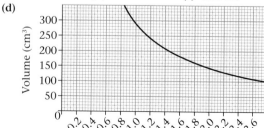

Exercise 17L

1 (a) $y = 45$ (b) $x = 7$
2 (a) $Q = 6.75$ (b) $m = 5$
3 $a = 11, b = 224$
4 (a) $y = 36$ (b) $x = 9$
5 (a) $y = 6x^3$ (b) $y = 384$ (c) $x = 5$
6 (a) $33.5\,\text{cm}^3$ to 3 s.f.   (b) $5.00\,\text{cm}$ to 3 s.f.
7 C. $Q = 0.4h^3$

Exercise 17M

1 (a) $y = 6$ (b) $x = 16$
2 (a) $y = 0.8$ (b) $x = 2$
3 (a) $y = 384$ (b) $x = 4$
4

t	2	4	1	0.5
V	9	2.25	36	144

5 (a) $Wm^2 = 60$ (b) 1.67 to 3 s.f.
   (c) $m = 2$
6 (a) $10\,\text{cm}$ (b) $4\,\text{cm}$
7 (a) 18.75 newtons   (b) $15\,\text{cm}$
8 $4 : 1$

Examination style questions

1 £3500
2 $126°$
3 B
4 60

Chapter 18

Exercise 18A

1 (a)
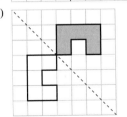

  (b)
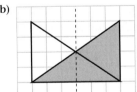

  (c)

(d)
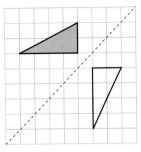

2 (a)
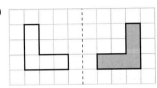

  (b)
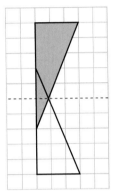

  (c)
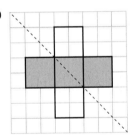

3 (a) and (b)
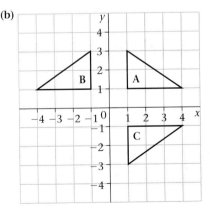

4

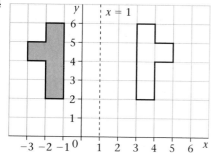

5

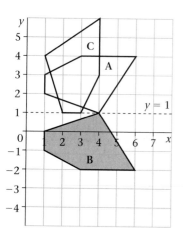

Exercise 18B

1 (a)

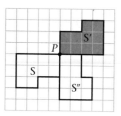

(b)

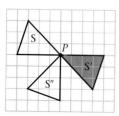

(c)

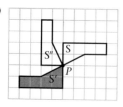

2 (a), (b) and (c)

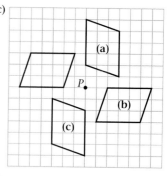

3 (a), (b) and (c)

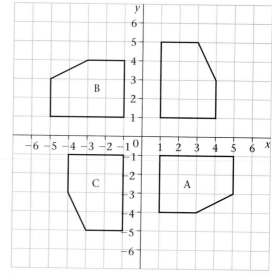

4 90° clockwise about the point $(-3, 4)$

Exercise 18C

1 (i) (a)

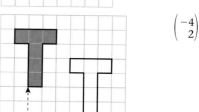

$\begin{pmatrix} 3 \\ -2 \end{pmatrix}$

(b)

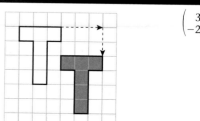

$\begin{pmatrix} -4 \\ 2 \end{pmatrix}$

(c)

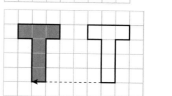

$\begin{pmatrix} -5 \\ 0 \end{pmatrix}$

(d) 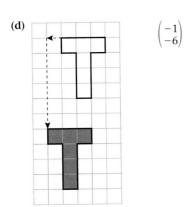 $\begin{pmatrix} -1 \\ -6 \end{pmatrix}$

(e) 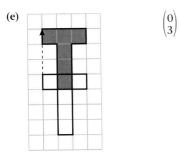 $\begin{pmatrix} 0 \\ 3 \end{pmatrix}$

(ii) (a) 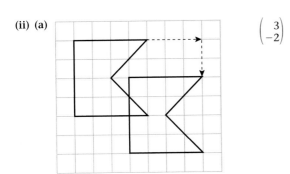 $\begin{pmatrix} 3 \\ -2 \end{pmatrix}$

(b) 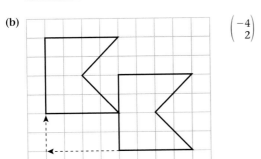 $\begin{pmatrix} -4 \\ 2 \end{pmatrix}$

(c) 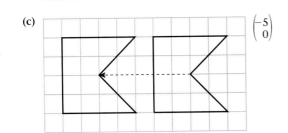 $\begin{pmatrix} -5 \\ 0 \end{pmatrix}$

(d) 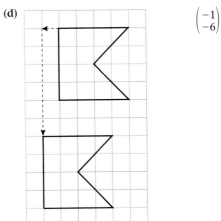 $\begin{pmatrix} -1 \\ -6 \end{pmatrix}$

(e) 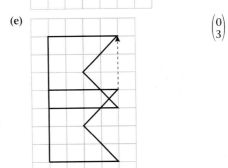 $\begin{pmatrix} 0 \\ 3 \end{pmatrix}$

2 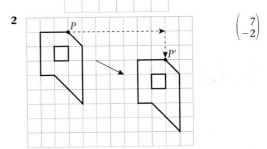 $\begin{pmatrix} 7 \\ -2 \end{pmatrix}$

3 A to B $\begin{pmatrix} -7 \\ 5 \end{pmatrix}$ A to C $\begin{pmatrix} -2 \\ 6 \end{pmatrix}$ A to D $\begin{pmatrix} 1 \\ 5 \end{pmatrix}$ A to E $\begin{pmatrix} -6 \\ -1 \end{pmatrix}$

4 (a)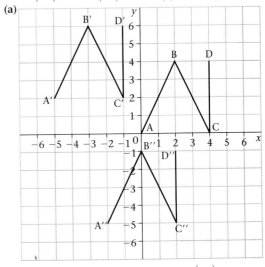

(b) $(0, -1)$ **(c)**   translation by $\begin{pmatrix} -2 \\ -5 \end{pmatrix}$

Exercise 18D

1 (a) scale factor 2
  (b) scale factor 3
  (c) scale factor 2
  (d) scale factor 4

2 (a)

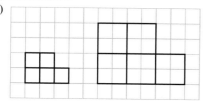

  (b)

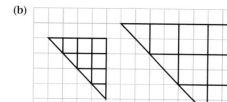

  (c)

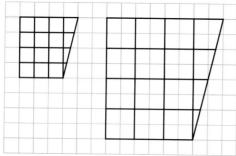

3 (a)

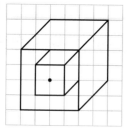

  (b)

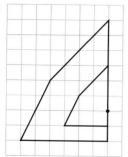

4

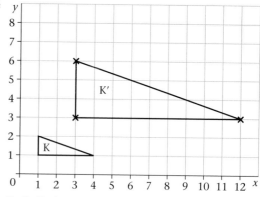

(3, 3), (3, 6) and (12, 3)

5

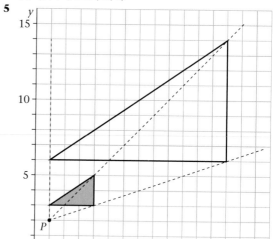

6 (a) $\frac{1}{3}$ (c) (0, 17)

7

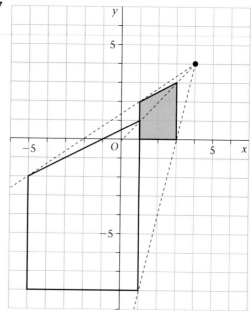

8

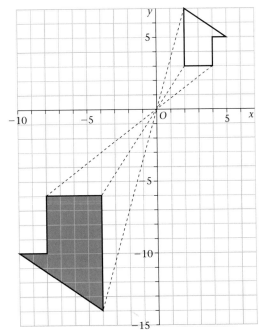

Exercise 18E

1

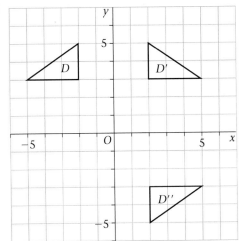

180° rotation clockwise or anticlockwise about the origin

2

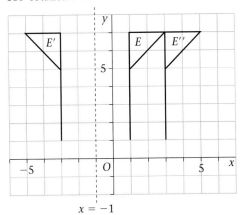

$x = -1$

translation vector $(2, 0)$

3

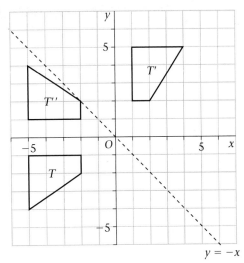

$y = -x$

reflection in the x-axis ($y = 0$)

4 (a)(b)(c)

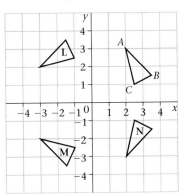

(d) reflection in the x-axis ($y = 0$)

Examination style questions

1 (a)

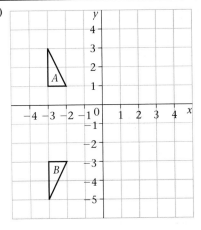

(b)

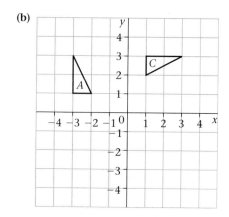

2 (a) (i) A
  (ii) E
  (iii) B
 (b) reflection in $x = -2$

3

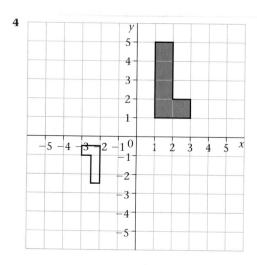

4

5 (a) rotation 90° clockwise (or 270° anticlockwise) about (0, 2)
  (b)

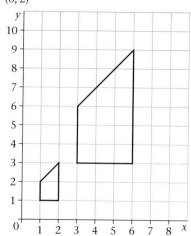

Chapter 19

Exercise 19A

1 (a) C and F   **(b)** D and H
2 various answers
3 (a) A and B   **(b)** A and C
  (c) A, B and C   **(d)** A and B
4 (a) SSS   **(b)** SAS
  (c) No – the 60° angle is included in one triangle, but not in the other triangle.
    (An alternative answer might be "not necessarily", since both triangles could be equilateral triangles, in which case they would be congruent.)
  (d) ASA   **(e)** RHS   **(f)** RHS

Exercise 19B

1 $a = 18, b = 15$ **2** no
3 $c = 4.5$ cm, $d = 5$ cm
4 (a) T   **(b)** F   **(c)** F
  (d) F   **(e)** T   **(f)** T
5 13.5 m
6 $x = 6.7$ cm, $y = 5.3$ cm both to 2 s.f.

Exercise 19C

1 216 cm²   **2** 6.32 cm to 3 s.f.
3 (a) 22.5 cm²   **(b)** 12.5 cm²
4 141.032 cm²   **5** 78.2 cm² to 3 s.f.
6 11.0 cm to 3 s.f.
7 (a) 125 cm³   **(b)** 4 : 25

Examination style questions

1 $p = 9$ cm, $q = 8$ cm
2 10.8 m   **3** 2 cm
4 (a) 5.4 cm   **(b)** 45°
5 1620 ml   **6** 74.25 cm³

Chapter 20

Exercise 20A

1 (a) mean 6, median 6, mode 6, range 11
   (b) mean 22.85, median 22.0, mode none, range 9.5
   (c) mean $\frac{29}{32}$ or 0.90625, median $\frac{5}{8}$ or 0.625, mode $\frac{1}{4}$,
       range $\frac{9}{4}$ or 2.25
   (d) mean 158 to 3 s.f., median 161, mode 180, range 56
2 mean 65 (to the nearest integer), median 64, mode 64
3 mean 6, mode 8
4 £9948.33
5 The (original) mean is £87. The new value is £59.
6 157.68 cm
7 (a) 22.2 to 3 s.f.   (b) 23
   (c) 24   (d) 12
8 456678 or 466668
9 mean 16.75°C, mode 21°C, median 17.5°C, range 30°C
10 (a) mode = 137, median = 152, range = 46, mean = 148
   (b) mode = 161, median = 158.5, range = 49,
       mean = 155.7 (4 s.f.)
   (c) mode = 155, median = 155, range = 53,
       mean = 151.6
   All answers in cm.

Exercise 20B

1 mode 38, median 32.5, mean 30.9. The median best represents the data.
2 mode 70, median 54.5, mean 50.2 to 3 s.f. The median is best, as 9 and 12 are extreme values.
3 mean 3.33 to 3 s.f., mode 4, median 3.5. Values are integers.
4 It is not clear which average is being used.

Exercise 20C

1 (a) 90
   (b)

10	0	0
20	23	460
30	48	1440
40	16	640
50	2	100
60	1	60

   mean 30 mph, median 30 mph, mode 30 mph
2 mean 2.2, median 2, mode 1
3

Score	1	2	3	4	5	6	7	8	9	10
Frequency	2	7	6	7	9	10	9	13	7	5

   mean 5.96, median 6, mode 8, range 9

Exercise 20D

1

0–2	12	1	12
3–5	18	4	72
6–8	31	7	217
9–11	11	10	110
Totals	72		411

estimated mean 5.71 to 3 s.f., modal class 6–8, median class 6–8

2

3–4	5	3.5	17.5
5–6	8	5.5	44.0
7–8	12	7.5	90.0
9–10	5	9.5	47.5
11–12	2	11.5	23.0
Totals	32		222.0

estimated mean 6.9375, modal class 7–8, median class 7–8

3

50–64	12	57	684
65–79	0	72	0
80–94	11	87	957
95–109	9	102	918
110–124	15	117	1755
125–139	33	132	4356
140–154	28	147	4116
155–169	41	162	6642
170–184	16	177	2832
	165		22 260

estimated mean 135 to 3 s.f., median class 140–154, modal class 155–169
4 mean 10.4 to 3 s.f., median class 5–9, modal class 0–4
5 estimated mean 1.36 to 3 s.f.

Exercise 20E

1 (a)

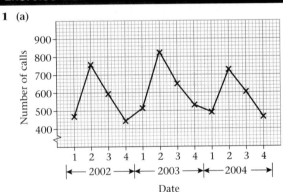

   (b) 2002:1–2002:4 566   2002:2–2003:1 578
       2002:3–2003:2 594   2002:4–2003:3 608
       2003:1–2003:4 630   2002:2–2004:1 624
       2003:3–2004:2 601   2003:4–2004:3 590
       2003:1–2003:4 573   to the nearest integer
   (c)

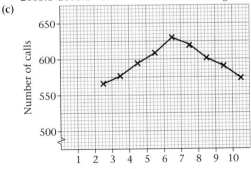

   (d) Number of calls increased until the 2nd quarter of 2003, then decreased.

2 (a)

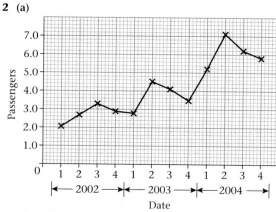

(b)
2002:1–2002:4	2.78
2002:2–2003:1	2.95
2002:3–2003:2	3.38
2002:4–2003:3	3.58
2003:1–2003:4	3.73
2002:2–2004:1	4.33
2003:3–2004:2	4.98
2003:4–2004:3	5.50
2003:1–2003:4	6.08 to 3 s.f.

(c)

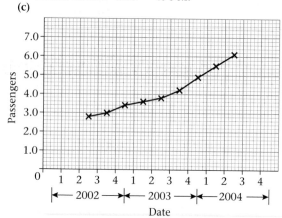

(d) There is an upward trend throughout.

Examination style questions

1 (a) 42   **(b)** 35 years
2 1.7
3 (a)

0–9	1
10–19	3
20–29	7
30–39	4
40–49	2
50–59	7
60–69	5
70–79	3
80–89	0
90–99	1

(b) 20–29 and 50–59   **(c)** 44.5

4 3.8 cm
5 (a) 50 to less than 60
  (b) 58 mph
6 (a) 7.6 minutes
  (b) $6 < t \leqslant 8$
7 (a) The figures will be smoothed.
  (c) £24.90

Chapter 21

Exercise 21A

1 (a)

Mark (%)	Cumulative frequency
≤10	1
≤20	3
≤30	7
≤40	14
≤50	19
≤60	27
≤70	29
≤80	31
≤90	32
≤100	32

(b)

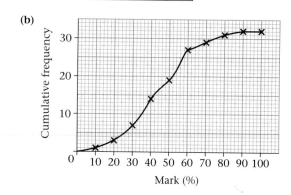

2

Marks	Cumulative frequency
≤10	1
≤20	7
≤30	15
≤40	30
≤50	47
≤60	71
≤70	93
≤80	108
≤90	117
≤100	120

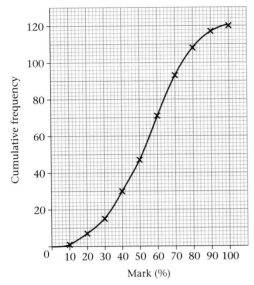

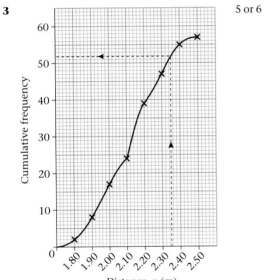

3
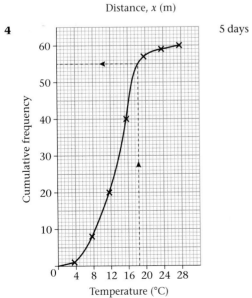

5 or 6

4
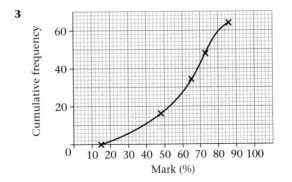

5 days

5

Time, t (min)	Cumulative frequency	Frequency (f)
$0 < t < 20$	4	4
$20 < t < 40$	11	7
$40 < t < 60$	21	10
$60 < t < 80$	35	14
$80 < t < 100$	47	12
$100 < t < 120$	50	3

29 pupils

Exercise 21B

1 (a) 3.4 km   (b) 2.1 km
   (c) 5.5 km   (d) 3.4 km
2 (a) 5 hrs 12 mins   (b) 3 hrs 54 mins
   (c) 7 hrs 12 mins   (d) 3 hrs 42 mins
   (e) 45%
3 (a) 53%   (b) 43%
   (c) 63%   (d) 20%
4 (a) 22 500 miles   (b) 18 750 miles
   (c) 27 750 miles   (d) 9000 miles
   (e) 17%

Exercise 21C

1   ; 316 hours

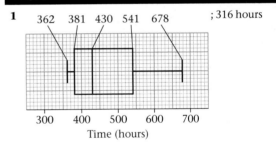

2

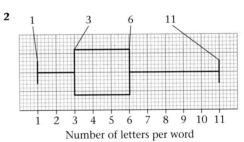

3

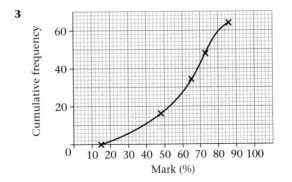

4 (a)

Waiting time (t) minutes	Cumulative frequency
⩽1	0
⩽2	4
⩽3	27
⩽4	70
⩽5	128
⩽6	165
⩽7	176
⩽8	179
⩽9	179
⩽10	180

(b)

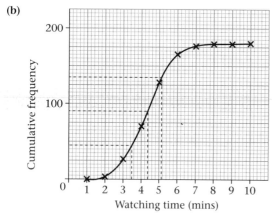

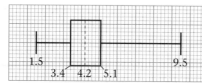

(c) estimated median waiting time = 4.2 minutes

5 (a)

Age a (years)	Cumulative frequency
⩽ 10	41
⩽ 20	179
⩽ 30	347
⩽ 40	539
⩽ 50	665
⩽ 60	750
⩽ 70	789
⩽ 80	800

(b)

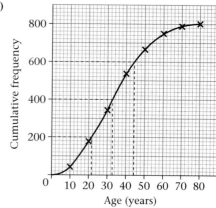

(c) estimated median age of members = 32 years

(d)

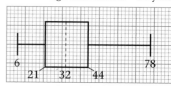

Exercise 21D

1

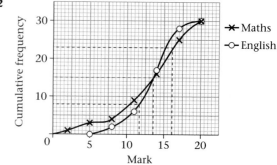

The range of lengths completed are approximately the same.
Swimmer B has a smaller IQR.
The median lengths completed is greater for swimmer A than swimmer B.

2

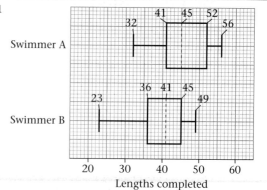

The spread of marks (given by the IQR) is much greater for Maths than English.
The range of marks is much greater for Maths than English.
The median marks for Maths and English are identical.

3

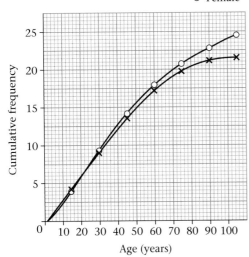

★ Male
-○- Female

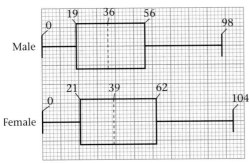

The median age for females is 39 years compared to 36 years for males.
The IQR is slightly greater for females (42 years) than males (38 years).
A greater number of females live longer than males.

Examination style questions

1

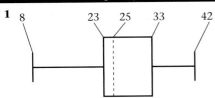

8   23 25 33   42

2 (a) The boys did slightly better and were less spread out.
(b) 27
3 11 and 19 minutes
4 (a)

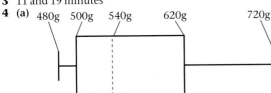

480g 500g 540g 620g   720g

(b) 120 g
(c) (i) 0 **(ii)** 20
5 (a) (i) 100 **(ii)** 13
(b) (i) George is more consistent (IQR of 8 − Brian's IQR = 13)
(ii) Brian is better (median of 100 − George's median = 107)

Chapter 22

Exercise 22A

1 (i) $r = 10$ cm **(ii)** $s = 12.75$ cm
(iii) $t = 8.2$ cm **(iv)** $u = 6.0$ cm
2 3.8 m
3 26.9 m
4 26 km
5 (a) $XY = 3.9$ cm
(b) $QR = 11.8$ cm
6 2.9 cm
7 $c = 5$ cm and $d = 12$ cm
8 $w = 3.75$ m
9 21.4 cm

Exercise 22B

1 $AB = 13$ units
2 $XY = 7.2$ units
3 (a) $GH = 6.7$ **(b)** $CD = 8.5$
(c) $WV = 12.8$ **(d)** $FE = 7.2$
(e) $XY = 11.7$ **(f)** $AB = 5.7$
(g) $RS = 6.4$ **(h)** $MN = 3.2$

Exercise 22C

1

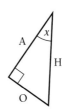

2 (a) 0.4540 **(b)** 4.0108
(c) 0.9511 **(d)** 0.0699
(e) 0.0872 **(f)** 1.0000
(g) 1.0000 **(h)** 0.3640
(i) −0.9659 **(j)** 0.0000
(k) 0.9848 **(l)** −0.5948
3 (a) $x = 49.6°$ **(b)** $y = 55.3°$
(c) $\theta = 66.8°$ **(d)** $A = 19.1°$
(e) $B = 47.5°$ **(f)** $\phi = 87.3°$

Exercise 22D

1 (i) $z = 41°$ **(ii)** $x = 48°$ **(iii)** $y = 38°$ **(iv)** $w = 33°$
2 (i) 59.5°, 30.5° **(ii)** 37.4°, 52.6° **(iii)** 49.3°, 40.7°

Exercise 22E

1 (i) $a = 9.5$ cm **(ii)** $b = 7.3$ cm
(iii) $c = 9.6$ cm **(iv)** $d = 12.7$ cm
2 (i) 23.0, 4.9 **(ii)** 30.0, 33.1
(iii) 12.2, 20.8 **(iv)** 18.2, 27.2

Exercise 22F

1 $x = 6$ cm
2 $y = 15$ cm
3 $b = 6.9$ cm
4 (a) 10 cm   (b) 0.8

Exercise 22G

1 (a) $\sqrt{149}$ or 12.2 cm   (b) 268 cm^2
2 (a) 3.0 cm   (b) 69.9 cm^3
3 328.62 cm^2

Exercise 22H

1 (a) $FC = 9.0$ m   (b) $FA = 7.5$ m   (c) 11.2 m
2 6813 m
3 19.6 m
4 7.1°

Exercise 22I

1 98.5 m, 207 m
2 13.2 m
3 (a) 2578 m   (b) 619 kph
4 400 m

Exercise 22J

1 13 cm
2 (a) $EC = 6.6$ cm   (b) 64.8°
3 (a) 12.7 m   (b) 35.3°
   (c) 30.8°   (d) 22.5°

Exercise 22K

1 (a) (4, 0, 0); (4, 5, 0); (0, 5, 0); (4, 0, 2); (4, 5, 2); (0, 5, 2);
        (0, 0, 2)
   (b) (i) (0, 2.5, 0)   (ii) (0, 5, 1)   (iii)(2, 5, 0)
        (iv) (4, 5, 1)
   (c) (i) (2, 2.5, 0)   (ii) (2, 2.5, 2)   (iii)(2, 5, 1)
2 (a) (3, 0, 0); (0, 0, 4); (0, 5, 4)
   (b) (i) (1.5, 0, 0)   (ii) (0, 2, 0)   (iii)(3, 2.5, 0)
   (c) (i) (1.5, 2.5, 0) (ii) (1.5, 0, 2)   (iii)(3, 2.5, 2)
   (d) (1.5, 2.5, 2)
   (e) (i) 5   (ii) $\sqrt{41}$ (iii)$\sqrt{34}$
        (iv) $\sqrt{50}$ (v) $\sqrt{50}$
3 (b) (i) 5   (ii) $5\sqrt{2}$ or 7.07

Examination style questions

1 177 m   2 230 m   3 56°
4 40°   5 21°

Chapter 23

Exercise 23A

1 12
2 $x = 226°$
3 $x = 36°$
4 $x = 38°, y = 104°$
5 $x = 59°, y = 31°$
6 $x = 50°, y = 40°$

Exercise 23B

1 $x = 52°$
2 $x = 56°$
3 $x = 214°, y = 146°$
4 $x = 28°, y = 56°$
5 $x = 90°, 49°$
6 $x = 52°, y = 52°, z = 76°$
7 $x = 22.5°$
8 $x = 47°, y = 43°$
9 $x = 39°, y = 51°$
10 $x = 86°, y = 47°$
11 $w = 45°, x = 32°, y = 58°, z = 64°$
12 $x = 38°, y = 32°, z = 38°$

Exercise 23C

1 $x = 46°, y = 24°$
2 $x = 101°, y = 79°$
3 $x = 17°, y = 45°$
4 $x = 43°, y = 70°, z = 67°$
5 $x = 25°, y = 51°, z = 70°$
6 $x = 61°, y = 39°, z = 39°$
7 $x = 43°$
8 $x = 40°, y = z = 42°$
9 $x = 53°$
10 $x = 36°; y = 72°$
11 $x = 29°$
12 $x = 75°, y = 37.5°$

Exercise 23D

1 $x = 35°$
2 $x = 20°, y = 36°$
3 $x = 33°, y = 57°$
4 $x = 53°, y = 37°, z = 106°$
5 $x = 93°, y = 112°$
6 $x = 54°, y = 54°$
7 $x = 53°, y = 64°$
8 $x = 18°, y = 72°$
9 $x = 23°, y = 34°$
10 $x = 27°, y = 93°$
11 $x = 58°, y = 32°, z = 26°$
12 $x = 84°, y = 42°, z = 6°$

Exercise 23E

1 $x = 43°$
2 $x = 57°$
3 $x = 53°, y = 37°$
4 $x = 55°, y = 35°$
5 $x = 50°, y = 65°, z = 50°$
6 $x = 59°, y = 78°$
7 $x = 68°, y = 44°, z = 44°$
8 $x = 51°, y = 78°, z = 78°$
9 $x = 36°$
10 $w = 47°, x = 86°, y = 47°, z = 86°$

Exercise 23F

1 $x = 69°, y = 121°$
2 $x = 112°, y = 124°$
3 $x = 28°$
4 $x = 26°$
5 $x = 38°$
6 $x = 100°, y = 46°$
7 $x = 67°, y = 111°$
8 $x = 63°, y = 117°, z = 54°$
9 $x = 39°, y = 81°$
10 $x = 52°, y = 128°, z = 48°$

Examination style questions

1 (a) 4 cm
(b) 7 cm
2 (a) The triangle PQR is isosceles, so the angle QPR is $\left(\dfrac{180° - 116°}{2}\right) = 32°$, and QPR and QSR are equal.
(b) (i) 116°   **(ii)** 42°

Chapter 24

Exercise 24A

(a) $(x + 2)(x - 2)$
(b) $(x + 5)(x - 5)$
(c) $(x + 8)(x - 8)$
(d) $(y + 6)(y - 6)$
(e) $(n + 10)(n - 10)$
(f) $(t + 1)(t - 1)$
(g) $(a + 11)(a - 11)$
(h) $(7 + x)(7 - x)$
(i) $(9 + x)(9 - x)$
(j) $(x + y)(x - y)$
(k) $(a + b)(a - b)$
(l) $(2x + 9)(2x - 9)$
(m) $(3x + 4)(3x - 4)$
(n) $(5x + 2)(5x - 2)$
(o) $(3x + y)(3x - y)$
(p) $(4a + b)(4a - b)$
(q) $(3x + 5y)(3x - 5y)$
(r) $4(3n + m)(3n - m)$
(s) $5(2p + 3q)(2p - 3q)$
(t) $3(5t + 4w)(5t - 4w)$

Exercise 24B

1 (a) $(a + 1)(a + 9)$
(b) $(x + 1)(x + 3)$
(c) $(n + 1)(n + 7)$
(d) $(x + 3)(x + 5)$
(e) $(y + 2)(y + 4)$
(f) $(p + 3)(p + 4)$
2 (a) $(x - 1)(x + 5)$
(b) $(k - 1)(k + 3)$
(c) $(r - 1)(r + 8)$
(d) $(z - 2)(z + 6)$
(e) $(d - 2)(d + 5)$
(f) $(x - 2)(x + 3)$
3 (a) $(b + 1)(b - 5)$
(b) $(x + 1)(x - 7)$
(c) $(g + 1)(g - 10)$
(d) $(m + 2)(m - 5)$
(e) $(t + 2)(t - 4)$
(f) $(f + 3)(f - 4)$
4 (a) $(q - 3)(q - 4)$
(b) $(y - 3)(y - 5)$
(c) $(x - 3)(x - 8)$
(d) $(a - 1)(a - 11)$
(e) $(x - 2)(x - 8)$
(f) $(h - 3)(h - 3)$
5 (a) $(t - 2)(t + 7)$
(b) $(z + 2)(z - 9)$
(c) $(p + 4)(p + 5)$
(d) $(d + 4)(d - 9)$
(e) $(n - 1)(n + 24)$
(f) $(x + 4)(x - 5)$

Exercise 24C

(a) $(2x + 3)(x + 1)$
(b) $(3x + 1)(x + 5)$
(c) $(2x + 1)(x + 1)$
(d) $(3t + 2)(t + 1)$
(e) $(2y + 3)(2y + 1)$
(f) $(4x + 3)(x + 5)$
(g) $(4x + 5)(x + 3)$
(h) $(6a + 1)(a + 4)$
(i) $(2x - 1)(x + 3)$
(j) $(3x - 2)(x + 1)$
(k) $(2x + 1)(x - 5)$
(l) $(2x - 3)(2x + 1)$
(m) $(2x - 1)(x - 1)$
(n) $(3x - 5)(x - 2)$
(o) $(4x - 3)(2x - 3)$
(p) $2(2x + 1)(x - 7)$
(q) $3(2x - 5)(x + 6)$
(r) $2(3x - 2)(x + 6)$
(s) $2(3x - 4)(2x - 3)$
(t) $3(4x - 7)(2x + 1)$

Exercise 24D

1 (a) $t = \pm 3$
(b) $x = \pm 5$
(c) $y = \pm 4$
(d) $z = \pm 7$
(e) $a = \pm 4$
(f) $x = \pm 2$
(g) $x = -3, x = 7$
(h) $y = 7, y = -5$
(i) $x = 1, x = -5$
(j) $x = 1, x = 9$
(k) $t = 2, t = -8$
2 (a) $x = 2.73, x = -0.73$
(b) $x = 1.76, x = 6.24$
(c) $x = 0.41, x = -2.41$
(d) $x = -0.17, x = -5.83$
3 Have to find the square root of a negative number $(\sqrt{-9})$.

Exercise 24E

1 (a) $t = 0, t = -5$
(b) $x = 0, x = -6$
(c) $n = 0, n = 9$
(d) $x = 0, x = 8$
(e) $z = 0, z = 2$
(f) $a = 0, a = 10$
(g) $b = 0, b = 7$
(h) $y = 0, y = -1$
(i) $x = 0, x = 1$
2 (a) $b = -1, b = -9$
(b) $x = -1, x = -4$
(c) $m = 1, m = -2$
(d) $t = 2, t = 6$
(e) $x = 3, x = -5$
(f) $y = 6$
(g) $z = 6, z = -4$
(h) $a = 7, a = -2$
(i) $x = -5, x = -6$
(j) $x = -7$
3 (a) $x = 6, x = -1$
(b) $x = 4, x = -3$
(c) $x = 2$
(d) $x = 2, x = -5$
(e) $x = 2, x = 3$
(f) $x = 9$
(g) $x = 5, x = -6$
(h) $x = 3, x = -8$
(i) $x = 6, x = 9$

Exercise 24F

1 (a) $t = 0, t = \frac{1}{2}$
(b) $x = 0, x = -\frac{2}{3}$
(c) $y = 0, y = \frac{5}{2}$
(d) $n = 0, n = \frac{1}{3}$
(e) $z = 0, z = \frac{5}{3}$
(f) $a = 0, a = -\frac{3}{5}$
2 (a) $x = -\frac{1}{2}, x = -3$
(b) $x = \frac{1}{2}, x = -2$
(c) $x = -\frac{1}{3}, x = 3$
(d) $a = \frac{2}{3}, a = -4$
(e) $y = \frac{5}{2}, y = -2$
(f) $k = -\frac{1}{2}, k = -\frac{3}{2}$
(g) $z = \frac{1}{4}, z = -2$
(h) $d = \frac{2}{3}, d = 4$
(i) $r = \frac{1}{3}, r = -\frac{3}{2}$
(j) $p = -\frac{1}{2}$
3 (a) $x = \frac{1}{3}, x = -2$
(b) $x = \frac{1}{3}, x = 3$
(c) $x = \frac{1}{2}, x = -3$
(d) $x = -3, x = 4$
(e) $x = \frac{1}{2}$
(f) $x = 5, x = -7$
(g) $x = -\frac{2}{3}, x = \frac{1}{2}$
(h) $x = \frac{3}{2}, x = -\frac{1}{6}$

Exercise 24G

1 (a) $x = -0.38, -2.62$
(b) $x = -0.44, -4.56$
(c) $x = 0.65, -4.65$
(d) $d = 3.45, -1.45$
(e) $y = 3.37, -2.37$
(f) $k = 0.84, 7.16$
(g) $x = 0.39, -3.89$
(h) $x = 3.28, 1.22$
(i) $x = 0.18, -1.85$
(j) $p = 0.54, -2.29$
(k) $y = 1.22, -0.82$
(l) $t = -0.12, -2.13$
(m) $z = 0.18, -1.85$
(n) $p = 1.40, -0.24$
2 (a) $x = 0.33, x = -1.5$
(b) $0.33\ldots = \frac{1}{3}, -1.5 = -\frac{3}{2}$, could get these from factorising: $(3x - 1)(2x + 3)$
3 (b) $b^2 - 4ac = -23$, can't square root a negative number

Exercise 24H

(a) $b^2 - 4ac = 1, 2$ solutions
(b) $b^2 - 4ac = 37, 2$ solutions
(c) $b^2 - 4ac = -12, 0$ solutions
(d) $b^2 - 4ac = 0, 1$ solution
(e) $b^2 - 4ac = 25, 2$ solutions
(f) $b^2 - 4ac = -8, 0$ solutions
(g) $b^2 - 4ac = 73, 2$ solutions
(h) $b^2 - 4ac = -20, 0$ solutions
(i) $b^2 - 4ac = 9, 2$ solutions
(j) $b^2 - 4ac = 0, 1$ solution

Exercise 24I

1 (a) $(x + 5)^2 - 20$
(b) $(x + 2)^2 - 6$
(c) $(x - 6)^2 - 16$
(d) $(y + 1)^2 + 8$
(e) $(x - \frac{3}{2})^2 - \frac{13}{4}$
(f) $(y + \frac{5}{2})^2 + \frac{3}{4}$
2 (a) $(x - 3)^2 + 2$ $p = 3, q = 2$
(b) $(x - 2)^2 - 5$ $p = 2, q = -5$
(c) $(x + 7)^2 - 41$ $p = -7, q = -41$
(d) $(x - 1)^2 + 4$ $p = 1, q = 4$
(e) $(x - \frac{7}{2})^2 - \frac{33}{4}$ $p = \frac{7}{2}, q = -\frac{33}{4}$
(f) $(x + \frac{1}{2})^2 - \frac{25}{4}$ $p = -\frac{1}{2}, q = -\frac{25}{4}$
3 (a) $x = -0.68, -7.32$
(b) $x = 9.69, 0.31$
(c) $y = 0.24, -12.24$
(d) $x = 4.45, -0.45$
(e) $x = 16.25, -0.25$
(f) $z = -1.59, -4.41$
(g) $y = 4.56, 0.44$
(h) $x = 1.19, -4.19$
4 (a) $x = -2 \pm \sqrt{3}$
(b) $x = 6 \pm \sqrt{20}$ or $6 \pm 2\sqrt{5}$
(c) $x = 4 \pm \sqrt{22}$
(d) $x = -3 \pm \sqrt{10}$
(e) $x = 1 \pm \sqrt{5}$
(f) $x = -4 \pm \sqrt{14}$

5 (a) $x = 5.24, 0.76$ (b) $x = 9.47, 0.53$
  (c) $x = -0.84, -7.16$ (d) $x = 0.24, -8.24$
  (e) $x = 6.46, -0.46$ (f) $x = 0.30, -3.30$
6 $(x + 3)^2 + 1 = 0 \Rightarrow (x + 3)^2 = -1$, cannot square root -1
7 $2[(x + 2)^2 - \frac{13}{2}] = 0$ $x = 0.55, -4.55$ to 2 d.p.
8 $x = \dfrac{-b \pm \sqrt{b^2 - 4ac}}{2a}$

Exercise 24J

1 $7, -13$
2 11 m by 16 m
3 11 and 14
4 $x = 4.30, 0.697$
5 7.29, 12.29, 14.29
6 $x = 1.5$
7 $x = 7.24$
8 $(2\frac{1}{2}, 15), (-3, 4)$
9 $x = 0.70, 4.30$
10 $x = 3.90, -0.90$

Examination style questions

1   (a) (i) $(x - 8)(x + 1)$ (ii)
    $x = 8, x = -1$
2 $(x - 2)(x + 8)$
3 $(y - 2)(y - 7)$
4 $x = 10.48, -0.48$
5 $x^2 - 2x - 8 = 0$
6 (a) $a = 3, b = 4$
  (b) 4

Chapter 25

Exercise 25A

1

x	-4	-3	-2	-1	0	1	2	3	4
y	48	27	12	3	0	3	12	27	48

$x = 0$

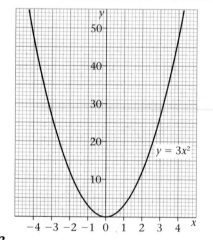

2

x	-4	-3	-2	-1	0	1	2	3	4
y	-32	-18	-8	-2	0	-2	-8	-18	-32

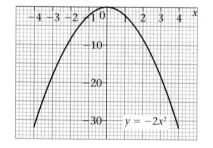

The minus sign inverts the graph.

3

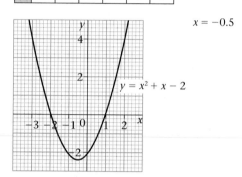

4 (a)

x	-3	-2	-1	0	1	2
y	4	0	-2	-2	0	4

$x = -0.5$

(b)

x	-5	-4	-3	-2	-1	0	1	2
y	11	5	1	-1	-1	1	5	11

$x = -1.5$

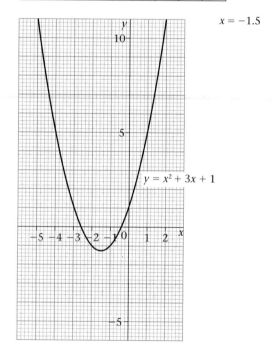

(c)

x	−3	−2	−1	0	1	2	3	4
y	27	15	7	3	3	7	15	27

$x = 0.5$

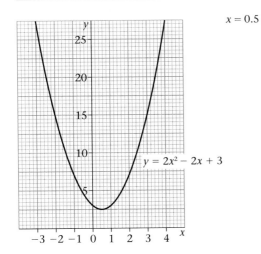

$y = 2x^2 - 2x + 3$

(d)

x	−2	−1	0	1	2	3	4
y	7	−3	−9	−11	−9	−3	7

$x = 1.0$

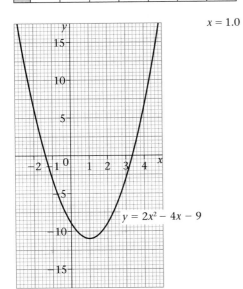

$y = 2x^2 - 4x - 9$

5

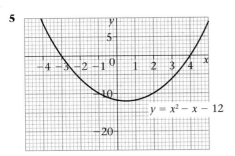

$y = x^2 - x - 12$

$x = -3$ and $x = 4$

6

x	−3	−2	−1	0	1	2	3	4
$3x^2$	27	12	3	0	3	12	27	48
$-7x$	21	14	7	0	−7	−14	−21	−28
-6	−6	−6	−6	−6	−6	−6	−6	−6
y	42	20	4	−6	−10	−8	0	14

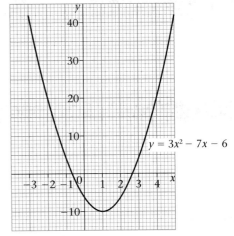

$y = 3x^2 - 7x - 6$

$(-0.7, 0)$ and $(3, 0)$

Exercise 25B

1

x	−3	−2	−1	0	1	2	3
$y = -x^3$	27	8	1	0	−1	−8	−27

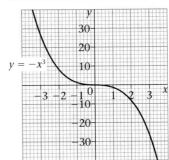

$y = -x^3$

2

x	−2	−1	0	1	2	3
x^3	−8	−1	0	1	8	27
$-3x^2$	−12	−3	0	−3	−12	−27
$-5x$	10	5	0	−5	−10	−15
y	−10	1	0	−7	−14	−15

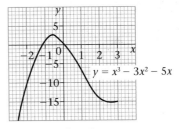

$y = x^3 - 3x^2 - 5x$

$x = -1.2$ and $x = 0$

3

x	−3	−2	−1	0	1	2	3
x^3	−27	−8	−1	0	1	8	27
$-3x$	9	6	3	0	−3	−6	−9
-2	−2	−2	−2	−2	−2	−2	−2
y	−20	−4	0	−2	−4	0	16

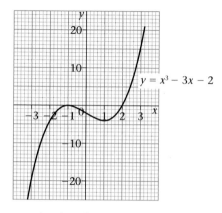

$x = -1$ and $x = 2$

4

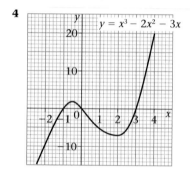

$x = -1$, $x = 0$ and $x = 3$

5 (a)

x	−3	−2	−1	0	1	2	3
y	−11.5	−2	1.5	2	2.5	6	15.5

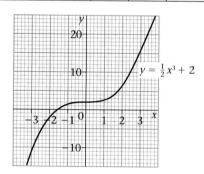

(b)

x	−3	−2	−1	0	1	2	3
y	−64	−27	−8	−1	0	1	8

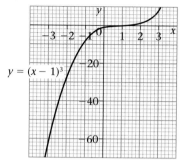

(c)

x	−3	−2	−1	0	1	2	3
y	−21	−6	−1	0	3	14	39

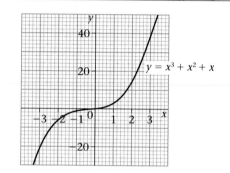

Exercise 25C

1 (a)

x	−3	−2	−1	−0.5	0.5	1	2	3
y	0.7	1	2	4	−4	−2	−1	−0.7

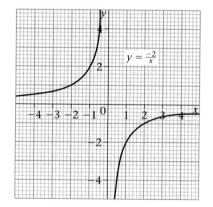

(b)

x	−3	−2	−1	−0.5	0.5	1	2	3
y	1.7	2.5	5	10	−10	−5	−2.5	−1.7

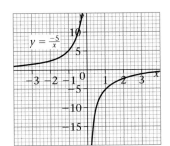

(c)

x	−3	−2	−1	−0.5	0.5	1	2	3
y	5	5.5	7	10	22	1	2.5	3

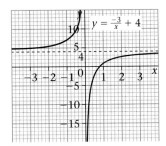

(d)

x	−3	−2	−1	−0.5	0.5	1	2	3
y	−4.7	−4	−2	2	−14	−10	−8	−7.3

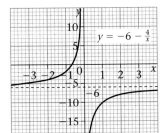

(e)

x	−3	−2	−1	−0.5	0.5	1	2	3
y	−2	−3	−6	−12	12	6	3	2

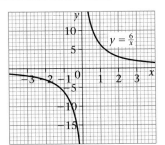

(f)

x	−3	−2	−1	−0.5	0.5	1	2	3
y	−1.3	−2	−4	−8	8	4	2	1.3

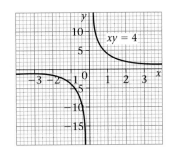

2

x	−3	−2	−1	−0.5	−0.2	0.2	0.5	1	2	3
y	1	1.5	3	6	15	−15	−6	−3	−1.5	−1

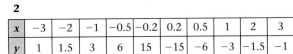

$x = 0,\ y = 0$

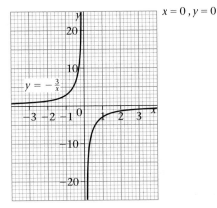

3

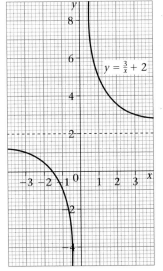

It is reflected in the *y*-axis and shifted up by 2 units.

4

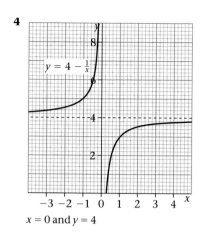

$x = 0$ and $y = 4$

Exercise 25D

1

x	-2	-1	0	1	2	3	4
$y = (1.5)^x$	0.4	0.7	1	1.5	2.3	3.4	5.1

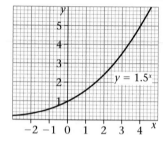

2

x	-0.5	-0.2	-0.1	0	0.1	0.2	0.5
y	0.1	0.4	0.6	1	1.6	2.5	10

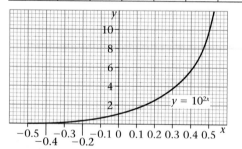

3

x	-3	-2	-1	0	1	2
y	37.0	11.1	3.3	1	0.3	0.1

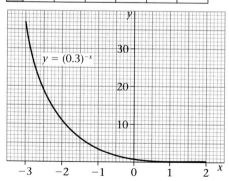

4 160 000
5 0.958 g

Exercise 25E

1

x	0.1	0.2	0.5	1	2	3	4	5
y	9.9	4.8	1.5	0	-1.5	-2.7	-3.8	-4.8

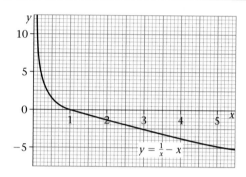

2

x	0.1	0.2	0.5	1	2	3	4	5
$4/x$	40	20	8	4	2	1.3	1	0.8
$-3x$	-0.3	-0.6	-1.5	-3	-6	-9	-12	-15
y	39.7	19.4	6.5	1	-4	-7.7	-11	-14.2

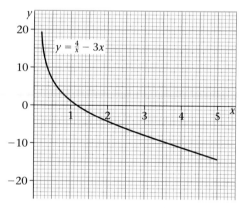

3

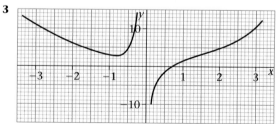

Exercise 25F

1

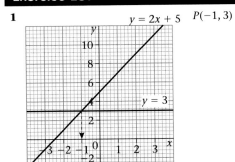

$y = 2x + 5$ $P(-1, 3)$
$y = 3$

2

$y = 4x + 9$ $Q(-0.5, 7); x = -2.3$
$y = 3$

3

x	-2	-1	0	1	2
y	5	2	1	2	5

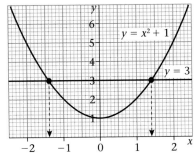

$y = x^2 + 1$
$y = 3$

coordinates of intersection $(-1.4, 0)$, $(1.4, 0)$

4

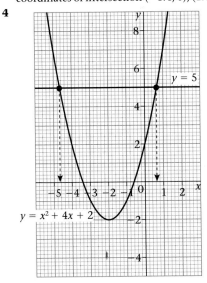

$y = 5$
$y = x^2 + 4x + 2$

(b) $x = -4.7$ and $x = 0.6$
(c) $x^2 + 4x - 3 = 0$
(d) $x = -3.4$ and $x = -0.6$

5 Students own graphs
  (a) $x = -2, x = 1; x = -2.6, x = 1.6$
  (b) $x = -4.2, x = 0.2; x = -3.4, x = -0.6$
  (c) $x = -1.4, x = 3.4; x = -0.4, x = 2.4$
  (d) $x = -2.6, x = -0.4; x = -4.5, x = 1.5$
  (e) $x = -0.7, x = 5.7; x = 0.4, x = 4.6$
  (f) no solution; $x = -2.5, x = 3.5$
  (g) $x = -2.9, x = 0.9; x = -3.4, x = 1.4$
  (h) $x = -1.3, x = 3.3; x = -0.2, x = 2.2$

6

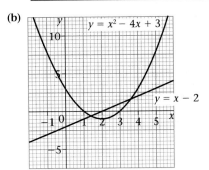

$y = x^2 - 2x - 4$
$y = -x + 2$

$x = -1.3$ and $x = 3.3$, $x = -2$ and $x = 3$

7 (a)

x	-1	0	1	2	3	4	5
y	8	3	0	-1	0	3	8

(b)

$y = x^2 - 4x + 3$
$y = x - 2$

(c) $x = 1$ and $x = 3$
(d) $x = 1.4$ and $x = 3.6$

8

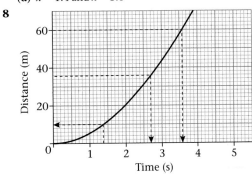

(a) 10–12 m
(b) 2.6–2.7 s
(c) 3.4–3.5 s

9

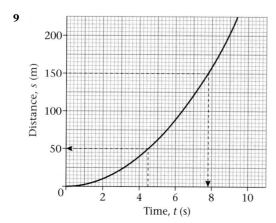

(a) 50–52 m
(b) 7.7–7.8 s

10

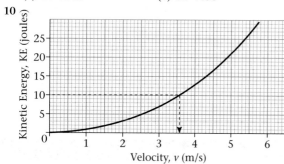

$v = 3.5$–3.6 m s

Exercise 25G

1 (a) Q (b) Cu (c) R
 (d) Cu (e) E (f) L
 (g) Co (h) Q (i) Co
2 (a) C (b) D (c) A
 (d) E (e) B
3 (a) R (b) L (c) E
 (d) Q (e) Q (f) R
 (g) Cu (h) Co (i) L

4

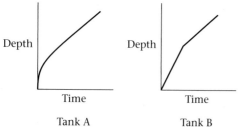

5

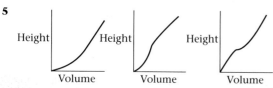

Examination style questions

1

x	−1	0	1	2	3
y	−3	−1	1	3	5

$(-0.5, -2)$

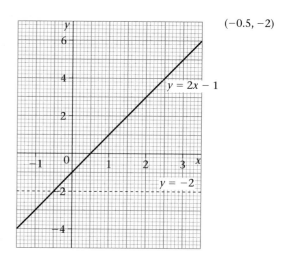

2 $(-1, -3)$
3 (a)

x	−2	−1	0	1	2	3
y	15	5	−1	−3	−1	5

(b)

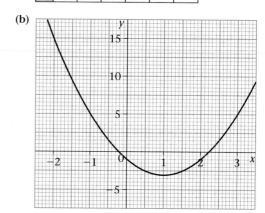

(c) (i) The graph intersects with the x-axis at $(2.2, 0)$.
 (ii) -0.2
4 (a)

x	0	1	2	3	4
y	1	0.8	0.64	0.51	0.41

(b)

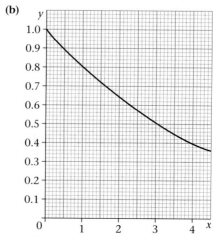

(c) 1.23
5 (d) Graph **(d)** Height increases slowly then more quickly.

6 (a)

x	-3	-2	-1	0	1	2	3	4	5	6
y	14	6	0	-4	-6	-6	-4	0	6	14

(b)

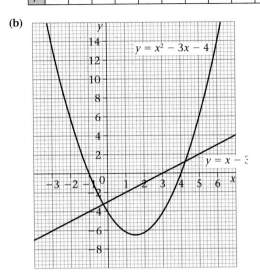

$y = x^2 - 3x - 4$

$y = x - 3$

(c) $-1, 4$
(d) $-0.2, 4.2$ to 1 d.p.

Chapter 26

Exercise 26A

1 (a) $0.\dot{6}$ recurring
(b) 0.8 terminating
(c) $0.\dot{4}$ recurring
(d) $0.4\dot{5}$ recurring
(e) 0.875 terminating
(f) $0.58\dot{3}$ recurring
(g) 0.24 terminating
(h) 0.1875 terminating
(i) $0.8\dot{3}$ recurring
(j) $0.\dot{7}1428\dot{5}$ recurring
2 (a) T   **(b)** R   **(c)** T   **(d)** R   **(e)** T
  (f) T   **(g)** R   **(h)** R   **(i)** T   **(j)** T
3 (a) $0.\dot{7}, 1$ **(b)** $0.\dot{7}\dot{2}, 2$ **(c)** $0.\dot{7}1428\dot{5}, 6$
  (d) $0.\dot{5}1\dot{8}, 3$ **(e)** $1.\dot{2}7\dot{9}, 3$
4 (a) yes   **(b)** yes   **(c)** can't tell
5 (a) no   **(b)** 1
  (c) multiples of 3   **(d)** prime numbers
  (e) You can not generalise from a small selection.

Exercise 26B

1 (a) $\frac{9}{20}$ **(b)** $\frac{3}{8}$ **(c)** $\frac{1}{200}$
2 (a) $\frac{5}{9}$ **(b)** $\frac{8}{9}$ **(c)** $\frac{7}{33}$
  (d) $\frac{5}{11}$ **(e)** $\frac{7}{11}$ **(f)** $\frac{320}{999}$
  (g) $\frac{208}{333}$ **(h)** $\frac{526}{3333}$ **(i)** $\frac{1235}{9999}$
  (j) $\frac{1}{6}$ **(k)** $\frac{1}{22}$ **(l)** $6\frac{14}{333}$
  (m) $2\frac{16}{99}$ **(n)** $15\frac{4}{45}$ **(o)** $4\frac{65}{1998}$
3 Using the method for converting recurring decimals into fractions gives $\frac{9}{9} = 1$

Exercise 26C

1 (a) I   **(b)** R   **(c)** R   **(d)** R
  (e) I   **(f)** R   **(g)** R   **(h)** I
  (i) R   **(j)** I   **(k)** R   **(l)** I
  (m) I   **(n)** I   **(o)** I   **(p)** R
2 $\sqrt{2}$ cm   **3** $\sqrt{13}$ cm
4 $\sqrt[3]{124}$ **5** $\dfrac{4 \pm \sqrt{60}}{2}$ or $2 \pm \sqrt{15}$

Exercise 26D

1 (a) 3   **(b)** 1   **(c)** $\sqrt{5}$
2 (a) $\sqrt{10}$ **(b)** $\sqrt{3}$ **(c)** $\sqrt{40}$
  (d) $\sqrt{4}$ **(e)** $\sqrt{21}$ **(f)** $\sqrt{6}$
3 (a) $6\sqrt{15}$ **(b)** $2\sqrt{5}$ **(c)** $2\sqrt{3}$
  (d) $4\sqrt{3}$ **(e)** $8\sqrt{85}$ **(f)** $12\sqrt{66}$
4 (a) $\sqrt{(25 \times 2)}$ **(b)** $\sqrt{(16 \times 2)}$ **(c)** $\sqrt{(4 \times 6)}$
  (d) $\sqrt{(9 \times 3)}$ **(e)** $\sqrt{(9 \times 5)}$ **(f)** $\sqrt{(36 \times 2)}$
  (g) $\sqrt{(100 \times 2)}$ **(h)** $\sqrt{(100 \times 10)}$
5 (a) $5\sqrt{2}$ **(b)** $4\sqrt{2}$ **(c)** $5\sqrt{3}$ **(d)** $4\sqrt{3}$
  (e) $3\sqrt{6}$ **(f)** $10\sqrt{3}$ **(g)** $2\sqrt{41}$ **(h)** $7\sqrt{6}$
6 (a) $\dfrac{4\sqrt{2}}{5\sqrt{2}}$ **(b)** $\dfrac{\sqrt{6}}{4}$ **(c)** $\dfrac{\sqrt{3}}{\sqrt{7}}$ **(d)** $\dfrac{4\sqrt{3}}{7\sqrt{2}}$
7 (a) $5\sqrt{2}$ **(b)** $5\sqrt{5}$ **(c)** $\sqrt{3}$
  (d) $\sqrt{6}$ **(e)** $8\sqrt{7}$ **(f)** $3\sqrt{2}$

Exercise 26E

1 (a) $\dfrac{\sqrt{2}}{2}$ **(b)** $\dfrac{\sqrt{11}}{11}$ **(c)** $2\sqrt{3}$ **(d)** $4\sqrt{2}$
2 (a) $\dfrac{\sqrt{3}}{6}$ **(b)** $\dfrac{\sqrt{7}}{35}$ **(c)** $\dfrac{\sqrt{5}}{2}$ **(d)** $\dfrac{\sqrt{3}}{4}$
3 (a) $\dfrac{\sqrt{15}}{5}$ **(b)** $\dfrac{\sqrt{15}}{6}$ **(c)** $\frac{1}{2}$ **(d)** $\frac{2}{3}$
4 (a) 2   **(b)** 2   **(c)** $2\sqrt{3}$ **(d)** 2
5 (a) $\dfrac{\sqrt{6}}{6}$ **(b)** $\dfrac{4\sqrt{15}}{5}$ **(c)** $\dfrac{\sqrt{30}}{10}$ **(d)** $\frac{1}{3}$

Exercise 26F

1 $\sqrt{2}$ **2** $7\sqrt{2}$
3 $3 \pm 3\sqrt{2}$ **4** $9\sqrt{3}$
5 (a) $\pi, 2\sqrt{3}$ **(b)** various
6 18π
7 (a) $\dfrac{2\sqrt{7}}{7}$ **(b)** $13\sqrt{2}$ **(c)** $2\sqrt{10}$
8 (a) $20 + 4\sqrt{3}$ units   **(b)** $28 + 10\sqrt{3}$ square units
9 (a) $\frac{229}{500}$ **(b)** $\frac{7}{9}$ **(c)** $\frac{50}{111}$
10 (a) $6\sqrt{2} + 4\sqrt{3}$ units   **(b)** $6\sqrt{6}$ square units
  (c) $\sqrt{30}$ units
11 $\frac{4}{9} + \frac{5}{9} = \frac{9}{9} = 1$ **12** $\sqrt{10}$
13 2   **14** AB $= \sqrt{23}$
15 (a) 5   **(b)** 27   **(c)** 24
  (d) 11   **(e)** 10   **(f)** 2

Examination style questions

1 (a) $a = 5$ **(b)** 18
2 $\frac{16}{33}$ **3 (a)** $\frac{14}{33}$ **(b)** $\frac{49}{66}$
4 (a) $3\sqrt{5}$
  (b) $(\sqrt{3} + \sqrt{12})^2 = 3 + 12 + 2\sqrt{3}\sqrt{12} = 15 + 2\sqrt{3} \times 2\sqrt{3}$
                            $= 15 + 12 = 27$

Chapter 27

Exercise 27A

1 (a) 53° or 127°   (b) 74° or 254°   (c) 63° or 297°
   (d) 130° or 310°   (e) 194° or 346°   (f) 150° or 210°
2 $x = 42°$ and 138°   **3** $x = 60°$ and 300°
4 $x = -116°$ and 64°
5 (a) 322°   (b) 142° and 218°   (c) 52° and 128°
6 (a) $x = -113°, -67°, 247°, 293°$
   (b) $x = -337°, -23°, 23°, 337°$
7 $x = 75.5°$ and 284.5°   **8** $x = -34.8°$ and $-145.2°$

Exercise 27B

1 $a = 2\sqrt{2}$ $b = \sqrt{2}\sqrt{3} = \sqrt{6}$
2 1.5 m   **3** $5\sqrt{2}$ m
4 (a) $0.9\sqrt{3}$ m   (b) 1.8 m
5 $\sin x = \dfrac{\sqrt{7}}{4}$, $\tan x = \dfrac{\sqrt{7}}{3}$
6 $\sin x = \dfrac{7}{\sqrt{170}}$, $\cos x = \dfrac{11}{\sqrt{170}}$

Exercise 27C

1 (a) 23.1 cm²   (b) 15.2 cm²
2 41.03°   **3** 75.5 cm²
4 19.8°, 160.2°   **5** 249.4 cm²

Exercise 27D

1 (a) 5.8 cm²   (b) 417.3 cm²   (c) 10.4 cm²
2 (a) 117.6 cm²   (b) 65.3 cm

Exercise 27E

1 $c = 4.86$ cm   **2** $Q = 26.8°$
3 (a) $a = 6.8$ cm, $c = 7.4$ cm
   (b) $b = 102.8$ cm , $c = 80$ cm
4 $Y = 15.5°$ or 108.5°
5 (a)

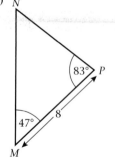

   (b) MNP = 50°   (c) 7.64 km
6 (a)

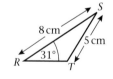

   (b) 55.5° and 124.5°

Exercise 27F

1 $b = 8.76$ cm   **2** 110.8 m
3 $C = 55.3°$ **4** $A = 119.1°$
5 34.4°, 67.4°, 78.2°   **6** 95 cm

Examination style questions

1 (a)

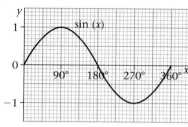

   (b) 113°   (c) -0.92
2 14.5 cm² to 3 s.f.   **3** $30\sqrt{2}$ cm or 42.4 cm to 3 s.f.
4 28.7 km to 3 s.f.

Chapter 28

Exercise 28A

1 $(-2, -1), (3, 4)$ **2** $(-1, 3), (4, 8)$
3 $(1, 4), (5, 20)$ **4** $(-2, 3), (1.5, -0.5)$
5 $(-1, -3)$ **6** $(-2.5, -8), (4, 5)$
7 $(-2, -8), (6, 32)$ **8** $(-3, 1), (9, 5)$

Exercise 28B

1 $(2, -5), (5, -2)$ **2** $(-1, -1), (0.2, 1.4)$
3 $(-1, 3), (3, 1)$
4 $\left(-1.5 + \sqrt{\frac{41}{2}},\ 1.5 + \sqrt{\frac{41}{2}}\right), \left(-1.5 - \sqrt{\frac{41}{2}},\ 1.5 - \sqrt{\frac{41}{2}}\right)$
   or $(1.70, 4.70), (-4.70, -1.70)$ to 3 s.f.
5 $\left(1 - \sqrt{\frac{14}{2}},\ 1 + \sqrt{\frac{14}{2}}\right), \left(1 + \sqrt{\frac{14}{2}},\ 1 - \sqrt{\frac{14}{2}}\right)$
   or $(-0.871, 2.87), (2.87, -0.871)$ to 3 s.f.
6 $\left(0.4 - \sqrt{\frac{79}{5}},\ -0.2 - 2\sqrt{\frac{79}{5}}\right), \left(0.4 + \sqrt{\frac{79}{5}},\ -0.2 + 2\sqrt{\frac{79}{5}}\right)$
   or $(1.38, -3.76), (2.18, 3.36)$ to 3 s.f.
7 1 point of intersection   **8** 0 point of intersection

Exercise 28C

1 $\dfrac{x + 2}{3}$ **2** cannot be simplified

3 $\dfrac{2x}{x - 5}$ **4** $\dfrac{2a}{3b}$

5 cannot be simplified   **6** cannot be simplified

7 $\dfrac{b + 2}{a}$ **8** $\frac{2}{3}$

9 cannot be simplified   **10** $\dfrac{1}{x - 3}$

11 $x + 6$ **12** $\dfrac{y - 5}{y - 3}$ **13** $\dfrac{x + 6}{x - 2}$

14 $\dfrac{m + 1}{5m + 3}$ **15** $\dfrac{2x + 3}{3x + 5}$

Exercise 28D

1 $\dfrac{a^2}{12}$ **2** $\dfrac{5b}{2a}$ **3** qr **4** $\dfrac{z}{y}$

5 $\dfrac{y(y + 2)}{6}$ **6** $\frac{1}{12}$ **7** $4ab$ **8** $\dfrac{x + 5}{4}$

9 $\frac{1}{4}$ **10** $2(m - 3)$

11 $\dfrac{3a(a - 2)^2}{(a - 1)(a + 1)^2(a + 3)}$ **12** $\dfrac{(y - 4)^2(y + 4)\ (y - 5)}{2y(y - 1)^2}$

Exercise 28E

1 $\dfrac{13x}{15}$ **2** $\dfrac{3b + 4a}{12ab}$ **3** $\dfrac{10y - 3x}{2xy}$

4 $\dfrac{9x - 8}{12x^2}$ **5** $\dfrac{3(4y + 1)}{10xy}$ **6** $\dfrac{7x + 11}{12}$

7 $\dfrac{13x + 6}{12}$ **8** $\dfrac{x + 16}{6}$ **9** $\dfrac{1 - 2y}{15}$

10 $\dfrac{16 - 11m}{10}$ **11** $\dfrac{7x - 13}{x^2 + 2x - 3}$ **12** $\dfrac{2x + 5}{x^2 + 7x + 12}$

Exercise 28F

1 $x = 7$ **2** $x = 3$
3 $x = 23$ **4** $x = 9$
5 $x = 1$ **6** $x = -8$
7 $x = 0.5$ or $x = 4$ **8** $x = -1$ or $x = 8$
9 $x = -0.5$ or $x = 5$ **10** $x = 2$ or $x = 6$
11 $x = -7$ or $x = 2$ **12** $x = -\frac{1}{12}$ or $x = 2$

Exercise 28G

1 $y = \dfrac{d - b}{2m}$ **2** $y = \dfrac{4}{p - q}$

3 $y = \dfrac{w + x}{a - b}$ **4** $y = \dfrac{kb - wa}{w - k}$

5 $y = \dfrac{md + ch}{ha - m}$ **6** $y = \dfrac{h(1 - k)}{k + 1}$

7 $y = \dfrac{2d + 3e}{d - e}$ **8** $y = \dfrac{a}{m^2 - 2}$

9 $y = \dfrac{4w^2m}{3 + 4w^2}$ **10** $y = \dfrac{gT^2}{4k^2}$

11 $y = \pm\sqrt{\dfrac{f - 2e}{k + h}}$ **12** $y = \pm\sqrt{\dfrac{((hw - ma)}{(2m + 3h))}}$

Examination style questions

1 (a)

x	-2	-1	0	1	2	3	4	5	6
y	10	3	-2	-5		-5	-2	3	10

(b)(d)

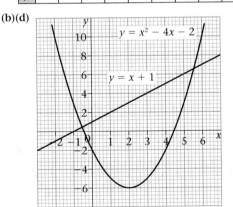

(c) $x = -\frac{1}{2}$ or $4\frac{1}{2}$ (d) $x = -\frac{1}{2}$ or $5\frac{1}{2}$

2 (a) $c = \sqrt{\dfrac{E}{m}}$ (b) $m = \dfrac{2E}{2gh + v^2}$

3 $\dfrac{x - 4}{3x - 2}$ **4** $-\frac{1}{3}$ or $-0.\dot{3}$

Chapter 29

Exercise 29A

1 (a) (i) 19   (ii) 12
      (iii) 3   (iv) 3.25
  (b) (i) 5   (ii) 1, 3
2 (a) (i) 1   (ii) 3
      (iii) 15   (iv) 136
  (b) (i) 0.5, 1   (ii) $-1.5, 3$

Exercise 29B

1 (a) translation (0, 5)
   (b) translation (0, -4)
   (c) translation (-1, 0)
   (d) translation (3, 0)
2 (a) $y = x^2 + 2$
   (b) $y = (x + 6)^2$
   (c) $y = x^2 - 5$
   (d) $y = (x + 4)^2$

Exercise 29C

Some of these exercises have multiple answers. For example, a transformation mapping $y = x^2$ onto $y = (2x)^2 = 4x^2$ could be a one-way stretch parallel to the x-axis scale factor 1/2 or a one-way stretch parallel to the y-axis scale factor 4. The most likely answers are given here.

1 (a) one-way stretch from $y = 0$ (x-axis) parallel to $x = 0$ (y-axis) of scale factor 3
   (b) one-way stretch from $x = 0$ (y-axis) parallel to $y = 0$ (x-axis) of scale factor 1/3

2 (a) (i) $y = 4x^2$ (ii) $y = 2.5x^2$ (iii) $\dfrac{y = x^2}{5}$

   (b) (i) $y = \left(\dfrac{x}{4}\right)^2$ (ii) $y = (4x)^2$ (iii) $y = (1.5x)^2$

Exercise 29D

Some of these questions have multiple answers.

1 (a) (i)

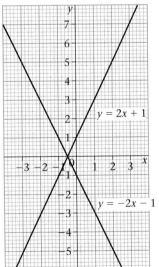

     (ii) $y = -2x - 1$

(b) (i)

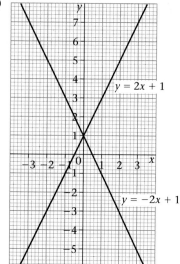

(ii) $y = -2x + 1$

2 (a) (i)

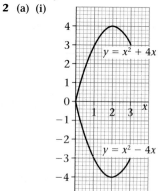

(ii) $y = -x^2 + 4x$
(b) (ii) $y = x^2 + 4x$

3 (a) (i) $y = -3x + 2$ **(ii)** $y = -3x - 2$
   (b) (i) $y = 4x - 1$ **(ii)** $y = 4x + 1$
   (c) (i) $y = -x^2 - 6x$ **(ii)** $y = x^2 - 6x$
   (d) (i) $y = -x^3 - 4$ **(ii)** $y = -x^3 + 4$
   (e) (i) $y = -2x^2 + 5$ **(ii)** $y = 2x^2 - 5$
   (f) (i) $y = -x^2 - 3x + 1$ **(ii)** $y = x^2 - 3x - 1$

4 (a) translation by $(0, 2)$

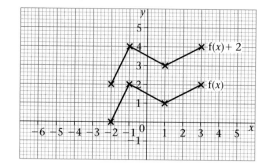

(b) one-way stretch from $y = 0$ (x-axis) parallel to $x = 0$ (y-axis) of scale factor 2

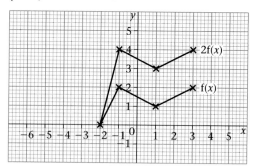

(c) reflection in $y = 0$ (x-axis)

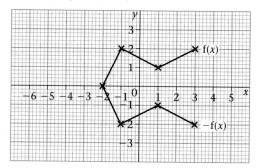

(d) translation by $(-3, 0)$

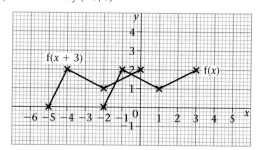

(e) (1) one-way stretch from $y = 0$ (x-axis) parallel to $x = 0$ (y-axis) of scale factor 2
    (2) translation $(0, -3)$

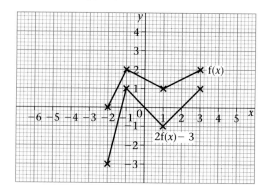

(f) (1) reflection in $y = 0$ (x-axis) (2) translation by $(0, 4)$
or (1) translation by $(0, -4)$ (2) reflection in $y = 0$
(x-axis)

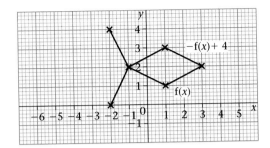

Exercise 29E

Some of these exercises have multiple answers. For example, where a transformation involves a number of steps, it will sometimes (but not always) be possible to change the order without affecting the outcome. The most likely answers are given here.

1 (a) $y = (x - 2)^2 - 7$; translation by $(2, -7)$
(b) $y = (x - 3)^2 + 2$; translation by $(3, 2)$
(c) $y = (x + 1)^2 + 5$; translation by $(-1, 5)$
(d) $y = (x + 5)^2 - 6$; translation by $(-5, -6)$
2 (a) translation by $(-2, -3)$
(b) translation by $(3, 2)$
(c) (1) one-way stretch from $y = 0$ (x-axis) parallel to $x = 0$ (y-axis) of scale factor 2
(2) translation by $(0, 3)$
(d) (1) translation by $(3, 0)$ (2) one-way stretch from $y = 0$ (x-axis) parallel to $x = 0$ (y-axis) of scale factor 2
(e) (1) translation by $(-2, 0)$ (2) one-way stretch from $y = 0$ (x-axis) parallel to $x = 0$ (y-axis) of scale factor 3
(f) (1) reflection in $y = 0$ (x-axis) (2) translation by $(0, 3)$
(g) (1) reflection in $y = 0$ (x-axis) (2) one-way stretch from $y = 0$ (x-axis) parallel to $x = 0$ (y-axis) of scale factor 3
(h) (1) reflection in $y = 0$ (x-axis) (2) one-way stretch from $y = 0$ (x-axis) parallel to $x = 0$ (y-axis) of scale factor 2
(3) translation by $(0, -3)$
3 (a) (1) translation by $(0, 2)$ (2) one-way stretch from $y = 0$ (x-axis) parallel to $x = 0$ (y-axis) of scale factor 2 (3) translation by $(0, -1)$
(b) (1) translation by $(-4, 0)$ (2) one-way stretch from $y = 0$ (x-axis) parallel to $x = 0$ (y-axis) of scale factor 3 (3) translation by $(0, -2)$
(c) (1) translation by $(3, 0)$ (2) one-way stretch from $y = 0$ (x-axis) parallel to $x = 0$ (y-axis) of scale factor 4 (3) translation by $(0, 7)$
(d) (1) translation by $(4, 0)$ (2) one-way stretch from $y = 0$ (x-axis) parallel to $x = 0$ (y-axis) of scale factor 2 (3) reflection in $y = 0$ (x-axis) (4) translation by $(0, 5)$
or (1) translation by $(4, 0)$ (2) one-way stretch from $y = 0$ (x-axis) parallel to $x = 0$ (y-axis) of scale factor 2 (3) reflection in $y = 2.5$
4 (a) $y = -(x + 2)^2 + 5$
(b) $y = 2(x + 3)^2 + 6$
(c) $y = -2(x - 4)^2$
(d) $y = -\left(\dfrac{x}{2}\right)^2 - 3$ **or** $y = -\left(\dfrac{x^2}{4}\right) - 3$
(e) $y = -3(x + 5)^2 + 2$

Exercise 29F

Some of these questions have multiple answers. For example, the answer for 1**(a)** could be translation by $(180°, \ 0)$ or translation by $(-180°, \ 0)$. The answers given here are the obvious ones.

1 (a) (ii) reflection in $y = 0$ (x-axis)
(iii) period: $360°$;
(b) (ii) one-way stretch from $x = 0$ (y-axis) parallel to $y = 0$ (x-axis) of scale factor $\frac{1}{2}$
(iii) period: $180°$;
(c) (ii) translation by $(0, -2)$
(iii) period: $360°$;
(d) (ii) one-way stretch from $y = 0$ (x-axis) parallel to $x = 0$ (y-axis) of scale factor 2
(iii) period: $360°$;
(e) (ii) one-way stretch from $x = 0$ (y-axis) parallel to $y = 0$ (x-axis) of scale factor 2
(iii) period: $720°$;
(f) (ii) translation by $(90°, 0)$
(iii) period: $360°$;
(g) (ii) (1) one-way stretch from $y = 0$ (x-axis) parallel to $x = 0$ (y-axis) of scale factor $\frac{1}{2}$ (2) translation by $(0, 1)$
(iii) period: $360°$;
(h) (ii) (1) reflection in $y = 0$ (x-axis) (2) translation by $(90°, 1)$ **or** translation by $(270°, 1)$
(iii) period: $360°$;

Exercise 29G

1 (a) $326°$
(b) $146°, 214°$
2 (a) $109°$
(b) $251°, 289°$
3 (a) $138°$
(b) The period will be half that of the original graph.
(c) (i) $21°, 69°, 201°, 318°$

Examination style questions

1 (a)

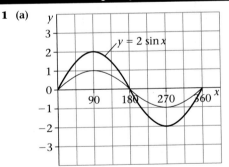

(b)

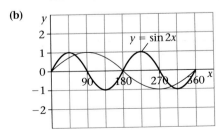

(c)

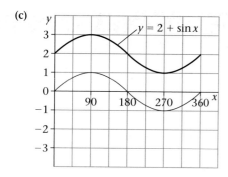

2 A $y = x^2 - 3$ B $y = x^2 + 3$
   C $y = -x^2$ D $y = -x^2 + 3$

3

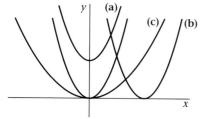

Chapter 30

Exercise 30A

1 **(a)**

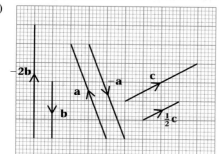

(b) $3\mathbf{a} = \begin{pmatrix} -6 \\ 15 \end{pmatrix}$ $-4\mathbf{b} = \begin{pmatrix} 0 \\ 12 \end{pmatrix}$ $-1\frac{1}{2}\mathbf{c} = \begin{pmatrix} -6 \\ -3 \end{pmatrix}$

2 **(a)**

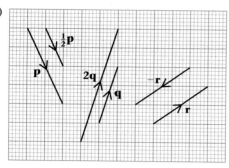

(b) $5\mathbf{r} = \begin{pmatrix} -15 \\ -10 \end{pmatrix}$ $-2\mathbf{q} = \begin{pmatrix} -2 \\ -6 \end{pmatrix}$ $1\frac{1}{2}\mathbf{p} = \begin{pmatrix} 3 \\ -6 \end{pmatrix}$

Exercise 30B

1 $\mathbf{d} + \mathbf{e} = \begin{pmatrix} -2 \\ -1 \end{pmatrix}$ $\mathbf{f} - \mathbf{d} = \begin{pmatrix} 3 \\ 4 \end{pmatrix}$ $2\mathbf{f} + \mathbf{e} = \begin{pmatrix} 0 \\ 8 \end{pmatrix}$

$\mathbf{d} - 3\mathbf{e} = \begin{pmatrix} -10 \\ 7 \end{pmatrix}$ $\mathbf{d} + \mathbf{e} + \mathbf{f} = \begin{pmatrix} -3 \\ 4 \end{pmatrix}$ $3\mathbf{f} + 2\mathbf{d} = \begin{pmatrix} -11 \\ 17 \end{pmatrix}$

$4\mathbf{e} - 3\mathbf{d} = \begin{pmatrix} 20 \\ -11 \end{pmatrix}$ $2\mathbf{d} - \mathbf{e} - 2\mathbf{f} = \begin{pmatrix} -8 \\ -6 \end{pmatrix}$

2

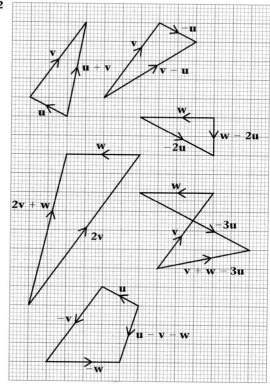

Exercise 30C

1 **(a)**

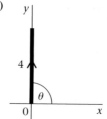

$|OA| = 4$
$\theta = 90°$

(b)

$|OA| = \sqrt{3^2 + 5^2}$
$= \sqrt{34}$
$\theta = \tan^{-1}\left(\frac{5}{3}\right)$
$= 59.0°$